T0394079

Dynamics of Blood Cell Suspensions in Microflows

Dynamics of Blood Cell Suspensions in Microflows

Edited by
Annie Viallat
Manouk Abkarian

CRC Press is an imprint of the
Taylor & Francis Group, an **informa** business

CRC Press
Taylor & Francis Group
6000 Broken Sound Parkway NW, Suite 300
Boca Raton, FL 33487-2742

International Standard Book Number-13: 978-1-138-03205-7 (Hardback)

Library of Congress Cataloging-in-Publication Data

Names: Viallat, Annie, editor.
Title: Dynamics of blood cell suspensions in microflows / edited by Annie Viallat, Manouk Abkarian.
Description: Boca Raton : CRC Press, [2020] | Includes bibliographical references and index.
Identifiers: LCCN 2019043063 | ISBN 9781138032057 (hardback) | ISBN 9781315395142 (ebook)
Subjects: LCSH: Blood flow. | Microcirculation. | Blood cells.
Classification: LCC QP105 .D96 2020 | DDC 612.1/181--dc23
LC record available at https://lccn.loc.gov/2019043063

Visit the Taylor & Francis Web site at
http://www.taylorandfrancis.com

and the CRC Press Web site at
http://www.crcpress.com

Contents

Preface

Blood microcirculation is the archetype of a multitasking system with exceptional performance. It provides simultaneously an impressive number of essential physiological functions for human and animal organisms. These include nutrient supply and waste disposal, oxygen delivery, homeostasis, temperature regulation, active monitoring against exogenous attacks and transport of immunity soldiers and even a system of self-repair of the entire network in case of injury.

From a mechanical point of view, it is an extraordinary and highly admirable system. Imagine: Four to five litres of blood - a concentrated suspension of three types of cells of varying sizes, shapes and stiffness, also carrying proteins and molecules - circulating in a closed interconnected hierarchical network feeding each cell of each organ. Its supply: a single pump. Its length: some 80,000 km!

Evolution has succeeded in meeting the impressive specifications required to ensure that physiological functions are performed with a very high degree of performance, reliability and durability. Our best researchers and engineers are nowhere near being able to design and manufacture such a system with such demanding specifications. In fact, we are just beginning to understand some of the multiple physical mechanisms behind the complex behaviors involved in the functioning of the entire vascular system.

From the point of view of physics, if we consider only the traffic aspect, the efficient and sustainable circulatory performance is a tour de force. Indeed, on the one hand, the suspension is concentrated (45% in volume) and complex with three types of cells: leukocytes (also called white blood cells), viscoelastic spherical balls of 10μm, platelets, small rigid discs of 2 μm, and red blood cells. A red blood cell is a very deformable biconcave disc filled with a hemoglobin solution, 2.5 μm thick and 8 μm in diameter. For every one white blood cell, there are about 50 platelets and 1000 red blood cells. On the other hand, the vascular network has a very long and complex geometry, composed of arteries distributing blood, a thin intricate branched network of capillaries (of diameter as small as half the red cell diameter) with numerous bifurcations and veins collecting blood. The circulation must be very robust: blood cells must never (for at least 80 years and more!) form aggregates that can clog large portions of blood vessels, regardless of flow conditions, even in bifurcations where shear gradients are strong, and although the cells interact strongly with each other and with the walls of blood vessels at such a high concentration. Also, the mechanical properties of the cells must provide them excellent robustness and deformability to allow them to slip and squeeze into the smallest capillaries. Damage caused by white and red blood cells lysis in blood vessels can induce a deleterious systemic vascular inflammatory response.

The individual rheological properties of the different blood cells give specific physical properties to blood and induce specific behaviors to blood flow, both at individual and collective cellular scales. For example, the mechanical properties and the high deformability of red blood cells are responsible for their large diversity of regimes of motion in microflows. Collective cell behavior is even more complex. Margination, for example, is a very peculiar spatial organisation of red cells, leukocytes and platelets in venules and arterioles. Margination maintains leukocytes and platelets at the immediate vicinity of the vascular endothelium, where they must adhere in case of inflammation or wound. This phenomenon takes its origin in the mechanical properties, size and shapes, which are different for the three cell types and induce different responses when they interact with each other and with the vessel walls.

With the emergence of microfluidic techniques, it is now possible to generate in-vitro blood micro-flows and to exploit these specific physical properties for novel biomedical applications such as cell sorting.

Blood microcirculation has been studied over many years by different scientific communities, from physiologists to biomechanics engineers and immunologists. There are therefore several books on the subject of blood micro-flows, but most books have a viewpoint focused on one type of cell or on one specific function. For instance, many studies concerning hemorheology or mechanics of the microcirculation are dedicated to red cell rheology and aggregation and do not address questions on platelets and leukocytes.

In this book, we propose to illustrate how a physical approach can successfully describe, understand and predict several behaviors of blood flow and blood cells that are directly linked to important physiologic functions. Moreover, we also want to draw attention to the underlying innovative potential of the physical approach for novel applications in medicine. Issues of biological regulation and activity are not well developed in this book. Thrombosis and platelet activity are the exception, as a complete chapter is devoted to them. In particular, the book does not go into detail on the regulation of blood vessels: vaso-constriction, vaso-dilation, transition from pusatile to constant flow, etc....

The reader will see that some concepts are repeated in several chapters, and therefore redundancies exist, but often they are used under different lighting and angles of approach.

In the first chapter, the reader will find basic and relevant concepts that are useful for all chapters. Chapters that use and deepen a given concept are indicated throughout this introductory chapter. An effort is made to explain some important concepts "with simple words".

Chapter 2 deals with rigid suspensions. It details the different involved interactions, introduces formal concepts of fluid mechanics and shows that the understanding of the behavior, structure and rheology of suspensions of identical rigid spheres has considerably progressed. In contrast, these results highlight that the understanding of flow behavior of suspensions of particles of different sizes, shapes and deformabities like blood is still in its infancy.

In Chapter 3, the deformability of the suspended objects is introduced, which is an essential element of red blood cells. The suspended objects are considered as quasi-spherical and with a viscosity equal to that of the suspending fluid. A theoretical approach is proposed both in bulk and in confined flows. This approach highlights the phenomena of shear-induced diffusion and hydrodynamic cross-stream migration. It describes the red blood cell-free layer. Indeed, despite the simplifications, the theory succeeds in capturing characteristic features of blood flows, such as separation of the different particles and margination.

Chapter 4 proposes a multi-scale computational approach that focuses on the behavior of suspensions of particles mimicking red blood cells and platelets, thus of different deformabilities and sizes, with a red blood cell's viscosity still remaining equal to that of the external fluid. The chapter discusses how red blood cells play an important role in platelet adhesion to blood vessel walls and ultimately influence hemostasis and "blood clotting". Platelet margination is predicted and the cross-stream distribution of red blood cells in capillaries is described, with the presence of the cell free layer in which platelets accumulate. Effects of hematocrit, channel height, viscosity ratio and capillary number on the flow structure are discussed.

These chapters illustrate the enormous progress made over the past decades in theoretical and computational approaches, which now make it possible to deal with a large number of cells and thus to be able to predict the behavior of concentrated suspensions.

The next three chapters introduce experimental approaches and thus show the inherent complexity of the "real world".

Chapter 5 is dedicated to the most abundant element of blood, the red blood cell. The mechanical properties of the simplest cell in the human body are detailed and the dynamic under flow of individual red blood cells is experimentally and theoretically described. This very rich dynamic shows how much the softness of a capsule, even without the active responsiveness that characterises living objects, generates a complexity that has long been unsuspected. The consequence is that the rheology of a suspension of red blood cells, and therefore of blood, is still far from being quantitatively understood.

In Chapter 6, complexity is further added. Indeed, we know that red blood cells can aggregate together through different mechanisms that are discussed in the chapter. The aggregation of red blood cells is an important phenomenon, particularly in the context of certain diseases such as sickle cell disease, for which the adhesion of red blood cells is increased, while their deformability is reduced. In general, an inflammatory state, even limited, activates cells and cell walls and makes them more adhesive. This makes the risk of clogging vessels important.

In Chapter 7, we come back to platelets, but by including and discussing all their biological complexity, in particular when they are activated and create a clot. The interactions of platelets with reactive surfaces, upon which they tether, translocate, roll and eventually become firmly adhered, are presented. The chapter describes both the external physical environment and intraplatelet dynamics that guide platelet motion, deformation and binding. Several single-scale and multi-scale continuum and particle-based models developed to describe and predict phenomena leading up to thrombus formation are explored.

In the following chapters, the flow geometry, which was a straight capillary until now, becomes more complex. Networks typical of the micro-vasculature are considered in the next two chapters.

Chapter 8 is dedicated to blood flow in the network architecture of the micro-vasculature. The chapter highlights the current understanding of the physical ingredients involved in the couplings between network architecture and blood flow dynamics. After giving key elements about the architecture of the micro-vasculature, the chapter describes the behavior of blood in simple components of a network and finally discusses the heterogeneity of blood flow at the scale of micro-vasculature.

Chapter 9 introduces white blood cells into the microcirculation. These circulating cells leave the bloodstream when an inflammatory signal, the signature of a local infection, reaches them. The cascade process from the recruitment until the arrest of a white blood cell on the vascular wall, thanks to specific adhesion molecules, is briefly described. The chapter is mainly dedicated to the physical mechanisms driving the circulation of white blood cells in the pulmonary network, whose capillary diameter is smaller than that of white blood cells, and to their blockage in the lungs due to the impairment of their mechanical properties in the case of acute inflammation.

Chapters 10 and 11 are dedicated to microfluidic applications in the fields of diagnosis, cell sorting and mechanical analysis. These applications are based on the specific behaviors of the different cells in different external fields and different regimes of flow.

In Chapter 10 in particular, a new flow regime is exploited, characterised by inertial effects associated with high flow velocities, which are not in the physiological range of blood microcirculation. The theoretical background of inertial microfluidics that is important for applications in the field of hematology is summarised. It is shown how to take advantage of the properties of this type of flow in two main applications of inertial microfluidics in hematology: separations and single cell analysis.

In Chapter 11, multiple microfluidic applications are described, which also use other external forces created by an electric field, for example for dielectrophoresis applications, an acoustic field or a magnetic field. The chapter first introduces a few general physical principles applied in microfluidic platforms and their application in blood component separation. Then the chapter focuses on two diseases, malaria and cancer, and the two microfluidic platforms that have been developed.

Finally, the last chapter presents an opening on the blood of other animal species and discusses the specificities of these blood types, which might be induced by adaptation to the environment. This chapter presents some extraordinary blood suspensions with behaviors that would be very pathologic for human blood. It gives also an overview of the most important properties. It shows that native animal blood offers properties for nearly every situation thanks to its rheological "fingerprint" and highlights the huge diversity of the various bloods found in Nature.

We have identified a unified set of topics involving physical issues that govern important features in blood microcirculation. We believe that the topics we have chosen are timely and highlight active research areas. They also bridge to applications in biology and medicine. We hope that this book will provide the reader with some useful elements to understand the dynamic behavior of red blood cells, leukocytes and platelets in vascular microcirculation.

We would like to thank all the contributors to this book. Thank you for accepting this work, thank you for your patience and thank you for the quality of your contributions. We wish to warmly thank our colleagues for their intellectually stimulating discussions, help and kindness. We also express our gratitude to our home French institutions. We give our heartfelt thanks to our families, who are our most precious support, for their patience and love.

A repository of colour images can be downloaded from https://www.crcpress.com/9781138032057

Manouk Abkarian and Annie Viallat

Contributors

Manouk Abkarian
CBS, Univ. of Montpellier
Montpellier, France

Viviana Clavería
CBS, Univ. of Montpellier
Montpellier, France

Philippe Connes
Univ. of Lyon
Lyon, France

Jules Dupire
L'Oreal Recherche and Innovation
Aulnay sous Bois, France

Stany Gallier
ArianeGroup
Nice, France

Michael D. Graham
Univ. of Wisconsin-Madison
Madison, Wisconsin, USA

Emmanuèle Helfer
Aix Marseille Univ, CNRS, CINAM
Marseille, France

Kuan Jiang
Mechanobiology Institute
National University of Singapore, Singapore

Wonhee Lee
KAIST
Daejon, Republic of Korea

Elisabeth Lemaire
Institut de physique de Nice
Nice, France

Chwee Teck Lim
Dept. of Biomedical Engineering
National University of Singapore, Singapore

Etienne Loiseau
Aix Marseille Univ, CNRS, CINAM
Marseille, France

Sylvie Lorthois
IMFT, CNRS, Univ. of Toulouse
Toulouse, France

Simon Mendez
IMAG, Univ. of Montpellier
Montpellier, France

Qin M. Qi
Stanford University
Stanford, California, USA

Eric S.G. Shaqfeh
Stanford University
Stanford, California, USA

Annie Viallat
Aix Marseille Univ, CNRS, CINAM
Marseille, France

Christian Wagner
Saarland University
Saarbrücken, Germany

Ursula Windberger
University of Medicine of Vienne
Vienna, Austria

CHAPTER 1

Blood in flow. Basic concepts

Etienne Loiseau, Annie Viallat

Aix-Marseille Univ, CNRS, CINAM, Marseille, France

Manouk Abkarian

CBS, CNRS, INSERM, Univ Montpellier, Montpellier, France

CONTENTS

IN THIS INTRODUCTORY CHAPTER, we will present how questions about the microcirculation of blood and blood cells are addressed by physicists. There are two levels of reading in this chapter. The first part of the chapter, from 1.1 to 1.6, for "dummies", introduces and comments upon the fundamental concepts and quantities at the basis of the physical approach. The second part of the chapter, from 1.7, labeled with a sign *, develops a more formal approach that will give technical tools to the most expert and to those who want to get to the heart of the physics and mechanics. Experts in fluid mechanics can skip this chapter. Throughout this section, we indicate in which chapters the concepts presented are developed and discussed in a more specific way. Readers eager to read more on how the laws of hydrodynamics introduced in this chapter are theoretically established, as well as on the different order of magnitude related to the blood circulation, may refer to excellent books on the subject that we used to prepare this introductory section [1,4,6–9,11].

1.1 BLOOD AND ITS MICROCIRCULATION

This section briefly describes the components of blood, the geometry of flow in the microvascular system and defines the notion of suspension and its specificity.

1.1.1 Composition

Human blood is a suspension of cells in an aqueous solution called plasma.

Plasma is a solution of 10% proteins, ions, nutrients, wastes and dissolved gases in 90% w/w water. Among the molecules found in plasma there are hormones, antibodies, and clotting factors. Plasma without clotting factors is called serum.

The cellular components of blood include red blood cells (erythrocytes), platelets (thrombocytes), and white blood cells (leukocytes)

Red blood cells (RBCs) represent more than 90% of the cells in blood. The volume fraction in whole blood of RBCs ranges between 40% and 50%. They are responsible for transporting oxygen from the lung to the tissues, and carbon dioxide from tissues to the lung. In humans, RBCs are biconcave disk-shaped, with a diameter of $\sim 7.5\mu m$ and and a thickness of $\sim 2.5\mu m$. Mature RBCs do not contain a nucleus at odds with many other species as will be discussed in Chapter 12. Human RBCs membrane is a double shell. The outer shell is a lipid bilayer and the inner shell is a 2D spectrin cytoskeleton. Specific membrane proteins bind the two layers together. RBCs are filled with a solution of hemoglobin, the key protein used in oxygen transport. RBCs have an average life span of 120 days. Old or damaged RBCs are eliminated in the liver and spleen, and new ones are produced in the bone marrow. Red blood cells are specifically studied in Chapters 3, 5, 4, 6 and 12.

White blood cells (WBCs) make up less than 1% of the cells in blood. The two types of WBCs found in blood flow are neutrophils and monocytes They are responsible for the immune response. They recognise and neutralise invaders such as bacteria. They are nucleated cells, rather spherical, with a diameter ranging from 8 to 20 μm, depending on the type. They have a large reserve of membrane stored in microvillii. They circulate for a limited time in the bloodstream (a few days maximum) and new cells are produced primarily in the bone marrow. The dynamics of WBCs is discussed in Chapter 9.

Platelets, around 5% of the cells in blood, are cell fragments engaged along with the coagulation factors in the formation of a blood clot. Platelets are small discs with a diameter of about 2.5 μm. They are produced when large cells called megakaryocytes break into pieces, each one making 2000 - 3000 platelets as it comes apart. Chapter 7 is dedicated to the behavior of platelets, which is also discussed in Chapter 4.

1.1.2 The microcirculation

The microcirculation is composed of vessels of different sizes classified into arterioles, capillaries and post-capillary venules. Each type of vessel has a unique function and structure. Arterioles, of about 100 μm in size, are responsible for blood delivery to tissues and regulate the rate of delivery. They branch into a myriad of capillaries that are responsible for exchange between blood and tissue. Blood capillaries are the smallest vessels; their diameters are similar in size to RBCs ($\sim 2 - 8\mu m$) and the speed of RBCs is several hundred microns per second in this region. Then the blood flows out of the capillaries through post-capillary venules. The venules are responsible for blood return to the heart. They are in general parallel to the arterioles in organisation. The diameter of the post-capillary venules ranges between 8-30 μm and the RBCs' velocity can reach a few millimeters per second. They are important for macromolecular exchange, vascular resistance and immunological response.

The problem of describing blood flow in such a complex network is discussed in Chapter 8.

The main question that must be answered is how these blood cells and whole blood circulate in the microcirculation network under the action of the heart. The heartbeat exerts an upstream to downstream pressure drop in the vessels. This pressure drop generates blood flow. We therefore try to describe how the velocity of plasma and blood cells, their local trajectory and density, their position in the vessels, their deformations as well as the blood flow resistance and the energy dissipated vary with the pressure drop, and, for example, the diameter of the vessels. Physicists have developed methods to answer these questions valid for ordinary fluids, but also, more recently, for suspensions such as blood.

1.1.3 Definition of a suspension and continuum assumption for blood cell suspensions

Particule suspensions are part of a large class of complex fluids which span a large diversity of systems going from clay suspensions, cements, paints, lavas from volcanoes, bacterial biofilms or the one that is at the heart of this book, blood cell suspensions. By definition, a suspension is a discontinuous dispersion of particles surrounded by a continuous liquid phase. Therefore, blood can be considered as a cell suspension made of RBCs, platelets and WBCs, which are deformable particles and are suspended in an aqueous solution of proteins and ions called the plasma.

When working with such a suspension, one important problem is the ability to define local variables necessary to characterise the movement of this complex fluid. For simple fluids like water, variables such as density, pressure, temperature and velocity are defined thanks to the hypothesis of continuity. In this case, there exists an elementary volume much larger than the scale of the molecular fluctuations over which to perform adequate averaging but which remains still much smaller than the scale of macroscopic variations of the flow one would like to describe. In the case of water for instance, this scale can be as small as few nanometers. But in the case of a suspension like blood, the validity of such assumption is not obvious since particles are several microns in size, classifying blood as a non-Brownian suspension for which thermal fluctuations of cells' positions play a little role in their large scale flow. As we will see throughout the different chapters of this book, there are many situations where the "granularity" of the suspension has to be considered in order to understand general aspects of blood flow in the microcirculation, for example when blood flows in vessels of size comparable to individual cell size, while in other situations such as hemorheology studies, blood is considered as an effective continuous complex fluid.

1.2 RESPONSE OF A COMPLEX FLUID TO A MECHANICAL STRESS. AN INTRINSIC CHARACTERISTIC

In this section, it is pointed out that fluids can exhibit a wide variety of behaviors, sometimes very complex when subjected to mechanical stress. These behaviors depend on the structure of the fluid and, in particular, in the case of suspensions when particles can interact with each other. We introduce the notion of viscosity and elasticity. The flow of a fluid through a tube, induced by the application of mechanical stress, is first characterised by its rate of deformation. This is an intrinsic response of the fluid, involving an important quantity, the viscosity. The notion of viscosity is easily understood by considering the simple case of pure shear stress.

Consider the flow of a fluid between two flat parallel plates (see Fig. 1.1). The lower plate is stationary, and the upper plate of surface area S is moved horizontally by applying a force F. The distance h between them is constant and the fluid is assumed to adhere to both plates (fluid velocity equal to 0 at the lower plate wall and U and the upper plate wall). The movement of the upper plate along the direction of the applied force results in the development of a velocity gradient in the fluid, which is the strain rate, also called shear rate, $\dot{\gamma}$, and is a constant in the case of simple shear flow.

$$\dot{\gamma} = |\frac{du_x}{dy}| = \frac{U}{h}, \tag{1.1}$$

where $\mathbf{u_x}(y)$ is the fluid velocity at a distance y of the lower plate.

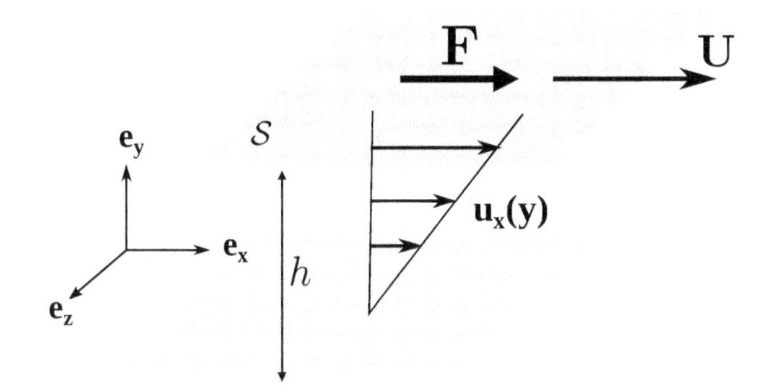

FIGURE 1.1
Velocity field in a simple shear driven by the motion of the upper plate.

1.2.1 Newtonian fluids

For an ordinary fluid, which is called a Newtonian fluid, the relationship between the applied stress τ (applied tangential force F per unit plate area S) and the observed shear rate is linear and defines the viscosity η such as:

$$\tau = \frac{F}{S} = \eta\dot{\gamma}. \tag{1.2}$$

The viscosity, whose unit is the Poiseuille (1 Poiseuille= 1 Pa s) measures the shear stress that must be applied to drive the plate at a speed U and to achieve a given shear rate. The work done by the shear stress supplies the energy dissipated by viscous friction when the fluid laminae parallel to the plates slide over each other in order to flow. This energy loss can be calculated. The mechanical power per unit volume is equal to $\tau\dot{\gamma}$ or $\eta\dot{\gamma}^2$. Thus, it increases as the square of the velocity gradient and is proportional to the fluid viscosity. It is the viscosity of the fluid that is at the origin of the dissipation of kinetic energy produced at large scales. The viscosity measures the slipping resistance between fluid layers.

1.2.2 Non-Newtonian fluids or complex fluids

While simple fluids like water and almost all gases, follow a Newtonian behavior, a large body of materials, called complex fluids like blood cell suspensions exhibit more complicated macroscopic properties. One interesting discussion of past and current attempts to develop the necessary constitutive theories to describe complex fluids can be found in the book of Gary Leal [11]. The author discusses the success of describing the non-Newtonian behavior of complex fluids by defining appropriate and general constitutive equations. He specifies how current theories are indeed either restricted to very specific simple flows like generalised shear flows or to very dilute solutions or suspensions for which the micro-scale dynamics is known and dominated by the motion and deformation of single, isolated objects such as particles, macromolecules or cells. In fact, there is currently no model that is known to provide quantitatively accurate or even qualitatively reliable descriptions of real complex fluids of dense suspensions such as blood for a wide spectrum of flows except the highly simplistic flows of conventional rheometers. Past continuum mechanics approaches were capable of obtaining Newtonian-fluid-like approximation in the simple flows, which we will describe briefly in this chapter. But these do not help the prediction of the behavior of cells in more complex flows such as the one of the microcirculation for instance. As we will see for example in Chapter 2 for particles and in Chapter 3 for capsules, for a dilute suspension case, a complete and exact fluid dynamics description of the motion of individual particles is somewhat possible, and the structural evolution of the suspension and the relationship with its macroscopic rheology can in principle be derived by a rigorous statistical averaging procedure. As we will see in Chapter 5, this single particle dynamics is just being fully solved for RBCs and should allow such a rigorous approach for dilute blood suspensions in the near future. Meanwhile, in the case of concentrated suspensions of rigid or deformable non-spherical particles, the multi-particle fluid dynamics problem is impossible to solve; only numerical approaches such as the one presented in Chapter 4 are available to describe the complex situation or ask new questions about for instance margination of WBCs (see Chapters 9 and 4) or platelets (see Chapter 7). Theoretically and more generally, only phenomenological models such as the suspension balance model or the diffusive flux models are available to describe general particle suspensions as presented in Chapter 2, with moderate successes in describing certain observations such as the shear-induced migration of spherical particles.

Complex fluids such as blood cell suspensions disclose macroscopic rheological properties which include non-linearity, elasticity and time dependent effects.

Non-linear behavior - For these fluids, the relation between stress and shear rate is not linear anymore. The shear viscosity may be $\dot{\gamma}$-dependent, $\eta(\dot{\gamma}) = \tau/\dot{\gamma}$. We can distinguish the shear-thinning fluids whose viscosity $\eta(\dot{\gamma})$ decreases when the shear-rate increases. Shear-thinning is generally observed when the fluid is made of interacting particles. At low shear rates, particles adhere to each other and resist the flow. As the shear rate increases, the particles align and slide over each other; they no longer stick together. Blood belongs to this category. The shear-thickening fluids, on the contrary, exhibit a viscosity $\eta(\dot{\gamma})$ which increases when the shear-rate increases. Corn starch is a typical example of such a fluid. Yield stress fluids have the property to start flowing provided that the shear stress overcomes a critical threshold. Some colloidal suspensions exhibit this behavior and tooth paste is a good example.

Time dependent effects - A thixotropic fluid exhibits time dependent shear thinning properties. These fluids are very viscous at rest. They exhibit a decrease in viscosity and flow over time under a mechanical stress such as a shear stress. The longer the fluid undergoes shear stress, the lower its viscosity. This results from a modification of the internal microstructure of the fluid with characteristic times comparable to the observation times. An example is tomato ketchup. Blood's thixotropy is still under debate since the measure of the yield stress is very difficult [2,13].

1.2.3 Elasticity and viscoelasticity

Though the strain rate characterises a fluid behavior under the action of mechanical stress, the response of a solid elastic is characterised by its strain. This intrinsic response depends on the elastic modulus of the material. The simple Hooke's law, generally valid at small deformations and forces, models the response of elastic solids. Let us consider a pulling force F exerted along one axis to a spring of initial length L_0. Hooke's law is $F = k\Delta L$, where

k is a characteristic of the spring and ΔL is the variation of the spring length. Similarly to fluids under a shear flow, one calculates the stress, $\tau = \frac{F}{S}$, where S is the surface area of the material on which the force is exerted. ΔL is renormalized by the initial length, $\epsilon = \Delta L/L_0$ and the relation between stress and strain reads

$$\tau = G\epsilon, \tag{1.3}$$

where G is the Young modulus of the material. This law applies to any elastic object submitted to traction forces (and for small deformations). For instance a similar law exists for a shear stress, which defines the shear modulus, G_s, as the ratio of the shear stress to the shear strain (Fig 1.2). It also applies when a material is bent and it is thin like in the case of membrane, a bending modulus is introduced and characterises the capacity of the material to bend under a given distribution of torques. This modulus scales for thin elastic materials like the product Gh^3.

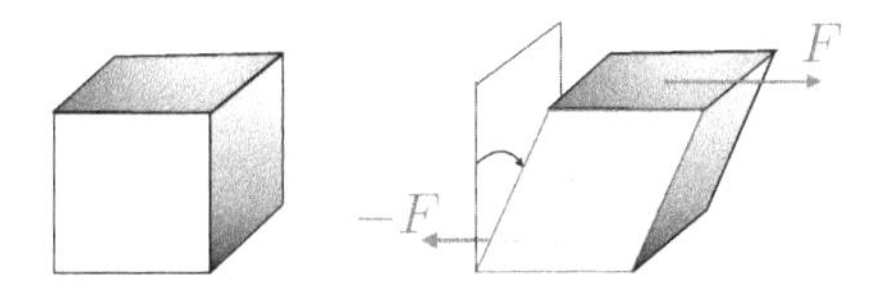

FIGURE 1.2
Generalized Hooke's law for shear

Blood cells such as red and white blood cells are not fluid material and have a non-zero shear, Young and bending moduli. In the blood flow, they are subjected to extensional and shear stresses. But red and white blood cells are not perfect elastic solids. Under a mechanical stress blood cells deform. Their shape change affects the behavior of blood in the microvasculature. The specific mechanical properties of red and white blood cells and their dynamics in microflows will be studied in Chapters 3, 5, 6, 9. As many other materials such as polymers and rubber, they present a viscoelastic behavior. The response to deformation of a viscoelastic fluid has both a viscous aspect (stress proportional to the rate of deformation) and an elastic aspect (stress proportional to the deformation). The elastic behavior dominates at short time-scale while the viscous behavior dominates at longer time-scale. Indeed, if the loading time is shorter than a characteristic time of the material, the microstructure does not have time to deform significantly and there is an elastic response. On the other hand, when the solicitation time is greater than the characteristic time, we observe a viscous response.

A classical approach to describe these viscoelastic fluids is to apply an oscillatory strain field. A simple model to describe a viscoelastic fluid in the limit of small deformations has been proposed by Maxwell, based on a linear behavior of the material. One general representation of the law describing the behavior of such a Maxwellian fluid is to look at the phenomenological behavior of a representative toy model of the material properties which associates a virtual elastic "spring" with a viscous "dashpot" in series (see Fig. 1.3).

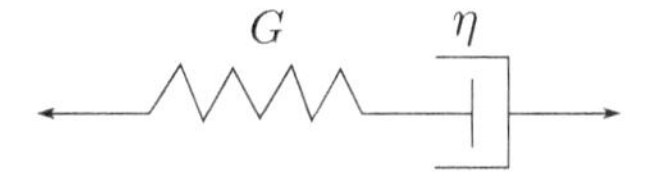

FIGURE 1.3
Maxwell model of a viscoelastic fluid.

In this case, the elastic and viscous strains add-up while the elastic stress and viscous stress are equal,

$$\begin{cases} \sigma = \sigma_S = G\epsilon_S \\ \sigma = \sigma_D = \eta\dot{\epsilon}_D \\ \epsilon = \epsilon_S + \epsilon_D \end{cases}, \tag{1.4}$$

where the subscript D indicates the stress/strain in the dashpot and the subscript S indicates the stress/strain in the spring. G and η are material coefficients homogeneous to an elastic modulus and a viscosity respectively. By taking the derivative of strain with respect to time we obtain the governing equation of the Maxwell's model,

$$\frac{1}{G}\frac{d\sigma}{dt} + \frac{\sigma}{\eta} = \frac{d\epsilon}{dt}, \tag{1.5}$$

associated to a characteristic relaxation time $t_c = \eta/G$. This characteristic time describes the deformation delay of a Maxwell fluid when a mechanical stress is applied to it

As stated earlier, the determination of relaxation times can be performed by subjecting the material to an oscillating deformation in a rheometer. If the imposed strain is sinusoidal and of low amplitude, $\epsilon = \epsilon_0 exp(i\omega t)$, the stress response is also sinusoidal, with a phase shift δ with respect to the strain, $\sigma = \sigma_0 exp(i\omega t + \delta)$. Stress and strain are connected via a complex shear modulus G^*,

$$\sigma(t) = G^*(\omega)\epsilon(t), \text{ and } G^* = G' + iG'', \tag{1.6}$$

with G' the storage modulus, associated to the elastic response of the material and G'' the loss modulus, associated to the viscous response. The complex shear modulus is easily calculated for the Maxwell's model,

$$G' = G\frac{\omega^2 t_c^2}{1 + \omega^2 \tau^2}, G'' = G\frac{\omega t_c}{1 + \omega^2 t_c^2}. \tag{1.7}$$

When $G' > G''$, the response of a Maxwell's fluid is governed by elasticity. It occurs at high frequencies (short time scales) when $\omega > 1/t_c$. When $G'' > G'$ the fluid has a viscous behavior. This occurs at low frequencies (long time scales). A fluid therefore may behave like a solid when the stress is applied during a short time and behave like a liquid and flow when observed at longer time scale [(i)]. Measurements of G' and G'' of blood are shown and discussed in Chapter 12 for human and animal blood.

A similar approach in which the association of a dashpot and a spring in parallel models a viscoelastic solid, called the Kelvin-Voigt solid. Such an approach can be used to model the behavior of RBCs' membrane mechanical behavior (see Chapter 5) and white blood cells' cytoplasm rheology (see Chapter 9).

1.3 DYNAMICS OF VISCOUS FLUIDS

Solving a fluid dynamics problem consists of the calculation of flow velocity, pressure, density, and temperature of the flowing fluid, as functions of space and time. In suspensions, the trajectories, local densities of the different particles are also of great interest. For effective continuous fluids, these physical quantities must be a solution of three fundamental

[(i)] This is also the case of water; think about how water is pretty rigid when your belly flops in a pool

laws, called conservation laws: conservation of mass, conservation of linear momentum, and conservation of energy. In addition to these laws, there is the law of behavior of the fluid (Newtonian, non-Newtonian, etc.) and boundary conditions experienced by the material during its flow. These conditions state how the variables of the problem are given in certain regions of space and at certain given times, for instance zero velocity on the walls. All these equations are complex and many problems remain unsolved or partially solved. Researchers are obliged to use numerical methods to predict flows. Several simplifications of the equations are however often possible. A first simplification is made when the fluid is considered incompressible, i.e. its density remains constant during movement. This is the case of blood.

In this section we first briefly comment upon the equations that apply for Newtonian incompressible fluids. Then, we focus on a simpler problem for which inertial effects (accelerations) can be neglected and we focus on the flow of a Newtonian fluid in a tube, whose geometry is relevant for blood vessels. More formal descriptions of the laws of motion will be described later in the chapter.

1.3.1 Newtonian fluids: Navier-Stokes equation

By applying the fundamental law of dynamics for an incompressible Newtonian fluid, we obtain the Navier-Stokes equation which, most often by an approximate resolution, allows describing the flow in many situations such as weather, ocean currents, water flow in a pipe. Navier-Stokes equations are differential equations that describe the force balance at a given point within a fluid. These forces involve stresses that are surface forces but also body forces that apply to the fluid volume, such as gravity and inertial accelerations.

$$\rho\left(\frac{\partial \mathbf{u}}{\partial t} + (\mathbf{u}.\nabla)\mathbf{u}\right) = \rho\mathbf{f} - \nabla p + \mu\nabla^2\mathbf{u}, \tag{1.8}$$

where ρ is the fluid density, $\mathbf{u}$ is the velocity field, ∇ is the gradient vector $(\frac{\partial}{\partial x}, \frac{\partial}{\partial y}, \frac{\partial}{\partial z})$ and ∇^2 is the operator $\frac{\partial}{\partial x^2}(\cdot) + \frac{\partial}{\partial y^2}(\cdot) + \frac{\partial}{\partial z^2}(\cdot)$ (called also the Laplacian).

The Navier Stokes equation contains 4 terms:

1- $\frac{\partial \mathbf{u}}{\partial t}$ is the acceleration of a fluid particle (temporal variation of the velocity u for a time-dependent flow). It vanishes for stationary flow.

2- $\mathbf{u}.\nabla\mathbf{u}$ is the spatial variation of the velocity of a fluid particle along the flow.

These two latter terms describe the acceleration of the fluid particles in the laboratory frame.

3- $\rho\mathbf{f} - \nabla p$ are the volume and pressure forces applied to the fluid. ∇p is the pressure gradient that exists because the fluid is incompressible.

4- $\mu\nabla^2\mathbf{u}$ is the viscous contribution. It represents the viscosity forces due to deformation of fluid elements.

These equations are still complex and most of the time their resolution requires a numerical approach.

1.3.2 Stokes flow

It is interesting to compare the different terms of the Navier Stokes equation to see if some terms may be neglected in specific situations. In steady state there is only one acceleration term, called the inertial term. It is of the order of $|\rho\mathbf{u}.\nabla\mathbf{u}| \sim \frac{\rho U^2}{L}$ and the viscous friction term is of the order of $|\mu\nabla^2\mathbf{u}| \sim \frac{\mu U}{L^2}$, where L is the characteristic size of the flow (typically the diameter of the tube in which the fluid flows) and U is the order of magnitude of the fluid velocity. The ratio of the two terms compares the inertial effects with the viscous effects and defines the Reynolds number:

$$Re = \frac{\rho U L}{\mu} \tag{1.9}$$

A flow associated with a low Reynolds number, $Re \ll 1$, is called a Stokes flow. In this case, the viscous effects dominate and the inertial effects are negligible. The nonlinear terms in the Navier-Stokes equations vanish and, for a steady flow, we obtain the Stokes equations which link the pressure field and the velocity field,

$$\mu \nabla^2 \mathbf{u} = \nabla p - \rho \mathbf{f}, \tag{1.10}$$

with the associated incompressibility condition $\nabla \cdot \mathbf{u} = 0$.

The Reynolds number of blood flowing in microvessels is generally in the range 10^{-3} to about 1. Therefore, effects of fluid inertia may be neglected in many cases, particularly in smaller microvessels. However, higher Reynolds numbers can be achieved in microfluidic devices, allowing other flow properties to be used. These developments are presented in Chapters 10 and 11.

Low Reynolds number flows (viscous flows) are counter-intuitive flows because on the human scale, for example when swimming in a pool, the Reynolds number is of the order of 10^4 and inertial effects dominate. The particularities of this flow are superbly illustrated in a video presented by Sir Taylor himself (techtv.mit.edu/collections/ifluids/videos/32604-low-reynolds-number-flow). These flows have remarkable characteristics, which result from the fact that the Stokes equations are linear. One striking property is instantaneity. The velocity responds instantaneously to the force and the boundary conditions. A sphere falling in an unbound fluid reaches its terminal velocity at once. Inversely, when a propelling force applying on an object stops, the object immediately stops. Another property is flow reversibility which is the result of the time reversibility of Stokes equations and their linearity. If a velocity field on the boundary of the Stokes flow is reversed, then the velocity field everywhere in the fluid is reversed. By reversal of the boundary conditions, each fluid particle will go back to its initial position. This is the reason why mixing two fluids is difficult in microfluidic applications. This also explains why swimming at low Reynolds number requires a technique which is not time-reversible.

1.3.3 Flow in a tube

Let us consider the flow of a Newtonian fluid at low Reynolds number in a circular cylindrical tube of constant cross section. This is a simple geometry that represents the flow in blood vessels (see Fig.1.4). Because of this simple geometry, the Stokes law can be analytically solved and yields the Poiseuille law. The Poiseuille's law states the relationship between the flow rate, the viscosity of the fluid, the pressure drop along the tube, the length L and radius R of the tube.

The velocity of the fluid is parallel to the tube axis (referred as z) and is a function of the radial position r in the tube. The velocity at the wall is equal to zero. For an upstream to downstream pressure drop equal to $\Delta P = P_i - P_o$, the velocity profile is given by

$$u(r) = \frac{\Delta P}{4 \eta L}(R^2 - r^2) \tag{1.11}$$

Note that the velocity profile is a parabola and that the pressure drop varies linearly along the tube.

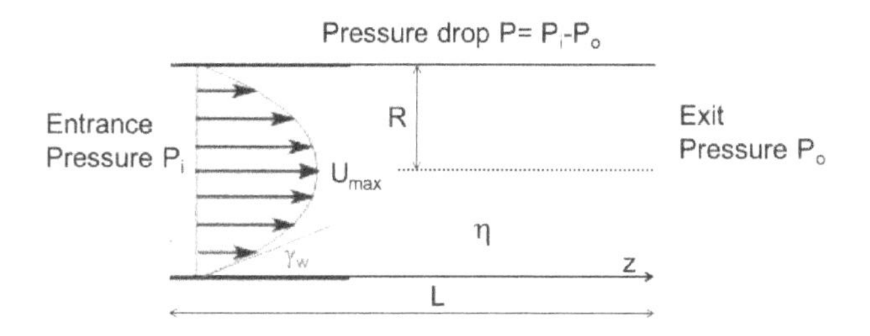

FIGURE 1.4
Poiseuille flow in a tube.

The mean velocity u_m, the maximal velocity at the center of the tube u_{max}, and the shear strain rate of the fluid at the tube wall $\dot{\gamma}_w$ are equal to

$$u_m = \frac{\Delta P R^2}{8\eta L} \tag{1.12}$$

$$u_{max} = \frac{\Delta P R^2}{4\eta L} \tag{1.13}$$

$$\dot{\gamma}_w = \frac{\Delta P R}{2\eta L}. \tag{1.14}$$

By integration on a transverse section, the pressure drop can be related to the volumetric flow rate of liquid, here noted Q, by the Poiseuille-Hagen law:

$$\frac{\Delta P}{Q} = \frac{8\eta L}{\pi R^4} \tag{1.15}$$

This relationship shows that there is a resistance to the flow, which can be defined as the ratio between pressure drop and flow rate, and is noted $\mathcal{R}_{\mathcal{H}}$. $\mathcal{R}_{\mathcal{H}}$ is dependent on the fluid viscosity and the tube geometry (L and R). For a given tube, therefore, this ratio gives a measure of the fluid viscosity. This expression is formally the analog of the electrokinetic Ohm's law between voltage difference and current, $\Delta V = \mathcal{R}i$ (see Fig. 1.5).Therefore, by analogy, the expression $\mathcal{R}_{\mathcal{H}} = \frac{8\eta L}{\pi R^4}$ is called the hydrodynamic resistance [(ii)]. This relation shows that the hydrodynamic resistance increases considerably as space scales decrease. For instance 10% of radius variation results in 40% in flow rate change. This shows that a very small constriction in a blood vessel drastically changes blood stream. For a tube of section area A and perimeter Π, a good estimation of the hydrodynamic resistance is $\mathcal{R}_{\mathcal{H}} = 2\eta L \frac{\Pi^2}{A^3}$.

1.3.4 Association of tubes in series and in parallel

The vascular network is complex with successive vessels and many branches. Based on the analogy between electrical and hydraulic circuits, it is possible to show that nodes and mesh laws exist, similarly to electrokinetics. In a node the sum of flow rate is equal to zero. For a mesh, we have the equivalent of a Kirchhoff's law. Hydraulic circuits can be treated as electric circuits (see Fig. 1.6). Two resistances $\mathcal{R}_1$ and $\mathcal{R}_2$ placed in series are equivalent to the resistance $\mathcal{R}_1 + \mathcal{R}_2$. Two resistances in parallel are equivalent to a resistance equal to $\frac{\mathcal{R}_1\mathcal{R}_2}{\mathcal{R}_1+\mathcal{R}_2}$.

These points will be used in Chapter 8.

[(ii)] But unlike electric resistance it has not a unit name

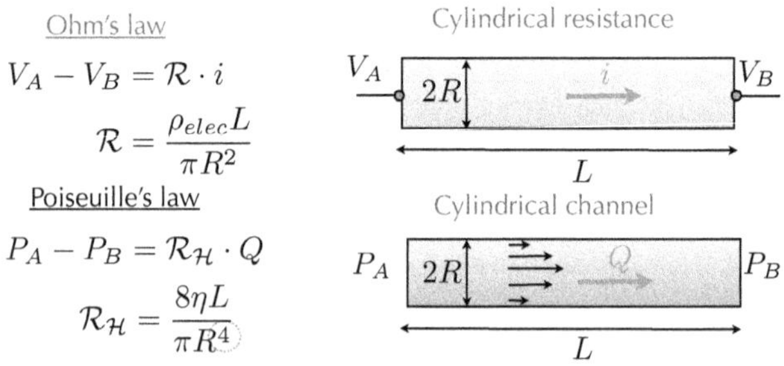

Ohm's law
$V_A - V_B = \mathcal{R} \cdot i$
$\mathcal{R} = \frac{\rho_{elec} L}{\pi R^2}$
Poiseuille's law
$P_A - P_B = \mathcal{R}_{\mathcal{H}} \cdot Q$
$\mathcal{R}_{\mathcal{H}} = \frac{8\eta L}{\pi R^4}$
Cylindrical resistance
V_A
$2R$
i
V_B
L
Cylindrical channel
P_A
$2R$
Q
P_B
L

FIGURE 1.5
Definition of the hydrodynamic resistance and its analogy with electricity.

Resistances in series
$\mathcal{R}_{eq} = \mathcal{R}_1 + \mathcal{R}_2$
$\mathcal{R}_1$
$\mathcal{R}_2$
Resistances in parallel
$\frac{1}{\mathcal{R}_{eq}} = \frac{1}{\mathcal{R}_1} + \frac{1}{\mathcal{R}_2}$
$\mathcal{R}_1$
$\mathcal{R}_2$

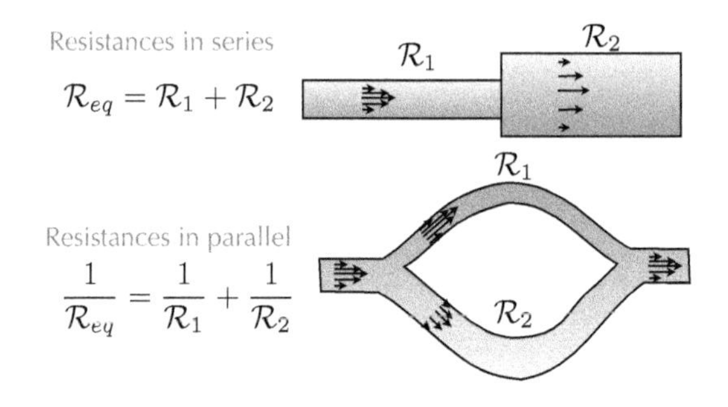

FIGURE 1.6
Equivalent hydrodynamic resistances in series and in parallel.

1.4 FORCES ACTING ON PARTICLES MOVING IN A FLUID

As we have seen, blood is a suspension of cells in plasma. The cells are deformable objects that disturb the flow which would exist in their absence. Velocity and pressure are changed close to the cells. Also, cells interact with each other and with the walls of the vessels. These interactions induce very specific behaviors of the cells, that depend on their shape, size and deformability. The determination of the general flow behavior of the suspension and of the specific cell behavior (deformation, trajectories, positions...) is a tricky problem, at the heart of all the chapters of this book. In this section, we give the basic concepts required to build general solutions for suspension flows. We first examine how a single rigid sphere disturbs simple flows at low Reynolds number, briefly evoke the behavior of solid ellipsoids in shear flow, and then focus on the effect of the vicinity of a wall to the behavior of a flowing particle.

1.4.1 A rigid sphere in translational, rotational and straining flows

Let us consider a rigid sphere in the three simple flows: a translational flow, a rotational flow and a straining flow. These flows are illustrated in Fig 1.7.

Let us first consider a sphere of radius R fixed in a uniform fluid of velocity $\mathbf{U}_\infty$ along the axis $\boldsymbol{x}$ and pressure p_∞. Its fixed position disturbs the fluid. The disturbance pressure $p-p_\infty$ away from the translating sphere varies as r^{-2} and the disturbance of the dominant portion of the velocity varies as r^{-1}, where r is the radial distance from the sphere. The important point is that the velocity's disturbance is long-ranged. The velocity will be changed at a pretty long distance from the sphere. Looking at the pressure, an overpressure applies on the front of the sphere while a depression applies on the back. A force $\boldsymbol{F}_d$ is exerted by the

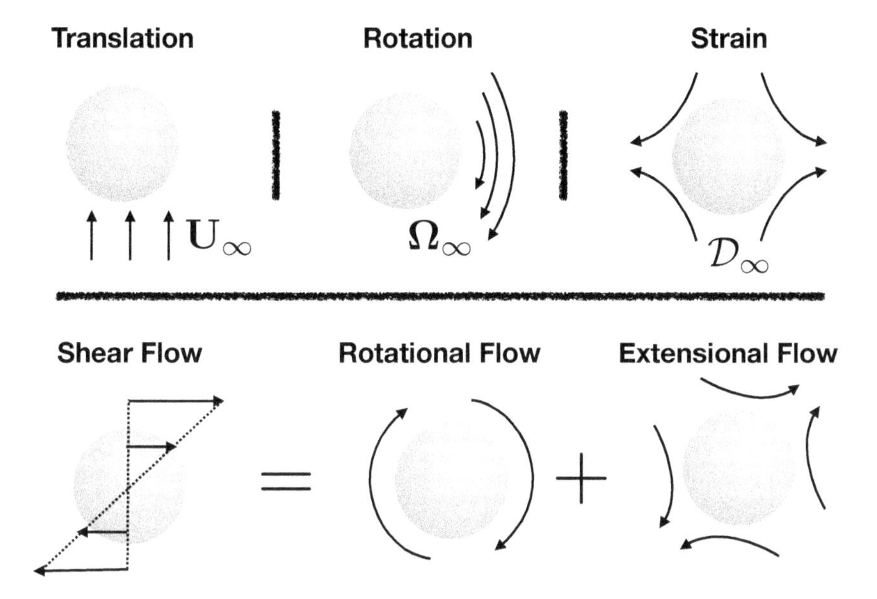

FIGURE 1.7
Top, from left to right: a rigid sphere in a translational, rotational and straining flow. Bottom: decomposition of a shear flow in the sum of a rotational flow and an extensional/straining flow

uniform flow around the sphere, due to the viscous friction. This force is called the drag force and is given by the Stokes law:

$$\boldsymbol{F}_d = 6\pi\eta R\boldsymbol{U}_\infty. \tag{1.16}$$

Note that the force that applies to the sphere is proportional to the fluid velocity. Also note that the Stokes drag force for a rigid sphere moving with velocity $\mathbf{U}_{sphere}$ in an otherwise static fluid bath is $\boldsymbol{F} = -6\pi\eta R\boldsymbol{U}_{sphere}$.

Let us now consider the same sphere, which is maintained fixed in a rotational fluid Ω_∞. No pressure is induced by the presence of the sphere. The disturbance velocity decays as r^{-2}. A hydrodynamic torque is exerted on the sphere held fixed. The Stokes law in rotational flow writes as

$$\boldsymbol{T}_d = 8\pi\eta R^3\boldsymbol{\Omega}_\infty. \tag{1.17}$$

Finally, let us consider the same sphere, which is maintained non-deformable in a straining fluid $\boldsymbol{\mathcal{D}}_\infty$ along the axis $\boldsymbol{x}$. The dominant part of the disturbance velocity decays as r^{-2}. At large distance from the sphere, the disturbance velocity is one order smaller than in the case of flow due to a sphere in a translational flow. The disturbance pressure is very short range since it decays as r^{-3}. A surface traction is applied over the sphere surface as a result of the resistance of the rigid particle to the straining motion. The symmetric part of the first moment of this surface traction is called the stresslet $\boldsymbol{S}$.

$$\boldsymbol{S} = \frac{20\pi}{3}\eta R^3\boldsymbol{\mathcal{D}}_\infty. \tag{1.18}$$

The stresslet is linear in $\boldsymbol{\mathcal{D}}_\infty$ and in R^3. As the stresslet results from the resistance of the particle to the strain, it depends on the deformability of the sphere and will be discussed in more detail in Chapter 2 for a rigid sphere and in Chapters 3 and 5 for deformable capsules and cells.

To study the movement of a sphere in a shear flow, it is advisable to consider this flow as the sum of a rotational flow and a straining flow (see Fig 1.7). The flow caused by a rigid sphere held fixed in the flow field $\boldsymbol{U}_\infty = (\dot{\gamma}y, 0, 0) = (\boldsymbol{\Omega}_\infty + \boldsymbol{\mathcal{D}}_\infty)\boldsymbol{x}$. The disturbance flow generated by the sphere is only due to its resistance to the straining component of the shearing flow. Indeed, no disturbance is created by a freely-rotating sphere displaying a solid-body rotation. This point has consequences in the fact that deformable particles have a behavior under flow that is different from rigid spheres.

1.4.2 A rigid ellipsoid in a shear flow: Jeffery's orbits

In 1922, Jeffery [10] was interested in the movement in a viscous shear flow of an isolated ellipsoid, whether elongated (prolate) or flattened (oblate) suspended in a Newtonian fluid. He showed that in addition to the translational movement with the suspending fluid, the ellipsoid is subjected to a rotational periodic movement, called tumbling, whose characteristics depend on the shear rate, the aspect ratio of the ellipsoid and its initial orientation. The trajectory of the end of the direction vector is an ellipse, called Jeffery's orbit. Its orientation with respect to the shear plane is characterised by a constant C. He showed that ellipsoids of revolution rotate in a linear shear field with a period

$$T = \frac{2\pi}{\dot{\gamma}}\left(\frac{a_2}{a_1} + \frac{a_1}{a_2}\right) \tag{1.19}$$

where a_1 and a_2 are respectively the major and the minor axis of the ellipsoid of revolution. Jeffery's orbits are referred to in Chapters 2, 3, 4, 5, 7, and 10.

1.4.3 Flowing particles in interaction with a static wall. The lift force

Under certain conditions, the presence of a solid wall close to a particle can generate its migration across the flow away from the wall in Stokes flow. This phenomenon is of major importance for the behavior of blood cells close to vessel walls.

In a shear flow at zero Reynolds number, a neutrally buoyant rigid particle in dilute suspension does not generally migrate across the flow. For spherical particles, this result follows from the reversibility of Stokes equations. A reversal of the flow direction would result in the opposite migration, which would be a contradiction. For a rigid ellipsoidal or disk-shaped particle that tumble in a shear flow, the same type of argument applies.

A deformable particle, however, migrates away from the wall in simple shear flow even at zero Reynolds number(Fig. 1.8 A). The origin of this migration lies in the interaction of the force dipole generated by the deformed particle and its image in the wall. This has been shown for deformable particles such as drops, vesicles and red blood cells and also for macromolecules and can be calculated by considering the stresslet applying on the particle. A paramount example is the drift away from the capillary walls of red blood cells leading to the formation of a cell-free layer at the capillary walls. The average steady inclination of a cell during tank-treading and its non-sphericity are two sufficient ingredients to induce migration away from walls. The lift velocity is a function of the shear rate and the distance from the substrate but also of the particle properties such as shape, volume, and deformability.

Additionally, if the velocity gradient is not constant, as in Poiseuille flow, a deformable particle can undergo lateral migration even in the absence of hydrodynamic wall effects (Fig. 1.8 B)

When inertial effects are present, even if they are weak, another migration process takes place, which also concerns spherical objects in bounded and unbounded flows such as shear

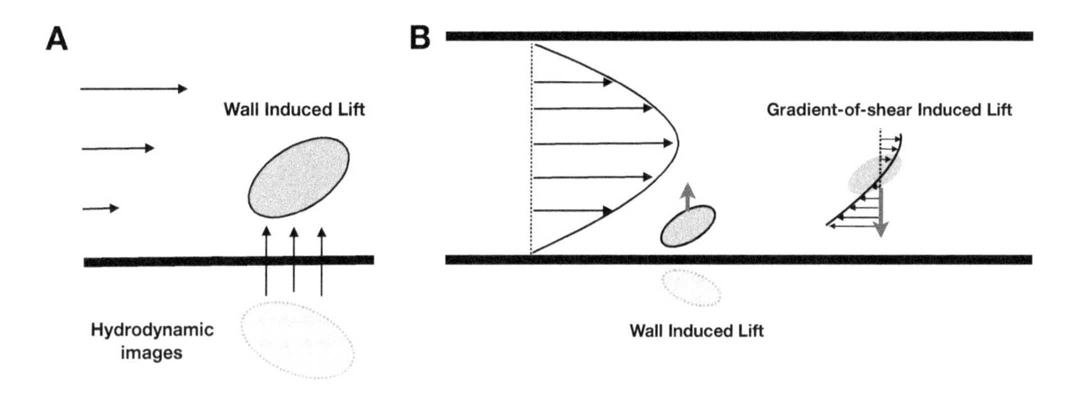

FIGURE 1.8
Schematic illustrating the viscous lift for a deformable particle flowing at low Reynolds number. A: Wall-induced viscous lift in shear flow. It results from the fore-aft asymmetry of the particle in the shear flow. B: Velocity gradient-induced viscous lift in a Poiseuille flow experienced by a deformable particle. It is induced by the flow profile curvature.

flow or straining flow. Wall effects and profile curvature however provide the dominant migration mechanisms when the Reynolds number of the mean flow is small. Typically, a neutrally buoyant sphere translates across streamlines of a Poiseuille flow to an equilibrium position 60% of the way from the centerline to the tube walls. This lateral migration originates mainly from the interaction of the induced disturbance flow, i.e. the stresslet, with the shear field and the wall. It has to be noted that axisymmetric, but non-spherical, particles generally show the same behavior in all shear flows as a sphere provided that the relevant quantities are time averaged to remove any periodic lateral motion associated with the periodic rotation of the particle.

The so-called inertial microfluidics, which is developed in Chapter 10, takes advantage of these specificities, for example, so that the different blood cells follow different trajectories in ad-hoc flows, for applications in cell sorting.

1.5 RHEOLOGY OF SUSPENSIONS

The behavior of a suspension in a flow is complex because, as we have seen, on the one hand the suspension can have a non-Newtonian rheology, and on the other hand, the particles composing the suspension disturb the flow by inducing very specific characteristics such as particle depletion near a wall.

A first issue is to describe the rheology of dilute suspensions and to define a suspension's viscosity. Einstein showed that the bulk shear viscosity η of a suspension of rigid, inertia-free, and neutrally buoyant mono-sized spheres in a Newtonian fluid is

$$\eta = (1 + 2.5\phi)\eta_f. \tag{1.20}$$

where $\phi = \frac{4\pi}{3}Na^3$ is the volume fraction occupied by the particles and η_f is the viscosity of the suspending fluid.

The suspension viscosity increases because the particles resist deformation. Newtonian suspensions shear-thin or thicken only at rather high volume fractions $\phi > 30-40\%$, because of particle interactions. Einstein's equation can be extended to account for effects of sphere interactions by adding higher order volume-fraction terms, such that

$$\eta = (1 + 2.5\phi + c\phi^2 + ...)\eta_f. \tag{1.21}$$

Today, there is no general agreement for the calculation of coefficients for the second and higher terms and different values for the ϕ^2 coefficient, c, have been proposed.

To get a deeper insight about the strain rate-stress relationship for rigid suspension, let us consider a suspension of rigid and electrically uncharged spheres dispersed in a Newtonian fluid and subjected to a shear flow. A schematic of such a system is represented in Figure 1.9, with the physical parameters to be considered.

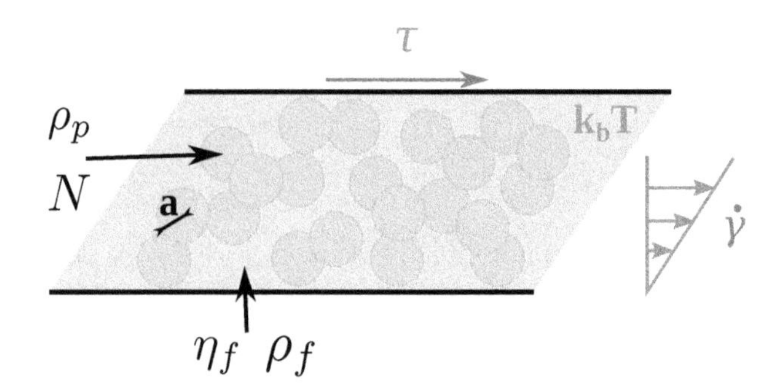

FIGURE 1.9
Characteristic parameters of a suspension of rigid spheres under a shear flow. ρ_p and a are the density and the radius of the spherical particles, and n is the number of particles per unit volume. The suspending fluid has a density ρ_f and a viscosity η_f.. $\dot{\gamma}$ and τ are the shear rate and shear stress of the shear flow. k_bT is the thermal energy.

In the case of isodense suspensions $\rho_p = \rho_f = \rho$ and with, a, ρ, n , η_f, k_BT, $\dot{\gamma}$, τ, two important dimensionless parameters can be defined:

- The particle Reynolds number,

$$Re_p = \frac{\rho_f a^2 \dot{\gamma}}{\eta_f} \tag{1.22}$$

This number is the Reynolds number at the scale of the particle (length scale equal to the particle size). For small particle Reynolds numbers, the inertia of the microscopic flow is negligible at the scale of the particle, and we refer to these as Stokesian suspensions.

- The Peclet number,

$$Pe = \frac{6\pi\eta_f a^3 \dot{\gamma}}{k_B T} \tag{1.23}$$

The Peclet number estimates the contribution of Brownian motion on the displacement of the particles. From the Stokes-Einstein equation which gives the diffusion coefficient of a particle, $D = K_bT/6\pi\eta_f a$, we can express a characteristic time for a particle to diffuse over its size, $t = 6\pi\eta_f a^3/K_bT$. Hence, the Peclet number can be seen as the ratio between the characteristic time scale of particle diffusion with the characteristic time scale associated to the shear flow $1/\dot{\gamma}$. At higher Peclet numbers, as is the case for blood, the Brownian motion of the particles is negligible compared to the shear deformation and we speak of non-Brownian suspensions.

In the regime of high Peclet numbers (non-Brownian suspensions) the rheology does not depend on temperature, thus the effective shear viscosity $\eta_{eff} = \frac{\tau}{\eta_f \dot{\gamma}}$ is not a function of Pe. In addition, for low Reynolds numbers (viscous regime) the inertia of the fluid is negligible and η_{eff} does not depend on Re_p. So for high Pe and low Re_p regime, the relationship between stress and shear rate is then linear and completely determined by the function $\eta_s(\phi)$,

$$\tau = \eta_f \eta_{eff}(\phi)\dot{\gamma}, \tag{1.24}$$

the suspension is Newtonian, with a viscosity $\eta_f \eta_{eff}(\phi)$ and $\eta_{eff}(\phi)$ is independent with regard to the particle size. Outside the range of high Pe and low Re_p, suspensions exhibit more complex rheological behaviors such as shear thinning and shear thickening.

However, when the spheres are non longer rigid but have a viscosity η_s and a surface tension γ_s, the possibility of deforming the spheres makes the problem even more complex and a new quantity, the capillary number Ca, is involved. $Ca = \frac{\eta_f v}{\gamma_s}$ measures the relative effect of the viscous force and the surface tension, where v is a characteristic velocity. For small Ca, the deformation of spheres is controlled by surface tension, viscous spheres remain spherical in the suspension and their deformation can be neglected. But for large Ca, viscous spheres are deformed by the hydrodynamic stress in the suspension. The orientation of deformed spheres in the flow then changes as well as their mutual interactions and the viscosity of the suspension is then modified.

1.6 RHEOLOGY OF BLOOD

In this section the most known features of blood rheology are introduced. We show that indeed, the deformability of the cells composing the blood suspensions generate non-linear effects such as shear thinning and two classical effects known as the Fåhraeus and the Fåhraeus - Lindqvist effects.

1.6.1 Blood, a shear thinning fluid

In the seventies, Chien and collaborators [3] found that blood is a shear thinning fluid. It is illustrated in Figure 1.10. At low shear rates, one can see how the viscosity versus shear rate curve of whole blood is higher than that of the same suspension treated to avoid aggregation, showing the importance of the structures formed by blood cells. At high shear rates, the viscosity of a suspension of RBCs stiffened by glutaraldehyde treatment is higher than that of whole blood, revealing how this second region of the curve is dominated by the deformability of RBCs.

In fact, as it is described in several chapters of this book, at low shear rates, RBCs aggregate and stack into three-dimensional structures called rouleaux, which are piles of RBCs analogous to piles of coins. Fibrinogen and globulin proteins present in the plasma promote this aggregation as will be discussed in Chapter 6. At a higher shear rate, for example for physiological shear rates (typically $\gg 100\ s^{-1}$), rouleaux fragment and cells circulate individually. This leads to a decrease in blood viscosity. At high shear, the RBCs elongate and align in the direction of flow, facilitating their movement relative to each other and further reducing blood viscosity (the process will be challenged in Chapter 5).

1.6.2 RBC-free layer

In confined microflows, for example in capillaries, blood cells interact with capillary walls. This generates structured flows and specific micro-rheological properties of blood that will be discussed and further developed in most chapters dedicated to RBCs. A major point is the tendency of RBCs to migrate towards the centre of the vascular vessels of the microcirculation. This is due to the lift forces experienced by RBCs flowing nearby a wall

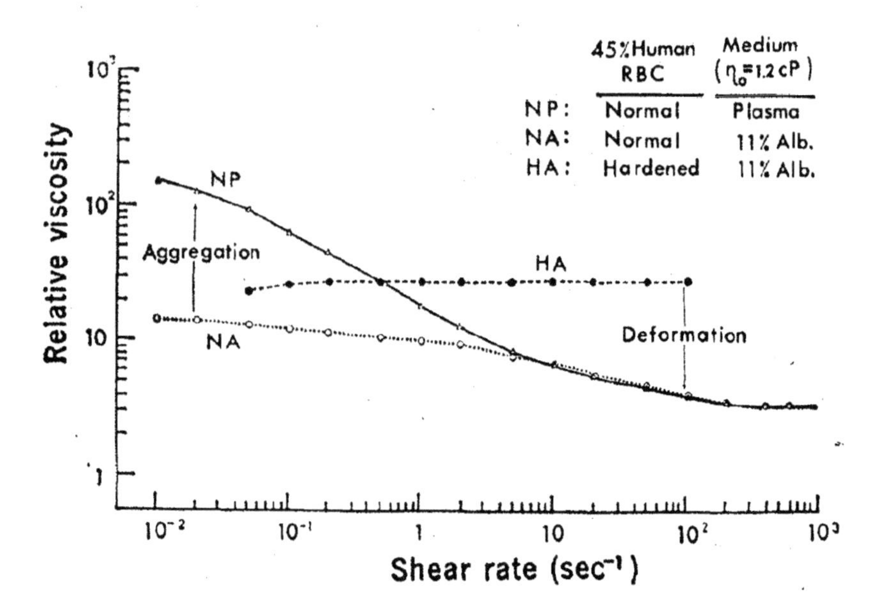

FIGURE 1.10
Relative apparent viscosity versus shear rate for three types of RBC suspensions at 45% hematocrit. The curve NP refers to normal blood. The curve NA refers to RBCs suspended in an albumin solution lacking fibrinogen and globulins. The absence of these proteins which favour aggregation results in a decrease of viscosity at low shear rates. The curve HA refers to hardened RBCs suspended in the same medium as NA. The difference between NP and HA curves at higher shear rates indicate the contribution of cell deformability. From [3].

described in the preceding section and results in the formation of a thin RBC-free layer at wall vessels. Fåhraeus and Lindqvist discovered that the cell-free layer plays a major role in the two rheological effects.

1.6.3 The Fåhraeus Effect and the Fåhraeus-Lindqvist Effect

The RBC volume fraction is called hematocrit. In the microcirculation it varies between 10-30 %, compared to arterial hematocrit of $\sim$ 50%,. The Fåhraeus effect is the decrease of hematocrit in blood flowing in a narrow tube (called tube hematocrit) compared to that in the blood entering and leaving the tube (called discharge hematocrit) [5]. The hematocrit decreases as the tube diameter decreases from 500 μm to 10 μm and reaches a minimum. The reason is that, due to the lift force, there is a cell-free layer of plasma against the vessel walls of a few micrometers, which is not negligible compared to the characteristic diameters of the vessels. Thus RBCs concentrate in the core of the vessel and their average velocity is therefore larger than that of blood. The hematocrit was shown to be inversely proportional to the average blood velocity [18]. Upon further decrease in tube diameter, RBCs form single files in the tube and the hematocrit increases sharply.

The Fåhraeus-Lindqvist effect is an effect associated with the Fåhraeus effect that describes how blood viscosity changes with the tube diameter. For tube diameters ranging between 10 and 300 μm blood viscosity decreases as the tube's diameter decreases. For tube diameters smaller than 6 μm, RBCs flow in single file and fill the cross section of the tube. The friction between the cell and the wall results in a large pressure drop across the cell, leading to an apparent large viscosity. Between 6 and 10 μm, a thick depletion cell-free layer is formed near the tube wall. The friction then decreases, the cell-free layer

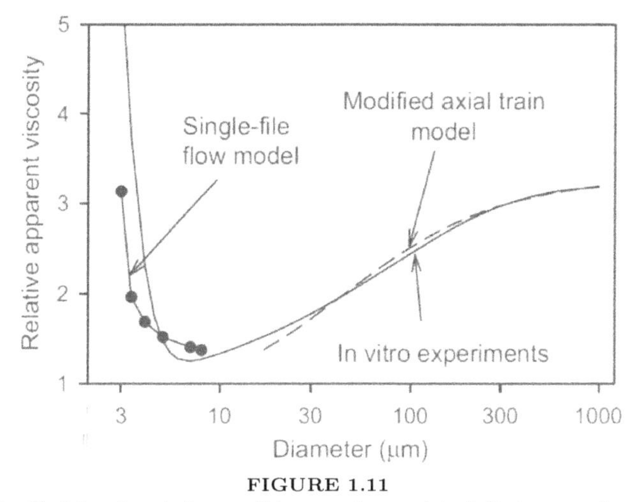

FIGURE 1.11
Fåhraeus-Lindqvist effect in glass tubes: solid curve is empirical fit to experimental data, dots are theoretical predictions, dashed curve is two-phase model with cell-free layer width 1.8 μm (from Secomb [17]).

has a locally reduced viscosity and the average result is a reduction of the relative blood viscosity. When the tube diameter increases beyond 10 μm, the cells form multiple files and the ratio of the depletion layer to the tube diameter starts decreasing. Cells flowing at different radial positions in the tube interact with their neighbours leading to a complex anisotropic shear-induced diffusion. A two-phase empirical model considering a core region containing RBCs associated with a core viscosity and a plasma layer close to the walls associated with a plasma viscosity allows estimating the relative viscosity and hematocrit of the suspension (Figure 1.11) as discussed further in Chapter 8. The presence of a plasma layer decreases the local viscosity in the region near the wall. This effect progressively decreases upon increasing the tube diameter [15, 16].

1.7 *RELEVANT CONCEPTS IN CONTINUUM MECHANICS FOR BLOOD FLOW

1.7.1 Density and hematocrit

One important parameter in any approach to describe fluid flow is the spatial and instantaneous distribution of matter within the volume of interest, which is measured with the material mass per unit volume of fluid often noted as $\rho(x, y, z, t)$. This mass "concentration" is a scalar field and is estimated for blood to be around 1.06×10^3 kg/m^3.

A related quantity for suspensions is the volume fraction of a species of particles. It is defined as the ratio of the total added volume of that particular species of elements over the entire volume of the suspension considered. Since one abundant species of particles for blood is RBCs, their volume fraction has a special terminology used in this book (and by clinicians) which is the "hematocrit" (it literally means *separated blood*, since it is measured by centrifuging the cells in a microcapillary tube).

As for many suspensions, the volume fraction of blood has often a more straightforward rheological meaning that the local mass concentration, though local mass concentration plays a role for instance during sedimentation flows (since RBCs have a cytoplasm density of about 1.09-1.1×10^3 kg/m^3) or inertial flows of the macrocirculation.

1.7.2 Simple idea about deformation and strain

If one pulls on both ends of an elastic rod with different forces parallel to an axis x, it will stretch and move in this direction and its total length L will increase from its resting length L_0 as shown in Figure 1.12A. The deformation of the rod implies both a change of position and a local change of length. Indeed, if we note the position of one end of the rod with respect to the origin of the coordinate system, we could try to define deformation as the rate $F = (x(t) - x_0)/x_0$. However, we easily see that $F = L/L_0 - 1 + \Delta x(t)/L_0$, which is a mixture of both the intrinsic stretch of the rod L/L_0 and its displacement.

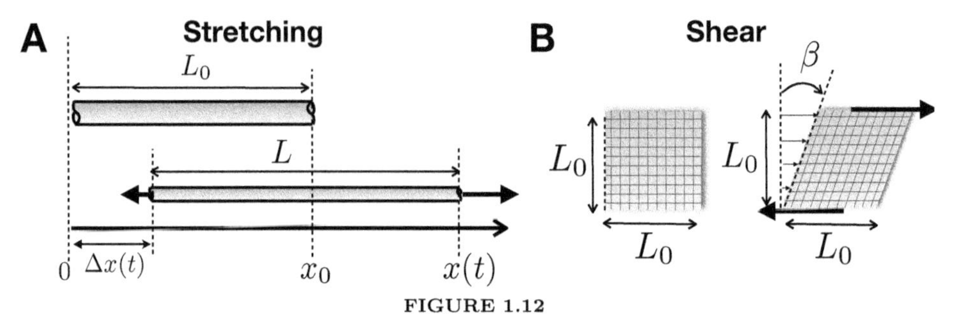

FIGURE 1.12
(A) a rod being deformed in time and (B) a square cross-section material body being sheared.

As we will see in the next section, a true measure of stretching must remain intrinsic and involve only the stretching ratio $\lambda \equiv L/L_0$ or inversely $\Lambda \equiv L_0/L$. The quantification of the stretching will give different numbers depending upon whether we look at the deformation from the current deformed state of the rod relative to its initial length or inversely, its actual length relative to its initial one. Another measure of the intrinsic deformation of the rod, is to introduce a relative percentage of stretch in length, what we will call later the strain and again there are two ways of defining it, one relative to the current deformed state and one from the initial unstressed state:

$$e = \frac{L^2 - L_0^2}{2L_0^2} = \frac{1}{2}\left(1 - \lambda^2\right), \text{ or } E = \frac{L^2 - L_0^2}{2L^2} = \frac{1}{2}\left(\Lambda^2 - 1\right). \tag{1.25}$$

e is called the Green-Lagrange strain, while E is the Euler-Almansi strain. For small deformations ϵ such as $L = L_0 + \epsilon$, both measures are equal: $|e| = |E| = \epsilon$. But these two measures are different for finite deformations which especially is the case in fluid mechanics where flow implies inevitably large deformations.

This simple example shows how the measure of strain implies to carefully look at the local changes of length, which in formal vocabulary is called the metric of the material. However other types of deformation keep local lengths constant but induce changes in the local orientation of material elements and relative material displacement. In Figure 1.12B, an example is showed of such deformation called a shear strain. The square piece of material represented in the figure is sheared by a tangential force applied on its surface displacing each material element from a quantity proportional to the height. The surface area of this square element did not change but an angle β quantifies the amount of shear deformation induced in the material.

1.7.3 Deformation field and formal measure of strain in a body

In continuum mechanics, any local displacement is the composition of a rigid-body displacement (translation and/or rotation) with the intrinsic modification of the distances of the material elements in the body. If $\mathbf{x^0}$ represents the position of the reference configuration of the material elements, and $\mathbf{x}$ the position in the current deformed configuration of these elements, expressed in another reference frame (translated and/or rotated from the initial one), one can write a functional relationship between these vectors (Figure 1.13):

$$\mathbf{x} = \chi(\mathbf{x^0}, t), \tag{1.26}$$

where t stands for time and χ is a continuously differentiable functional with respect to $\mathbf{x^0}$[iii].

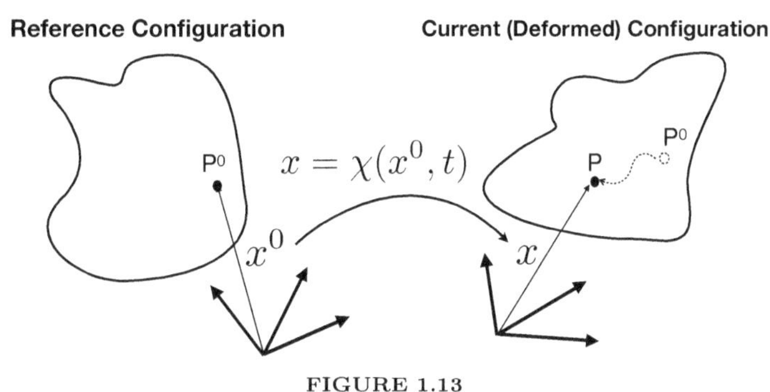

FIGURE 1.13
A schematic representation of two different configurations of states of material elements.

The fundamental measure of deformations of the material is called the deformation gradient tensor $\mathbf{F}$, which relates the infinitesimal differential material elements in the current state of deformation of the body with the one of its reference state, such as :

$$d\mathbf{x} = \mathbf{F}d\mathbf{x^0}, \text{ with } F_{ij} \equiv \frac{\partial x_i}{\partial x_j^0}. \text{ (iv)} \tag{1.27}$$

It is a generalization of the stretch ratio we defined roughly in the previous paragraph. $\mathbf{F}$ combines both rigid body translations and rotations as well as deformations[v]. Since rigid-body motions induce no change in distance between any two material points of the body, we need to define an intrinsic entity which better describes the inner deformations of the material, independently from its local displacements. As also used in Chapter 5 in the

---

[iii] The differentiability of χ ensures that there exists an inverse transformation bringing back the elements from the current state to the reference one.

[iv] We use both boldface notation and index notation for vectors and tensors throughout the book. The index in Latin represents one component of the mathematical entity considered. For instance, the list of three numbers (x_1, x_2, x_3) is denoted by the symbol x_i, where the index i ranges from 1 to 3. Similarly, the list X_{ij} represents the nine following numbers $(X_{11}, X_{12}, X_{13}, X_{21}, X_{22}, X_{23}, X_{31}, X_{32}, X_{33})$ which will be put in matrix form when representing a tensor in a given coordinate frame. Moreover, we adopt Einsteins' convention that if a subscript appears twice in the same term, then it means a summation over that subscript. Therefore, X_{ii} represents in fact the sum $X_{11}+X_{22}+X_{33}$ and is by definition the trace of the tensor, while $X_{ij}x_j$ means the sum $X_{i1}x_1 + X_{i2}x_2 + X_{i3}x_3$. Finally, we use the dot notation to indicate either an inner product between two vectors or between a vector and a tensor, such as :$\mathbf{x}\cdot\mathbf{y}$ is the scalar $x_1y_1+x_2y_2+x_3y_3$ while $\mathbf{x}\cdot\mathbf{X}$ is the vector $(X_{11}x_1 + X_{12}x_2 + X_{13}x_3, X_{21}x_1 + X_{22}x_2 + X_{23}x_3, X_{31}x_1 + X_{32}x_2 + X_{33}x_3)$.

[v] Reversibility of the deformations implies that $\mathbf{F}\mathbf{F}^{-1} = \mathbf{F}^{-1}\mathbf{F} = \mathbf{I}$, where $\mathbf{I}$ is the identity matrix given in index notation by the Kronecker delta δ_{ij}.

special case of membranes, one can write the variations of the local lengths of materials, what is called a "metric" of the current elastic state $(ds)^2 = |d\mathbf{x}|^2$ of the body, in comparison to the reference configuration which we identify with a 0 superscript, $(ds^0)^2 = |d\mathbf{x^0}|^2$.

Classically, two choices exist which do not lead to the same quantities as we presented in the former section in the case of the stretching of a simple rod. We define the Green-Lagrange strain tensor e_{ij} which expresses only local deformations relative to the reference configuration :

$$(ds)^2 - (ds_0)^2 = dx_k dx_k - dx_k^0 dx_k^0 \equiv 2e_{ij} dx_i^0 dx_j^0, \text{ with } e_{ij} = \frac{1}{2}\left(\frac{\partial x_k}{\partial x_i^0}\frac{\partial x_k}{\partial x_j^0} - \delta_{ij}\right). \quad (1.28)$$

Inversely, the same variation is given by the Euler-Almansi strain tensor E_{ij} when deformations are relative to the current state:

$$(ds)^2 - (ds_0)^2 = dx_k dx_k - dx_k^0 dx_k^0 \equiv 2E_{ij} dx_i dx_j, \text{ with } E_{ij} = \frac{1}{2}\left(\delta_{ij} - \frac{\partial x_k^0}{\partial x_i}\frac{\partial x_k^0}{\partial x_j}\right). \quad (1.29)$$

These tensors are symmetric, in that $E_{ij} = E_{ji}$ and $e_{ij} = e_{ji}$. The Euler-Almansi strain tensor corresponds to the so-called Eulerian representation of the deformations while the Green-Lagrange strain tensor is the Lagrangian representation. The latter is suited for elastic descriptions of large deformations where one has to follow the material elements of the medium and describe their current elastic state, while the Eulerian representation describes current deformations in the deformed state, and only the velocity field and its local gradients are important. It is essential in the description of fluid behaviors. Any full strain-stress relationship seeks the proper representation for both viscous terms that are usually expressed in an Eulerian view and elastic terms which are Lagrangian in nature (see example of Chapter 5)[(vi)].

For any deformation, one can find three mutually orthogonal directions along which the material elements undergo a pure stretch (i.e. for which $\mathbf{F}$ is diagonal). These directions are called principal axes of deformation and the associated stretches are called principal stretches. They are respectively eigenvectors and eigenvalues of the strain tensor. In these principal axes of deformation for which $\frac{\partial x_i}{\partial x_i^0} \equiv \lambda_i$, we have each of the principal components of $\mathbf{e}$ which are given by the diagonal terms $e_{ii,(\text{no sum})} = \frac{1}{2}(\lambda_i^2 - 1)$, where the λ_i physically describe the fractional elongations or contractions along the three local principal axes of strain of the infinitesimal neighborhood of a point in the material as used to describe membrane deformations in Chapter 5 (see also Fig. 5.2 in the same chapter).

1.7.4 Velocity field

As stated in the previous paragraph, care should be taken in the expression of the strain rate in the material, especially for fluids, where the important parameter of its deformation is not the strain but the rate of strain.

Following Ogden [12], the Lagrangian velocity field of a material element, which is identified by the position $\mathbf{x^0}$ in the reference configuration, is denoted $\mathbf{U}(\mathbf{x^0}, t)$. It is defined

---

(vi) One can easily demonstrate that the strain field in the Lagrangian representation can be obtained from the Eulerian form by the simple transformation in matrix notation:

$$\mathbf{E} = {}^{\mathbf{T}}\mathbf{FeF}. \quad (1.30)$$

by:

$$\mathbf{U}(\mathbf{x^0}, t) \equiv \frac{\partial \chi(\mathbf{x^0}, t)}{\partial t}. \tag{1.31}$$

In the Eulerian description, however, the velocity field of a material point occupying a position $\mathbf{x}$ at time t is written $\mathbf{u}(\mathbf{x}, t)$ and we must have $\mathbf{U}(\mathbf{x^0}, t) \equiv \mathbf{u}(\mathbf{x} = \chi(\mathbf{x^0}, t), t)$. The spatial gradient tensor associated to $\mathbf{u}$ writes as $L_{ij} \equiv \partial u_i / \partial x_j$ and defines the local kinematics in the current (deforming) configuration of the material. It relates the infinitesimal differential material element $d\mathbf{x}$ of length ds to its temporal variation such as[(vii)(viii)]:

$$\frac{D(d\mathbf{x})}{Dt} = \mathbf{L} d\mathbf{x}, \tag{1.34}$$

where $D(\cdot)/Dt$ represents the material derivative following the vector $d\mathbf{x}$.

As noted for the deformation gradient tensor, the derivative of this tensor is also not suited to describe local flow components in time. One approach to find the good entity is to calculate the time derivative of the square of the metric which describes the rate of strain in the instantaneous or deformed coordinate system and it is expressed in matrix notation as [(ix)]:

$$\frac{d}{dt}((ds)^2 - (ds_0)^2) = 2\ {}^T d\mathbf{x}\ \boldsymbol{\mathcal{D}}\ d\mathbf{x} \tag{1.36}$$

where $\boldsymbol{\mathcal{D}} = \frac{1}{2}\left(\mathbf{L} + {}^T\mathbf{L}\right)$ is the Eulerian strain-rate tensor and represents the symmetric part of $\mathbf{L}$, while the antisymmetric component $\boldsymbol{\Omega} = \frac{1}{2}\left(\mathbf{L} - {}^T\mathbf{L}\right)$ expresses the rotations of the current configuration of material elements. $\boldsymbol{\mathcal{D}}$ is in fact the classical quantity used in fluid mechanics to describe flows. It is the true measure of the local deformation rate in Eulerian representation and not the time derivative of $\mathbf{E}$.

1.7.5 Acceleration in Eulerian representation and spatial derivative

Accelarations are tricky entities to calculate in fluid and continuum mechanics. In fact, while in a Lagrangian representation following material elements during their deformations, acceleration could be defined as in Newtonian mechanics as $\mathbf{a} \equiv \frac{D\mathbf{U}}{Dt}$, where $\mathbf{U}$ is the Lagrangian velocity of the moving material elements. The Eulerian form, which represents

(vii) Since $\mathbf{x^0}$ and t are independent variables, we have:

$$\dot{\mathbf{F}} = \mathbf{L} \cdot \mathbf{F}. \tag{1.32}$$

(viii)

$$\dot{F}_{ik} = \frac{\partial}{\partial t}\left(\frac{\partial \chi_i}{\partial x_k^0}\right) = \frac{\partial \dot{\chi}_i}{\partial x_k^0} = \frac{\partial u_i}{\partial x_j}\frac{\partial x_j}{\partial x_k^0} = L_{ij} F_{jk}. \tag{1.33}$$

(ix)

$$\begin{aligned}
\frac{d}{dt}((ds)^2 - (ds_0)^2) &= {}^T d\dot{\mathbf{x}} d\mathbf{x} + {}^T d\mathbf{x} d\dot{\mathbf{x}} \\
&= {}^T d\mathbf{x^0}\ {}^T\dot{\mathbf{F}} d\mathbf{x} + {}^T d\mathbf{x} \dot{\mathbf{F}} d\mathbf{x^0} \\
&= {}^T d\mathbf{x^0}\ {}^T\mathbf{F}\ {}^T\mathbf{L} d\mathbf{x} + {}^T d\mathbf{x} \mathbf{L} \mathbf{F} d\mathbf{x^0} \\
&= 2\ {}^T d\mathbf{x} \left[\frac{1}{2}\left({}^T\mathbf{L} + \mathbf{L}\right)\right] d\mathbf{x} \\
&= 2\ {}^T d\mathbf{x}\ \boldsymbol{\mathcal{D}}\ d\mathbf{x},
\end{aligned} \tag{1.35}$$

how the velocity field of the flow is changing spatially relative to the observer looking at the fluid in movement, can be written with respect to the spatial (Eulerian) representation as:

$$\mathbf{a} = \frac{\partial \mathbf{u}}{\partial t} + (\mathbf{u} \cdot \nabla)(\mathbf{u}). \tag{1.37}$$

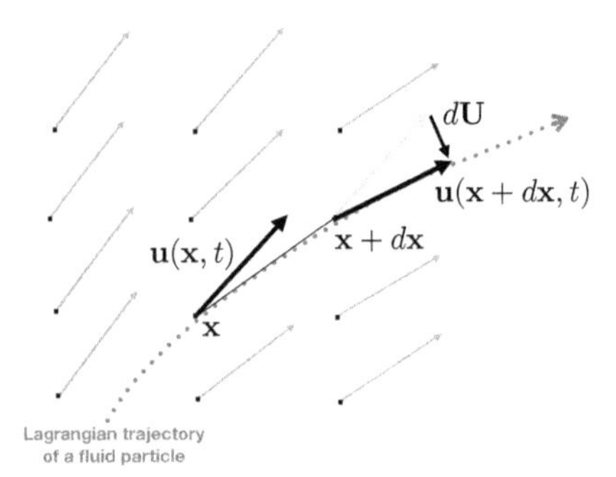

FIGURE 1.14
Definition of the Eulerian acceleration field.

To demonstrate this equality one has to understand the local meaning of such a relationship. In Figure 1.14, one can see the velocity field of a fluid in a fixed frame, the Eulerian frame. At each point on the grid the velocity $\mathbf{u}$ of the fluid is measured, but this velocity does not represent the velocity of the same particle of fluid but many different ones that acquired this velocity at a given place in the grid. If one follows a particle of fluid, one will see it moving through this grid acquiring the velocity $\mathbf{U}(t) = \mathbf{u}(\mathbf{x}, t)$, where $\mathbf{x} = (x_1, x_2, x_3)$ is the position of the specific points of the grid where the particle is at time t. The differential variation of velocity $D\mathbf{U}$ of this Lagrangian particle of fluid should be given by:

$$D\mathbf{U} = \mathbf{u}(x_1 + d\mathbf{x}, t + dt) - \mathbf{u}(\mathbf{x}, t).$$

Using a simple Taylor expansion, one obtains:

$$\begin{aligned} D\mathbf{U} &= \mathbf{u}(\mathbf{x}, t) + dx_1 \frac{\partial \mathbf{u}(\mathbf{x}, t)}{\partial x_1} + dx_2 \frac{\partial \mathbf{u}(\mathbf{x}, t)}{\partial x_2} + dx_3 \frac{\partial \mathbf{u}(\mathbf{x}, t)}{\partial x_3} + \frac{\partial \mathbf{u}(\mathbf{x}, t)}{\partial t} dt + \cdots - \mathbf{u}(\mathbf{x}, t) \\ &= \frac{\partial \mathbf{u}(\mathbf{x}, t)}{\partial t} dt + (\mathbf{u}(\mathbf{x}, t) \cdot \nabla)(\mathbf{u}(\mathbf{x}, t)) dt, \end{aligned} \tag{1.38}$$

where the gradient operator $\nabla(\cdot)$ is the vector $(\partial(\cdot)/\partial x_1, \partial(\cdot)/\partial x_2, \partial(\cdot)/\partial x_3)$. Finally the acceleration is:

$$\frac{D\mathbf{U}}{dt} = \frac{\partial \mathbf{u}(\mathbf{x}, t)}{\partial t} + (\mathbf{u}(\mathbf{x}, t) \cdot \nabla)(\mathbf{u}(\mathbf{x}, t)). \tag{1.39}$$

In fact, this expression is the result of a general result about material derivative which will present the same functional form for any variable of the problem. The material derivative is indeed related to the local spatial derivatives using the chain rule for derivation. Following Robertson [8], let us take the example of another variable than the velocity, for instance the density field. Let us define the density of the material $\bar{\rho}$ at a point $\mathbf{x^0}$ in the Lagrangian representation of the reference configuration and $\hat{\rho}(\mathbf{x})$ its Eulerian representation in the current configuration at a point $\mathbf{x}$. By definition, $\bar{\rho}(\mathbf{x^0}) = \hat{\rho}(\mathbf{x})$. If we would like to know its time derivative with respect to the variable in the Eulerian representation, we just need to decompose this function and apply the chain rule of derivation:

$$\frac{D\bar{\rho}(\mathbf{x^0}, t)}{Dt} \equiv \frac{\partial \bar{\rho}(\mathbf{x^0}, t)}{\partial t} = \frac{\partial \hat{\rho}(\mathbf{x}, t)}{\partial t} + u_i \frac{\partial \hat{\rho}(\mathbf{x}, t)}{\partial x_i}, \tag{1.40}$$

which in vectorial notation gives us:

$$\frac{D(\cdot)}{Dt} = \frac{\partial(\cdot)}{\partial t} + (\mathbf{u} \cdot \nabla)(\cdot). \tag{1.41}$$

1.7.6 Conservation of mass and the incompressibility condition

As depicted in Figure 1.15, let us consider the domain of volume V of total mass M delimited by the surface S at time t that will deform into another domain of volume V' and surface S' at a time $t + dt$, sweeping a volume $\Delta V = V' - V$. The total mass variation in time of V can be written as a difference:

$$\frac{DM}{Dt} = \lim_{dt \to 0} \frac{1}{dt} \left[\int_{V'} \rho(\mathbf{x}, t + dt) dv - \int_{V} \rho(\mathbf{x}, t) dv \right] \tag{1.42}$$

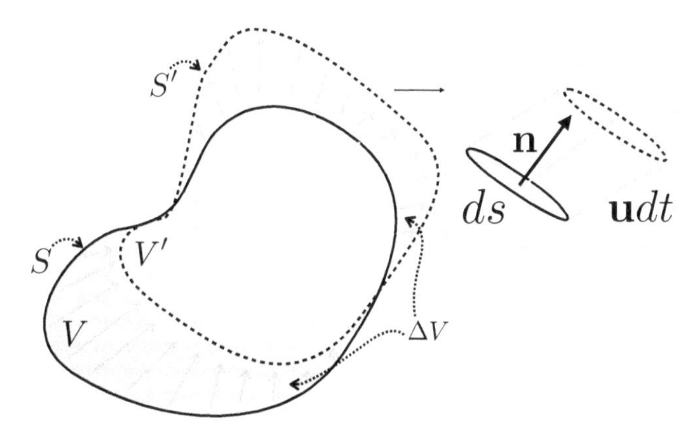

FIGURE 1.15
Deformation of a volume V of material of surface S into another volume V' of surface area S' for an elementary time step dt.

The integral over V' can be decomposed into an integral over V and ΔV, leading to:

$$\frac{DM}{Dt} = \lim_{dt \to 0} \frac{1}{dt} \left[\int_{V} \left[\rho(\mathbf{x}, t + dt) - \rho(\mathbf{x}, t)\right] dv + \int_{\Delta V} \rho(\mathbf{x}, t + dt) dv \right]. \tag{1.43}$$

The first term on the right hand side is simply: $\int_V \frac{\partial \rho}{\partial t} dv$. The second term however represents the amount of matter that entered and left the initial volume crossing the surface S which transformed the domain into the final deformed domain of volume V'. This net flux of matter is given then by:

$$\int_S (\rho \mathbf{u} \cdot \mathbf{n}) ds, \tag{1.44}$$

where $\mathbf{n}$ is the normal vector field around surface differential elements ds. Here, the differential flux of mass crossing a particular surface element ds per unit time dt is simply given by $(\rho \mathbf{u} \cdot \mathbf{n})$ (it is the differential cylindrical volume swept by the local deformation in Figure 1.15).

By analogy with electrostatics, this surface integral can be transformed into a volume integral of the divergence of the vector field $(\rho \mathbf{u} \cdot \mathbf{n})$, using Gauss's theorem (also called Green-Ostrogradski's theorem):

$$\int_S (\rho \mathbf{u} \cdot \mathbf{n}) ds = \int_V \nabla \cdot (\rho \mathbf{u}) dv. \tag{1.45}$$

We obtain finally the following integral relationship for the density (referred to as the Reynolds transport theorem in the literature):

$$\frac{D}{Dt}\left(\int_V \rho dv\right) = \int_V \left[\frac{\partial \rho}{\partial t} + \nabla \cdot (\rho \mathbf{u})\right] dv, \tag{1.46}$$

If no matter is destroyed or produced during fluid motion, the principle of mass conservation states that the total amount of matter $M = \left(\int_V \rho dv\right)$ should remain constant. Therefore $DM/Dt \equiv 0$. This leads to the continuity equation:

$$\frac{\partial \rho}{\partial t} + \nabla \cdot (\rho \mathbf{u}) = 0, \tag{1.47}$$

which expresses a local form of the mass conservation. Since density variations are often negligible for liquids such as water or suspensions such as blood, ρ can be considered as constant and the former equation reduces further to a condition on the divergence of the velocity field:

$$\nabla \cdot \mathbf{u} = 0, \tag{1.48}$$

or in a cartesian coordinate system:

$$\frac{\partial u_x}{\partial x} + \frac{\partial u_y}{\partial y} + \frac{\partial u_z}{\partial z} = 0. \tag{1.49}$$

This is the so-called incompressibility condition which is a strong constraint on the velocity field.

1.8 *NOTION OF TRACTION FORCES, STRESS TENSOR AND BODY FORCES

Deformation and flow of continuous materials are produced by internal and external forces. As we will see in the next section, in order to apply Newton's second law to determine the way these material elements move under force, we have to generalize the concept of force applied to point masses to a continuous medium. One important notion to understand is the one of stress.

1.8.1 Simple idea about stress

Let us go back to our rod with a circular cross-section of radius R and total length L_0 as shown in Figure 1.16. If we pull on it with two forces F applied at the center of each cross-section of both ends and directed orthogonally to these cross-sections, the rod will elongate to a new length L. The question now is how much more force one should apply to a second rod made of the same material but twice as big in its cross-section radius to reach the same length L? The idea of stress is all about these type questions. Indeed, the important parameter to consider is not the absolute value of the force but the ratio of the force to the cross-section area of the rod, which is called a stress. In the case of Figure 1.16 the force will need to be four times larger because the cross-sectional area of the second rod is 4 times larger too.

1.8.2 The formal notion of traction forces

Let us consider indeed a body $\mathcal{B}$ and imagine a small surface $\mathcal{S}$ of material inside enclosing virtually an inner part from the rest as represented in Figure 1.17A. This part of the body

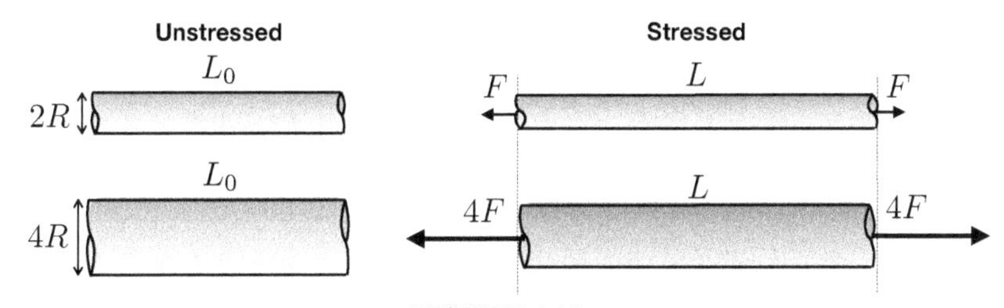

FIGURE 1.16
Stretching of two rods of different cross-section radii.

is interacting with the outer neighbouring elements through the imaginary interface $\mathcal{S}$. If we consider now an elementary element ds of this surface (Figure 1.17A), the neighbouring material elements literally "pull" on $\mathcal{S}$ through ds with a force $\mathbf{T}$. It can also be seen as the force produced by $\mathcal{S}$ over the rest of the body through ds. The intensity of this force changes with the surface ds but reaches a stable definite value when ds goes to zero. This force per unit area is called the traction force or stress vector. It is a vector field that can be defined at any point of $\mathcal{S}$. The stress vector $\mathbf{T}$ has a normal component $\mathbf{T_n}$ in direction of the normal $\mathbf{n}$ to ds and its norm is the familiar notion of mechanical pressure p such as $\mathbf{T_n} = -p\mathbf{n}$ (the negative sign indicates that pressure is a compressive force acting on $\mathcal{S}$). Like pressure the unit of the stress vector norm is in Pascal (Pa). This pressure is called the isostatic pressure and it is a generalization of the notion of the hydrostatic (or thermodynamic) pressure in a liquid. Noteworthy, the tangential components are called the shear components.

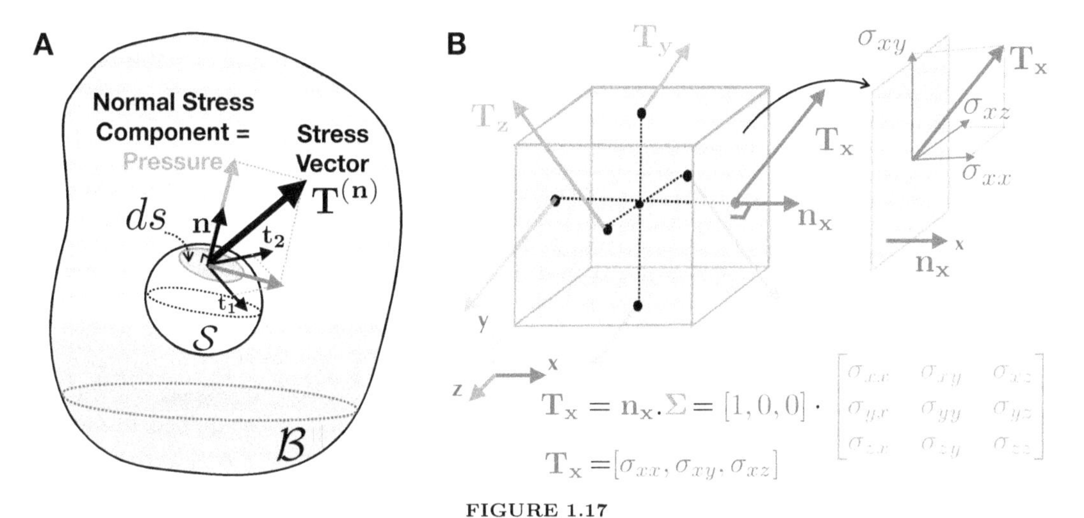

FIGURE 1.17
(A) Definition of the stress vector acting over a differential surface element ds inside of a body $\mathcal{B}$. (B) Representation of the traction forces acting over the faces of parallelipiped element, and the operational definition of the stress tensor.

1.8.3 The formal notion of the stress tensor

The stress vector is defined when the normal to an elementary surface is given around a point of interest. Therefore, changing the orientation of this differential surface around a given point shows that an infinite number of traction vectors $\mathbf{T}^{(\mathbf{n})}$ can be associated with each orientation with a normal $\mathbf{n}$. In fact, one can associate a basis of vectors at that point with the normal $\mathbf{n} = \mathbf{t_3}$ being orthogonal to the interface and two other mutually orthogonal vectors tangent to the surface $\mathbf{t_1}$ and $\mathbf{t_2}$ (see Figure 1.17A), such as $\mathbf{T}^{(\mathbf{n})}$ can be decomposed on this local basis:

$$\mathbf{T}^{(\mathbf{n})} = T_1^{(\mathbf{n})}\mathbf{t_1} + T_2^{(\mathbf{n})}\mathbf{t_2} + T_3^{(\mathbf{n})}\mathbf{t_3} \tag{1.50}$$

According to Cauchy's fundamental stress theorem (see Fung [7]) by knowing the stress vectors on three mutually perpendicular planes, the stress vector on any other plane passing through that point can be found through coordinate transformation equations. This implies that $\mathbf{T}$ and $\mathbf{n}$ are directly related with each other by a tensorial entity called the stress tensor $\mathbf{\Sigma}$ such as:

$$\mathbf{T} = \mathbf{n} \cdot \mathbf{\Sigma} \tag{1.51}$$

In a cartesian coordinate system, the components of $\mathbf{\Sigma}$ are simply given by the following matrix:

$$\mathbf{\Sigma} = \begin{bmatrix} \sigma_{xx} & \sigma_{xy} & \sigma_{xz} \\ \sigma_{yx} & \sigma_{yy} & \sigma_{yz} \\ \sigma_{zx} & \sigma_{zy} & \sigma_{zz} \end{bmatrix} \tag{1.52}$$

An example of how to use such an entity is represented in Figure 1.17B. In the figure, a rectangular parallelepiped continuum is represented in a cartesian reference frame $(\mathbf{x}, \mathbf{y}, \mathbf{z})$ whose vectors are parallel to the edges of the parallelepiped. As shown in Figure 1.17B, If one considers the surface with an outer normal $\mathbf{n_x}$ pointing in the positive direction of the $\mathbf{x}$-axis, the stress vector acting on this surface will be simply given by:

$$\mathbf{T_x} = \mathbf{n_x} \cdot \mathbf{\Sigma} = (\sigma_{xx}, \sigma_{xy}, \sigma_{xz}). \tag{1.53}$$

Noteworthy, the component σ_{xy} is therefore the y component of the traction force exerted on the unit surface with a normal in the direction x. Moreover, for bodies in equilibrium and in the absence of external torques, the resultant moment around any element in equilibrium must vanish leading to the important property that the stress tensor is symmetric and therefore:

$$\sigma_{ij} = \sigma_{ji}. \tag{1.54}$$

1.8.4 Body forces: example of gravity

In the former paragraph, we saw that deformations in a continuous medium are driven by surface forces called stresses that act on surface elements and can be transmitted through the outer surface of the body by forces like hydrodynamic friction forces (as we will see later). But a second type of forces are long-range actions that can be applied to volume elements of the body. Such forces are called body forces and common examples are gravity and electromagnetic forces (see applications in Chapter 11). The total force $\mathbf{f_b}$ acting on a given elementary volume dV can then be expressed using a force density $\mathbf{b}$ such as $\mathbf{f_b} = \mathbf{b}dV$.

In the special case of gravity $\mathbf{b} = \rho\mathbf{g}$, where $\mathbf{g}$ is the acceleration of gravity.

1.9 *CONSERVATION OF LINEAR MOMENTUM AND THE EQUATIONS OF MOTION

Newton's second law of motion states that in an inertial frame of reference, the material rate of change of the linear momentum $\mathcal{P}$ of a body is equal to the resultant of the forces $\mathcal{F}$ applied to the body:

$$\frac{D\mathcal{P}}{Dt} = \mathcal{F}. \tag{1.55}$$

Since the total linear momentum inside of the volume V of a body is equal to:

$$\mathcal{P} = \int_V \rho \mathbf{u} \, dV, \tag{1.56}$$

its material derivative can be expressed using the same equality to express the continuity condition (*i.e.* the transport theorem) but component by component:

$$\frac{D\mathcal{P}_i}{Dt} = \int_V \left[\frac{\partial(\rho u_i)}{\partial t} + \frac{\partial}{\partial x_j}(u_j \, \rho u_i) \right] dV. \tag{1.57}$$

The integrand can be developed in:

$$\frac{\partial(\rho u_i)}{\partial t} + \frac{\partial}{\partial x_j}(u_j \, \rho u_i) = u_i \left(\frac{\partial \rho}{\partial t} + \frac{\partial(\rho u_j)}{\partial x_j} \right) + \rho \left(\frac{\partial u_i}{\partial t} + u_j \frac{\partial u_i}{\partial x_j} \right). \tag{1.58}$$

The first term of the right hand side vanishes because of mass conservation, while the second is exactly the acceleration we defined earlier.

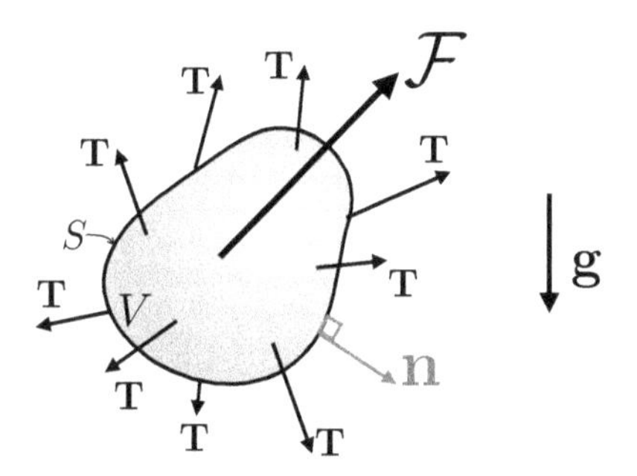

FIGURE 1.18
Total traction forces acting on a volume element.

Moreover, the total force $\mathcal{F}$ acting at the center of mass of the body is the sum of all the traction forces acting over the surface S and the body force $\mathbf{b}$ integrated over the volume V (Figure 1.18), which writes as:

$$\mathcal{F} = \int_S \mathbf{T} \, dS + \int_V \rho \mathbf{b} \, dV, \tag{1.59}$$

Remembering the relationship between traction forces and the stress tensor, $\mathbf{T} = \mathbf{n} \cdot \boldsymbol{\Sigma}$, and transforming the resulting surface integral of the first right hand side into a volume integral using Gauss's theorem, leads to:

$$\mathcal{F} = \int_V \nabla \cdot \boldsymbol{\Sigma} \, dV + \int_V \rho \mathbf{b} \, dV, \tag{1.60}$$

Finally, applying Newton's second law, we get an integral momentum equation. Since it is true for any arbitrary volume element $\mathcal{V}$, it reduces to the general and local law of motion for a deformable body:

$$\underbrace{\rho \frac{\partial \mathbf{u}}{\partial t} + \rho(\mathbf{u} \cdot \nabla)(\mathbf{u})}_{\text{Acceleration Term}} = \underbrace{\nabla \cdot \boldsymbol{\Sigma}}_{\text{Traction Forces Resultant}} + \underbrace{\rho \mathbf{b}}_{\text{Body Forces Resultant}}. \tag{1.61}$$

1.10 *BOUNDARY CONDITIONS

One important element necessary to solve the partial differential equations of eqns. 1.61 are the boundary conditions experienced by the material during its flow. These conditions state how some variables of the problem are given in certain regions of space and at certain given times.

1.10.1 Fluid-Solid interfaces: impermeability and no-slip conditions

One important assumption in fluid mechanics is that a fluid cannot cross the surface of an impermeable solid. This implies that the normal velocity component must vanish there. Meanwhile, for the tangential component, it is customary to assume that both fluid and solid velocities remain equal, the so-called "no-slip" condition. Since often the walls are immobile, then the fluid velocity is zero there. For motions of a fluid around moving solid boundaries, these conditions are sufficient to determine completely a solution of the equations of motion, provided the motion of the solid boundaries is specified.

1.10.2 Fluid-Fluid interfaces

Paraphrasing Gary Leal [11]: "in problems involving two fluids separated by an interface, [the former] conditions are not sufficient because they provide relationships only between the velocity components in the fluids and the interface shape, all of which are unknowns. The additional conditions necessary to completely determine the velocity fields and the interface shape come from a force equilibrium condition on the interface."

1.10.2.1 Kinematic condition

Let us consider two fluids (I) and (II) separated by a continuous interface with a unit normal vector field $\mathbf{n}$ pointing for instance toward the fluid phase (I) (Figure 1.19). Without mass transfer between the two phases, the velocity fields in fluids (I) and (II) must be continuous across the interface. By analogy to the no-slip condition at a rigid boundary one can write the continuity of the tangential components such as:

$$\mathbf{u}^{(I)} - \left(\mathbf{n} \cdot \mathbf{u}^{(I)}\right) \mathbf{n} = \mathbf{u}^{(II)} - \left(\mathbf{n} \cdot \mathbf{u}^{(II)}\right) \mathbf{n}. \tag{1.62}$$

For the normal component, it should also remain continuous across the interface. For both fluids, this normal component of velocity must equal the velocity of the interface normal to itself. One difficulty is therefore in this case to provide the explicit equation of motion of the interface [11].

Moreover, since we have two unknown vector fields $\mathbf{u}^{(I)}$ and $\mathbf{u}^{(II)}$, we need twice as many boundary conditions to describe such a situation. To solve problems in this case, one need to write a boundary condition connecting the state of stress in each fluid at the interface.

1.10.2.2 Surface tension

Because of thermodynamic reasons, the immiscibility of two fluids induces a normal stress jump or pressure difference even without relative movement of the fluid elements. In fact, the interface supports a continuous tension force field acting tangentially to it. This "surface tension" often denoted by the Greek letter γ is isotropic and constant over the entire interface for simple liquids such as water in air (its value is in fact 72 mN/m in this case). It has units of a tension force in N/m and represents the in-plane force per unit length $d\mathbf{f}$ any line of fluid elements dl will be submitted to because of this effect (Figure 1.19B). γ induces a pressure or "normal stress" jump across the interface which depends locally on the mean curvature of the interface at that point. The pressure jump is given by the so-called Young-Laplace equation. If at a specific point of the interface, one writes the pressure difference between the two fluids as $\Delta p = p^{(I)} - p^{(II)}$ and the mean curvature $H = \frac{1}{2}\left(\frac{1}{R_1} + \frac{1}{R_2}\right)$, with R_1 and R_2 the local principal radii of curvature (Figure 1.19A), then the Young-Laplace equation writes as:

$$\Delta p = 2H\gamma = \gamma\left(\frac{1}{R_1} + \frac{1}{R_2}\right) \tag{1.63}$$

If we consider a virtual curve $\mathcal{C}$ enclosing a piece of the interface $\mathcal{S}$, the total action of the tangential surface tensions $d\mathbf{f}$ acting along $\mathcal{C}$ has a net normal component producing the overpressure on the side of fluid (II) as shown in Figure 1.19B.

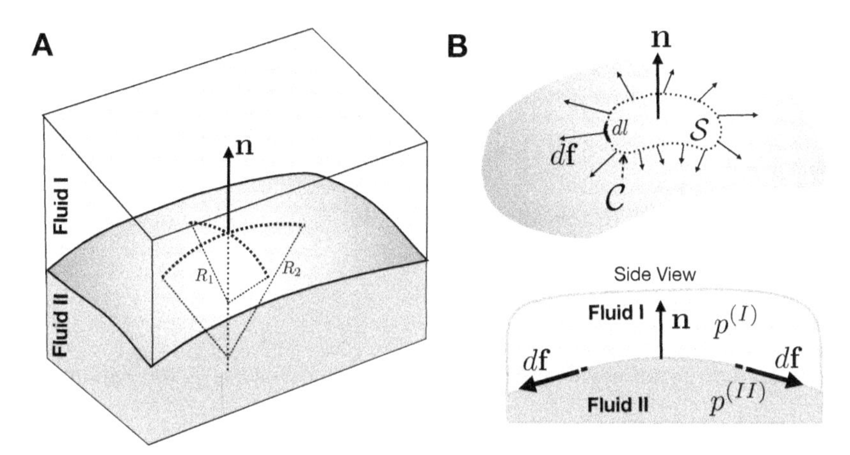

FIGURE 1.19
(A) Sketch of a curved interface between two immiscible fluids. The local normal at the interface is shown as well as the two principal radii of curvature. (B) Top drawing: A surface element of this interface with a virtual closed curve $\mathcal{C}$ of surface $\mathcal{S}$, over which surface tension is pulling. Bottom drawing: same element in side view.

Interestingly, this law can be analogous for elastic membranes which can develop non-isotropic in-plane tension forces. In two principal directions (1) and (2) tangent to the surface, one could then write a generalized Laplace law such as:

$$\Delta p = \frac{T_1}{R_1} + \frac{T_2}{R_2}, \tag{1.64}$$

T_1 and T_2 are respectively the tensions in direction (1) and (2). Such a relationship is easy to use for quasi-spherical interfaces such as spherical cell membranes (see for instance in Chapter 9). It also shows that if an interface is not spherical, it could imply that the membrane elements bear an anisotropic tension field.

1.10.2.3 Stress balance at the interface between two flowing fluids

If one considers the situation of Figure 1.19B of the interface of two flowing fluids, the total force equilibrium of the traction forces due to the flows acting on each side of the interface can be written in an integral form, taking into account the interfacial tension:

$$\int_{\mathcal{S}} \left(\Sigma^{(I)} - \Sigma^{(II)}\right) \cdot \mathbf{n} ds + \int_{\mathcal{C}} \gamma \mathbf{t} dl = 0, \tag{1.65}$$

where $\Sigma^{(I)}$ and $\Sigma^{(II)}$ are the bulk stress tensor respectively in fluid (I) and (II), and $\mathbf{t}$ the vector field tangent to $\mathcal{S}$ and orthogonal to the curve $\mathcal{C}$. The curvilinear integral on the right hand side of this equation can be transformed into a surface integral using Stokes' theorem, such as:

$$\int_{\mathcal{C}} \gamma \mathbf{t} dl = \int_{\mathcal{S}} \nabla_{\mathcal{S}}(\gamma) ds - \int_{\mathcal{S}} \gamma \mathbf{n}(\nabla \cdot \mathbf{n}) ds, \tag{1.66}$$

where $\nabla_{\mathcal{S}} = \nabla - \mathbf{n}(\nabla \cdot \mathbf{n})$ is the surface projected component of the gradient operator ∇ in the local plane of the interface. These tensile forces contribute a net force in the tangential direction that is locally proportional to the gradient of γ, plus a net force normal to the interface that is proportional to γ times the curvature of the interface since $2H \equiv \nabla \cdot \mathbf{n}$[x].

Combining the two former surface integral equations one obtains the following relation:

$$\int_{\mathcal{S}} \left[\left(\Sigma^{(I)} - \Sigma^{(II)} \right) \cdot \mathbf{n} + \nabla_{\mathcal{S}}(\gamma) - \gamma \mathbf{n}(\nabla \cdot \mathbf{n}) \right] ds = 0, \tag{1.67}$$

which is true for any arbitrary surface element $\mathcal{S}$, leading to the normal stress balance relation for the interface between two fluids:

$$\left(\Sigma^{(I)} - \Sigma^{(II)} \right) \cdot \mathbf{n} + \nabla_{\mathcal{S}}(\gamma) - \gamma \mathbf{n}(\nabla \cdot \mathbf{n}) = 0. \tag{1.68}$$

Let us discuss separately in the following sections the normal components and the tangential one of this stress condition.

1.10.2.4 Normal stress balance

Taking the inner product of eqns. 1.68 with $\mathbf{n}$ we get the normal stress balance:

$$\left[\left(\Sigma^{(I)} - \Sigma^{(II)} \right) \cdot \mathbf{n} \right] \cdot \mathbf{n} = \gamma(\nabla \cdot \mathbf{n}) = \gamma \left(\frac{1}{R_1} + \frac{1}{R_2} \right), \tag{1.69}$$

which is a dynamical version of the Young-Laplace equation. It relates the curvature of the interface through the divergence term with the normal stress components, which gather both the thermodynamic pressure component associated to the no-flow condition with the dynamic pressure or normal stress associated to the two flows on both side of the interface.

1.10.2.5 Tangential stress balance

In this case, there will be, in general, two tangential components, which we obtain by taking the inner product with the two orthogonal unit tangent vectors that are normal to $\mathbf{n}$. If we denote these unit vectors as $\mathbf{t_i}$ (with $i = 1$ or 2), the "shear-stress" balance can be written symbolically in the form,

$$\left[\left(\Sigma^{(I)} - \Sigma^{(II)} \right) \cdot \mathbf{n} \right] \cdot \mathbf{t_i} + \nabla_{\mathcal{S}}(\gamma) \cdot \mathbf{t_i} = 0. \tag{1.70}$$

We see that the shear- (*i.e.*, tangential-) stress components are discontinuous across the interface whenever there is a gradient of surface tension on the surface. Such gradients can appear when there are temperature gradients over the interface or surface active molecules such as surfactants or proteins with variable spatial concentrations, both of which locally change the value of γ. This discontinuity of shear stress due to surface tension effects is

---

[x] The latter equality is a classical result of surface geometry.

often called the Marangoni effect in the literature. If $\nabla_{\mathcal{S}}(\gamma) = \mathbf{0}$, the tangential stress must be continuous when crossing the interface.

Noteworthy, if an elastic membrane replaces this interface, the same discontinuity of stress components holds but one has to consider the balance of elastic forces in the membrane to calculate its reaction on the fluid. By analogy to the fluid case, the stress jump at the interface will be balanced by the divergence of the traction force in the plane of the membrane (which will play the same role as surface tension) and of the transverse shear component acting in the cross section of the membrane which are related to the bending moments present across the thickness and due to membrane curvature (see discussion in Pozrikidis [14] and a simple case in Chapter 5).

1.11 *CONSTITUTIVE EQUATIONS

To close the system of governing equations we provided earlier for the deformation and flow of any material, we need to select the constitutive equations describing the behavior of the particular material of interest. These equations must express the relationship between stress and strain.

1.11.1 Viscometric flows: the example of simple shear flow

It can be shown that the mechanical behavior of a chosen (but arbitrary) incompressible simple fluid is completely determined in some characteristic flows, called viscometric flows. The most famous one is the simple shear flow, sometimes called Couette flow. In this case, the velocity field is steady, fully developed and unidirectional. The magnitude of the velocity depends linearly on the spatial component of the axis perpendicular to the direction of flow. As shown in Figure 1.20, one simple configuration is to sandwich the fluid to be sheared in between two parallel plates, one of which is moving parallel to the other. If cartesian coordinates x_i are chosen such that the flow direction is parallel to the x_1-axis (see Figure 1.20), then the orientation and origin of the coordinate system can be chosen such that the velocity field can be written as:

$$\mathbf{u} = u_x(y)\boldsymbol{e_x}. \tag{1.71}$$

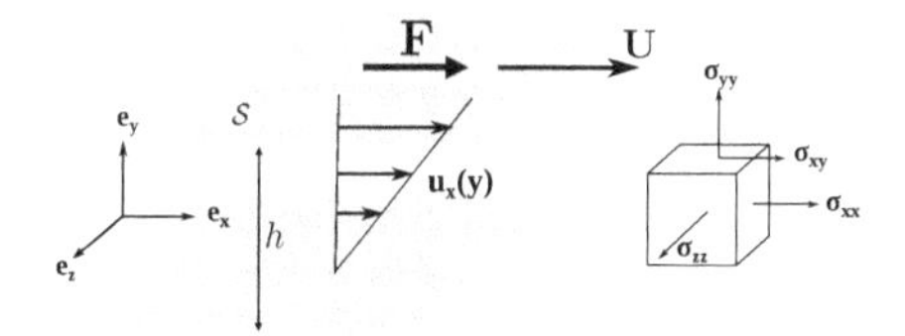

FIGURE 1.20
Velocity field in a simple shear driven by the motion of the upper plate. On the right side, a schematic of the components of the stress tensor on a volume unit.

The rate of deformation in this flow is simply given by:

$$\boldsymbol{\mathcal{D}} = \frac{1}{2}\left(\frac{\partial u_i}{\partial x_j} + \frac{\partial u_j}{\partial x_i}\right) \tag{1.72}$$

$$= \frac{1}{2}\begin{pmatrix} 0 & \dot{\gamma} & 0 \\ \dot{\gamma} & 0 & 0 \\ 0 & 0 & 0 \end{pmatrix}, \tag{1.73}$$

where

$$\dot{\gamma} = |\frac{\partial u_x}{\partial y}| = \frac{U}{h}, \tag{1.74}$$

is called the shear rate and is constant throughout the bulk, since the velocity field is linear in space. The characteristic time scale of the shear flow is then defined by $1/\dot{\gamma}$.

In the case of simple shear, $\mathbf{\Sigma}$ can be expressed as in cartesian coordinates,

$$\mathbf{\Sigma} = \begin{pmatrix} \sigma_{xx} & \sigma_{xy} & 0 \\ \sigma_{xy} & \sigma_{yy} & 0 \\ 0 & 0 & \sigma_{zz} \end{pmatrix}, \tag{1.75}$$

where the diagonal components of the tensor are the normal stress and the non-diagonal components of the shear stress. We note τ, the norm of the shear stress $\tau = |\sigma_{xy}| \equiv \frac{F}{S}$. τ represents therefore the force F per unit area S necessary to drag the plate at the velocity U, as depicted in Figure 1.20.

Viscometric flows include steady, fully developed flow in a straight pipe of constant circular cross section (called Poiseuille flow) and steady, unidirectional flow between two concentric circular cylinders driven by the rotation of one or both of the cylinders about their common axis (Couette flow) (see Figure 1.21). Most rheometers used to measure blood rheology are designed to generate such viscometric flows.

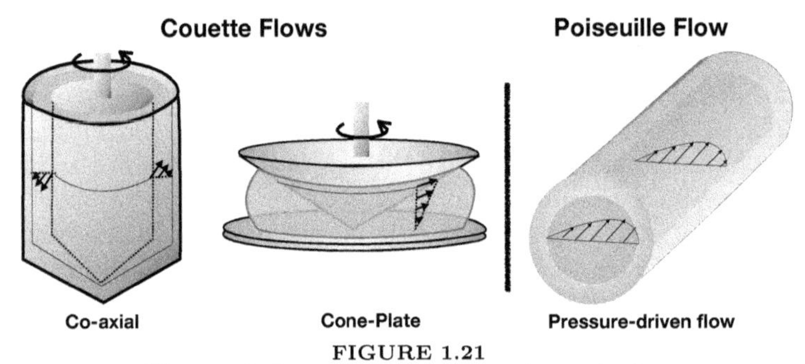

FIGURE 1.21
Viscometric flows produced in rheometers and tubes.

1.11.2 The notion of viscosity and the Newtonian fluid

Only three material functions are necessary to describe viscometric flows such as shear flows and to therefore characterize the rheology of the material in flow. We will refer to these three functions as viscometric functions and the viscosity $\eta(\dot{\gamma})$ is one of them and is defined as:

$$\eta(\dot{\gamma}) \equiv \frac{\tau(\dot{\gamma})}{\dot{\gamma}} \tag{1.76}$$

The definition of viscosity goes back to Newtonian physics itself. A Newtonian viscous fluid is indeed a fluid for which the shear stress $\mathbf{\Sigma}$ is everywhere linearly proportional to the strain rate $\boldsymbol{\mathcal{D}}$, with a coefficient of proportionality called the dynamic viscosity of the fluid. This parameter has units of Pa.s. It is typically equal to 10^{-3} Pa.s and it is constant for water at 20°C. Another unit often used in hemorheology is the centi Poise, denoted cP with $1cP = 10^{-3}$ Pa.s. The viscosity of blood is known to remain in a range between 1-100 cP and we will see that this depends strongly on the hematocrit, the concentration of macromolecules in plasma and the shear rate applied (temperature plays also a role but we will not discuss it here).

For a Newtonian incompressible fluid the stress-strain relationship is therefore formally written:

$$\boldsymbol{\Sigma} = -p\mathbf{I} + 2\eta\boldsymbol{\mathcal{D}}; \tag{1.77}$$

here we decomposed the stress tensor in its isotropic part related simply to pressure ($\mathbf{I}$ being the identity tensor) and its deviatoric component associated to $\boldsymbol{\mathcal{D}}$.

The flow properties of numerous simple Newtonian fluids such as water, oils, or plasma are in fact independent of the flow velocity and therefore $\eta(\dot{\gamma}) = \eta$. Indeed, the time scale associated to the microscopic displacement of the molecules is much shorter than the characteristic time scale, $1/\dot{\gamma}$, of the flow. This means that the shear flow does not modify the microstructure of a Newtonian fluid.

In contrast, if the microscopic time scale is of the same order of $1/\dot{\gamma}$, the microscopic structure of the fluid can be affected by the shear flow and such fluids do not verify anymore Newton's law in a certain range of shear rates. They are called non-Newtonian fluids.

1.11.3 The normal stress differences

Two other viscometric functions can be defined in the framework of simple shear flow:

$$\psi_1(\dot{\gamma}) \equiv \frac{N_1(\dot{\gamma})}{\dot{\gamma}^2}, \text{ and } \psi_2(\dot{\gamma}) \equiv \frac{N_2(\dot{\gamma})}{\dot{\gamma}^2}, \tag{1.78}$$

where N_1 and N_2 are the first and second stress differences respectively and are writen as:

$$N_1(\dot{\gamma}) \equiv \Sigma_{11} - \Sigma_{22} \tag{1.79}$$

$$N_2(\dot{\gamma}) \equiv \Sigma_{22} - \Sigma_{33}. \tag{1.80}$$

While for Newtonian fluids these two functions are identically zero, they are functions of the shear rate for complex fluids and in particular for suspensions. N_1 measures the difference in normal stress between the flow and the direction of the gradient, orthogonal to the flow. A positive value corresponds to the fluid forcing the plates apart in Figure 1.20, which is the case for most viscoelastic fluids but it is not true for suspensions or colloidal dispersions at high shear rates, where N_1 can become negative. The presence of a first stress difference in a fluid can have counterintuitive effects as described for instance in Chapter 2. In the case of N_2, which is the normal stress difference in the plane perpendicular to the flow direction, its value is always negative and much smaller than N_1 for suspensions (as discussed in Chapter 2). However, the measure of normal stress differences in suspensions remains scarce, though essential to fully characterize suspension fluid flow. There are only a few experimental studies of these properties for non-Brownian suspensions of hard spheres [9], and lack of data for colloidal suspensions, not speaking about blood cell suspensions for which these measurements have yet to be realized.

1.12 *THE NAVIER-STOKES EQUATIONS OF FLUID MOTION

The combination of equations 1.61 and 1.77 leads to the famous Navier-Stokes equations describing the flow of an incompressible Newtonian fluid [xi]:

$$\rho\frac{D\mathbf{u}}{Dt} = \rho\left(\frac{\partial\mathbf{u}}{\partial t} + (\mathbf{u}\cdot\nabla)\mathbf{u}\right) = -\nabla p + \eta\nabla^2\mathbf{u}. \tag{1.81}$$

[xi] Here without a body force which will add a term $\rho\mathbf{b}$ in the right hand side of the equations

The right hand side of these equations has two separated terms: $\nabla(p)$ which is the gradient of pressure present in the fluid and $\eta\nabla^2\mathbf{u}$ is the viscous contribution and is nothing else than the Laplacian of the velocity field. These terms specify that for viscous fluids, flow is possible only if a gradient of pressure is present in the fluid and/or boundaries apply viscous traction to each layer of material and by a diffusion-like process transmitting the momentum from the boundaries to the bulk thanks to friction for instance on solid moving walls [(xii)]; one example being the configuration of the Couette flow we saw earlier.

From the left hand side of the equations, two terms are present. The first one describes the temporal variation of the local velocity in the fixed laboratory frame (Eulerian representation), while the second is called the convective acceleration and represents the spatial variation of velocity.

1.12.1 Dimensional analysis and the Reynolds number

Osborn Reynolds was the first one to characterize in 1883 channeled flows in a large range of velocities and pressure gradients and discovered that at least three different regimes of flow are present that have been coined in time laminar, transitory and turbulent flows depending on a non-dimensional number built empirically by Reynolds and called the Reynolds number Re. When $Re \lesssim 1$ the flow is defined as laminar while for $Re \gtrsim 10^3$ the flow can become turbulent.

This number can be deduced by proper dimensional analysis of the Navier-Stokes equations. In fact one can rewrite the equations by non-dimensionalizing the variables with typical scales of the flow. Let's consider a fluid of fixed density ρ and viscosity η for which we define the kinematic viscosity $\nu = \eta/\rho$, flowing at a typical velocity U and considering a typical size L over which the highest gradients of velocity are observed (typically the radius of a vessel or the size of a particle), and a typical time $\mathcal{T}$ of variations of the velocity field (the period for oscillatory flows for instance). Then one can write a new set of variables:

$$\tilde{\mathbf{u}} = \mathbf{u}/U,\ \tilde{\mathbf{x}} = \mathbf{x}/L,\ \tilde{t} = t/\mathcal{T},\ \tilde{p} = p/(\eta U/L) \tag{1.83}$$

and obtain a new set of Navier-Stokes equations non-dimensionalized:

$$Re\left(Str\frac{\partial\tilde{\mathbf{u}}}{\partial\tilde{t}} + (\tilde{\mathbf{u}}\cdot\tilde{\nabla})\tilde{\mathbf{u}}\right) = -\tilde{\nabla}\tilde{p} + \tilde{\nabla}^2\tilde{\mathbf{u}},^{(\mathrm{xiii})} \tag{1.84}$$

where:

$$Str = \frac{L/U}{\mathcal{T}} \text{ and } Re = \frac{\rho U L}{\eta} = \frac{UL}{\nu} \tag{1.85}$$

are respectively the Strouhal number and the Reynolds number. Once non-dimensionalized, we can see that Re and Str are measures of the relative magnitude of the different terms in the Navier-Stokes equations. So Str is a ratio of two time scales: L/U which represents the typical time of convection of the fluid and $\mathcal{T}$ the typical time scale of local variations of the velocity. A small Strouhal number defines a flow which is quasi-stationary. In the case of the Reynolds number, it also measures the ratio of two time scales: the convective time

[(xii)] Indeed without pressure gradient, the equations become a diffusion equation for $\mathbf{u}$:

$$\frac{D\mathbf{u}}{Dt} = \frac{\eta}{\rho}\Delta\mathbf{u} = \nu\Delta\mathbf{u}, \tag{1.82}$$

with a coefficient of diffusion $\nu = \eta/\rho$ which is called the kinematic viscosity and which characterizes the ability for the fluid to diffuse linear momentum thanks to viscous effects.

[(xiii)] Note that $\tilde{\nabla} = L\nabla$ and $\tilde{\nabla}^2 = L^2\nabla^2$.

scale L/U and the viscous diffusive time scale of fluid momentum transport L^2/ν. A low Reynolds number means therefore that transport of momentum in the fluid is dominated by viscous effects. Re is in fact given by the ratio of the advective term over the viscous one in the dimensional Navier-Stokes equations:

$$|\rho \mathbf{u}.\nabla \mathbf{u}| \sim \frac{\rho U^2}{L}, |\eta \nabla^2 \mathbf{u}| \sim \frac{\eta U}{L^2}. \tag{1.86}$$

Therefore,

$$\frac{|\rho \mathbf{u}.\nabla \mathbf{u}|}{|\eta \nabla^2 \mathbf{u}|} \sim \frac{\rho U L}{\eta} = Re \tag{1.87}$$

Laminar flows, which are typical of the microcirculation are in fact characterized by a low Reynolds number of the order of 10^{-2}, while in the large arteries of the body, the flows are at large Reynolds numbers $Re \sim 10^2 - 6 \times 10^3$ where inertial effects dominate and even turbulence could occur. Noteworthy, the pulsatile nature of blood flow due to the contraction of the heart is at a frequency of the order of few Hz and therefore a characteristic time scale of variation. In this case, Str can be estimated for arterioles around $Str \sim 10^{-1}$. Non-stationary effect will not be inertial in nature at these frequencies.

1.12.2 Inertial flows: the example of Dean's flow

For steady flows, and for Reynolds numbers higher than unity but far from turbulent regimes $Re \ll 10^3$, unidirectional straight flows are somewhat stable and it appears that inertial effects are absent from the flow. Indeed, the convective acceleration term $\mathbf{u} \cdot \nabla(\mathbf{u})$ is identically zero for instance for unidirectional flows if the flow velocity does not change in the direction of the flow. However, this is no more true for instance if the fluid is flowing through a curvilinear channel. The fluid experiences then centrifugal acceleration forces directed radially outward leading to the formation of two counter-rotating vortices known as Dean vortices in the top and bottom halves of the channel as shown in Figure 1.22. They will induce helicoidal trajectories for the fluid particles. The magnitude of these secondary flows is quantified by a dimensionless Dean number De given by:

$$De = Re\sqrt{\frac{R}{R_c}} \tag{1.88}$$

where R_c is the radius of curvature of the channel and R its cross-sectional radius. The inertia-induced spiral motion is noticeable when the Dean number is larger than 1.

On top of being important for the understanding of blood flow in large arteries, this effect is relevant to many technological problems and found in the last decade surprising applications in cell sorting and manipulation in what is called now inertial microfluidics. It will be described in further detail in Chapter 10.

1.12.3 Calculation of the Poiseuille flow

The pressure-driven motion of blood in small vessels inspired Poiseuille in 1840 to describe this flow as the one of a Newtonian fluid in straight tubes of circular cross-sections. A simple approach of stress balance can be used to deduce Poiseuille's original result.

In Figure 1.23, we consider a Newtonian fluid flowing unidirectionally in a tube of circular cross-section of radius R. We isolate virtually a cylindrical portion of radius $r < R$ and length L of this flowing fluid. Writing the equilibrium of the total tangential force

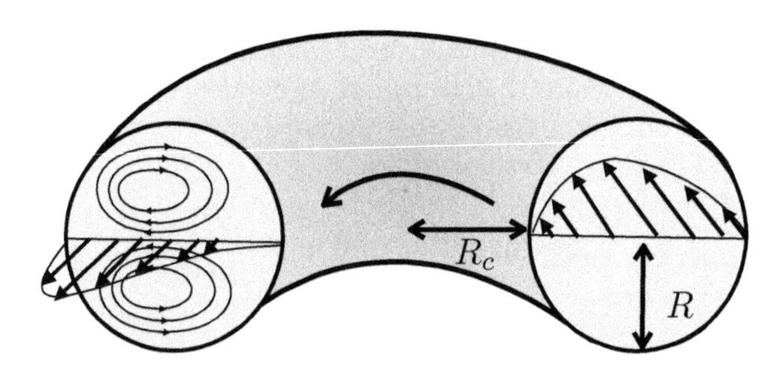

FIGURE 1.22
Secondary flows as Dean vortices of a fluid flowing in curved pipe.

balance of this fluid element in cylindrical coordinates, one obtains a simple equality:

$$P_+ \times S_+ - P_- \times S_- + \sigma_{rz} \times S_{syl} = P_+ \pi r^2 - P_- \pi r^2 + \sigma_{rz} 2\pi r L = 0. \tag{1.89}$$

Since for a Newtonian fluid, $\sigma_{rz} = \eta \frac{\partial u_z(r)}{\partial r}$ and defining the pressure difference $\Delta P = P_+ - P_-$, we have the following differential equation:

$$\frac{\partial u_z(r)}{\partial r} = -\frac{\Delta P}{L} \frac{r}{2\eta}. \tag{1.90}$$

A simple integration with the no-slip boundary condition of $u_z(R) = 0$, leads to the famous Poiseuille flow, showing that the velocity profile is parabolic in a circular vessel (as represented in Figure 1.23):

$$u_z(r) = \frac{\Delta P R^2}{4\eta L}\left(1 - \frac{r^2}{R^2}\right). \tag{1.91}$$

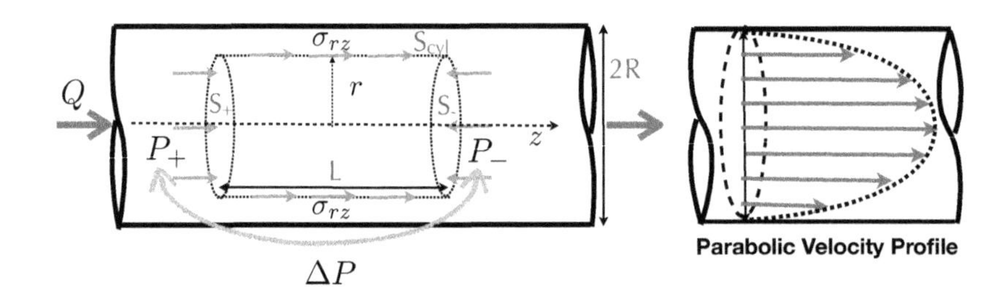

FIGURE 1.23
(Left) Stress balance on a fluid element in flow. (Right) Parabolic velocity profile called Poiseuille flow.

Bibliography

[1] Dominique Barthès-Biesel. *Microhydrodynamics and Complex Fluids.* CRC Press, 2012.

[2] D. Bonn, M.M. Denn, L. Berthier, T. Divoux, and S. Manneville. Yield stress materials in soft condensed matter. *Rev. Mod. Phys.*, 89:1–40, 2017.

[3] S. Chien. Shear dependence of effective cell volume as a determinant of blood viscosity. *Science*, 168:977–979, 1970.

[4] Shu Chien. Biophysical behavior of red cells in suspensions. *The Red Blood Cell*, 2(4):1031–133, 1975.

[5] R. F Acta Astronauticahraeus. Die strömungsverhältnisse und die verteilung der blutzellen im gefässystem. *Journal of Molecular Medicine*, 7(3):100–106, 1928.

[6] Y.C. Fung. *Biomechanics: Mechanical Properties of Living Tissues.* Springer, New York, 2 edition, 1993.

[7] Y.C. Fung. *First Course in Continuum Mechanics for Physical and Biological Engineers and Scientists.* Prentice Hall, Englewood Cliffs, New Jersey, 3rd edition, 1994.

[8] G.P. Galdi, R. Rannacher, A.M. Robertson, and S. Turek. *Hemodynamical Flows. Modeling, Analysis and Simulation.* Birkh´auser, Basel–Boston–Berlin, 2000.

[9] G.P. Galdi, R. Rannacher, A.M. Robertson, and S. Turek. *A Physical Introduction to Suspension Dynamics.* Cambridge University Press, Cambridge, UK, 2012.

[10] G. B. Jeffery. The motion of ellipsoidal particles immersed in a viscous fluid. *Proc. R. Soc. Lond. A*, 102(715):161–179, 1922.

[11] G.L. Leal. *Advanced Transport Phenomena: Fluid Mechanics and Convective Transport.* Cambridge University Press, 2007.

[12] R.W. Ogden. *Non-Linear Elastic Deformations.* Dover Publications, 1997.

[13] C. Picart, J.-M. Piau, H. Galliard, and P. Carpentier. Human blood shear yield stress and its hematocrit dependence. *J. Rheol.*, 42:1–12, 1998.

[14] C. Pozrikidis. *Modeling and Simulation of Capsules and Biological Cells.* Boca Raton: Chapman & Hall/CRC, 2003.

[15] A.R. Pries, D. Neuhaus, and P. Gaehtgens. Blood viscosity in tube flow: dependence on diameter and hematocrit. *American Journal of Physiology-Heart and Circulatory Physiology*, 263(6):H1770–H1778, 1992.

[16] A.R. Pries and T.W. Secomb. Microvascular blood viscosity in vivo and the endothelial surface layer. *American Journal of Physiology-Heart and Circulatory Physiology*, 289(6):H2657–H2664, 2005.

[17] T. W. Secomb. Mechanics and computational simulation of blood flow in microvessels. *Med. Eng. and Phys.*, 33:800–804, 2011.

[18] S.P. Sutera, V. Seshadri, P.A. Croce, and R.M. Hochmuth. Capillary blood flow: Ii. deformable model cells in tube flow. *Microvascular research*, 2(4):420–433, 1970.

CHAPTER 2

Dynamics of suspensions of rigid particles

Stany Gallier

ArianeGroup, Centre de Recherches du Bouchet

Elisabeth Lemaire

Institut de physique de Nice (InPhyNi)

CONTENTS

2.1 INTRODUCTION

SUSPENSIONS of rigid particles in low Reynolds number flows are ubiquitous in industry (food transport, cosmetic products, civil engineering, etc.) or natural flows (such as mud or lava flows) to mention but a few. This wide occurrence of suspensions has fostered significant research in the past years. Considering non-deformable particles is obviously not a clever choice to describe blood suspensions. However suspensions with rigid particles are simpler systems that have been studied for a long time, so that the current understanding might be a bit more advanced compared to suspensions of deformable particles. The objective of this first chapter is therefore to provide the reader with an overview of the current knowledge concerning the physics and rheology of suspensions of rigid particles. This might appear as a preliminary step before moving to more complex suspensions having deformable or non-spherical particles. Throughout this chapter, particles are assumed to be homogeneous and rigid spheres.

2.2 BASIC CONCEPTS OF SUSPENSION PHYSICS

2.2.1 Interactions in suspensions

The complexity of suspensions is partly related to the wide variety of interactions between particles, which motivates some brief discussions in this section. Before moving to interactions, let us recall that the key parameter controlling the physics of suspensions is the volume fraction ϕ of particles. To give a simple physical picture, a very dilute suspension ($\phi \to 0$) may be seen as a viscous fluid whereas a very dense suspension would behave more like a dry granular material. For a disordered suspension of rigid monodisperse spheres, the maximum volume fraction is $\phi_{max} \approx 0.64$ [90, 91]. This value differs however in the case of non-spherical, polydisperse or deformable particles. The importance of the volume fraction can be understood in terms of an average gap h between particles since we have a scaling $h/a \propto (\phi/\phi_{max})^{-1/3} - 1$ with a the particle radius. As will be seen in Sec. 2.3, suspensions cannot flow at a particle fraction ϕ^* below ϕ_{max} and this value of ϕ_{max} must be seen as a geometrical parameter related to a specific arrangement of particles. For high volume fractions, the gap h is extremely small which favors close particle interactions. The situation where h is significantly greater than particle radius a can be considered as dilute and ϕ is typically a few % only. Conversely, the case $h \to 0$ where direct particle interactions play a major role delineates the regime of dense suspensions, typically above 30–40 % .

Interactions in suspensions can be classified based on some dimensionless numbers—as described in the following—using the particle radius a as the relevant length scale.

Inertia

Fluid inertia at the particle scale is defined by the particle Reynolds number $Re = \rho_f \dot{\gamma} a^2/\eta_f$ where $\dot{\gamma}$ is the shear rate and η_f the fluid viscosity. A small Reynolds number means that Stokes equations prevail at particle scale and fluid inertia is negligible. Note that this does not imply that the whole flow is Stokesian and motion at large scales can be inertial. Particles have their own inertial time scale τ_p so that we can define a non-dimensional number based on the ratio between τ_p and the flow time scale $\dot{\gamma}^{-1}$, which defines the Stokes number $St = \dot{\gamma}\tau_p$. For low Re, we have $\tau_p = m/6\pi\eta_f a$ (with m the mass of the particle), so that $St = \rho_p \dot{\gamma} a^2/\eta_f$ with ρ_p the particle density[(i)]. If $St \ll 1$, the particle dynamics adapts instantaneously to the fluid motion. For the usual case of moderate density mismatch ($\rho_p \approx \rho_f$), Re and St are commensurate which is not surprising since both quantities define inertia to some extent. A suspension which is not inertial ($Re \to 0$ and $St \to 0$) can be described by the steady Stokes equations, which has some consequences on the flow at microscale. Notably, the fluid and particle motion is linear and reversible. Particles adapt instantaneously to the fluid and particle motion can be seen as a succession of quasi-steady configurations.

[(i)] The exact expression obtained is $St = 2\rho_p \dot{\gamma} a^2/9\eta_f$ but we drop the 2/9 factor here for simplicity.

Brownian motion

Brownian interactions can affect the motion of the smallest particles due to thermal agitation. The influence of thermal agitation relative to shear is defined by the Péclet number $Pe = 6\pi\eta_f\dot{\gamma}a^3/kT$ where T is the temperature and k the Boltzmann's constant. It can be seen as the ratio between a Brownian time scale $6\pi\eta_f a^3/kT$ and the flow time scale $\dot{\gamma}^{-1}$. The Pe=0 case means that the flow is completely governed by the thermal motion whereas the $Pe \to \infty$ limit implies that the shear-driven motion overwhelms the Brownian motion.

Colloidal interactions

Small particles are likely to be subject to specific attractive or repulsive interactions. Attractive forces are generally of van der Waals type (dispersive adhesion) whose intensity depends on the interparticle distance h and material nature (through the Hamaker constant A_H ~10^{-20}–10^{-21} J). Repulsive electrostatic forces arise due to particle electric charge which, in turn, can attract counterions from the fluid. This results in an electrical double layer at the particle-fluid interface, with a typical thickness of a few nanometers (Debye length). Both attractive and repulsive interactions are gathered into the DLVO (Derjaguin-Landau-Verwey-Overbeek) theory. The relative effect of the double-layer repulsion compared to the attractive Van der Waals force is given by the dimensionless number $\varepsilon\Psi_0^2 a/A_H$ [16,17] with ε the fluid permittivity ($\sim 10^{-10}$ F/m) and Ψ_0 the electrical surface potential ($\sim 10^{-3}$ V). For a=10 μm, this number is typically of order one, meaning that repulsion and attraction have similar intensities for such particles. But since the magnitude of colloidal interactions is strongly linked to particle and fluid materials, they have to be estimated carefully case-by-case. Let F_{coll} be a scale for colloidal forces; then the dimensionless numbers aF_{coll}/kT and $F_{coll}/\eta_f\dot{\gamma}a^2$ describe the importance of colloidal forces relative to Brownian motion and viscous forces respectively. Note that there can be other related interactions, like additional repulsion due to polymer brushes grafted on particle surface or depletion forces for polymeric fluids when the interparticle distance h is smaller than the size of the polymer chain (see Figure 2.1).

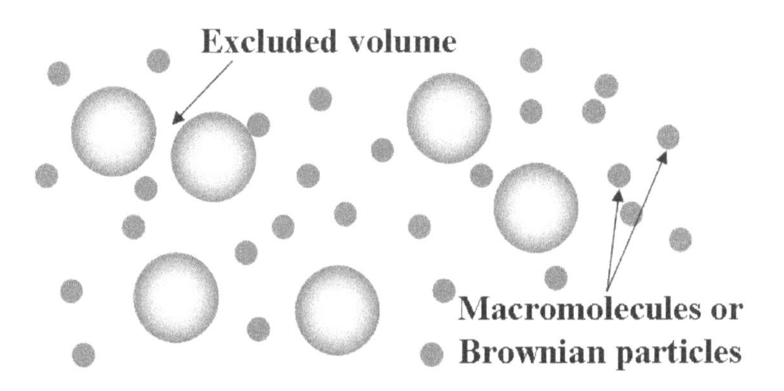

FIGURE 2.1
Particles dispersed in a solution of macromolecules or a dispersion of smaller Brownian particles may experience attractive depletion forces that come from excluded volume effects.

Hydrodynamic interactions

Viscous forces primarily manifest as a drag for particles moving in a fluid and this drag force scales as $\eta_f \dot{\gamma} a^2$ for low Reynolds numbers. If this force prevails, the regime is termed hydrodynamic or viscous. Reciprocally, particles exert a force on the fluid which is transmitted to neighboring particles. Those hydrodynamic interactions are many-body and long-ranged. For an isolated particle subject to an external force, the resulting fluid perturbation (referred to as a Stokeslet) decays as r^{-1} and can therefore extend over large distances r. There are short-range hydrodynamic interactions as well when two particles come in near contact. Consider two spheres approaching each other along their line of centers (squeeze flow problem). Then, the interstitial fluid in the gap becomes more and more difficult to drain due to viscous effects as $h \to 0$. An intense pressure, scaling as h^{-2}, locally develops in the gap and hinders further motion. This pressure gives birth to a force referred to as lubrication force. For the squeeze flow problem, the lubrication force is $6\pi\eta_f a^2 \dot{h}/h$ (with $\dot{h}$=dh/dt) and is singular at contact (h=0). Lubrication hampers contact for two approaching spheres ($\dot{h} < 0$) but becomes an attractive force when particles separate ($\dot{h} > 0$). The h^{-1} singularity of lubrication force should theoretically prevent any direct contact but this relation only holds for infinitely rigid and perfectly smooth particles. In reality, actual contacts can occur for instance through particle surface roughness and this prevents any singularity. Note that for fluid drops, the lubrication force is weaker and diverges only as $h^{-1/2}$, so that a contact is now possible in finite time [50]. The normal component of lubrication force (squeeze flow) is singular as h^{-1} but tangential force (shear flow) or torque is also singular as $\ln h$. Exact analytical expressions for lubrication interactions between a pair of spheres are available in the specialized literature [50].

Contact forces

We may move from viscous to contact-dominated regimes when contact forces between particles commensurate with viscous forces. This limit is identified by the Leighton number Le defined as $Le = \eta\dot{\gamma}a/(h\sigma_n)$ [16, 17]. It measures the relative importance of viscous effects compared to a normal stress σ_n applied to particles (confining pressure, gravity, etc.). If $Le \ll 1$, contacts prevail and there is a continuous and extended network of contacting particles. For higher Le, this network tends to fragment and disappear under shear. A correct *a priori* estimation of σ_n (hence, Leighton number) is not straightforward for volume-imposed density-matched suspensions. Recent studies using particle-based simulations [32] have shown that contact and viscous contributions to viscosity become similar for $\phi \approx 40$ %. This value can therefore be considered as the onset of a regime driven by contacts. Actual contacts generally involve friction between particles and friction forces are also found to strongly affect suspension dynamics, structure and rheology [32, 63].

For highly sheared suspensions, there could be collisions between particles. By collision, we mean that the contact time is very short and that the life-time of connected networks is too small to transmit significant stresses. This collisional regime (or Bagnoldian regime) is defined by the Bagnold number as Ba=$\rho_p\dot{\gamma}ha/\eta_f$ [16]. For $a \sim 10$ μm and $h \sim 10^{-2}a$, we find a small $Ba \sim 10^{-3}$ even for $\dot{\gamma} \sim 10^3$ s^{-1} meaning that a collisional regime is not expected in typical blood flows.

2.2.2 Interactions in blood flows

The above numbers are useful for hard-sphere or more complex suspensions, like blood flows. We hereafter provide some estimates for blood flows assuming a suspending fluid of density $\rho_f \sim 10^3$ kg/m^3 and viscosity $\eta_f \sim 10^{-3}$ Pa.s, which are typical values for blood plasma. The particles (here, red blood cells) have a radius $a \sim 3$–4 μm and typical shear rates are in the range $\dot{\gamma} \sim 10^2$–10^3 s^{-1}.

We find $Re \sim 10^{-3}$–10^{-2} and $Pe \sim 10^5$–10^6. It is generally assumed that a suspension is non-Brownian if it has a Pe larger than 10^3, so that blood flows can be considered as non-Brownian and Stokesian suspensions. As mentioned previously, we do not expect a collisional regime (small Bagnold numbers) but a predominance of contact forces is possible since the volume fraction of red blood cells can be high, up to $\phi \approx 50$ %.

There are however some additional specificities in blood flows, notably linked to the deformability of particles which can alter the hydrodynamic interactions. Red blood cells are often considered as a viscous fluid (of viscosity η_{in}) enclosed in a deformable elastic membrane with modulus E, so that two additional non-dimensional numbers arise: the viscosity ratio $\lambda = \eta_{in}/\eta_f$ and an analogue of a capillary number defined as $Ca = \eta_f \dot{\gamma} a/E$. This capillary number may be regarded as a dimensionless shear rate indicating how the membrane is likely to deform under the prescribed shear rate. Typical values for λ and Ca are given in Chapter 2 of this book.

At last, although plasma is a quasi Newtonian fluid, it contains macromolecules that may create attractive forces between red blood cells that are presumably involved in erythrocyte aggregation. Aggregation can either come from depletion forces or be induced by a bridging of the red blood cells by macromolecules.

2.2.3 Hydrodynamics of a single particle

Before moving to many-particle systems, the simple relevant case of a single particle in a Stokes flow is considered. We here assume a linear flow so that the prescribed fluid velocity $\mathbf{u}^\infty$ reads

$$\mathbf{u}^\infty = \mathbf{U}^\infty + \boldsymbol{\Omega}^\infty \times \mathbf{x} + \boldsymbol{\mathcal{D}}^\infty \cdot \mathbf{x} \tag{2.1}$$

where $\mathbf{U}^\infty$ is a uniform flow velocity, $\boldsymbol{\Omega}^\infty$ the fluid rotation vector, $\mathbf{x}$ the position with respect to the particle center, and $\boldsymbol{\mathcal{D}}^\infty$ the rate-of-strain tensor which is defined as the symmetrical part of the velocity gradient: $\boldsymbol{\mathcal{D}}^\infty = \frac{1}{2}(\nabla \mathbf{u}^\infty + \nabla^t \mathbf{u}^\infty)$. Because of the linearity of the Stokes equations, the superposition principle applies and the motion of a particle in a flow described by Eq. (2.1) is the sum of the elementary motions of a particle in a uniform translation $\mathbf{U}^\infty$, rotation $\boldsymbol{\Omega}^\infty$ and straining $\boldsymbol{\mathcal{D}}^\infty \cdot \mathbf{x}$ as depicted in Figure 2.2. For each of those three elementary motions, analytical solutions of the Stokes equations are available for the velocity field $\mathbf{u}(\mathbf{x})$ or pressure field $p(\mathbf{x})$. By superposition, the global velocity and pressure fields are obtained by summing the analytical solutions for each motion. Such analytical developments are not recalled here and they can be obtained for instance using integral representations based on Green's functions. More information can be found in classical textbooks [37, 50, 85].

We now present the classical concepts of hydrodynamic force $\mathbf{F}^h$ and hydrodynamic torque $\mathbf{T}^h$ but also the notion of stresslet $\mathbf{S}$ which is useful for rheology.

Starting from an infinitesimal surface dS oriented with normal vector $\mathbf{n}$ on the particle surface S_p, the hydrodynamic force $\mathbf{F}^h$ is defined as

$$\mathbf{F}^h = \int_{S_p} \boldsymbol{\Sigma} \cdot \mathbf{n} \, dS \tag{2.2}$$

where $\boldsymbol{\Sigma}$ is the fluid stress tensor and with the integral performed on the particle surface S_p. This hydrodynamic force is readily understood as the sum of the infinitesimal force $\boldsymbol{\Sigma} \cdot \mathbf{n} \, dS$ applied by the fluid on the particle surface. For a rigid sphere of radius a in a translational

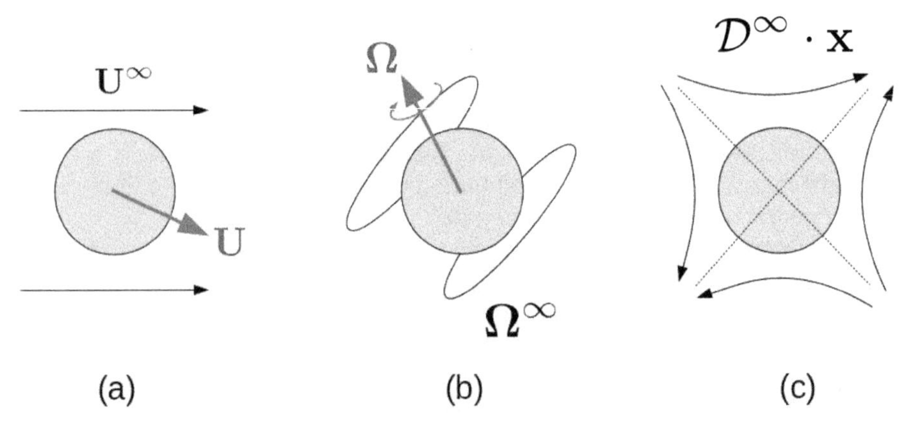

FIGURE 2.2
Particle motion in a linear flow as the sum of motions in a uniform translation (a), pure rotation (b), and pure straining (c).

motion at velocity $\mathbf{U}$ in a uniform ambient field $\mathbf{U}^\infty$ (see Figure 2.2a), the classical result is the Stokes law which reads

$$\mathbf{F}^h = 6\pi\eta_f a(\mathbf{U}^\infty - \mathbf{U}) \tag{2.3}$$

One way to derive this expression is to use the analytical expressions for the velocity and pressure fields in order to obtain an expression for the fluid stress $\mathbf{\Sigma}$ (using a constitutive law for the fluid, supposed Newtonian here) and perform the integral Eq. (2.2). In absence of inertia, we recall that the sum of forces applied on the particle is zero, which gives $\mathbf{F}^h+\mathbf{F}^e=0$ where $\mathbf{F}^e$ represents the external forces on the particle (gravity, collision, Brownian, electrostatic, etc.). Without any external forces ($\mathbf{F}^e=0$), we have $\mathbf{F}^h=0$ from which $\mathbf{U}=\mathbf{U}^\infty$. In the case of a non-rigid fluid spherical drop of viscosity η_{in} and for small capillary number Ca, the Stokes law becomes

$$\mathbf{F}^h = 6\pi\eta_f a\frac{2+3\lambda}{3+3\lambda}(\mathbf{U}^\infty - \mathbf{U}) \tag{2.4}$$

with λ the viscosity ratio $\lambda=\eta_{in}/\eta_f$. The solid sphere corresponds to the case $\lambda \to \infty$.

The hydrodynamic torque is similarly defined as

$$\mathbf{T}^h = \int_{S_p} \mathbf{x} \times (\mathbf{\Sigma} \cdot \mathbf{n})\, dS \tag{2.5}$$

where $\mathbf{x}$ is the position with respect to the center of gravity of the particle. For a rigid sphere of radius a rotating at velocity $\mathbf{\Omega}$ in an ambient flow with uniform rotation $\mathbf{\Omega}^\infty$ (see Figure 2.2b), the hydrodynamic torque is found to be

$$\mathbf{T}^h = 8\pi\eta_f a^3(\mathbf{\Omega}^\infty - \mathbf{\Omega}) \tag{2.6}$$

In absence of external torques, we get $\mathbf{T}^h=0$, which yields a rotational velocity of the particle $\mathbf{\Omega}=\mathbf{\Omega}^\infty$. Note that the relation Eq. (2.6) still holds for a viscous spherical drop at small Ca.

To define the stresslet, let us first write the force dipole tensor D_{ij} as

$$D_{ij} = \int_{S_p} \Sigma_{ik} n_k x_j\, dS \tag{2.7}$$

It can be further decomposed into its symmetric and antisymmetric part as $D_{ij}=S_{ij}+T_{ij}$ where the symmetric part S_{ij} is the stresslet and T_{ij} is called the rotlet. The stresslet is therefore formally defined as

$$S_{ij} = \frac{1}{2}\int_{S_p} [\Sigma_{ik}n_k x_j + \Sigma_{jk}n_k x_i]\, dS \tag{2.8}$$

This is the usual expression for rigid particles and an extended relation for non-rigid particles will be presented in the next chapter.

The stresslet has a less obvious physical meaning since it is not directly related to particle motion as forces and torques can be. Yet it has a profound effect on rheology as will be seen in the following section since it represents an additional stress associated with the presence of the particle that resists the flow deformation. This is depicted in the following sketch Figure 2.3 showing a particle in a straining flow. For clarity, we restrict our discussion to the xy component. Noting $f_i=\Sigma_{ik}n_k$, we have by definition $D_{xy}=\int_{S_p} f_x y\, dS$ and $D_{yx}=\int_{S_p} f_y x\, dS$ as illustrated in Figure 2.3(a)-(b). Now, the stresslet S_{xy} and rotlet T_{xy} respectively read $S_{xy}=\frac{1}{2}(D_{xy}+D_{yx})$ and $T_{xy}=\frac{1}{2}(D_{xy}-D_{yx})$ as pictured in Figure 2.3(c)-(d). This sketch clearly shows that the rotlet conveys the same information as the hydrodynamic torque acting on the particle(ii) while the stresslet actually acts as deforming the particle. The stresslet is hence connected to how the particle resists the imposed straining motion that tends to deform it.

For a rigid sphere in a straining flow, we have

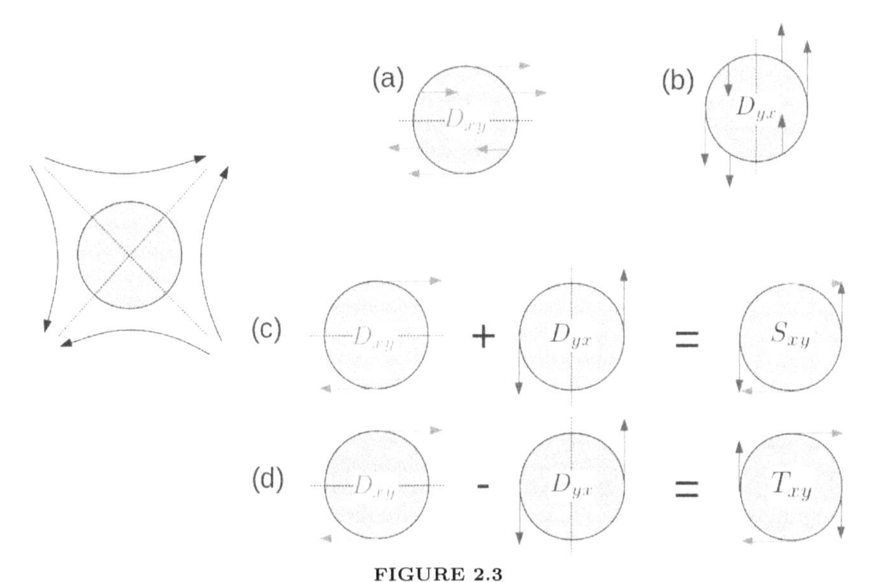

FIGURE 2.3
A particle in a straining flow with schematic repartition of D_{xy} (a), D_{yx} (b), S_{xy} (c) and T_{xy}(d).

$$S_{ij} = \frac{20}{3}\pi\eta_f a^3 \mathcal{D}^{\infty}_{ij} \tag{2.9}$$

Here again, this result can be obtained from the analytical flow field solution plugged into Eq. (2.8). For a spherical viscous drop at $Ca=0$, we have

$$S_{ij} = \frac{20}{3}\pi\eta_f a^3 \frac{2+5\lambda}{5+5\lambda}\mathcal{D}^{\infty}_{ij} \tag{2.10}$$

(ii)We have $T_{ij} = -\frac{1}{2}\epsilon_{ijk}T^h_k$ with ϵ_{ijk} the Levi-Civita tensor (permutation tensor).

The relations given by Eqs. (2.4), (2.6), and (2.10) hold for spherical particles, either viscous or rigid ($\lambda \to \infty$), but in the limit of small Ca. For the more general case of arbitrary-shaped particles, there are still linear relations between ($\mathbf{F}^h$, $\mathbf{T}^h$, $\mathbf{S}$) and ($\mathbf{U}^\infty$,$\boldsymbol{\Omega}^\infty$, $\mathcal{D}^\infty$) due to the linearity of the Stokes equations but those relations are more complex. In particular, there can be some couplings between translation, rotation, and straining that are absent for spheres for which forces depend only on translational velocity (Eq. (2.3)), torques only on rotational velocity (Eq. (2.6)), and stresslets only on rate-of-strain (Eq. (2.9)). The general relation writes

$$\begin{pmatrix} \mathbf{F}^h \\ \mathbf{T}^h \\ \mathbf{S} \end{pmatrix} = \mathcal{R} \begin{pmatrix} \mathbf{U}^\infty - \mathbf{U} \\ \boldsymbol{\Omega}^\infty - \boldsymbol{\Omega} \\ \mathcal{D}^\infty \end{pmatrix} \tag{2.11}$$

where the matrix $\mathcal{R}$ is called the resistance matrix. This resistance matrix is available analytically for shapes other than spheres, such as ellipsoids for instance [50]. Equation (2.11) can be used to obtain the motion of a non-spherical particle, as the well-known Jeffery orbits which describe the motion of a rigid ellipsoid in a shear flow [46] (also see Chapter 2 and Chapter 6 in this book). Generalization to any axisymmetric particles is provided by Bretherton [9].

2.2.4 Particle stress and rheology

A major objective of rheology is to relate the stress Σ_{ij} with the strain rate. For a shear flow of a Newtonian fluid, the stress tensor Σ_{ij} reads

$$\Sigma_{ij} = -p\delta_{ij} + 2\eta_f \mathcal{D}_{ij} \tag{2.12}$$

where p is the pressure, δ_{ij} the Kronecker tensor and $\mathcal{D}_{ij}$ the local rate-of-strain tensor which—as seen previously—is defined as

$$\mathcal{D}_{ij} = \frac{1}{2}\left(\frac{\partial u_i}{\partial x_j} + \frac{\partial u_j}{\partial x_i}\right) \tag{2.13}$$

The tensor $\mathcal{D}_{ij}$ describes the deformation of a fluid element and its volume variation. Low Reynolds suspensions are generally incompressible and this tensor is traceless $\mathcal{D}_{ii} = 0$. The coefficient η_f is the shear viscosity which is a constant for a Newtonian fluid.

Suspension flows are well-known to be non-Newtonian. As will be addressed further in Sec. 2.4, non-Newtonian effects can involve shear-dependent viscosity, normal stresses, or time-dependent viscosity so that Eq. (2.12) might not be suitable for a suspension. In his seminal works, Batchelor [3] formally introduced the so-called *particle stress tensor* Σ^p_{ij} which represents that additional stress due to the presence of particles. Following Batchelor's ideas, the suspension stress Σ^s_{ij} may be split into the stress Σ^f_{ij} of the suspending fluid and this additional Σ^p_{ij} contribution from particles as

$$\Sigma^s_{ij} = \underbrace{\frac{1}{V}\int_{V_f} \Sigma_{ij}\, \mathrm{d}V}_{\Sigma^f_{ij}} + \underbrace{\frac{1}{V}\int_{V_p} \Sigma_{ij}\, \mathrm{d}V}_{\Sigma^p_{ij}} \tag{2.14}$$

where V_f (resp., V_p) is the fluid (resp., particle) volume and $V = V_f + V_p$.

Now consider a suspension of neutrally-buoyant spheres in a flow with a prescribed rate-of-strain $\mathcal{D}^\infty_{ij}$. Assuming a Newtonian suspending fluid gives $\Sigma_{ij} = -p\delta_{ij} + 2\eta_f\mathcal{D}_{ij}$ in the fluid phase, with η_f the fluid viscosity. The fluid contribution then reads

$$\Sigma_{ij}^f = \underbrace{-\frac{1}{V}\int_{V_f} pdV}_{\langle p\rangle_f}\ \delta_{ij} + \eta_f \frac{1}{V}\int_{V_f} 2\mathcal{D}_{ij}dV \tag{2.15}$$

where we have introduced an average fluid pressure $\langle p\rangle_f$. The second integral gives

$$\begin{aligned}\frac{1}{V}\int_{V_f} 2\mathcal{D}_{ij}dV &= \frac{1}{V}\int_{V} 2\mathcal{D}_{ij}dV - \frac{1}{V}\int_{V_p} 2\mathcal{D}_{ij}dV\\ &= \frac{1}{V}\int_{V} 2\mathcal{D}_{ij}dV - \frac{1}{V}\int_{S_p} (u_i n_j + u_j n_i)dS\\ &= 2\mathcal{D}_{ij}^{\infty} - \frac{1}{V}\int_{S_p} (u_i n_j + u_j n_i)dS\end{aligned} \tag{2.16}$$

with $\mathbf{n}$ is the outward-pointing normal vector on particle surface S_p.

The stress in the particles can be expressed using the following relation [3,37,50]

$$\int_{V_p} \Sigma_{ij}\ \mathrm{d}V = \int_{S_p} \Sigma_{ik} n_k x_j\ \mathrm{d}S - \int_{V_p} \frac{\partial \Sigma_{ik}}{\partial x_k} x_j\ \mathrm{d}V \tag{2.17}$$

where vector $\mathbf{x}$ is the position with respect to particle center. If particles are force-free, then $\partial\Sigma_{ik}/\partial x_k=0$ and the second integral vanishes. Combining all the equations above gives the suspension stress as

$$\Sigma_{ij}^s = -\langle p\rangle_f \delta_{ij} + 2\eta_f \mathcal{D}_{ij}^{\infty} + \Sigma_{ij}^p \tag{2.18}$$

with

$$\Sigma_{ij}^p = \frac{1}{V}\int_{S_p} \left[\Sigma_{ik} n_k x_j - \eta_f (u_i n_j + u_j n_i)\right]\ \mathrm{d}S \tag{2.19}$$

The first term is recognized as the force dipole D_{ij} defined by Eq. (3.34) while the second term is linked to the deformation of the particle. This term is zero for rigid particles but must be retained for deformable particles. The stresslet definition becomes slightly different for deformable particles and is shown to be [50]

$$S_{ij} = \frac{1}{2}\int_{S_p} \left[\Sigma_{ik} n_k x_j + \Sigma_{jk} n_k x_i\right]\ dS - \eta_f \int_{S_p} (u_i n_j + u_j n_i)\ dS \tag{2.20}$$

Using this definition and considering the usual case of torque-free particles (the rotlet T_{ij} is then zero, meaning that the force dipole comes only from the stresslet), then the extra particle contribution Σ_{ij}^p to the bulk stress can be viewed as the sum of the stresslets S_{ij} of all N suspended particles. Equation (2.19) can then be rewritten as

$$\Sigma_{ij}^p = \frac{1}{V}\sum_{k=1}^{N} S_{ij}^{(k)} \tag{2.21}$$

where $S_{ij}^{(k)}$ is the stresslet associated with particle k. Note that all the above relations hold irrespective of the particle shape.

The quantities usually considered in rheology can be fully defined from the suspension stress tensor Σ_{ij}^s. In particular, for a pure shear flow of shear rate magnitude

$\dot{\gamma} = (2\mathcal{D}_{ij}^{\infty}\mathcal{D}_{ij}^{\infty})^{1/2}$ in the plane (x,y) (see sketch in Figure 2.4), the suspension shear viscosity η_s is given as

$$\eta_s = \frac{\Sigma_{xy}^s}{\dot{\gamma}} = \eta_f + \frac{\Sigma_{xy}^p}{\dot{\gamma}} \tag{2.22}$$

For a rigid isolated force-free and torque-free sphere in a straining flow, we have from Eq. (2.9)

$$S_{ij} = \frac{20}{3}\pi\eta_f a^3 \mathcal{D}_{ij}^{\infty} \tag{2.23}$$

If particles are so dilute that they do not interact hydrodynamically, Eq. (2.21) gives for N rigid particles

$$\Sigma_{ij}^p = \frac{20}{3}\pi\eta_f a^3 \frac{N}{V}\mathcal{D}_{ij}^{\infty} = 5\phi\eta_f\mathcal{D}_{ij}^{\infty} \tag{2.24}$$

Combining Eq. (2.24) and Eq. (2.18) finally leads to

$$\Sigma_{ij}^s = -\langle p\rangle_f\delta_{ij} + 2\eta_f(1+\frac{5}{2}\phi)\mathcal{D}_{ij}^{\infty} \tag{2.25}$$

From which we can deduce the classical Einstein relation for the shear viscosity of a dilute suspension

$$\eta_s = \eta_f(1+\frac{5}{2}\phi) \tag{2.26}$$

This latter relation only applies for a dilute suspension of rigid spheres. Deformable or non-spherical spheres lead to different relations owing to the change in the stresslet expression. For a spherical viscous drop at Ca=0, the coefficient 5/2 in Eq. (2.26) becomes $(5\lambda+2)/(2\lambda+2)$. Also, for non-spherical particles, the stresslet depends on particle orientation which requires orientational averaging [50]. Some results and expressions can be found in [40, 50].

Although the suspending fluid is Newtonian, this is not the case for the suspension due to the additional Σ_{ij}^p contribution. In particular, Σ_{ij}^p has normal components, i.e. has non-zero Σ_{xx}^p, Σ_{yy}^p or Σ_{zz}^p components (see sketch in Figure 2.4 for the definitions of directions x, y, and z). This means for instance that a suspension sheared in a cell can push or pull apart the walls of the cell. The occurrence of normal stresses in suspensions can lead to the so-called anti-Weissenberg effect (or "rod-dipping") as it will be illustrated further (Figure 2.15). In rheology, the normal stress differences N_1 and N_2 are mostly used and defined as

$$N_1 = \Sigma_{xx}^s - \Sigma_{yy}^s \tag{2.27}$$

$$N_2 = \Sigma_{yy}^s - \Sigma_{zz}^s \tag{2.28}$$

Note that Σ_{ij}^p is in general not traceless. This means that there exists an additional pressure (also referred to as particle pressure) acting similarly as an osmotic pressure. The pressure in a suspension is therefore the sum of a fluid pressure and a particle pressure.

The particle stress defined by Eq. (2.19) only arises from hydrodynamic stress Σ_{ij} and there can be additional contributions from other non-hydrodynamic forces (Brownian, colloidal, contacts, etc.). In dense suspensions of non-Brownian rigid particles, a major contribution comes from the contact through a contact force $\mathbf{F}^c$ between particles. The resulting additional contribution to stress $\Sigma_{ij}^{p,(c)}$ (where superscript (c) refers to contact) is given by, assuming pairwise additivity

$$\Sigma_{ij}^{p,(c)} = \frac{1}{V}\sum_{pairs} F_i^c r_j \tag{2.29}$$

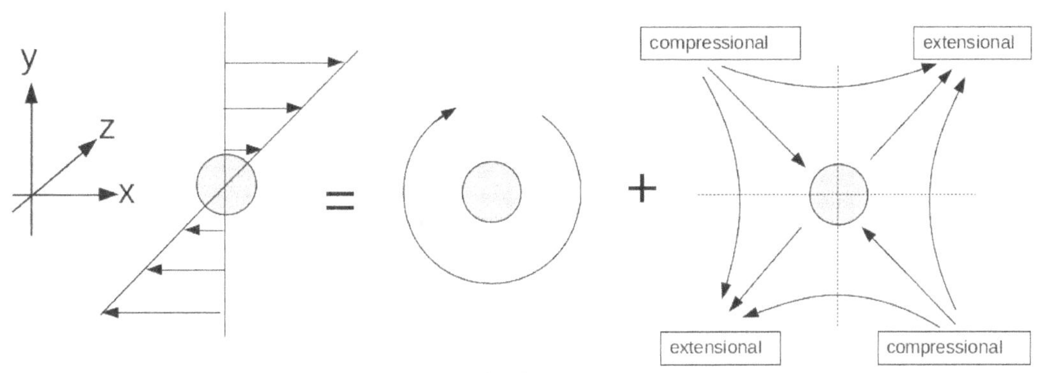

FIGURE 2.4
Sketch of a simple shear flow in the (x,y) plane. The directions x, y, and z respectively define the velocity, velocity gradient and vorticity directions. A shear flow can be decomposed as a rotation plus a strain flow. The straining flow involves compressional and extensional quadrants.

where the sum runs over all pairs of contacting particles and $\mathbf{r}$ is the vector joining the two centers of a pair. Note that $\Sigma_{ij}^{p,(c)}$ is in general not traceless and not symmetric due to possible contact torques. Numerical simulations suggest that this contact stress is the major contribution to the overall particle stress for dense suspensions, typically for ϕ above 40% [32] meaning that direct contacts are the main contributor to the viscosity of dense suspensions.

2.2.5 Microstructure of suspensions

The dynamics and rheology of suspensions are strongly related to the particle volume fraction ϕ. However ϕ is a global, low-order statistical quantity that misses fine microscale details, i.e. it does not bring any information on how particles are arranged at small scale—which is termed the microstructure. The relation between microstructure and rheology has received a growing interest recently and is instrumental in understanding and modeling suspensions. A simple and widely-used quantity to characterize the microstructure statistically is the pair-distribution function $g(\mathbf{x}_1, \mathbf{x}_2)$ which is related to the conditional probability $P(\mathbf{x}_1, \mathbf{x}_2)$ to find a particle at $\mathbf{x}_1$ given a test particle is at $\mathbf{x}_2$. For a spatially homogeneous system, statistics depend only on $\mathbf{r} = \mathbf{x}_1 - \mathbf{x}_2$ and we define $g(\mathbf{r}) = P(\mathbf{r})/n$ with n the average number density of particles. Hence, $g(\mathbf{r})$ defines the likelihood of finding a particle at position $\mathbf{r}$ with respect to a reference particle.

The first experimental determinations of $g(\mathbf{r})$ date back to the works of Husband *et al.* [41] in dilute regimes and Parsi and Gadala-Maria [77] in dense regimes of non-Brownian suspensions (ϕ=40% and 50% suspensions of polystyrene beads). The latter study has in particular revealed that, under shear flow, the microstructure was not fore-aft symmetric (with respect to flow direction) with more particles in the compressional quadrant and conversely a depleted region in the extensional quadrant. This means that the microstructure is anisotropic since there exists an angular dependence of $g(\mathbf{r})$. Figure 2.5 presents an experimental pair-distribution function for a sheared non-Brownian suspension ($Pe \sim 10^8$). It clearly illustrates this break of fore-aft symmetry with higher probability (dark red color) to find a pair of particles in the compressional quadrant ($xy < 0$) and lower probability in the extensional region ($xy > 0$). In particular, there is a marked depleted region close to the extension direction (at 45°), apparent as a "hole" in the distribution, meaning that finding particles oriented along this specific direction is low. The sketch in Figure 2.4 recalls the extensional and compressional regions arising from the straining component of a shear flow. Another feature is that the likelihood of finding a pair is strongly amplified only in a very thin region near contact ($\|\mathbf{r}\| \sim 2a$). This has been also confirmed by numerical simulations of non-colloidal or strongly sheared Brownian suspensions [29, 32, 73, 95, 105]. The anisotropy of the microstructure increases with the Péclet number and for vanishing Pe, an isotropic microstructure at equilibrium is expected because there are no preferred directions of thermal agitation [29, 73]. This anisotropic microstructure at finite Pe is induced by the flow and is experimentally found to invert upon a shear reversal [5, 77].

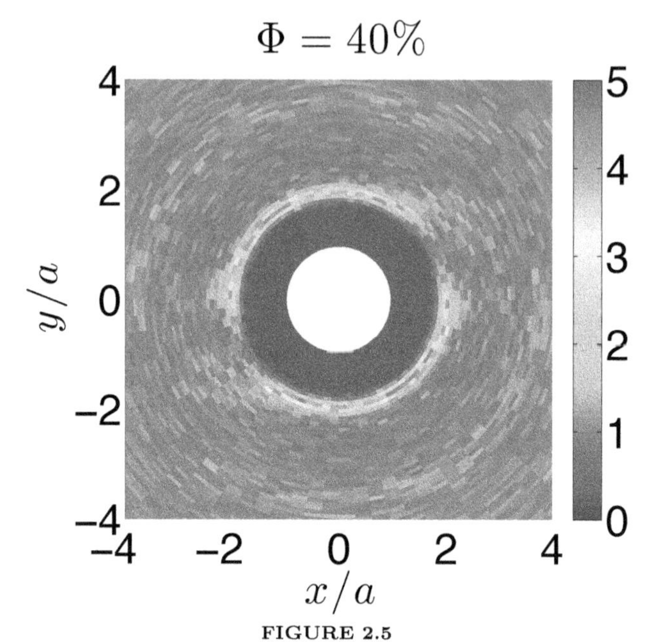

FIGURE 2.5
Experimental pair-distribution function $g(x, y)$ in the shear plane (x,y) for a ϕ=40 % suspension. Compression occurs for $xy < 0$. Particles are spherical PMMA of radius $a \sim$ 80–90 μm. The white disk is the reference particle. Reprinted from [5] with the permission of the author.

From a theoretical perspective, the dilute case was addressed by Batchelor and Green [4] who have predicted that $g(\mathbf{r})$ becomes singular at contact but still remains symmetric, a consequence of the reversibility of the motion of a pair of particles in a Stokes flow. Some more general works by Brady and Morris [8] have shown that $g(\mathbf{r})$ at contact follows the scaling law $g(\mathbf{r}) \sim Pe^{0.78}$, which gives a singular nature to the pure hydrodynamic limit $(Pe \to \infty)$.

This fore-aft anisotropy has profound consequences on suspension rheology. Intuitively, a $g(\mathbf{r})$ which is spherically symmetric indicates an equally-oriented distribution of particles and normal stress, thereby leading to zero normal stress differences and eventually a Newtonian behavior. Conversely, if $g(\mathbf{r})$ is not symmetric (as seen in Figure 2.5), there will be non-zero normal stress differences depending on how normal stress and particles are spread. This simple physical picture has been confirmed theoretically by Brady et al. [8, 29]. Normal stress differences therefore arise in a sheared suspension primarily because of the flow-induced anisotropy of particle arrangements. This anisotropy leads to a compressive normal stress of a pair in the compressional quadrant not being balanced by an equal tensile stress for pairs in recession (extensional region). As noted from Figure 2.5, the particle build-up is relatively constant in the compressional quadrant, from which we expect $\Sigma^p_{xx} \approx \Sigma^p_{yy}$, therefore a small value of $N_1 = \Sigma^p_{xx} - \Sigma^p_{yy}$. This first normal stress difference is thus strongly linked to the anisotropy in the shear plane (x,y), hence the microstructure. Conversely, the value of N_2 is larger (in magnitude) since interactions in the vorticity direction z are much weaker ($|\Sigma^p_{yy}| \gg |\Sigma^p_{zz}|$). This also makes N_2 less sensitive to microstructural details [32].

The physical reason for this microstructural anisotropy is believed to be connected to contacts between particles [6,20,23,86]. Contact forces are non-hydrodynamic (i.e., not proportional to velocity): they are repulsive in the compressional quadrant (to prevent particle overlaps) but they are not attractive in the extensional quadrant when particles separate. This just breaks down the fore-aft symmetry. Even the case of dilute suspensions (ϕ=5 %)—theoretically expected to have isotropic microstructures—has been experimentally found to show an anisotropy well predicted by a roughness-induced contact model [6].

All the above discussion is valid for rigid spheres with the striking result being the breaking of symmetry due to the orientation of pairs of particles along preferred directions. Note that we are here referring to a "configurational" microstructure since the relative position of pairs of particles is studied. In addition, deformable or non-spherical particles have a single-body microstructure associated with particle deformation or orientation. Both configurational and single-body microstructures are expected to be strongly coupled. Simulations by Clausen et al. [13] on deformable capsules at large Ca show that the shape of the pair-distribution function $g(\mathbf{r})$ becomes ellipsoidal, highlighting the role of deformation on particle interactions. The breaking of symmetry was however still apparent. Overall, this suggests that a more complex description of the microstructure is required and explains why the microstructure of suspensions with deformable or non-spherical particles is less known than for rigid particles.

2.2.6 Irreversibility in suspensions

A salient feature of suspensions is their irreversible and chaotic nature—diffusion or shear-induced migration are a manifestation of it, among others. At first glance, this can seem at odds with the linearity and reversibility of motion expected from Stokes equations. This irreversible nature has been clearly established by the experiments of Pine *et al.* [83] where particles are tracked during cycles of oscillatory shear. If the suspension is reversible, particles should return to their initial configuration after each cycle. But experiments have shown that when the strain amplitude exceeds a critical strain γ^*, they actually do not and drift away. The dynamics of the suspension becomes chaotic and the loss of reversibility is quite rapid, typically $\gamma^* \sim 1$ for a ϕ=40 % suspension. Figure 2.6 presents Pine *et al.*'s results through measurements of particle diffusivities, which clearly show this loss of reversibility for strain amplitudes above γ^*.

Another demonstration of this irreversible feature is the famous Taylor experiment[(iii)] but applied to a spherical cloud of particles as done by Metzger *et al.* [68]. The cloud is alternatively sheared back and forth and it loses its spherical shape and disperses progressively in the flow direction if the strain amplitude exceeds a certain threshold which depends on volume fraction. The chaotic nature of sheared suspensions has also been confirmed numerically [22, 25] through the Lyapunov exponent[(iv)] λ which grows linearly with ϕ and is in the range 0.05–0.5 for ϕ between 5% and 35% [25]. The source of this irreversibility

[(iii)] A droplet of dye is introduced into a viscous liquid between two concentric cylinders. When the inner cylinder is rotated, the droplet is stretched into a barely visible long filament. When the rotation of the cylinder is reversed, the filament astoundingly reforms the original spherical droplet, demonstrating reversibility.

[(iv)] The Lyapunov exponent λ measures the average rate of separation S of two initially neighboring trajectories in phase space $S(t) \sim S_0 \exp(\lambda t)$.

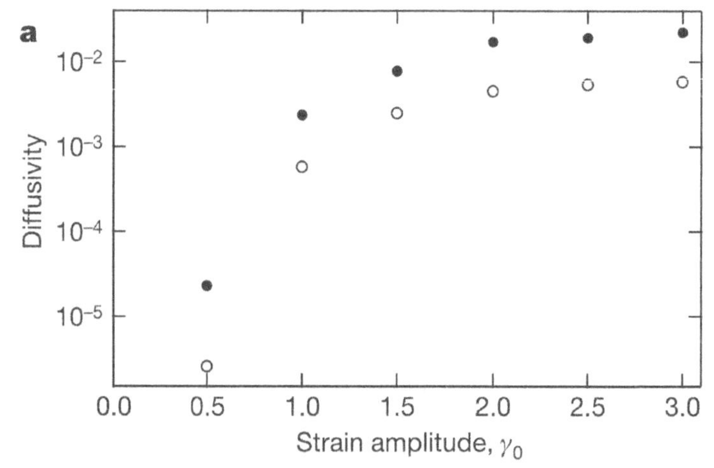

FIGURE 2.6
Experimental diffusivities D_{xx} (filled circles) and D_{zz} (open circles) as a function of the oscillatory strain amplitude γ_0 for a suspension of volume fraction ϕ=40 %. The x and z diffusivities both become negligible for $\gamma_0 < \gamma^* \sim 1$, showing a threshold in the suspension reversibility. Reprinted from [83] with the permission of Nature Publishing Group.

has been first thought to arise from many-body long-range hydrodynamic interactions since it is well known that the trajectories of three particles sedimenting are chaotic [45]. This view might be different for sheared suspensions due to the predominance of short-range interactions such as contacts or lubrication. We have mentioned in the previous section that contacts may break the Stokes symmetry and imply a loss of reversibility, which may identify interparticular forces to be at the microscopic origin of irreversibility [25]. Experimental studies on a two-particle system (therefore discarding any multi-body interactions) clearly demonstrate a strong relation between irreversibility and surface properties of particles [84]. Numerical works have also substantiated that irreversibility and chaos cannot significantly arise from long-range interactions or lubrication interactions [67, 69]. Recent experimental works have clearly confirmed contact forces—due to particle roughness—to be at the origin of irreversibility [79]. Despite being driven by reversible Stokes equations, a suspension can be irreversible and chaotic due to short-range non-hydrodynamic interactions, e.g. roughness-induced contacts. Hence, the behavior of the bulk suspension can be profoundly affected by forces acting on a scale well below particle size. Indeed, the typical roughness size for particles commonly used in suspension studies is in the range of 10^{-3}–10^{-2} the particle radius [6, 97].

2.3 VISCOSITY OF SUSPENSIONS

Particle volume fraction has often been considered as the unique parameter that controls the relative viscosity η_r of a suspension $\eta_r = \eta_s / \eta_f$. However, experimental measurements show a large data scatter when relative viscosity is plotted against particle volume fraction. This scatter may be explained either by experimental artifacts or by the involvement of other factors such as particle shape, polydispersity, interparticle or Brownian forces and type of flow.

2.3.1 Viscosity measurements

Conventional viscosity measurements are performed in simple shear flows (see Sec. 2.2). Ideally, simple shear flow is the flow that develops between two infinite parallel plates moving relatively one from each other. The continuum description assumption—on which rheometric measurements rely—is correct only if the suspension remains homogeneous and if the characteristic length scale of the flow is much larger than the particle size. However, measurements are often conducted under much less ideal conditions. The shear flows are either generated in rotational geometries (parallel plates, rotating cone-plate or rotating concentric cylinders) or produced by pressure drop in circular or rectangular channels. The characteristic dimension of the flow (gap width for torsional flows or rectangular capillaries and diameter for circular tubes) may be not very large compared to the size of the particles, in particular in the case of non-Brownian suspensions where the particles may have a diameter of several hundreds of micrometers. Furthermore, even though particles are much smaller than the flow gap, the continuum description fails near the walls. In the neighborhood of solid surface, the particle volume fraction cannot be the same as in the bulk because of excluded-volume effects and this particle depletion influences the flow. Generally, the particle fraction is lower near the wall than in the bulk and so is the viscosity, leading to a larger shear rate that is often described by an apparent wall-slip velocity of the suspension [44]. Wall slip velocity can be deduced from viscosimetric measurements carried out in various geometries (Couette, torsional or Poiseille flow) where the distance separating the walls is varied [71, 107]. Then, the bulk viscosity is deduced by extrapolating the measurements to an infinite gap. To minimize wall slip, rough surfaces can be used with a roughness size comparable to the particle size although the complete elimination of slipping cannot be guaranteed.

The only quantitative way to characterize apparent wall slip and to determine the viscosity of an unbounded suspension is to directly measure the velocity profile from which the viscosity is deduced at each position in the gap. Many local rheometry techniques have emerged at the end of the 1990s and in the early 2000s. Some of them are based on optical detection of the particles such as confocal microscopy for micronic or sub-micronic particles [34] and laser doppler velocimetry (LDV), particle tracking (PTV) or particle image velocimetry (PIV) for larger particles [7, 44, 98]. These techniques are powerful but limited to transparent suspensions. On the opposite, nuclear magnetic resonance (NMR) can be used for most suspensions with few restrictions on suspension characteristics [87].

Besides wall slip, shear induced particle migration (see Sec. 2.5) is a possible artifact that may invalidate macroscopic rheometric measurements. Once again, performing local rheometry is the only way to avoid errors that would result from macroscopic measurements carried out in a suspension where particle volume fraction is heterogeneous.

2.3.2 Concentration dependence of the viscosity

Before the development of local rheometry, it was held that the scattering observed in viscosity data was largely due to experimental errors or artifacts and some authors (see for instance [89]) chose to cover up the scatter with average curves rather than examine the other factors that may influence the rheological properties of suspensions. Most of the various expressions that have been proposed for predicting the concentration dependence of suspension viscosity are empirical even though some of them have theoretical foundations. Exact theoretical calculations starting from basic principles have been successful only for dilute suspensions so far. This is the case for the Einstein relation Eq. (2.26) that holds only for concentrations lower than 1 or 2 %. An extension to the second order has been proposed by Batchelor and Green [4] who found:

$$\eta_r = 1 + 2.5\phi + k\phi^2 \tag{2.30}$$

with k=6.95 for non-Brownian suspensions in a pure straining flow and k=6.2 for Brownian suspensions in the zero Péclet limit, whatever the flow. However this relation is a good approximation of the suspension viscosity only for $\phi \lesssim 5 - 10\%$.

For higher particle volume fractions, the concentration dependence of the relative viscosity is often expressed by the Maron-Pierce relation [64]:

$$\eta_r = \left(1 - \frac{\phi}{\phi^*}\right)^{-2} \tag{2.31}$$

or the Krieger-Dougherty relation [52]:

$$\eta_r = \left(1 - \frac{\phi}{\phi^*}\right)^{-[\eta]\phi^*} \tag{2.32}$$

where ϕ^* is the jamming fraction and $[\eta]$ is the intrinsic viscosity which is equal to 2.5 for spheres to recover the Einstein limit for dilute suspensions.

The viscosity diverges at $\phi = \phi^*$ and the value of ϕ^* is usually taken as a fitting parameter typically ranging between 0.53 and 0.74 for monodisperse spherical suspensions. It can significantly increase and possibly reach values close to 1 for polydisperse suspensions. Geometrical arguments are insufficient for predicting the value of ϕ^* for a system of particles and a general theoretical framework is still lacking. In particular, as already mentioned in Sec. 2.2, ϕ^* is different from the maximum volume fraction ϕ_{max} that would be measured from sedimentation. Rather, the exact value of ϕ^* seems to be controlled by the forces acting on the particles whether they are Brownian, colloidal or contact forces. Two results corroborating this idea are given hereafter. The first one is the influence of the particle friction coefficient on suspension viscosity. Recent numerical simulations [63, 78] have shown that for suspensions of non-Brownian and non-colloidal rough spheres, increasing the friction coefficient μ from 0 to 1 results in a decrease of the jamming fraction from 0.7 to 0.56 (Figure 2.7). The variation of ϕ^* with the Péclet number is a second example showing how interactions at the particle scale affect the jamming fraction. It has been observed (see for instance [99] or [102]) that as Pe increases, ϕ^* becomes larger and changes from $\phi^* \simeq 0.63$ in the zero Péclet number limit to $\phi^* \simeq 0.7$ in the high Péclet number limit. This dependence of ϕ^* with Pe has been interpreted as a disorder-order transition that facilitates the flow through the sliding of particle layers over each other [99]. This hypothesis has been confirmed by scattering experiments [101] and discrete numerical simulations based on a molecular dynamics method that accounts for many-body hydrodynamic interactions, known as the Stokesian Dynamics [29, 82].

Predicting the value of ϕ^* remains a current challenge which is important as it seems to be the main factor that governs suspension viscosity. In particular, plotting the relative suspension viscosity against the ratio of particle volume fraction to jamming fraction makes possible a collapse of all the data on almost a single curve, even for polydisperse suspensions [11] (see Figure 2.8).

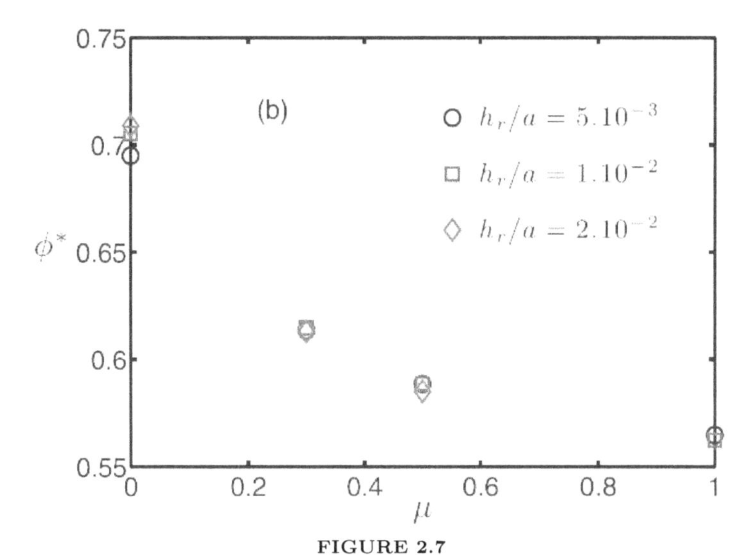

FIGURE 2.7
Variation of the jamming fraction ϕ^* against the friction coefficient μ for a non-Brownian non-colloidal suspension of rough spheres, for three different values of roughness size h_r/a (with a the particle radius). Jamming fraction mostly depends on friction while roughness size seems to have minor effects. Reproduced from [78] with the permission of the Journal of Rheology.

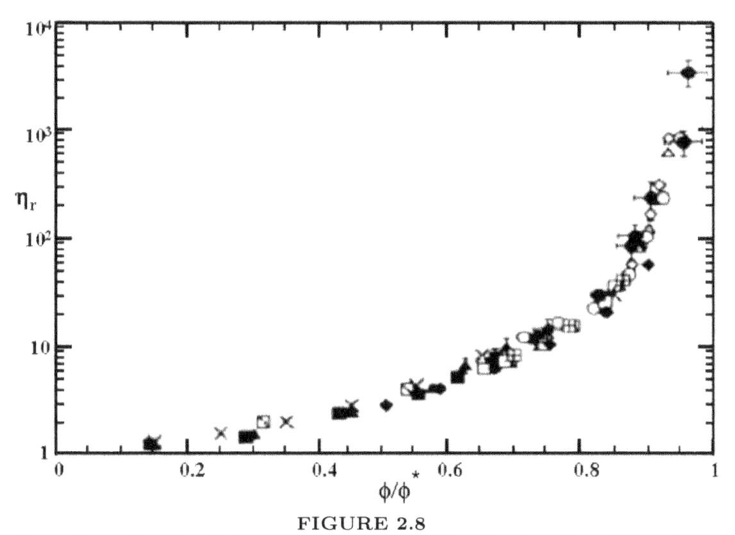

FIGURE 2.8
Relative viscosity of non-Brownian bimodal suspensions as a function of reduced particle volume fraction ϕ/ϕ^* for different size ratio and different small to large particles fraction ratio. Reproduced from [11] with the permission of the Journal of Rheology.

2.4 NON-NEWTONIAN EFFECTS

Most of natural (mud, blood, mucus...) and industrial (fresh concrete, polymers, paints...) materials exhibit non-Newtonian behaviors. Among them, are the suspensions. Following Barnes *et al.* [2], a fluid is termed non-Newtonian if—at constant pressure and temperature–shear viscosity depends on shear-rate and/or time, or if normal stress differences differ from zero in a simple shear. In the following, we address the shear-rate dependence of viscosity and the existence of normal stress differences.

2.4.1 Shear-rate dependence of viscosity

In a non-Newtonian fluid the relationship between the shear stress and the shear rate is not linear so that a constant viscosity coefficient cannot be defined but its value has to be specified for each value of the shear stress (or shear rate). Viscosity can either increase or decrease with increasing shear stress. In the former case, we talk about shear-thickening fluid while the latter is referred to as shear-thinning fluid. In most cases, shear-thinning or shear-thickening is associated with microsctructural changes inside the material. A paradigmatic example is electro or magnetorheological suspensions. When an electric (or a magnetic) field is applied, the particles of the suspension align themselves along the field direction (Figure 2.9). This structuring results in a profound modification of the flow properties of the suspension. Under high field, the suspension behaves like a solid gel. This "solid" state is characterized by the maximum stress that it can withstand which is called the yield stress. When the suspension is sheared with a stress higher than this yield stress, the chains of particles are progressively destroyed and the viscosity of the suspension decreases, illustrating the interplay between microstructure and rheological properties. But shear-dependent viscosity can stem from other reasons than microstuctural changes and generally shear-thinning or shear-thickening behaviors are expected each time non-hydrodynamic interactions are not negligible compared to hydrodynamic forces.

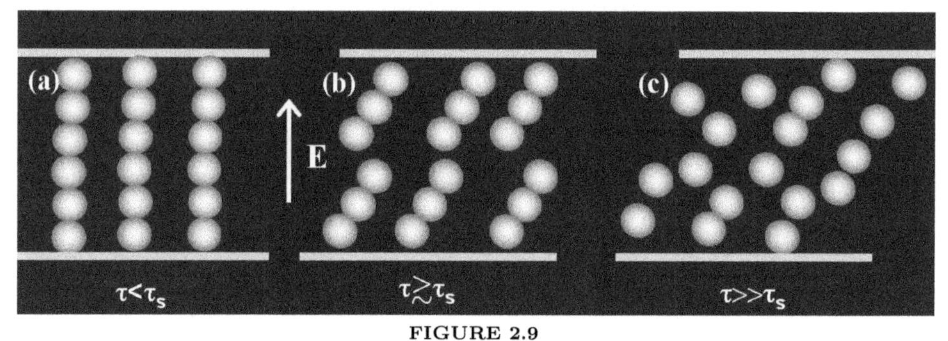

FIGURE 2.9
Illustration of the electrorheological effect. Under the application of an electric field **E**, the particles of a suspension form chains aligned in the field direction (a). The suspension behaves like a gel with a yield stress, τ_S. The chains that formed under the effect of the field must be broken for the suspension to flow (b) and, as the stress increases, the aggregates are progressively destroyed (c) and the viscosity of the suspension decreases.

Non-viscous behaviors (i.e. stress-dependent viscosity) are all the more pronounced as the particle volume fraction is large. In dilute to semi-dilute suspensions ($\phi \lesssim 10 - 20\%$), viscosity hardly varies with shear rate, contrary to the case of concentrated suspensions which may present large variations of the viscosity with shear rate. There is no universal shear rate-viscosity relationship but there is a general trend that when increasing shear rate, viscosity first shear-thins and reaches a plateau that is sometimes followed by a shear-thickening behavior and, for even higher shear rates, by a second shear-thinning regime. This general trend is illustrated schematically in Figure 2.10.

This quite complex behavior of viscosity originates from the non-hydrodynamic forces acting on the particles. They introduce a stress scale that is different from the shear stress associated with the flow. Consequently, the viscosity of such suspensions may depend on a dimensionless stress Σ_r given by the ratio between hydrodynamic forces and non-hydrodynamic forces F^c (that may be any type of Brownian, colloidal, contact or external forces): $\Sigma_r = \Sigma^s_{xy}a^2/F^c = \eta_s\dot{\gamma}a^2/F^c$. To give an overview of the viscosity behavior with shear rate, we will distinguish three types of suspensions: Brownian suspensions, colloidal suspensions and non-Brownian suspensions. Non-Brownian and non-colloidal suspensions

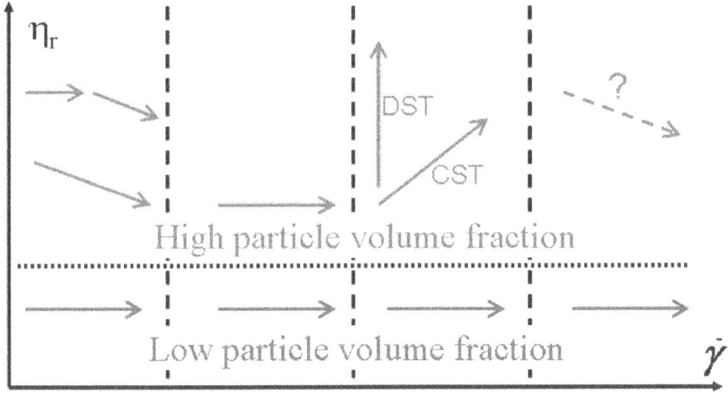

FIGURE 2.10
Schematic diagram of the shear rate-viscosity variation for a suspension. For low particle concentration, the viscosity is almost constant with shear rate. For higher concentrations, with increasing shear rate, the initial shear-thinning is followed by a shear-thickening which can be either continuous (CST) or discontinuous (DST). At even higher shear rates, a second shear-thinning behavior may appear.

are often confused probably because it is usual to classify the suspensions according to the size of the particles. Thus suspensions of particles smaller than 1 μm in size are often assimilated to both Brownian and colloidal suspensions. However, as explained previously, it is the ratio between hydrodynamic forces and colloidal (resp., thermal) forces which defines the boundary between colloidal (resp., Brownian) suspensions and non-colloidal (resp., non-Brownian) suspensions. Thus some suspensions may be Brownian—but non-colloidal—and are therefore referred to as Brownian hard sphere suspensions. Conversely, for particles of a few micrometers in diameter, the Brownian motion may be neglected while colloidal forces between particles have to be taken into account in describing the suspension behavior. In this last case, the suspension is non-Brownian but colloidal.

Brownian hard sphere suspensions

In the case of Brownian suspensions, the Péclet number Pe (see Sec. 2.2) measures the ratio between hydrodynamic forces $6\pi\eta_f a^2\dot{\gamma}$ and thermal forces kT/a. The role of thermal motion on Brownian suspension rheology has been carefully studied by Krieger [51] through experiments on suspensions of monodisperse particles for which the surface charge was adjusted to minimize other interparticle forces. They observed that the relative viscosities of suspensions with several particle radii and fluid viscosities, measured over a wide range of shear rates, superimpose when plotted against a reduced stress, $\Sigma_r = \Sigma^s_{xy}a^3/kT = Pe\ \eta_r/6\pi$ (see Figure 2.11). For particle volume fractions lower than 0.2, the viscosity is constant with shear rate but for higher particle fractions, viscosity shear-thins and, at $\phi = 0.5$, the low shear viscosity is about twice the viscosity measured in the high shear limit. As shear stress increases, the direct Brownian contribution to viscosity decreases while the hydrodynamic contribution remains mostly constant as shown by numerical simulations [82].

Numerical simulations reveal three well-defined regions where different rheologies and microstructures are observed. There is first a regime dominated by Brownian motion ($Pe \lesssim 1$) where, due to thermal diffusion, particles are almost isotropically distributed and the suspension viscosity shear-thins. When Brownian and hydrodynamic forces balance for $Pe \simeq 10$, particles tend to form a string-ordered structure along the flow direction. This flow-induced ordering persists over a range of Pe and the rheology remains unchanged. Finally, there is a hydrodynamically-dominated regime for $Pe > 200$ where shearing forces push particles into close contact. The suspension then shear-thickens owing to the increase

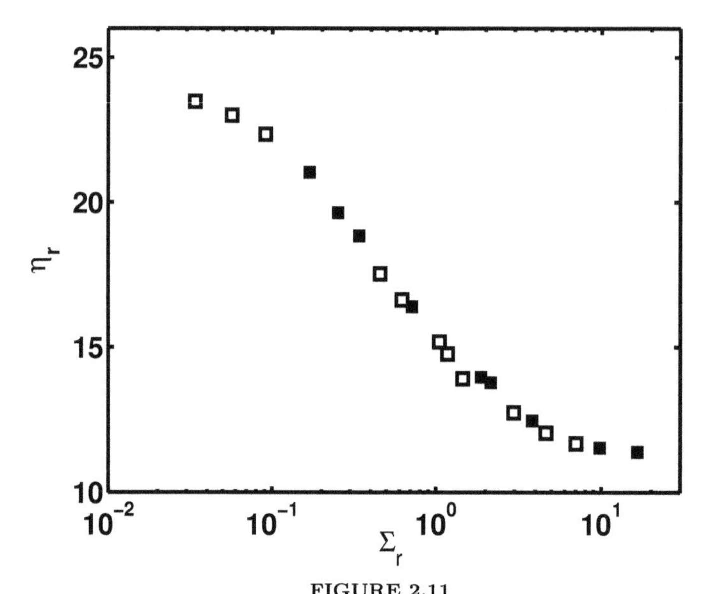

FIGURE 2.11
Relative viscosity as a function of the reduced stress $\Sigma_r = \Sigma^s_{xy} a^3/kT$ for eleven different suspensions with three different suspending media and five particle sizes ranging from 155 nm to 433 nm at $\phi = 0.5$ [51].

in the hydrodynamic stress caused by the formation of particle clusters (hydroclusters) in which lubrication dissipation prevails (see Figure 2.12). This cluster formation was the first attempt to explain shear-thickening in concentrated suspensions. There now exists another explanation for shear-thickening that is based on shear-induced frictional contacts between particles [63] which is probably more appropriate to describe shear-thickening in non-Brownian suspensions.

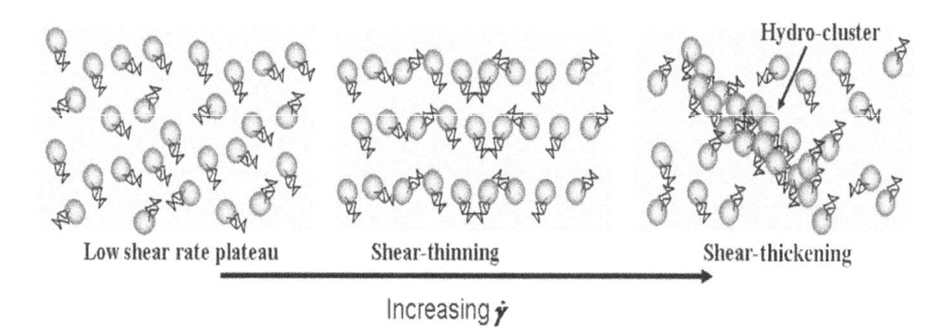

FIGURE 2.12
Schematic representation of the shear-rate-dependent microstructure in a suspension of Brownian hard spheres.

Colloidal suspensions

In addition to thermal forces, colloidal forces that may be electrostatic or steric in nature may be present and influence the rheological behavior of suspensions of micronic or submicronic particles. The contribution of colloidal forces on rheology is particularly discernible for particles in the $\mathcal{O}(1 - 10\ \mu m)$ range. In this size range, Brownian effects are quite weak and flow properties of the suspension are mainly controlled by the competition between colloidal and hydrodynamic forces. The rheology of colloidal suspensions is known to be very sensitive to the physicochemical properties of the particle interface. In most aqueous suspensions, aggregation that would result from attractive Van der Waals forces is prevented by longer-range electrostatic repulsion. The increase of viscosity caused by electrostatic forces is traditionally attributed to two distinct physical mechanisms. The first one, which is dominant for dilute suspensions, arises from the deformation by the shear flow of the ionic cloud that surrounds the particles. The second, which is all the more important as particle volume fraction grows, comes from the increase in the probability of interactions between particles resulting from the extent of the ionic clouds. This secondary electroviscous effect has been proposed as early as 1966 to be responsible for shear thinning observed in electrostatically stabilized suspensions [10]. The authors proposed the following qualitative picture: as two charged spheres approach one another in a simple shear flow, the repulsive electrostatic forces deflect their trajectories over a distance δ that is given by the balance between electrostatic and hydrodynamic forces (see Fig. 2.13). Hence, since viscous forces increase with shear rate while electrostatic forces remain constant, the trajectory offset and the viscosity must decrease with increasing shear rate.

Shear-thinning in floculated suspensions is even more pronounced and can be interpreted by the disruption of the flocs by the flow as shear stress increases. For a complete review of the influence of colloidal interactions on non-Newtonian behavior of suspensions, see [88]. It is indeed difficult to extract general behaviors for colloidal suspensions and, because of the diversity of possible interactions between particles, it is necessary to examine their behavior case by case.

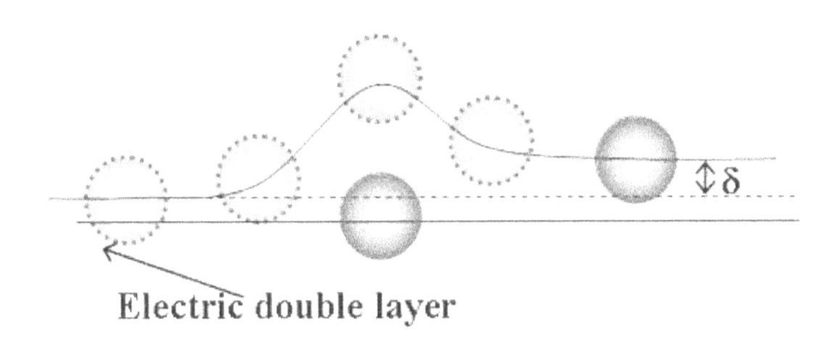

FIGURE 2.13
Schematic representation of the electroviscous effect.

Non-Brownian suspensions

Non-Brownian suspensions have received less interest than colloidal or Brownian suspensions, probably because they are expected to have simpler rheological behavior. In particular, it has long been assumed that particles in non-Brownian suspensions only interact via hydrodynamic forces. Under this hypothesis, there is only one stress scale Σ^s_{xy} which is fixed by the flow and no viscosity dependence with shear stress (or shear rate) is expected. In fact, non-Brownian suspensions exhibit various complex rheological behaviors meaning that the assumption of strictly hydrodynamic interactions between particles is not valid. The non-Newtonian behavior is all the more marked as particle volume fraction increases. In the general case, as the shear stress increases, the suspension switches from a shear-thinning behavior to a shear-thickening behavior with possibly a jump in viscosity. For still larger shear stress, the behavior is not well known but recent studies [27] show that a second shear-thinning regime appears as sketched in Figure 2.14. The non-hydrodynamic forces coming into play can be solid contact forces, possibly including friction, residual colloidal forces or a mix thereof. The discontinuous shear-thickening is a striking example of the interplay of colloidal and contact forces between particles and the viscosity jump is attributed to a transition from a lubricated non-frictional regime to a frictional one when the shear stress is high enough to overcome repulsive forces between particles [14, 63]. It should be noted that shear-thickening is only expected for suspensions where particles interact via colloidal repulsive forces that may be of steric or electrostatic origin. When those colloidal forces are truly negligible, non-Brownian suspensions no longer shear-thicken and only exhibit shear-thinning behavior whose origin is still unclear.

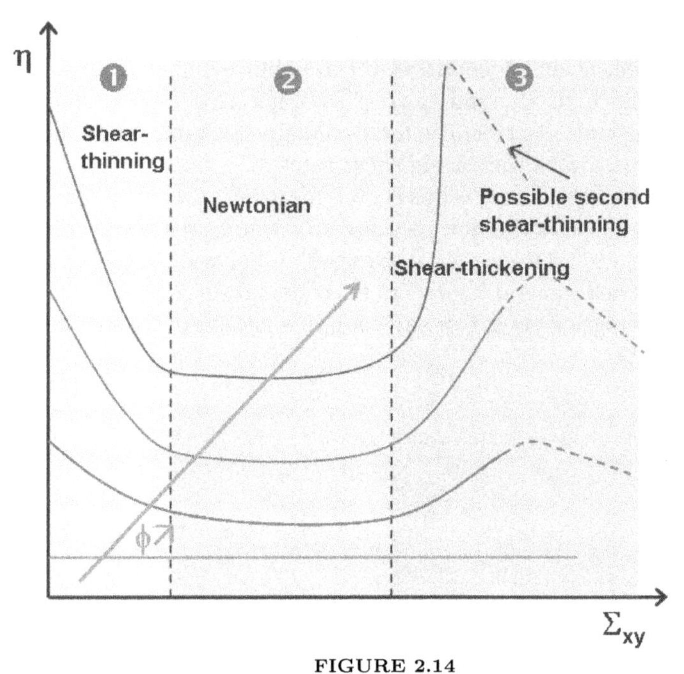

FIGURE 2.14
Schematic diagram of the viscosity behavior as a function of shear stress for non-Brownian suspensions. In the first zone, the suspension is shear-thinning due to colloidal or residual Brownian effects. In the second zone, the suspension behavior is quasi-Newtonian and the particles experience repulsive forces that prevent solid contacts between particles. In the third zone, the shear forces are high enough to overcome repulsive interactions between particles that come into contact.

2.4.2 Normal stress differences

Normal stress differences have been much less extensively studied than viscosity, especially from an experimental point of view. A spectacular manifestation of normal stress differences is the climbing effect of a viscoelastic fluid when stirred by a rotating rod. This effect is well-documented and known as the Weissenberg effect. On the contrary, when a spinning rod is introduced into a concentrated suspension, a rod dipping is observed (see Figure 2.15). This free surface deformation is not caused by inertial effects since it is observed even at very low Reynolds numbers but is a manifestation of normal stress differences that are different in sign and in magnitude from the normal stress differences exhibited by polymer solutions.

Normal stress differences are known to diverge for the same particle volume fraction ϕ^* as viscosity. Stokesian Dynamics simulations [29] indicate that, for suspensions of Brownian hard spheres, N_1 should be positive when the Péclet number is low enough for Brownian forces to dominate and that N_1 changes sign from positive to negative at the shear-thickening transition. In both shear-thinning and shear-thickening regimes, N_2 is found to be negative and about the same magnitude as N_1. The few experimental attempts carried out on suspensions of Brownian hard spheres have confirmed those results, almost qualitatively [19, 56].

FIGURE 2.15
Anti-Weissenberg effect observed in a suspension of PMMA spherical particles (diameter $80\mu m$) dispersed in a mixture of Ucon oil and water (viscosity 20 Pa.s) at $\phi = 0.52$. The rotating speed is equal to 10 rad.s^{-1} which leads to a Reynolds number as low as 10^{-2} meaning that inertial effects can be neglected. The suspension climbs down the rotating rod due to normal stress differences.

The case of suspensions of non-Brownian non-colloidal particles is more problematic. There is a consensus about the magnitude and the sign of N_2. Both numerical simulations [32, 95, 105] and experiments [18, 24, 96] indicate $N_2 < 0$ and $|N_2| \sim \eta_s(\phi/\phi^*)^2$. In contrast, the values of N_1 are very scattered. In all the numerical simulations [32, 95, 105], N_1 is negative but experiments report either negative N_1 [21,96,108], or almost zero N_1 [18] or positive N_1 of the order of $-N_2/2$ [24] or, at last, either positive or negative values depending on the size and polydispersity of particles [33].

2.4.3 Confinement effects

Confined suspensions are quite widespread and include rheometric experiments when the gap of the viscosimeter is not very large compared to the particle size, suspension flows in porous media or microfluidic devices, and of course blood flow in capillaries. In those situations, the interactions between particles and boundaries can play an important role and modify the behavior the suspension would have in an unbounded flow. Walls are known to induce a local ordering of the particles that tend to form layers in the neighborhood of the walls. The wall-induced layering is all the more marked and extended as particle volume fraction increases, as shown in Figure 2.16. For volume fractions of the order of 50%, the layered zone can extend over a distance of about 4 to 5 particle diameters. In those structured zones, the viscosity of the suspension is lower than in the bulk, which is reflected at a macroscopic scale by an apparent wall slip, in agreement with the empirical model of Jana *et al.* [44]. The layering with a hexagonal structure of the particles has been confirmed by numerical simulations [31, 106] which also show that such a particle ordering may dramatically change the value of N_1. In the layered zone, the magnitude of

$|N_1|$ is smaller than far from the walls and, for rough particles, the combined effects of wall-ordering and friction lead to positive values of N_1. The effect of walls on particle diffusion is also of importance: diffusion in the wall-normal direction changes from superdiffusion for particles close to the wall to subdiffusion for particles in the core of the channel [104].

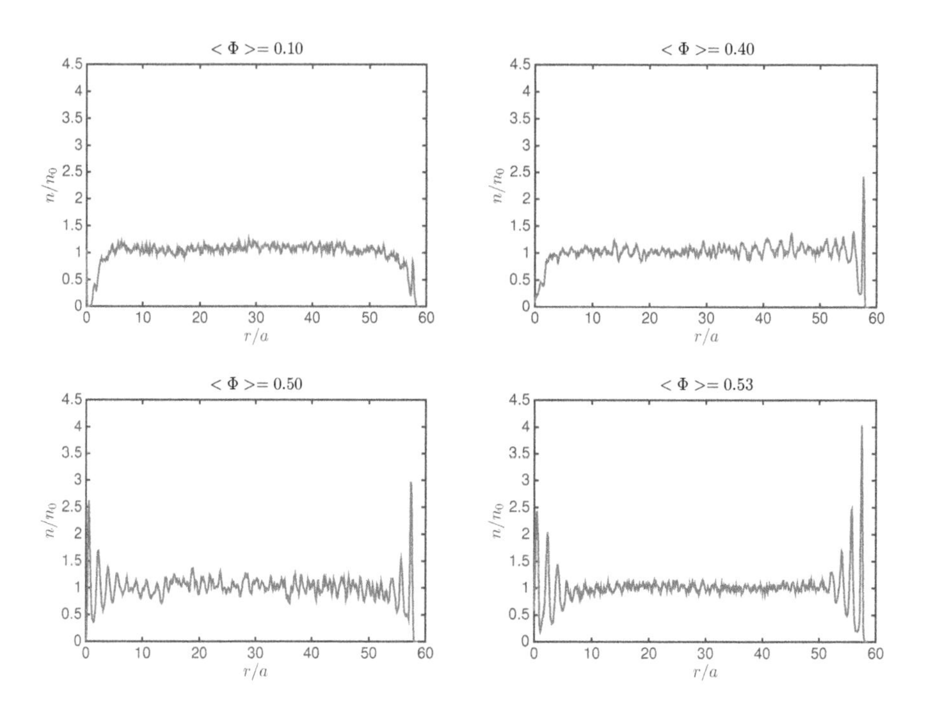

FIGURE 2.16
Concentration profile normalized by its mean value in the gap of a Couette rheometer as a function of the position r in the gap (normalized by particle radius a) for four particle volume fractions ϕ. Reproduced from [5] with the permission of the author.

2.5 SHEAR-INDUCED MIGRATION

2.5.1 Physical description

Migration of particles due to the flow can result from inertia (so-called Segré-Silberberg effect [65, 92]) or non-linearities in the flow field. We shall here solely focus of the latter process, which is the only migration mechanism for non-inertial force-free particles. The term migration here refers to a correlated motion of many particles (collective diffusion) and corresponds to a rate at which macroscopic spatial inhomogeneities relax. This differs from the self-diffusion of particles which is due to the stochastic nature of the trajectory of an individual particle, even in the absence of inhomogeneities. The fact that a particle exhibits a random walk motion is related to the chaotic nature of suspension dynamics as already discussed in Sec. 2.2.6. As pointed out by some works [58, 59], self-diffusion and migration are somewhat connected because they both result from interactions between particles. Like self-diffusion, the existence of migration can appear surprising based on the reversibility of motion predicted by Stokes equations but as pointed out previously, suspensions are, in effect, strongly irreversible. Shear-induced migration can result in strong heterogeneities in volume fraction and has therefore received much attention due to its relevance in many fields.

The first evidence of shear-induced migration dates back from the early 80's when Gadala-Maria and Acrivos [30] noticed a slow and continuous decrease of suspension viscosity in a Couette rheometer. This was experimentally studied in detail later on by Leighton and Acrivos [57] who proposed a theoretical framework describing migration. They found that particles tend to migrate from high shear rate to low shear rate regions of the flow. The collision frequency between particles being proportional to the local shear rate, particles may experience more interactions in high shear rate regions, eventually pushing them away toward more quiet zones where interactions are less frequent, finally leading to a macroscopic gradient in the volume fraction. Leighton and Acrivos have found that the migration flux $\mathbf{j}$ scales as $\dot{\gamma}a^2$ and that this flux is normal to the mean flow, i.e. across the streamlines of the bulk flow. Migration has been since experimentally evidenced in many non-linear flows[(v)], notably Poiseuille flows [38, 39, 62, 98] or wide-gap Couette flows [1, 57, 76, 81]. In torsional flows (parallel-disk rheometer), migration seems absent [12, 49] or very weak [66].

Figure 2.17 shows two images illustrating this migration in a Poiseuille flow from refractive index matching techniques. The left image is taken in the initial suspension before start-up of the flow while the right image is obtained after the suspension has been sheared until a fully-developed migration is reached. The migration is clearly visible as a higher concentration of particles (black disks in the figure) in the center of the pipe, i.e. regions of lower shear rates.

2.5.2 Migration modeling

Most migration models consider the suspension as a continuous medium governed by continuity and momentum equations. This continuum description is obtained through some averaging procedure not described further here [26, 43, 75]. The particle mass conservation leads to an equation on particle volume fraction ϕ as

$$\frac{\partial \phi}{\partial t} + \mathbf{u}_s \cdot \nabla \phi = -\nabla \cdot \mathbf{j} \qquad (2.33)$$

where $\mathbf{u}_s$ is the average suspension velocity given by solving a momentum equation. In absence of sedimentation or other external forces, the flux $\mathbf{j}$ in Eq. (2.33) describes the shear-induced migration flux. There are two main ways to model $\mathbf{j}$: the diffusive flux models and the suspension balance models. Many models are available in the literature (e.g., see Ref. [100]) but they fall into one of those two classes.

Diffusive flux models:
The ideas of Leighton and Acrivos [57] have been used to develop a phenomenological description of migration based on different diffusive terms, for instance the model from Phillips *et al.* [81]. The first term is a collision flux $\mathbf{j}_c$ which originates from the gradient in shear rate and particle concentration and, from scaling arguments by Leighton and Acrivos [57], is given as $\mathbf{j}_c = -K_c a^2 \phi \nabla(\phi\dot{\gamma})$. A second flux arises due to spatial variations of viscosity and reads $\mathbf{j}_\eta = K_\eta a^2 \phi^2 \dot{\gamma} \nabla \ln \eta$. This flux physically describes the tendency for particles to migrate more easily in less viscous regions. K_c and K_η are two empirically-determined diffusion coefficients. The total migration flux then reads $\mathbf{j}=\mathbf{j}_c+\mathbf{j}_\eta$. Later on, some *ad hoc* additional curvature-induced flux was added by Krishnan *et al.* [54] in order to fix the inability of the model to predict the absence of migration in torsional flows.

(v) Flows having a spatial gradient in the shear rate, i.e $\nabla\dot{\gamma} \neq 0$.

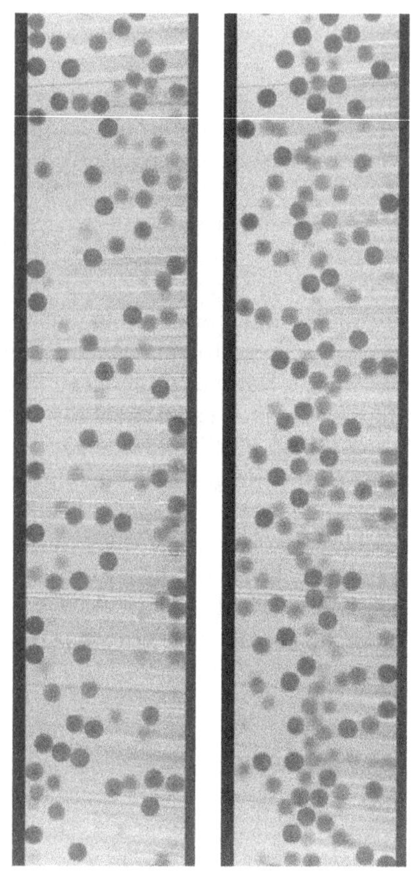

FIGURE 2.17
Images of a suspension at volume fraction ϕ=20% subjected to large oscillating displacements in a pipe. The left image corresponds to the initially mixed suspension and the right image to the suspension after 32 oscillations where fully-developed migration toward the centerline is observed. Reproduced from movie 1 of [98].

Suspension balance models:
Diffusive flux models are phenomenological models that are generally satisfactory for most flows but can fail in certain situations, such as curvilinear flows (e.g., parallel-plate or cone-and-plate flows). They can be improved by free parameter adjustments or by increasing the model complexity [28, 36, 49]. The suspension balance model attempts to relate the migration flux and the suspension rheology based on volume-averaged balances of mass and momentum for fluid and particle phase. Relying on the ideas of Jenkins and McTigue [47] and Nott and Brady [74], Morris and Boulay [72] developed the suspension balance model where the migration flux is given as

$$\mathbf{j} = \frac{2a^2}{9\eta_f} f(\phi) \nabla \cdot \mathbf{\Sigma}^p \tag{2.34}$$

where $f(\phi)$ is taken as the sedimentation hindrance function. This requires a model for $\mathbf{\Sigma}^p$ which in Ref. [72] is given as

$$\mathbf{\Sigma}^p = 2(\eta_s - \eta_f)\boldsymbol{\mathcal{D}}^\infty - \eta_n \dot{\gamma} \mathbf{Q} \tag{2.35}$$

where η_n is termed the normal viscosity and is given by [72] as

$$\eta_n = 0.75 \left(\frac{\phi}{\phi^*}\right)^2 \left(1 - \frac{\phi}{\phi^*}\right)^{-2} \tag{2.36}$$

and $\mathbf{Q}$ is a diagonal tensor which captures the anisotropy of the normal stresses ($\mathbf{Q}$=$\mathbf{I}$ for isotropic normal stresses). Simply speaking, at steady-state, the migration flux acts to balance a gradient in $\mathbf{\Sigma}^p$. There have been further recent works [60, 75] providing a different interpretation about the stress term in Eq. (2.34). Those works show that only the contact contribution $\mathbf{\Sigma}^{p,(c)}$ (and not the total particle stress $\mathbf{\Sigma}^p$) should be included in the divergence in Eq. (2.34). For dense systems, differences are expected to be limited due to the predominance of the contact contribution to the stress [32] but the relevance of Eq. (2.34) in dilute regimes remains an open question.

The backbone of suspension balance models is a constitutive equation for the particle stress. It is important to mention that formulating constitutive equations for continuum modeling of suspensions has a wide importance that extends beyond migration. The simple model of Eq. (2.35) with its explicit splitting between shear and normal stress contributions has some limitations since it becomes problematic to define $\mathbf{Q}$ for arbitrary flows despite some recent attempts [24,70]. More general constitutive equations for particle stress are still a topic of intense research [35,60,80,103].

A well-known limitation of such models is that they break down in regions of zero shear rate, such as the centerline in pressure-driven channel flows, and the predicted volume fraction ϕ at centerline goes up to its maximum value ϕ^*. This singular behavior can be easily understood by noting that in such flows at steady-state, we have a zero radial migration flux j_y=0 from Eq. (2.33), so that $\partial \Sigma^p_{yy}/\partial y$=0 from Eq. (2.34). Using the model given by Eq. (2.35), this suggests that $\eta_n \dot{\gamma}$ is a constant so that η_n must diverge in regions where $\dot{\gamma}$=0. This flaw is related to the assumption of considering the suspension as a continuum and such models are therefore unable to describe the flow at scales smaller than the particle size. Classical corrections consist in adding extra terms that account for the finite size of the particles. More details are given in the Chapter 6 of this book.

2.5.3 Segregation

Particles in actual suspensions are generally not monodisperse and have a size distribution. Similarly, a suspension is often composed of particles having different sizes. In those cases, migration can result in size segregation. It is generally accepted that the underlying physical mechanisms for migration and segregation are similar and due to interactions between particles in a non-linear flow. Because of the expected a^2 scaling, large particles are likely to migrate rapidly while small particles migrate slowly or not at all. Segregation has been evidenced experimentally, mostly in bidisperse configurations where two classes of particles are considered with fraction ϕ_{big} and ϕ_{small} (ϕ=ϕ_{big}+ϕ_{small}) and size ratio $\beta = a_{big}/a_{small}$. Available experimental data include Couette [1,53,54] or Poiseuille flows [42,62]. As a general rule, authors have found that when ϕ_{small} is small, large particles migrate while small particles do not. But for $\phi_{small}/\phi > 0.25$ there is a weak migration of small particles toward the center of a Poiseuille flow (ϕ=0.4, β=3.4) [62]. However, for ϕ=0.3, there does not seem to be any migration of fine particles whatever their fraction [62]. Similar results are obtained in Poiseuille flows by Semwogerere *et al.* [93] who have found migration of large particles and a small depletion of fine particles in the center (β=2). The effect is reverse when fine particles are in large concentration. This means that the idea that large particles migrate and small do not is not general and is case-dependent meaning that interactions between classes may exist. Those interactions possibly explain why it is generally not possible to scale the effective diffusivity based on an average of the diffusivity of each class [53].

Segregation modeling is basically an extension of existing migration models, notably diffusive flux models. The simplest models consider a migration model applied to the largest particles, assuming that small particles can be lumped into an equivalent fluid [54,62]. Such models are not universal and may not work in the case where there are interactions between classes. They are therefore restricted to large β and low ϕ_{small}/ϕ so that migration of large particles dominate. More complex models taking into account interactions between classes have been proposed recently [94, 100] with encouraging prediction capabilities. Yet, there are still some closure issues as well as a lack of rheological models (viscosity, hindrance function, etc.) for polydisperse suspensions.

2.5.4 On the role of deformability

All the above discussion addresses rigid spheres, which have been the subject of most studies on migration and segregation. Yet, migration in Poiseuille flows has been also reported for capsules, red blood cells, and drops indicating that migration also prevails for non-spherical or non-rigid particles. The study of such particles is somewhat less advanced but it is believed that most of the physics presented so far still applies. The suspension balance model enlightens the connection between migration and normal stresses which, in turn, are related to the fore-aft asymmetry of the microstructure as discussed previously. We have also underlined that this symmetry breaking arises from contacts or, more generally, to non-hydrodynamic interactions. Loewenberg and Hinch [61] performed simulations of collisions between two deformable drops in shear flows and found self-diffusivities larger than for rigid spherical particles. They concluded that deformation endows the drops with apparent short-range repulsive pairwise interactions that produce net cross-flow displacements just as in the case of rigid contacts. Simulations by Clausen et al. [13] on deformable capsules show that the microstructure remains anisotropic, suggesting normal stresses and a migration physics similar to rigid particles. All those results support that the physical mechanisms of shear-induced migration, as described previously for rigid particles, persist for deformable particles.

However, additional complexity arises and the deformability of particles generates additional lift forces that have been attributed to wall-particle interactions as well as a coupling between flow non-linearity and particle deformation [15,48] (see also Chapter 6). As far as we are aware, there are no detailed migration models for deformable particles such as the ones presented previously for rigid particles (diffusive flux models and suspension balance models). Because diffusive flux models are ad-hoc models fitted for rigid spheres, they are not expected to provide valuable insights into deformable particles. On the other hand, suspension balance models are based on mass/momentum balance and—although developed primarily for rigid spheres—they may be still valid provided new closure models for $\mathbf{\Sigma}^p$ are developed. For rigid spheres, the normal stress depends on volume fraction and is linear in shear rate but for non-rigid particles, it is also a function of particle deformation which itself depends on shear rate, thereby involving additional interactions. Indeed, simulations by Clausen et al. [13] show that the normal stresses Σ^p_{ii} can undergo rapid changes even with moderate levels of particle deformation. In addition, rigidity contrast is likely to cause segregation. Kumar and Graham [55] showed that stiffness contrast could play a role in margination (see Chapter 2). They studied the flow of a binary suspension composed of capsules with two different rigidities and showed the trend of the stiffer particles to marginate.

2.6 CONCLUSIONS

Dense suspensions of rigid particles are a complex physical domain which has significantly progressed in the recent years through experimental data and numerical simulations. A part of its complexity is nested in the variety of interactions between particles. Even the case of non-Brownian, non-colloidal and non-inertial suspensions of spheres present non-Newtonian features whereas only hydrodynamics and contacts are at play. Recent studies have highlighted the role of contacts that can be promoted by particle roughness. Roughness-induced contacts between particles explain the chaotic and irreversible nature of suspensions which manifests macroscopically as diffusion or migration. A key feature of suspensions is that its behavior can be driven by forces acting on a scale well below particle size. Contacts (or other non-hydrodynamic forces) affect the microstructure that presents a flow-induced angular distortion growing with Pe which, in turn, involves the occurrence of normal stresses. Yet, much less is known when particles are deformable although the above statements for rigid particles remain valid to some extent. Therefore the knowledge of suspensions of rigid particles is a prerequisite before moving to non-rigid particles. Deformability involves a subtle interplay between non-spherical shape, intricate dynamics, and additional interactions which makes deformable suspensions a quite open and challenging domain.

FURTHER READING

Guazzelli, E., and Pouliquen, O. (2018). Rheology of dense granular suspensions. *Journal of Fluid Mechanics*, 852, P1.

Denn, M. M., and Morris, J. F. (2014). Rheology of non-Brownian suspensions. *Annual Review of Chemical and Biomolecular Engineering*, 5: 203-228.

Guazzelli, E., and Morris, J. F. (2011). *A Physical Introduction to Suspension Dynamics*, volume 45. Cambridge University Press.

Bibliography

[1] J.R. Abbott, N. Tetlow, A.L. Graham, S.A. Altobelli, E. Fukushima, L.A. Mondy, and T.S. Stephens. Experimental observations of particle migration in concentrated suspensions: Couette flow. *Journal of Rheology*, 35(5):773–795, 1991.

[2] H.A. Barnes, J.F. Hutton, and K. Walters. *An Introduction to Rheology*, volume 3. Elsevier Science, 1989.

[3] G.K. Batchelor. The stress system in a suspension of force-free particles. *Journal of Fluid Mechanics*, 41(03):545–570, 1970.

[4] G.K. Batchelor and J.T. Green. The determination of the bulk stress in a suspension of spherical particles to order c^2. *Journal of Fluid Mechanics*, 56(03):401–427, 1972.

[5] F. Blanc. *Rhéologie et microstructure des suspensions concentrées non browniennes*. PhD thesis, Université Nice Sophia Antipolis, 2011.

[6] F. Blanc, F. Peters, and E. Lemaire. Experimental signature of the pair trajectories of rough spheres in the shear-induced microstructure in noncolloidal suspensions. *Physical Review Letters*, 107(20):208302, 2011.

[7] F. Blanc, F. Peters, and E. Lemaire. Particle image velocimetry in concentrated suspensions: Application to local rheometry. *Applied Rheology*, 21:23735, 2011.

[8] J.F. Brady and J.F. Morris. Microstructure of strongly sheared suspensions and its impact on rheology and diffusion. *Journal of Fluid Mechanics*, 348:103–139, 1997.

[9] F.P. Bretherton. The motion of rigid particles in a shear flow at low reynolds number. *Journal of Fluid Mechanics*, 14(2):284–304, 1962.

[10] F.S. Chan, J. Blachford, and D.A.I. Goring. The secondary electroviscous effect in a charged spherical colloid. *Journal of Colloid and Interface Science*, 22(4):378–385, 1966.

[11] C. Chang and R.L. Powell. Effect of particle size distributions on the rheology of concentrated bimodal suspensions. *Journal of Rheology*, 38(1):85–98, 1994.

[12] A.W. Chow, S.W. Sinton, J.H. Iwamiya, and T.S. Stephens. Shear-induced particle migration in couette and parallel-plate viscometers: Nmr imaging and stress measurements. *Physics of Fluids*, 6:2561, 1994.

[13] J.R. Clausen, D.A. Reasor, and C.K. Aidun. The rheology and microstructure of concentrated non-colloidal suspensions of deformable capsules. *Journal of Fluid Mechanics*, 685:202–234, 2011.

[14] J. Comtet, G. Chatté, A. Niguès, L. Bocquet, A. Siria, and A. Colin. Pairwise frictional profile between particles determines discontinuous shear thickening transition in non-colloidal suspensions. *Nature Communications*, 8, 2017.

[15] G. Coupier, B. Kaoui, T. Podgorski, and C. Misbah. Noninertial lateral migration of vesicles in bounded poiseuille flow. *Physics of Fluids*, 20(11):111702, 2008.

[16] P. Coussot and C. Ancey. Rheophysical classification of concentrated suspensions and granular pastes. *Physical Review E*, 59(4):4445, 1999.

[17] P. Coussot and C. Ancey. *Rhéophysique des pâtes et des suspensions*. EDP Sciences, 1999.

[18] E. Couturier, F. Boyer, O. Pouliquen, and E. Guazzelli. Suspensions in a tilted trough: second normal stress difference. *Journal of Fluid Mechanics*, 686:26–39, 2011.

[19] C.D. Cwalina and N.J. Wagner. Material properties of the shear-thickened state in concentrated near hard-sphere colloidal dispersions. *Journal of Rheology*, 58(4):949–967, 2014.

[20] F.R. Da Cunha and E.J. Hinch. Shear-induced dispersion in a dilute suspension of rough spheres. *Journal of Fluid Mechanics*, 309:211–223, 1996.

[21] S.-C. Dai, E. Bertevas, F. Qi, and R.I. Tanner. Viscometric functions for noncolloidal sphere suspensions with newtonian matrices. *Journal of Rheology*, 57(2):493–510, 2013.

[22] J. Dasan, T.R. Ramamohan, A. Singh, and P.R. Nott. Stress fluctuations in sheared stokesian suspensions. *Physical Review E*, 66(2):021409, 2002.

[23] R.H. Davis, Y. Zhao, K.P. Galvin, and H.J. Wilson. Solid–solid contacts due to surface roughness and their effects on suspension behaviour. *Philosophical Transactions of the Royal Society of London. Series A: Mathematical, Physical and Engineering Sciences*, 361(1806):871–894, 2003.

[24] T. Dbouk, E. Lemaire, L. Lobry, and F Moukalled. Shear-induced particle migration: Predictions from experimental evaluation of the particle stress tensor. *Journal of Non-Newtonian Fluid Mechanics*, 198:78–95, 2013.

[25] G. Drazer, J. Koplik, B. Khusid, and A. Acrivos. Deterministic and stochastic behaviour of non-brownian spheres in sheared suspensions. *Journal of Fluid Mechanics*, 460(1):307–335, 2002.

[26] D.A. Drew. Mathematical modeling of two-phase flow. *Annual Review of Fluid Mechanics*, 15(1):261–291, 1983.

[27] A. Fall, F. Bertrand, D. Hautemayou, C. Mezière, P. Moucheront, A. Lemaitre, and G. Ovarlez. Macroscopic discontinuous shear thickening versus local shear jamming in cornstarch. *Physical Review Letters*, 114(9):098301, 2015.

[28] Z. Fang, A.A. Mammoli, J.F. Brady, M.S. Ingber, L.A. Mondy, and A.L. Graham. Flow-aligned tensor models for suspension flows. *International Journal of Multiphase Flow*, 28(1):137–166, 2002.

[29] D.R. Foss and J.F. Brady. Structure, diffusion and rheology of brownian suspensions by stokesian dynamics simulation. *Journal of Fluid Mechanics*, 407(167-200):166, 2000.

[30] F. Gadala-Maria and A. Acrivos. Shear-induced structure in a concentrated suspension of solid spheres. *Journal of Rheology*, 24(6):799–814, 1980.

[31] S. Gallier, E. Lemaire, L. Lobry, and F. Peters. Effect of confinement in wall-bounded non-colloidal suspensions. *Journal of Fluid Mechanics*, 799:100–127, 2016.

[32] S. Gallier, E. Lemaire, F. Peters, and L. Lobry. Rheology of sheared suspensions of rough frictional particles. *Journal of Fluid Mechanics*, 757:514–549, 2014.

[33] C. Gamonpilas, J.F. Morris, and M.M. Denn. Shear and normal stress measurements in non-brownian monodisperse and bidisperse suspensions. *Journal of Rheology*, 60(2):289–296, 2016.

[34] C. Gao, S.D. Kulkarni, J.F. Morris, and J.F. Gilchrist. Direct investigation of anisotropic suspension structure in pressure-driven flow. *Physical Review E*, 81(4):041403, 2010.

[35] J.D. Goddard. A weakly nonlocal anisotropic fluid model for inhomogeneous stokesian suspensions. *Physics of Fluids*, 20(4):040601, 2008.

[36] A.L. Graham, A.A. Mammoli, and M.B. Busch. Effects of demixing on suspension rheometry. *Rheologica Acta*, 37(2):139–150, 1998.

[37] E. Guazzelli and J.F. Morris. *A Physical Introduction to Suspension Dynamics*, volume 45. Cambridge University Press, 2011.

[38] R.E. Hampton, A.A. Mammoli, A.L. Graham, N. Tetlow, and S.A. Altobelli. Migration of particles undergoing pressure-driven flow in a circular conduit. *Journal of Rheology*, 41:621, 1997.

[39] M. Han, C. Kim, M. Kim, and S. Lee. Particle migration in tube flow of suspensions. *Journal of Rheology*, 43:1157, 1999.

[40] J. Happel and H. Brenner. *Low Reynolds Number Hydrodynamics*. Kluwer Academic Publishers Group, 1983.

[41] D.M. Husband and F. Gadala-Maria. Anisotropic particle distribution in dilute suspensions of solid spheres in cylindrical couette flow. *Journal of Rheology*, 31(1):95–110, 1987.

[42] D.M. Husband, L.A. Mondy, E. Ganani, and A.L. Graham. Direct measurements of shear-induced particle migration in suspensions of bimodal spheres. *Rheologica Acta*, 33(3):185–192, 1994.

[43] R. Jackson. Locally averaged equations of motion for a mixture of identical spherical particles and a newtonian fluid. *Chemical Engineering Science*, 52(15):2457–2469, 1997.

[44] S.C. Jana, B. Kapoor, and A. Acrivos. Apparent wall slip velocity coefficients in concentrated suspensions of noncolloidal particles. *Journal of Rheology*, 39(6):1123–1132, 1995.

[45] I.M. Jánosi, T. Tél, D.E. Wolf, and J.A.C. Gallas. Chaotic particle dynamics in viscous flows: The three-particle stokeslet problem. *Physical Review E*, 56(3):2858, 1997.

[46] G.B. Jeffery. The motion of ellipsoidal particles immersed in a viscous fluid. *Proceedings of the Royal Society of London A*, 102(715):161–179, 1922.

[47] J.T. Jenkins and D.F. McTigue. Transport processes in concentrated suspensions: the role of particle fluctuations. In *Two Phase Flows and Waves*, pages 70–79. Springer, 1990.

[48] B. Kaoui, G.H. Ristow, I. Cantat, C. Misbah, and W. Zimmermann. Lateral migration of a two-dimensional vesicle in unbounded poiseuille flow. *Physical Review E*, 77(2):021903, 2008.

[49] J.M. Kim, S.G. Lee, and C. Kim. Numerical simulations of particle migration in suspension flows: Frame-invariant formulation of curvature-induced migration. *Journal of Non-Newtonian Fluid Mechanics*, 150(2):162–176, 2008.

[50] S. Kim and S.J. Karrila. *Microhydrodynamics: Principles and Selected Applications*, volume 507. Butterworth-Heinemann Boston, 1991.

[51] I. M. Krieger. Rheology of monodisperse latices. *Advances in Colloid and Interface Science*, 3(2):111–136, 1972.

[52] I.M. Krieger and T.J. Dougherty. A mechanism for non-newtonian flow in suspensions of rigid spheres. *Journal of Rheology*, 3:137, 1959.

[53] G. Krishnan and D.T. Leighton. Dynamic viscous resuspension of bidisperse suspensions—i. effective diffusivity. *International Journal of Multiphase Flow*, 21(5):721–732, 1995.

[54] G.P. Krishnan, S. Beimfohr, and D.T. Leighton. Shear-induced radial segregation in bidisperse suspensions. *Journal of Fluid Mechanics*, 321:371–393, 1996.

[55] A. Kumar and M.D. Graham. Segregation by membrane rigidity in flowing binary suspensions of elastic capsules. *Physical Review E*, 84(6):066316, 2011.

[56] M. Lee, M. Alcoutlabi, J.J. Magda, C. Dibble, M.J. Solomon, X. Shi, and G.B. McKenna. The effect of the shear-thickening transition of model colloidal spheres on the sign of n 1 and on the radial pressure profile in torsional shear flows. *Journal of Rheology*, 50(3):293–311, 2006.

[57] D. Leighton and A. Acrivos. The shear-induced migration of particles in concentrated suspensions. *Journal of Fluid Mechanics*, 181:415–439, 1987.

[58] A.M. Leshansky and J.F. Brady. Dynamic structure factor study of diffusion in strongly sheared suspensions. *Journal of Fluid Mechanics*, 527:141–169, 2005.

[59] A.M. Leshansky, J.F. Morris, and J.F. Brady. Collective diffusion in sheared colloidal suspensions. *Journal of Fluid Mechanics*, 597:305–341, 2008.

[60] D. Lhuillier. Migration of rigid particles in non-brownian viscous suspensions. *Physics of Fluids*, 21:023302, 2009.

[61] M. Loewenberg and E.J. Hinch. Collision of two deformable drops in shear flow. *Journal of Fluid Mechanics*, 338:299–315, 1997.

[62] M.K. Lyon and L.G. Leal. An experimental study of the motion of concentrated suspensions in two-dimensional channel flow. Part 2. Bidisperse systems. *Journal of Fluid Mechanics*, 363:57–77, 1998.

[63] R. Mari, R. Seto, J.F. Morris, and M.M. Denn. Shear thickening, frictionless and frictional rheologies in non-brownian suspensions. *Journal of Rheology*, 58(6):1693–1724, 2014.

[64] S.H. Maron and P.E. Pierce. Application of ree-eyring generalized flow theory to suspensions of spherical particles. *Journal of Colloid Science*, 11(1):80–95, 1956.

[65] J.P. Matas, J.F. Morris, and E. Guazzelli. Lateral forces on a sphere: Solid/liquid dispersions in drilling and production. *Oil & Gas Science and Technology*, 59(1):59–70, 2004.

[66] D. Merhi, E. Lemaire, G. Bossis, and F. Moukalled. Particle migration in a concentrated suspension flowing between rotating parallel plates: Investigation of diffusion flux coefficients. *Journal of Rheology*, 49(6):1429–1448, 2005.

[67] B. Metzger and J.E. Butler. Irreversibility and chaos: Role of long-range hydrodynamic interactions in sheared suspensions. *Physical Review E*, 82(5):051406, 2010.

[68] B. Metzger and J.E. Butler. Clouds of particles in a periodic shear flow. *Physics of Fluids*, 24:021703, 2012.

[69] B. Metzger, P. Pham, and J.E. Butler. Irreversibility and chaos: Role of lubrication interactions in sheared suspensions. *Physical Review E*, 87(5):052304, 2013.

[70] R.M. Miller, J.P. Singh, and J.F. Morris. Suspension flow modeling for general geometries. *Chemical Engineering Science*, 64(22):4597–4610, 2009.

[71] M. Mooney. Explicit formulas for slip and fluidity. *Journal of Rheology (1929-1932)*, 2(2):210–222, 1931.

[72] J.F. Morris and F. Boulay. Curvilinear flows of noncolloidal suspensions: The role of normal stresses. *Journal of Rheology*, 43:1213, 1999.

[73] J.F. Morris and B. Katyal. Microstructure from simulated brownian suspension flows at large shear rate. *Physics of Fluids*, 14(6):1920–1937, 2002.

[74] P.R. Nott and J.F. Brady. Pressure-driven flow of suspensions: simulation and theory. *Journal of Fluid Mechanics*, 275:157–199, 1994.

[75] P.R. Nott, E. Guazzelli, and O. Pouliquen. The suspension balance model revisited. *Physics of Fluids*, 23:043304, 2011.

[76] G. Ovarlez, F. Bertrand, and S. Rodts. Local determination of the constitutive law of a dense suspension of noncolloidal particles through magnetic resonance imaging. *Journal of Rheology*, 50:259, 2006.

[77] F. Parsi and F. Gadala-Maria. Fore-and-aft asymmetry in a concentrated suspension of solid spheres. *Journal of Rheology*, 31(8):725–32, 1987.

[78] F. Peters, G. Ghigliotti, S. Gallier, F. Blanc, . Lemaire, and L. Lobry. Rheology of non-brownian suspensions of rough frictional particles under shear reversal: A numerical study. *Journal of Rheology*, 60(4):715–732, 2016.

[79] P. Pham, B. Metzger, and J.E. Butler. Particle dispersion in sheared suspensions: Crucial role of solid-solid contacts. *Physics of Fluids*, 27(5):051701, 2015.

[80] N. Phan-Thien. Constitutive equation for concentrated suspensions in newtonian liquids. *Journal of Rheology*, 39(4):679–695, 1995.

[81] R.J. Phillips, R.C. Armstrong, R.A. Brown, A.L. Graham, and J.R. Abbott. A constitutive equation for concentrated suspensions that accounts for shear-induced particle migration. *Physics of Fluids A: Fluid Dynamics*, 4:30, 1992.

[82] T.N. Phung, J.F. Brady, and G. Bossis. Stokesian dynamics simulation of brownian suspensions. *Journal of Fluid Mechanics*, 313:181–207, 1996.

[83] D.J. Pine, J.P. Gollub, J.F. Brady, and A.M. Leshansky. Chaos and threshold for irreversibility in sheared suspensions. *Nature*, 438(7070):997–1000, 2005.

[84] M. Popova, P. Vorobieff, M.S. Ingber, and A.L. Graham. Interaction of two particles in a shear flow. *Physical Review. E, Statistical, Nonlinear, and Soft Matter Physics*, 75(6):066309, 2007.

[85] C. Pozrikidis. *Boundary Integral and Singularity Methods for Linearized Viscous Flow*. Cambridge University Press, 1992.

[86] I. Rampall, J.R. Smart, and D.T. Leighton. The influence of surface roughness on the particle-pair distribution function of dilute suspensions of non-colloidal spheres in simple shear flow. *Journal of Fluid Mechanics*, 339:1–24, 1997.

[87] J.S. Raynaud, P. Moucheront, J.C. Baudez, F. Bertrand, J.P. Guilbaud, and Ph. Coussot. Direct determination by nuclear magnetic resonance of the thixotropic and yielding behavior of suspensions. *Journal of Rheology*, 46(3):709–732, 2002.

[88] W.B. Russel. Review of the role of colloidal forces in the rheology of suspensions. *Journal of Rheology*, 24(3):287–317, 1980.

[89] I. R. Rutgers. Relative viscosity of suspensions of rigid spheres in newtonian liquids. *Rheologica Acta*, 2(3):202–210, 1962.

[90] G.D. Scott. Packing of spheres: packing of equal spheres. *Nature*, 188(4754):908–909, 1960.

[91] G.D. Scott and D.M. Kilgour. The density of random close packing of spheres. *Journal of Physics D: Applied Physics*, 2:863, 1969.

[92] G. Segré and A. Silberberg. Behaviour of macroscopic rigid spheres in poiseuille flow. Part 1. Determination of local concentration by statistical analysis of particle passages through crossed light beams. *Journal of Fluid Mechanics*, 14(01):115–135, 1962.

[93] D. Semwogerere and E.R. Weeks. Shear-induced particle migration in binary colloidal suspensions. *Physics of Fluids*, 20(4):043306, 2008.

[94] A. Shauly, A. Wachs, and A. Nir. Shear-induced particle migration in a polydisperse concentrated suspension. *Journal of Rheology*, 42(6):1329–1348, 1998.

[95] A. Sierou and J.F. Brady. Rheology and microstructure in concentrated noncolloidal suspensions. *Journal of Rheology*, 46:1031, 2002.

[96] A. Singh and P. R. Nott. Experimental measurements of the normal stresses in sheared stokesian suspensions. *Journal of Fluid Mechanics*, 490:293–320, 2003.

[97] J.R. Smart and D.T. Leighton. Measurement of the hydrodynamic surface roughness of noncolloidal spheres. *Physics of Fluids A: Fluid Dynamics*, 1(1):52–60, 1989.

[98] B. Snook, J.E. Butler, and E. Guazzelli. Dynamics of shear-induced migration of spherical particles in oscillatory pipe flow. *Journal of Fluid Mechanics*, 786:128–153, 2016.

[99] J.C. Van der Werff and C.G. De Kruif. Hard-sphere colloidal dispersions: The scaling of rheological properties with particle size, volume fraction, and shear rate. *Journal of Rheology*, 33(3):421–454, 1989.

[100] H.M. Vollebregt, R.G. Van Der Sman, and R.M. Boom. Suspension flow modelling in particle migration and microfiltration. *Soft Matter*, 6(24):6052–6064, 2010.

[101] N.J. Wagner and W.B. Russel. Light scattering measurements of a hard-sphere suspension under shear. *Physics of Fluids A: Fluid Dynamics*, 2(4):491–502, 1990.

[102] C.R. Wildemuth and M.C. Williams. Viscosity of suspensions modeled with a shear-dependent maximum packing fraction. *Rheologica Acta*, 23(6):627–635, 1984.

[103] K. Yapici, R.L. Powell, and R.J. Phillips. Particle migration and suspension structure in steady and oscillatory plane poiseuille flow. *Physics of Fluids*, 21(5):053302, 2009.

[104] K. Yeo and M.R. Maxey. Anomalous diffusion of wall-bounded non-colloidal suspensions in a steady shear flow. *EPL (Europhysics Letters)*, 92(2):24008, 2010.

[105] K. Yeo and M.R. Maxey. Dynamics of concentrated suspensions of non-colloidal particles in couette flow. *Journal of Fluid Mechanics*, 649(1):205–231, 2010.

[106] K. Yeo and M.R. Maxey. Ordering transition of non-brownian suspensions in confined steady shear flow. *Physical Review E*, 81(5):051502, 2010.

[107] A. Yoshimura and R.K. Prud'homme. Wall slip corrections for couette and parallel disk viscometers. *Journal of Rheology*, 32(1):53–67, 1988.

[108] I.E. Zarraga, D.A. Hill, and D.T. Leighton Jr. Normal stresses and free surface deformation in concentrated suspensions of noncolloidal spheres in a viscoelastic fluid. *Journal of Rheology*, 45(5):1065–1084, 2001.

CHAPTER 3

Blood as a suspension of deformable particles

Michael D. Graham

University of Wisconsin-Madison

CONTENTS

3.1 INTRODUCTION

RED BLOOD CELLS are very soft and are thus easily deformed as they flow. This deformability is of course essential for passing through capillary beds, but also plays a role in larger blood vessels. The latter situation is the focus of the present chapter, whose aim is to lay out the principles that govern the behavior of suspensions of deformable particles. Many of these principles are generic and here we will go back and forth between general arguments and issues that are specific to flowing blood. The first part of the chapter describes some of the basic equations that govern fluid motion and the stress in a flowing suspension, closely following the development in Ref. [28]. With this background in hand, we then briefly describe the motion of nonspherical and deformable particles in a flow. Then the key features of transport in flowing suspensions are described, specifically, momentum transport (the effective viscosity) and mass transport (shear-induced diffusion and migration of cells and solutes). Finally, some important effects of confinement in flowing suspensions are described, namely the development of concentration gradients across a channel during flow and the cross-stream segregation of different components of blood based on their physical properties, a phenomenon called margination.

3.2 MICROSCALE FLOW FUNDAMENTALS

3.2.1 Stokes equations and the Green's function

The relative importance of inertial and viscous effects in a flow is measured by the Reynolds number. The flow properties of a suspension are largely determined by this ratio when measured at the scale of the particles comprising it: i.e. the particle Reynolds number Re_p. A human red blood cell (RBC) is a biconcave discoid with radius a of about 4 μm. The viscosity η of plasma is around 1.2 mPa s and its density ρ is close to that of water. The viscosity of whole blood at high shear rates is about 5 mPa s [24, 25]. A typical shear rate $\dot{\gamma}$ experienced by a blood cell is $10^2 - 10^3$ s^{-1}. The particle Reynolds number is given by $\mathrm{Re}_p = \rho\dot{\gamma}a^2/\eta$, which for the conditions described here is $\sim 10^{-3} - 10^{-2}$. Since this is much smaller than unity, we see that viscous effects dominate, so the fluid motion on the cellular scale is governed by the Stokes equations

$$-\boldsymbol{\nabla} p + \eta \nabla^2 \boldsymbol{v} + \boldsymbol{f} = \boldsymbol{0}, \tag{3.1}$$

$$\boldsymbol{\nabla} \cdot \boldsymbol{v} = 0. \tag{3.2}$$

Here $\boldsymbol{v}$ and p are the fluid velocity and pressure, and $\boldsymbol{f}$ is a force density – all of these quantities are functions of position $\boldsymbol{x}$. The first equation here is momentum conservation and the second is mass conservation under the assumption of incompressibility, which is valid for blood flow. The stress tensor in the fluid is

$$\boldsymbol{\sigma} = -p\boldsymbol{\delta} + \eta\left(\boldsymbol{\nabla}\boldsymbol{v} + \boldsymbol{\nabla}\boldsymbol{v}^{\mathrm{T}}\right).$$

The linearity of the Stokes equations allows solutions to complex problems to be put together as sums of solutions to simpler ones. Perhaps the most important of these simpler problems is the flow generated by a point force (a delta function force density) exerted at some position in the flow domain. In general, the solution to a linear differential equation with a delta function forcing is called a *Green's function*. With this solution a number of other *fundamental solutions* can be generated. We describe some of these here for unbounded flow domains and illustrate how they form a basis for computing and understanding the dynamics of particles in flow.

Consider Stokes flow in an unbounded domain driven by a point force exerted at position $\boldsymbol{x} = \boldsymbol{x}_0$: i.e. $\boldsymbol{f}(\boldsymbol{x}) = \boldsymbol{F}\delta(\boldsymbol{x}-\boldsymbol{x}_0)$, with $p \to p_\infty$ and $\boldsymbol{v} \to 0$ as $r = |\boldsymbol{x} - \boldsymbol{x}_0| \to \infty$. For problems without a given reference pressure, we can take $p_\infty = 0$, since the velocity only depends on the pressure gradient. The solution to this problem is

$$\boldsymbol{v}(\boldsymbol{x}) = \mathbf{G}(\boldsymbol{x} - \boldsymbol{x}_0) \cdot \boldsymbol{F}, \tag{3.3}$$

where

$$\mathbf{G}(\boldsymbol{x} - \boldsymbol{x}_0) = \frac{1}{8\pi\eta|\boldsymbol{x} - \boldsymbol{x}_0|}\left(\boldsymbol{\delta} + \frac{(\boldsymbol{x} - \boldsymbol{x}_0)(\boldsymbol{x} - \boldsymbol{x}_0)}{|\boldsymbol{x} - \boldsymbol{x}_0|^2}\right) \tag{3.4}$$

is the free space Green's function for the Stokes equation. This tensor is also commonly called the *Stokeslet* or *Oseen-Burgers tensor* [13, 62]. Because the velocity field is divergence-free, the Green's function is as well: $\boldsymbol{\nabla} \cdot \mathbf{G} = \boldsymbol{0}$.

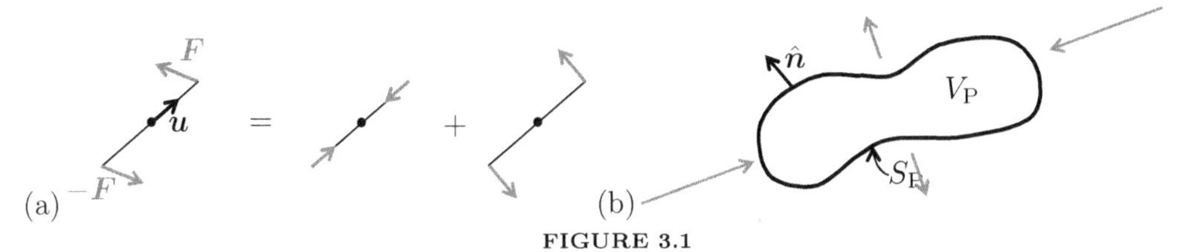

FIGURE 3.1
(a) A force dipole (left), showing its decomposition into symmetric (center) and antisymmetric (right) parts. (b) A force- and torque-free particle in a fluid and the force dipoles it exerts on the fluid (blue arrows).

By superposition of two Stokeslets, we can now construct a *force dipole* solution to the Stokes equations. The importance of this solution for suspensions will emerge below. A force dipole is a force density arising from equal and opposite forces applied at points $\boldsymbol{x}_0 \pm \boldsymbol{u}\, h/2$ as shown in Figure 3.1a:

$$\boldsymbol{f}_\mathrm{d}(\boldsymbol{x}) = \boldsymbol{F}\left(\delta(\boldsymbol{x} - (\boldsymbol{x}_0 + \boldsymbol{u}\, h/2)) - \delta(\boldsymbol{x} - (\boldsymbol{x}_0 - \boldsymbol{u}\, h/2))\right).$$

This density yields a velocity field

$$\boldsymbol{v}(\boldsymbol{x}) = (\mathbf{G}(\boldsymbol{x} - (\boldsymbol{x}_0 + \boldsymbol{u}\, h/2)) - \mathbf{G}(\boldsymbol{x} - (\boldsymbol{x}_0 - \boldsymbol{u}\, h/2))) \cdot \boldsymbol{F}.$$

In the limit $h \to 0$ with $\boldsymbol{F}_\mathrm{d} = \lim_{h\to 0} \boldsymbol{F} h$ held constant, this velocity field can be written

$$\boldsymbol{v}(\boldsymbol{x} - \boldsymbol{x}_0) = \boldsymbol{u} \cdot \boldsymbol{\nabla}_0 \mathbf{G}(\boldsymbol{x} - \boldsymbol{x}_0) \cdot \boldsymbol{F}_\mathrm{d} = -\boldsymbol{u} \cdot \boldsymbol{\nabla} \mathbf{G}(\boldsymbol{x} - \boldsymbol{x}_0) \cdot \boldsymbol{F}_\mathrm{d}. \tag{3.5}$$

In index notation:

$$v_i = \frac{\partial G_{ik}}{\partial x_{0j}} F_{\mathrm{d},k} u_j = -\frac{\partial G_{ik}}{\partial x_j} F_{\mathrm{d},k} u_j. \tag{3.6}$$

This flow field is called the Stokeslet doublet. Based on the structure of (3.5) we can see that this flow results from a force density

$$\begin{aligned} \boldsymbol{f}_\mathrm{d}(\boldsymbol{x} - \boldsymbol{x}_0) &= -\boldsymbol{F}_\mathrm{d}\boldsymbol{u} \cdot \boldsymbol{\nabla}\delta(\boldsymbol{x} - \boldsymbol{x}_0) \\ &= \boldsymbol{\nabla} \cdot (-\boldsymbol{u}\boldsymbol{F}_\mathrm{d}\delta(\boldsymbol{x} - \boldsymbol{x}_0)) \\ &= \boldsymbol{\nabla} \cdot \left(\mathbf{D}^\mathrm{T}\delta(\boldsymbol{x} - \boldsymbol{x}_0)\right), \end{aligned} \tag{3.7}$$

where

$$\mathbf{D} = -\boldsymbol{F}_\mathrm{d}\boldsymbol{u} \tag{3.8}$$

is called the force dipole tensor $\mathbf{D}$.

The dipole tensor can be decomposed as

$$\mathbf{D} = \mathbf{S} + \mathbf{R} + \frac{1}{3}(\mathrm{tr}\mathbf{D})\boldsymbol{\delta}, \tag{3.9}$$

where $\mathbf{S}$ is symmetric and traceless, and $\mathbf{R}$ is antisymmetric. The last term, the isotropic part of $\mathbf{D}$, drives no motion because $\boldsymbol{\nabla} \cdot \mathbf{G} = \mathbf{0}$. Figure 3.1a shows the force distributions corresponding to the symmetric and antisymmetric parts of $\mathbf{D}$. From the figure it is apparent that $\mathbf{R}$ corresponds to a point torque exerted on the fluid. The symmetric tensor $\mathbf{S}$ is called the *stresslet* and the antisymmetric tensor $\mathbf{R}$ is called the *rotlet*. The following section elaborates on how these solutions are relevant to the motion of a particle in a fluid.

3.2.2 Multipole expansion and the dipole for a force- and torque-free particle

Consider a particle with characteristic dimension a in an unbounded fluid in the Stokes flow regime. The flow induced by the presence of the particle is generated by the forces that the particle exerts on the fluid: it is an integral over the Stokeslets generated by the force each infinitesimal element of the particle surface exerts on the fluid. If the particle is centered at the origin, occupies the domain V_{P} and has surface S_{P} as shown in Figure 3.1b, then this statement can be written as follows:

$$v_i(\boldsymbol{x}) - v_{\infty,i}(\boldsymbol{x}) = -\int_{S_{\mathrm{P}}} G_{ij}(\boldsymbol{x}-\boldsymbol{x}')\,(\hat{n}_k\sigma_{kj}(\boldsymbol{x}'))\,dS(\boldsymbol{x}'). \tag{3.10}$$

Here $\boldsymbol{v}_\infty$ is the velocity the fluid would have in the absence of the particle, G_{ij} is the free-space Green's function derived above, $\hat{\boldsymbol{n}}$ is the unit normal pointing from the particle surface into the fluid and $(\hat{n}_k\sigma_{kj}(\boldsymbol{x}'))\,dS(\boldsymbol{x}')$ is the force exerted on the area element of particle surface located at position $\boldsymbol{x}'$ by the fluid.

Now consider a point $\boldsymbol{x}$ that is very far from the particle, so that $|\boldsymbol{x}| \gg |\boldsymbol{x}'|$ for all points $\boldsymbol{x}'$ on the particle surface ($|\boldsymbol{x}'| \lesssim a$). In this case we can Taylor expand the Green's function inside the integral to yield:

$$\begin{aligned} v_i(\boldsymbol{x}) - v_{\infty,i}(\boldsymbol{x}) = &-G_{ij}|_{\boldsymbol{x}'=0}\int_{S_{\mathrm{P}}} (\hat{n}_k\sigma_{kj}(\boldsymbol{x}'))\,dS(\boldsymbol{x}') \\ &+ \left.\frac{\partial G_{ij}}{\partial x_l}\right|_{\boldsymbol{x}'=\mathbf{0}} \int_{S_{\mathrm{P}}} x_l'\,(\hat{n}_k\sigma_{kj}(\boldsymbol{x}'))\,dS(\boldsymbol{x}') \\ &+ O\left(\left(\frac{a}{|\boldsymbol{x}|}\right)^3\right). \end{aligned} \tag{3.11}$$

This expression is called the *multipole expansion* of the velocity field induced by the presence of the particle. We now examine its structure.

Consider the integral in the first term on the right-hand side of this expression, which we will denote $\boldsymbol{F}_{\mathrm{drag}}$:

$$F_{\mathrm{drag},j} = \int_{S_{\mathrm{P}}} (\hat{n}_k\sigma_{kj}(\boldsymbol{x}'))\,dS(\boldsymbol{x}'). \tag{3.12}$$

This is simply the stress exerted on the particle by the fluid, integrated over the surface of the particle – the total drag force exerted on the particle by the fluid. By Newton's third law this is simply the negative of the force $\boldsymbol{F}$ that that the particle exerts on the fluid. It multiplies the free-space Green's function giving a Stokeslet contribution to the velocity field that just arises from the total force exerted by the particle on the fluid.

The next term contains the gradient of the Green's function, which we saw above is associated with point stresses and torques exerted on the fluid – it is the Stokeslet doublet. We denote the integral in this term as $\mathbf{D}$:

$$D_{jl} = \int_{S_{\mathrm{P}}} (\hat{n}_k\sigma_{kj}(\boldsymbol{x}'))\,x_l' dS(\boldsymbol{x}'). \tag{3.13}$$

This is the dipole tensor associated with the particle. Now we can write the multipole expansion in terms of the Stokeslet, stresslet and rotlet solutions introduced above:

$$v_i(\boldsymbol{x}) - v_{\infty,i}(\boldsymbol{x}) = -G_{ij}(\boldsymbol{x})F_{\mathrm{drag},j} + \frac{\partial G_{ij}}{\partial x_l}(\boldsymbol{x})D_{jl} + O\left(\left(\frac{a}{|\boldsymbol{x}|}\right)^3\right) \tag{3.14}$$

$$\approx -G_{ij}(\boldsymbol{x})F_{\mathrm{drag},j} + T_{lij}^{\mathrm{STR}}(\boldsymbol{x})S_{jl} + T_{lij}^{\mathrm{ROT}}(\boldsymbol{x})R_{jl}. \tag{3.15}$$

Here T_{lij}^{STR} and T_{lij}^{ROT} are the symmetric and antisymmetric parts of $\partial G_{ij}/\partial x_l$ with respect to indices j and l. Now consider a particle that is sufficiently small that its inertia is negligible, so any external forces or torques exerted on it are balanced by forces and torques the particle exerts in turn on the fluid. For a neutrally buoyant particle, or more generally a particle that is not subjected to an external force, the first term vanishes and the second dominates. Furthermore, if there are no external torques exerted on the particle it will exert no net torque on the fluid. Therefore, D_{jl} will be symmetric and only a stresslet flow field will be generated, so we have

$$v_i(\boldsymbol{x}) - v_{\infty,i} \approx T_{lij}^{\mathrm{STR}} S_{jl}.$$

Because it is symmetric, the particle stresslet $\mathbf{S}$ can be written in terms of its eigenvalues λ_i and orthonormal eigenvectors $\boldsymbol{w}_i$ as shown in Figure 3.1b:

$$\mathbf{S} = \lambda_1 \boldsymbol{w}_1 \boldsymbol{w}_1 + \lambda_2 \boldsymbol{w}_2 \boldsymbol{w}_2 + \lambda_1 \boldsymbol{w}_3 \boldsymbol{w}_3. \tag{3.16}$$

3.2.3 The stress in a suspension

Now consider a suspension of many neutrally buoyant torque-free particles. If the externally imposed deformations have a scale much larger than the distance between particles, then we can define effective properties of the suspension by considering quantities that are averaged over a volume V that is large compared to the distance between particles but small compared to the overall size of the flow system [4, 47]. Let there be N particles in the volume so the number density $n = N/V$. Each particle is given a label $\alpha = 1, 2, 3, \ldots, N$, and particle α occupies volume V_α with surface S_α. As above, $\hat{\boldsymbol{n}}$ denotes the unit normal pointing from a particle into the fluid. We assume that the boundary of the volume does not cut through any particles. The volume-averaged velocity gradient is

$$\Gamma_{ji} = \frac{1}{V} \int_V \frac{\partial v_i}{\partial x_j}\, dV \tag{3.17}$$

The integral can be split into a contribution from the suspending fluid and one from the particles:

$$\begin{aligned} \Gamma_{ji} &= \frac{1}{V} \int_{V - \sum_{\alpha=1}^{N} V_\alpha} \frac{\partial v_i}{\partial x_j}\, dV + \frac{1}{V} \int_{\sum_{\alpha=1}^{N} V_\alpha} \frac{\partial v_i}{\partial x_j}\, dV \\ &= \frac{1}{V} \int_{V - \sum_{\alpha=1}^{N} V_\alpha} \frac{\partial v_i}{\partial x_j}\, dV + \frac{1}{V} \sum_{\alpha=1}^{N} \int_{S_\alpha} v_i \hat{n}_j\, dS. \end{aligned} \tag{3.18}$$

The second expression arises from the first by use of the divergence theorem and is preferable because it avoids reference to the velocity gradient within the particle. Similarly, the volume-averaged stress is

$$\Sigma_{ij} = \frac{1}{V} \int_V \sigma_{ij}\, dV \tag{3.19}$$

$$= \frac{1}{V} \int_{V - \sum_{\alpha=1}^{N} V_\alpha} \left[-p\delta_{ij} + \eta \left(\frac{\partial v_i}{\partial x_j} + \frac{\partial v_j}{\partial x_i} \right) \right] dV + \frac{1}{V} \sum_{\alpha=1}^{N} \int_{V_\alpha} \sigma_{ij}\, dV \tag{3.20}$$

The second integral in this expression contains the stress inside the particles. While we cannot determine this without further information about the particles, we do know that if

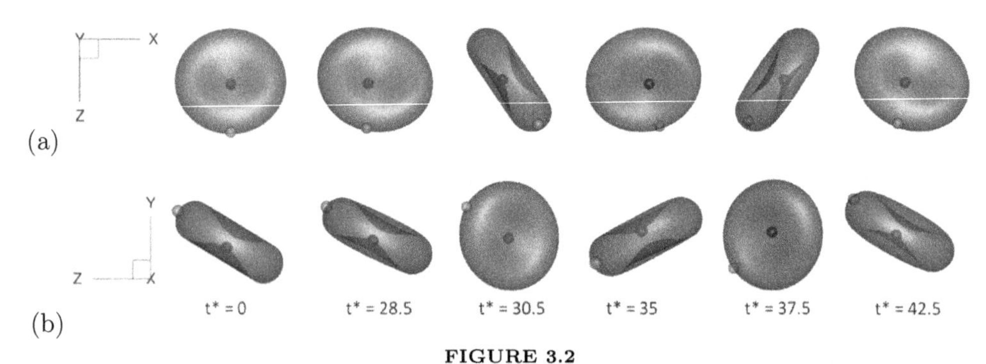

FIGURE 3.2
Time sequence images of a simulation of an RBC in kayaking (Jeffery orbit) motion from (a) top view (yz plane) and (b) side view (xy plane) respectively. Flow is in the x direction; the x-velocity varies linearly with y. The blue and green points are markers on the membrane surface that are initially on the symmetry axis and equator of the cell, respectively. Reproduced with permission from [70].

inertia is negligible within the particles, then the stress satisfies $\boldsymbol{\nabla} \cdot \boldsymbol{\sigma} = \mathbf{0}$. In this case

$$\frac{\partial}{\partial x_i}\sigma_{ij}x_k = \sigma_{kj}. \tag{3.21}$$

Therefore,

$$\int_{V_\alpha} \sigma_{ij}\, dV = \int_{V_\alpha} \frac{\partial}{\partial x_k}\sigma_{kj}x_i\, dV = \int_{S_\alpha} \hat{n}_k\sigma_{kj}x_i\, dS = D^\alpha_{ij} \tag{3.22}$$

where D^α_{ij} is precisely the dipole tensor for particle α that emerges from the multipole expansion; see (3.13). Incorporating (3.18) and (3.22) in (3.20) yields that

$$\Sigma_{ij} = -p_{\text{eff}}\delta_{ij} + \eta\left(\Gamma_{ij} + \Gamma_{ji}\right) + \Sigma_{\text{P},ij}. \tag{3.23}$$

Here p_{eff} is an effective pressure for the suspension, so the first two terms of (3.23) simply comprise a Newtonian stress based on the average velocity gradient. The last term, given by

$$\Sigma_{\text{P},ij} = n\left[\frac{1}{N}\sum_\alpha D^\alpha_{ij}\right] - n\eta\left[\frac{1}{N}\sum_\alpha \int_{S_\alpha} \left(u_i\hat{n}_j + u_j\hat{n}_i\right)\, dS\right]\, dV \tag{3.24}$$

is the particle-phase contribution to the average stress. The second term here simply cancels out the "extra" stress that comes from integrating the strain rate within the particle volume when evaluating the volume-averaged velocity gradient Γ_{ij} in (3.23). We have thus learned that we can understand the stress in a suspension by viewing the particles as force dipoles. We take advantage of this observation at several points in the discussion below.

3.3 DYNAMICS OF DEFORMABLE PARTICLES IN SHEAR FLOW

For perspective, we begin with a brief discussion of the dynamics of rigid particles in simple shear flow. Consider an imposed flow in the x-direction with velocity v_x that varies linearly with y and is independent of z: $v_x = \dot{\gamma}y$. These three directions will be called the flow, gradient and vorticity directions, respectively. During this flow a rigid spherical particle rotates with the local angular velocity of the surrounding fluid, while a spheroidal particle undergoes a time-periodic motion called a Jeffery orbit [36]. Each orbit is characterized by an orbit constant C that can take values in $0 \leq C \leq \infty$. When $C = 0$ the axis of revolution aligns with the vorticity direction for all time; the particle rolls like a log or a

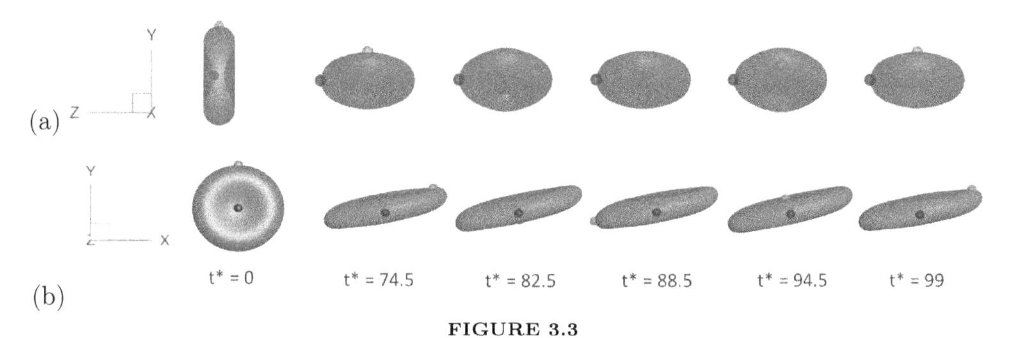

FIGURE 3.3
Time sequence images of a simulation of an RBC performing tank-treading motion during simple shear flow in (a) front view (yz plane) and (b) side view (xy plane). The front view is kept translucent to allow the green material point to be seen, while in the side view, the YZ plane is tilted by 4° to allow visualization of the dimple. Reproduced with permission from [70].

coin depending on whether the particle is prolate or oblate. For intermediate values of C, the particle undergoes what is often called a kayaking motion, during which the axis of revolution of the particle rotates like a kayak paddle about the vorticity axis. When $C = \infty$ the axis of revolution traces out a unit circle on the shear plane – the particle tumbles.

For an oblate spheroid, the more relevant shape for an RBC, the tumbling motion slows down as the axis of revolution approaches the gradient direction – a completely flat disk will rotate until its thin direction is aligned with the flow (then stay there until Brownian motion causes it to flip again). With regard to the discussion above regarding the particle stresslet and its role in the stress in a suspension, the shear stress arises from the yx component of the stresslet: for a particle aligned in the flow direction, this vanishes so a particle in this orientation does not contribute to the shear viscosity of the fluid. Only if the shape is nonaxisymmetric can more complex dynamics arise [80, 82]. In all rigid particle cases, the linearity of the Stokes equations implies that the dynamics are independent of shear rate – the shear rate $\dot{\gamma}$ can be absorbed into the time variable and thus scaled out of the problem.

For particles like red blood cells, which are not rigid, the dynamics can be much more complex. Now even if $\dot{\gamma}$ is absorbed into the time variable, it still appears in the problem via a quantity known as the capillary number and denoted Ca. This is a dimensionless group that measures the ratio of viscous stresses exerted by the surrounding fluid on the particle, which scale as $\eta\dot{\gamma}$, to elastic or interfacial stresses. For an RBC the elastic stress in the membrane can be estimated as G/a where G is the interfacial shear modulus of the cell membrane, so $\mathrm{Ca} = \eta\dot{\gamma}a/G$. For an RBC, $G \approx 4 \cdot 10^{-6}$ N/m [22, 30, 34, 50, 73]. In the circulation the capillary number is in the approximate range $0 \lesssim \mathrm{Ca} \lesssim 1$. The viscosity of the cytoplasm of an RBC, which we will denote η_{in}, is about five times as viscous as plasma [73], so the viscosity ratio $\lambda = \eta_{\mathrm{in}}/\eta \approx 5$.

Experiments in the past several decades have established that in shear flow at low shear rates, a suspended RBC behaves as a rigid body and undergoes a tumbling motion (here denoted TU) [26], while at higher shear rates in a sufficiently viscous fluid its orientation takes on a constant angle with respect to the flow direction and the membrane rotates about the interior in a so-called tank-treading (TT) motion [21, 26, 67, 72]. If the cell is oblique relative to the flow-gradient plane, then the tumbling can take the form of a Jeffery orbit with finite C; a simulation of one such "kayaking" orbit is shown in Figure 3.2 [70]. Figure 3.3 shows tank-treading [70]. In the case shown, the cell membrane reorients with respect to flow so that the blue marker ends up on the rim of the deformed cell. See also [12] and [17] for examples of this phenomenon.

To illustrate some qualitative features of the transitions that arise in shear flow of red blood cells, we briefly describe the model of Keller and Skalak [38]. These authors considered an ellipsoidal capsule in simple shear flow. The particle is characterized by an internal viscosity and two geometric ratios and one axis of the capsule is taken to be oriented with the vorticity direction of the flow. To simplify the model, the dynamics of the capsule membrane are not addressed; instead, the capsule is assumed not to change shape during flow, though it can change orientation. To determine an evolution equation for the capsule motion, the authors equated the rate of work done by the external fluid (plasma) with the rate of energy dissipation in the internal fluid (cytoplasm). The model can be written

$$\dot{\theta} = -A + \cos 2\theta \tag{3.25}$$

where θ is the orientation angle of the cell with respect to the flow direction and A is a parameter that is determined by the ellipsoid geometry and viscosity ratio. In particular, as λ increases, A does. (The shear rate can be absorbed into the time scaling in this model and thus plays no role in the dynamics – this is not the case in reality as we further discuss below). For small A, (3.25) has steady states that are solutions to $\cos(2\theta) = A$. At $A = 0$, these solutions are $\theta = \pi/4$ and $\theta = -\pi/4$. (There are two other solutions as well, but by symmetry they are equivalent to these). The solution $\theta = \pi/4$ is linearly stable – given an arbitrary initial condition, the particle aligns along the extensional axis of the rate of strain tensor. This is tank-treading: the fluid inside the particle is moving, but the overall orientation θ of the particle is not. The solution at $\theta = -\pi/4$ is linearly unstable; an initial condition that begins near this solution will move to the stable solution $A = \pi/4$. As A increases, these two solutions move closer together; when $A = 1$ they merge at $\theta = 0$ and for $A > 1$ there is no steady solution. Now $\dot{\theta} < 0$ for all time, so the particle tumbles in the $-\theta$ direction. The limit $A \to \infty$ corresponds to a rigid ellipsoid undergoing a Jeffery orbit.

More detailed observations have revealed a number of variations on these basic motions. Goldsmith and Marlow [26] observed a rolling motion of the RBC in which the axis of revolution of the RBC is oriented in the vorticity direction of shear flow. Bitbol [5] and Yao *et al.* [81] made very similar observations. Abkarian *et al.* [1] reported a swinging motion in which the orientation of the cell oscillates about a fixed angle while simultaneously tank-treading. Skotheim and Secomb [74] and Abkarian *et al.* [1] extended the Keller-Skalak model by introducing an elastic membrane to the ellipsoidal particle model and were able to predict an additional "swinging" motion, as well as intermittency during the transitions between different motions.

The Keller-Skalak model and its variants assume reflection symmetry across the flow-gradient plane, so out-of-shear plane RBC motions such as those observed by Dupire *et al.* [17] cannot be captured. Many such motions have been observed in some recent experimental studies *et al.* [16, 17]. These include, for example, a tilted rolling regime as well as a complex "flip-flopping" motion as shown in the simulation result in Figure 3.4 [70], which reproduces experimental observations in [17]. At present, simulations are not able to quantitatively capture all of the observed dynamics and the parameter regimes (Ca, λ) in which they reside; see references [2, 70, 79] and references within for further discussion of this issue, which revolves at least in part, around uncertainties regarding how to model the shear and bending elasticity of the RBC membrane. It is unclear to what extent subtle changes of dynamics in an isolated cell in simple shear are important during flow of a dense suspension of RBCs, though simulations of a mixture of flexible (tank-treading) and stiff (tumbling) RBCs indicate that the single-cell dynamics can still be discerned during flow of the mixture [84].

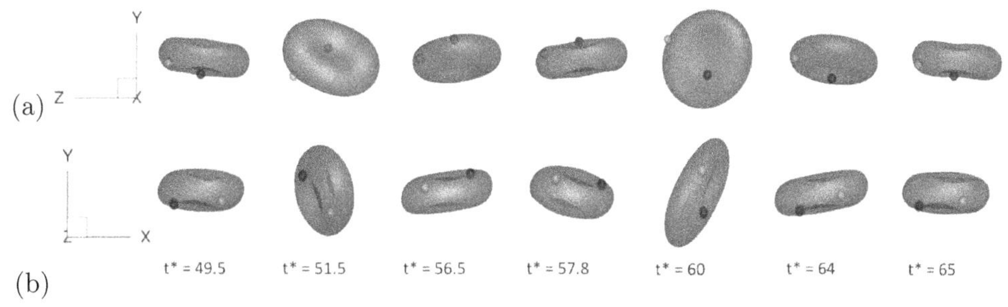

FIGURE 3.4
Time sequence images of a simulation of an RBC undergoing flip-flopping dynamics in (a) front view (yz plane) and (b) side view (xy plane). Reproduced with permission from [70].

3.4 TRANSPORT IN UNCONFINED SUSPENSIONS

3.4.1 Rheology

Macroscopic viscosity measurements of blood show that it is a shear-thinning fluid with a yield stress $\tau_0 \approx 4$ mPa [55, 78]. At 37 °C, the viscosity approaches a constant value $\eta_\infty \approx 3 - 5$ mPa s once $\dot{\gamma}_w \gtrsim 100$ s^{-1}. At 20 °C, the viscosity reaches a high-shear rate plateau of about $\eta_\infty \approx 10$ mPa s. These numbers depend somewhat on temperature and hematocrit. To a good approximation, the shear stress τ for blood undergoing shear flow can be related to the shear rate $\dot{\gamma}$ using the Casson model

$$\tau^{1/2} = \tau_0^{1/2} + (\eta_\infty \dot{\gamma})^{1/2}. \tag{3.26}$$

Blood is also somewhat viscoelastic, due to the deformability of the RBCs [77] – the quantity $\eta_{\text{in}} a/G$ is a characteristic time scale for relaxation of RBCs to their equilibrium shape. Plasma itself is generally considered to be Newtonian with a viscosity of about 1.2 mPa s, although very recent evidence suggests that it does have some viscoelasticity, with a relaxation time of about 2 ms [8]. Figure 3.5a shows the viscosity of blood and model RBC suspensions as a function of shear rate.

The yield stress of blood is small – any flow with $\dot{\gamma} \gtrsim 1 - 10$ s^{-1} is strong enough to exceed it. Nevertheless, during flow in a blood vessel the shear stress vanishes as the centerline is approached, so there will always be a region in which $\tau < \tau_0$. The origin of this yield stress is aggregation of RBCs into assemblages known as rouleaux, in which the cells are arranged like a stack of coins. Rouleau formation does not seem to be completely understood, but classical depletion flocculation (which is osmotic in origin [66]) due to the presence of the large number of albumin and globulin molecules and bridging flocculation via the protein fibrinogen both play a substantial role [59, 65].

Once the yield stress of blood is exceeded, the shear-thinning and plateau behavior are attributable to the combination of aggregation and the deformability of the RBCs under flow. Suspensions of chemically hardened RBCs (curve HA) in Figure 3.5a have approximately constant viscosity, consistent with the discussion above about the behavior of rigid particles in flow. Curve NA shows that a suspension of deformable RBCs without aggregation displays shear thinning to a plateau value. As shear rate increases the cells tank-tread and increasingly become aligned with the flow; the yx component of the stresslet and thus the RBC contribution to the viscosity decrease. The plateau at high shear rate (capillary number) reflects the fact that in tank-treading of cells and capsules, the orientation and shape tend to plateau when Ca $\gtrsim 1$ (see, e.g. [63]). Finally, in suspensions of aggregating RBCs (curve NP), there is substantial shear-thinning at low shear rates. This reflects the

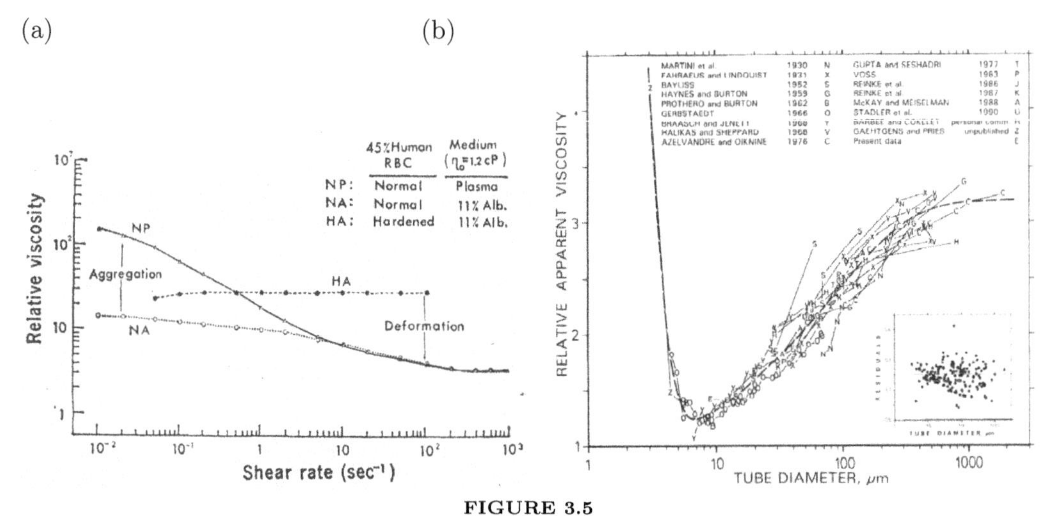

FIGURE 3.5
(a) Viscosity of blood (NP) as well as normal (NA) and hardened (HA) RBCs suspended in a buffer solution (Ringer's lactate) containing albumin. Hematocrit is constant in all three cases. Reproduced with permission from [11]. (b) Apparent viscosity vs. tube diameter for blood as measured by a large number of separate investigators. Reproduced with permission from [64].

orientation of the rouleaux with the flow. Now we think about the stresslet of the rouleaux rather than individual cells determining the stress – again, flow alignment reduces the yx component, thus decreasing the viscosity.

3.4.2 Shear-induced diffusion

Interactions between particles in a flowing suspension lead to a fluctuating component of the flow field that drives motions of the particles and fluid elements. The simplest way to view this phenomenon is that a shear flow drives suspended particles to collide with one another as illustrated schematically in Figure 3.6. Here δ_y and δ_z are the pre-collision pair offsets in the y and z directions and $\Delta_y^{\alpha\beta}(\delta_y, \delta_z)$ and $\Delta_z^{\alpha\beta}(\delta_y, \delta_z)$ are the cross-stream and cross-vorticity direction displacements of a particle of type α after collision with a particle of type β. Repeated collisions will lead to a diffusive behavior of the particles known as shear-induced diffusion [18,48,86]. In general this diffusion process is anisotropic so the diffusivity is a tensor; we will only be concerned with the component in the velocity gradient direction, which we will denote D_p. Solute transport can also occur in a suspension as fluid elements are moved around due to the presence of particles – we denote this diffusivity as D_f.

From dimensional analysis, shear-induced diffusivities scale as $a^2\dot{\gamma}$ [48]. Furthermore, if we take the diffusivity to scale as the displacement per collision times the frequency of collision, we arrive at the scaling result $D \sim a^2\dot{\gamma}\phi$, where D here is either D_p or D_f [14]. Experiments with spheres by Zarraga and Leighton [83], yield particle self-diffusivities (D_p) and fluid phase (passive solute) self-diffusivities (D_f) given approximately by

$$D_p = (0.0341\phi + 1.15\phi^3)a^2\dot{\gamma}$$

and

$$D_f = (0.0354\phi + 0.506\phi^3)a^2\dot{\gamma}.$$

For a 30% suspension of 10 micron spheres, a modest shear rate of 300 s^{-1} (commonly found in the circulation) leads to an effective diffusivity of $2 \cdot 10^{-6}$ cm^2/s, one to two orders of magnitude larger than the molecular diffusivities of many biomolecules and viruses. Because

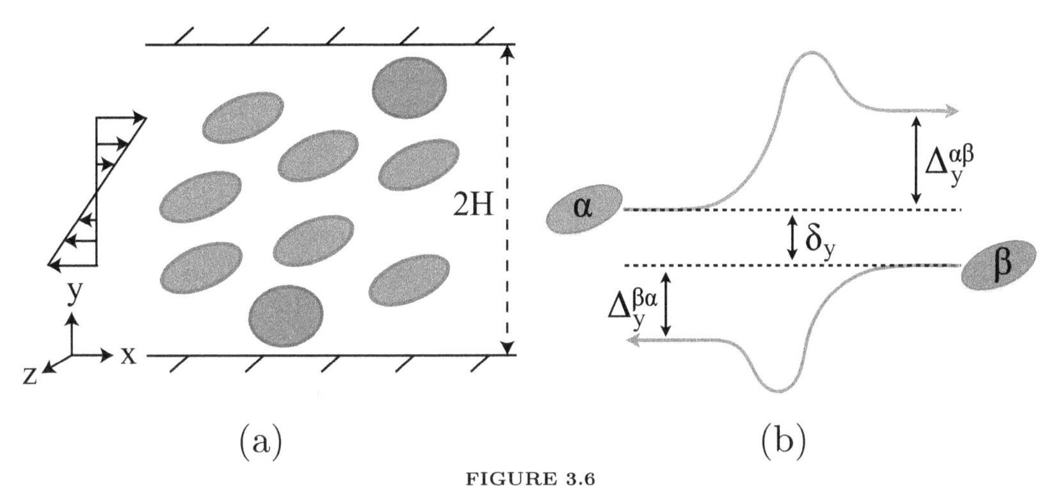

FIGURE 3.6
(a) Suspension of particles in a slit under simple shear flow. (b) Pair collision trajectories of particles of species α and β under simple shear flow, shown in a reference frame moving with the mean horizontal velocity of the particles. Reproduced with permission from [32].

of this increase in effective diffusivity, shear-induced diffusion has been shown to enhance adsorption to surfaces [39, 42, 52, 53].

The ϕ^1 scaling of D cited above relies on some nonhydrodynamic effect to induce nonzero net displacement in a pair collision, because for perfect hydrodynamic collisions of rigid spheres in Stokes flow, by symmetry and Stokes flow reversibility arguments there is no net displacement. In the latter case $D \sim a^2\dot{\gamma}\phi^2$ rather than $a^2\dot{\gamma}\phi$ [49]. For blood, either the nonsphericity or deformability of the cells suffices to break this symmetry.

3.5 CONFINED SUSPENSIONS

3.5.1 Nonuniform cell distributions in blood flow

The rheological properties described above are relevant for macroscopic flow geometries, i.e. those with scales much larger than that of a blood cell. An indication that blood flow at smaller scales displays features different from its bulk behavior is the Fahraeus-Lindqvist effect [19]: this is the observation that the apparent viscosity at high shear rates (beyond the shear thinning regime) determined in a capillary viscometer is a strong function of the capillary tube diameter, as shown in Figure 3.5b [64]. For large tubes, there is a plateau, but for diameters between 10 and 1000 μm, the apparent viscosity decreases substantially with decreasing tube size, before increasing sharply for tubes smaller than 10 μm. The increase for very small tubes is easily understood by recalling that the RBCs are about 8 μm across, but the behavior at larger tube diameters is more subtle.

This phenomenon is related to another classical observation in blood physiology called the Fahraeus effect [3, 78]: the measured hematocrit of blood *in* a tube – i.e. the true average volume fraction – can be less than the hematocrit inferred from measurements of the blood discharged by flow out *from* a tube. The ratio of "tube hematocrit" to "discharge hematocrit" displays a similar tube size dependence as the apparent viscosity. In particular, in a tube of 20 μm diameter the tube hematocrit (i.e. the true hematocrit within the tube) is about half the discharge hematocrit [3, 51, 78].

The resolution to this apparent paradox lies in the fact that the discharge hematocrit is weighted by the nearly parabolic velocity profile in the tube while the tube hematocrit is not. If the local concentration of RBCs in the tube were uniform, then the two measures would yield the same result, so the Fahraeus effect indicates that the concentration is not uniform and that the concentration of RBCs is higher near the center of the channel where the velocity is high than near the walls where the velocity is low. One reason for this concentration variation is the simple volume exclusion effect – cells cannot pass through the wall so there is always a region near the wall with a cell volume fraction lower than the nominal value. In this region the effective viscosity of blood is also smaller, resulting in a "lubricating layer" that leads in part to the Fahraeus-Lindqvist effect. Beyond this simple volume exclusion effect, there exist mechanisms that cause the concentration of RBCs to vary with cross-stream position. In particular, RBCs and indeed all deformable particles migrate away from walls during shear due to hydrodynamic interaction effects with the walls. The cell-depleted region near vessel walls is called a "cell-free layer" or "marginal layer"[41, 73]. We now discuss the mechanisms that underlie cross-stream migration and thus concentration variations during flow.

3.5.2 Cross-stream migration phenomena

As shown in Section 3.2.2, a force- and torque free particle in Stokes flow exerts a force dipole on the surrounding fluid. In the presence of confining walls, the velocity field generated by this dipole will induce a velocity at the position of the particle that will drive it to move with a velocity

$$\boldsymbol{U}(\boldsymbol{x}_0) = \boldsymbol{v}_\infty(\boldsymbol{x}_0) + \boldsymbol{U}^{\mathrm{W}}(\boldsymbol{x}_0) \tag{3.27}$$

where $\boldsymbol{U}^{\mathrm{W}} = \boldsymbol{v}^{\mathrm{W}}$ and $\boldsymbol{v}_\infty$ is the fluid velocity in the absence of the particle. The wall-normal component of $\boldsymbol{U}^{\mathrm{W}}$ is called the *migration velocity* and the tangential component the *slip velocity*. This velocity arises because the confining walls break the translation invariance symmetry of the unbounded domain and modify the Green's function from its form in an unbounded flow. For a force- and torque-free particle a distance y above a plane wall in a semi-infinite domain, where the modified Green's function is known analytically [6], the migration velocity, using a point dipole approximation for the particle, is

$$U_y^{\mathrm{W}} = \frac{3}{64\pi\eta y_0^2}\left(D_{xx} + D_{zz} - 2D_{yy}\right) = -\frac{9S_{yy}}{64\pi\eta y^2}, \tag{3.28}$$

[75]. On physical grounds, we can understand this result as follows: a force dipole consists of two point forces. In the presence of a wall, each point force generates a flow that tends to carry the other point force relative to the wall [37, 54].

For a rigid sphere in shear flow, $S_{yy} = 0$ by symmetry and no migration occurs. For deformable particles, this symmetry argument does not hold; suspended liquid drops [75], red blood cells [41] and polymer molecules [27] are all experimentally observed to migrate away from walls during flow. The direction of migration can be understood in terms of the components of $\mathbf{D}$; deformable objects tend to stretch and align with flow, leading D_{xx} to be positive and to be the dominant term in (3.28), as schematically illustrated in Figure 3.1b. The general form of the migration velocity is

$$\frac{U_y^{\mathrm{W}}}{\dot{\gamma}a} = \frac{a^2}{y^2}F(\mathrm{Ca}, \lambda), \tag{3.29}$$

where F also depends on the nature of the deformable particle. At low Ca, $F \sim \mathrm{Ca}^1$, while simulations for capsules and vesicles suggest that F increases more slowly than linearly Ca as Ca increases [58, 63]. For finite-sized particles, the a^2/y^2 scaling holds until the particle center is about $1.5a$ from the wall a from the wall [63].

The effect just described arises in any nontrivial flow, including simple shear, in which the velocity gradient is constant. In addition, if there is a spatial variation in the velocity gradient, as arises in pressure-driven flow, there is an additional migration effect that arises for deformable particles in the Stokes flow regime. This is not due directly to the presence of walls, but rather to the interaction of deformation and the change in velocity gradient (a rigid sphere in a pressure-driven flow displays no migration in the Stokes flow regime. Chan and Leal [10] and Helmy and Barthes-Biesel [29] found analytical expressions for the migration velocity of a drop in an unbounded plane Poiseuille flow in the limit of small deformation and the migration velocity of a capsule in unbounded pipe flow in the limit of $\mathrm{Ca}_w \ll 1$, respectively. Their results both indicate a linear dependence of the migration velocity on $\dot{\gamma}_w$, Ca_w, and the distance from the centerline of the channel, $1 - \frac{y}{C}$. In the droplet case, migration can be toward or away from the centerline, depending on viscosity ratio: migration is toward the wall for $0.1 < \lambda < 10$. The capsule case is more relevant to the present situation, and there it is found that migration is always toward the centerline. Near walls, this effect is much smaller than that due to the walls described above.

At any nonzero Reynolds number the symmetry argument for nonmigration of a rigid sphere in shear flow fails to hold. Experiments show that in plane Couette flow a sphere migrates to the centerline and in pressure-driven flow to a position intermediate between the wall and centerline, indicating the presence of an inertial lift force on the sphere. This phenomenon is often called the *Segré-Silberberg effect* [15, 68]. At low particle Reynolds numbers [33], the resulting wall-normal velocity due to this effect can be expressed

$$\frac{U_{\mathrm{lift},y}}{\dot{\gamma}a} = \mathrm{Re}_p G\left(\frac{y}{H}\right), \tag{3.30}$$

where $G(y/H)$ depends on the specific velocity profile chosen and is odd with respect to the channel or tube centerline $y = H$. For the physiologically relevant conditions of $\mathrm{Ca} \sim 1$, $\mathrm{Re}_p \sim 10^{-2}$ and for distances from the wall $y \sim a \sim 4\ \mu\mathrm{m}$, the migration of RBCs due to deformability is dominant over that due to inertia. For rigid particles however, the deformation-driven mechanism is absent so the Segre-Silberberg effect can be important even for small but nonzero Reynolds number.

The above effects are all present for single particles in a fluid. One additional cross-stream migration effect arises in suspensions, due to the same mechanism as shear-induced diffusion described in Section 3.4.2. There is a cross-stream particle velocity in the presence of gradients of either particle concentration or shear rate [49]; in either case a given particle will experience a higher rate of collisions from the side with higher shear rate or number density, and will thus be knocked preferentially to regions in the flow where these quantities are lower: particles will be driven down gradients in $\dot{\gamma}\phi$. This phenomenon shows up explicitly in the kinetic theory model that is described in the following section and suffices to cause number density gradients in pressure-driven flow even of rigid spherical particles in the Stokes flow regime, where the cross-stream migration mechanisms described above are not active [69].

3.5.3 Combined effects of migration and shear-induced diffusion – a simple model

We have introduced the phenomena of shear-induced diffusion and hydrodynamic migration. These can be integrated into a single model that predicts the distribution of cells across a channel, thus shedding light on the phenomena of cell-free layer formation and the distribution of leukocytes and platelets. These tend to be preferentially found near blood vessel walls, a widely observed phenomenon called *margination* [44,61].

In this model, the number density distributions for the various types of blood cells are determined by a kinetic master equation – a system of nonlinear integrodifferential equations – that captures the migration and collision effects ([45, 46, 58, 85]). We consider a dilute suspension containing N_s types of deformable particles with total volume fraction ϕ undergoing flow in a slit bounded by no-slip walls at $y = 0$ and $y = 2H$ and unbounded in x and z. Quantities referring to a specific component α in the mixture will have subscript α: for example n_α and a_α are the number density and characteristic particle size of component α. For clarity, we present the model for simple shear (plane Couette) flow and, consistent with the diluteness assumption, take the shear rate $\dot{\gamma}$ to be independent of the local number densities and thus independent of position. Poiseuille flow can be treated as well [32]. For the moment, we neglect molecular diffusion of the particles. Since the particles are deformable, they migrate away from the wall during flow with velocity $v_{\alpha m}(y)$ due to the various mechanisms described above.

While it is possible to perform direct simulations of the master equation [45, 46, 58], some additional assumptions reduce it to a form amenable to analytical progress [31, 32]. If the collisional displacements $\Delta_y^{\alpha\beta}$ and $\Delta_z^{\alpha\beta}$ are small and only occur for particles whose centers come very close together, then the model reduces to the coupled drift-diffusion equations

$$\frac{\partial n_\alpha}{\partial t} = -\frac{\partial}{\partial y}\left((v_{\alpha m} + v_{\alpha c})\, n_\alpha - \frac{\partial}{\partial y}(D_\alpha n_\alpha)\right). \tag{3.31}$$

Here $v_{\alpha c}$ is the collisional drift velocity of component α, while D_α is its short time shear-induced self-diffusivity. Additionally, if we assume that there is a "primary" component ($\alpha =$'p') and a "trace" component ($\alpha =$'t') such that $n_p \gg n_t$, then the primary component is unaffected by the trace, while the trace is driven by the primary. The condition $n_p \gg n_t$ is valid for blood, where RBCs outnumber platelets and white blood cells by one and three orders of magnitude respectively [7]. Now the collisional drift velocities and diffusivities become

$$v_{\alpha c} = -K_{\alpha c}\frac{\partial \dot{\gamma} n_p}{\partial y} \tag{3.32}$$

$$D_\alpha = K_{\alpha d}\dot{\gamma} n_p, \tag{3.33}$$

where $K_{\alpha c}$ and $K_{\alpha d}$ are integrals over the collision functions $\Delta_y^{\alpha p}(\delta_y, \delta_z)$ and $\Delta_z^{\alpha p}(\delta_y, \delta_z)$ [31,32]. Equations (3.32) and (3.33) correspond to the collisional migration described at the end of Section 3.5.2 and the shear-induced diffusion described in Section 3.4.2.

To describe the wall-induced hydrodynamic migration velocity, we assume that the wall hydrodynamic interaction effect described by (3.29) dominates. Superimposing the point-dipole approximations corresponding to each of the two walls [63] yields:

$$v_{\alpha m} = K_{\alpha m}\left(\frac{1}{y^2} - \frac{1}{(2H-y)^2}\right). \tag{3.34}$$

With these further idealizations, Eq. 3.31 becomes a pair of partial differential equations, which we present here in nondimensional form:

$$\frac{\partial \phi_p}{\partial t} = -\frac{\partial}{\partial y}\left[\kappa_{pm}\left(\frac{1}{y^2} - \frac{1}{(2\mathrm{C}-y)^2}\right)\phi_p - \kappa_{ppc}\frac{\partial \phi_p}{\partial y}\phi_p - \kappa_{ppd}\frac{\partial \phi_p^2}{\partial y}\right], \tag{3.35}$$

$$\frac{\partial \phi_t}{\partial t} = -\frac{\partial}{\partial y}\left[\kappa_{tm}\left(\frac{1}{y^2} - \frac{1}{(2\mathrm{C}-y)^2}\right)\phi_t - \kappa_{tpc}\frac{\partial \phi_p}{\partial y}\phi_t - \kappa_{tpd}\frac{\partial (\phi_p\phi_t)}{\partial y}\right]. \tag{3.36}$$

Here $\phi_p = n_p V_p$ and $\phi_t = n_t V_t$ are the volume fractions of the primary and trace components, where V_α is the volume per particle of component α, $\mathrm{C} = H/a_p$ is the confinement ratio, $\kappa_{pm} = \frac{K_{pm}}{\dot\gamma a_p^3}$, $\kappa_{ppc} = \frac{K_{ppc}}{V_p a_p^2}$, $\kappa_{ppd} = \frac{K_{ppd}}{V_p a_p^2}$, $\kappa_{tm} = \frac{K_{tm}}{\dot\gamma a_p^3}$, $\kappa_{tpc} = \frac{K_{tpc}}{V_p a_p^2}$, and $\kappa_{tpd} = \frac{K_{tpd}}{V_p a_p^2}$. Time t is nondimensionalized with $\dot\gamma^{-1}$ and y with a_p. For simplicity, we keep the symbols t and y for their nondimensionalized forms. For a single-component suspension of rigid particles ($N_s = 1, K_{\alpha m} = 0$) a model of similar form was proposed by [60] based on phenomenological arguments first proposed by [49].

Equations (3.35) and (3.36) can be solved to yield insights into the parameter dependence of the cell free layer and margination phenomena. The steady state concentration field for the primary component, $\phi_p(y)$, can be found by solving (3.35):

$$\phi_p = \begin{cases} 0, & y < l_d \\ \frac{2\mathrm{C}^2\bar\phi_p}{2\mathrm{C}(\mathrm{C}-l_d) - l_d(2\mathrm{C}-l_d)\ln\left(\frac{2\mathrm{C}-l_d}{l_d}\right)}\left(1 - \frac{l_d}{y}\frac{(2\mathrm{C}-l_d)}{(2\mathrm{C}-y)}\right), & y > l_d \end{cases}, \tag{3.37}$$

where $\bar\phi_p$ is the mean volume fraction of the primary component (i.e. the bulk hematocrit), $\eta_p = \frac{\kappa_{pm}}{\kappa_{ppc}+2\kappa_{ppd}}$, and l_d is the nondimensional cell-free or depletion layer thickness, which is implicitly given in dimensionless form by

$$\frac{\bar\phi_p \mathrm{C}}{\eta_p} = 2\frac{\mathrm{C}}{l_d}\frac{(1-\frac{l_d}{\mathrm{C}})}{(2-\frac{l_d}{\mathrm{C}})} - \ln\left(2\frac{\mathrm{C}}{l_d} - 1\right). \tag{3.38}$$

This result will remain valid as long as l_d is greater than the radius of the primary component. Otherwise the model would have to include an excluded volume force between the cells and the wall [32]. In the unconfined limit $\mathrm{C} \to \infty$, $l_d \to \eta_p/\bar\phi_p$, confirming the ϕ^{-1} dependence found earlier in scaling analyses [35, 46, 58, 63].

Equation (3.38) implies that $\bar\phi_p \mathrm{C}/\eta_p$ is a function of l_d/C, so given the single adjustable parameter η_p, all results should fall onto a master curve when plotted in terms of these two quantities. This master curve, as well as data from a number of simulations and experimental measurements for blood *in vitro* and *in vivo*, is shown in Figure 3.7. Fitted values of η_p lie in the range $0.36 < \eta_p < 0.85$. Remarkably, nearly all the data points collapse onto a single master curve, even for the experiments, which were performed in pressure-driven flow. This means that independent of the form of the theory, the relationship between $\bar\phi_p$ and l_d is set by only one adjustable parameter (η_p).

The concentration profile for the primary component given by the closed-form solution (Eq. 3.37) is illustrated as the black solid curve in Figure 3.8a. The profile is symmetric so

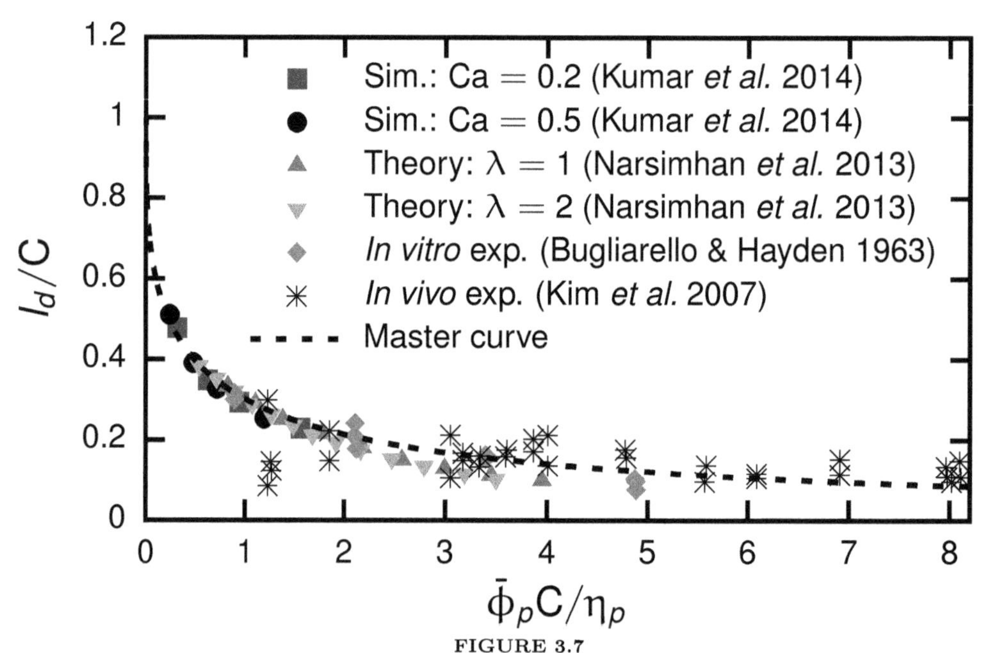

FIGURE 3.7
Master curve of cell-free layer thickness l_d (nondimensionalized with particle radius) *vs.* $\bar{\phi}_p$ from various sources; Eq. 3.38 is the black dashed curve. Blue and black symbols are from boundary integral simulations of pure suspensions of deformable capsules in simple shear flow [46]. Green and gray symbols are from and integrodifferential model for red blood cells in simple shear flow [58]. Red symbols are from *in vitro* experiments of blood in small glass capillaries in pressure-driven flow [9]. Asterisks are from *in vivo* experiments (star symbols) in arterioles in the rat cremaster muscle [40]. Best fit values of η_p for these data sets are $\eta_p = 0.65, 0.85, 0.51, 0.60, 0.45$ and 0.36, respectively. Reprinted from permission from [32].

only the bottom half of the channel $0 < y < \mathrm{C}$ is shown. The solution is zero within the cell-free layer $y < l_d$ and increases with increasing y, reaching a maximum at $y = \mathrm{C}$.

For the trace component, an analytical steady state solution can also be found:

$$\phi_t = \begin{cases} 0, & y < l_d \\ \phi_{tc}\left(\frac{\phi_p(y)}{\phi_{pc}}\right)^{\mathrm{M}}, & y > l_d \end{cases}, \tag{3.39}$$

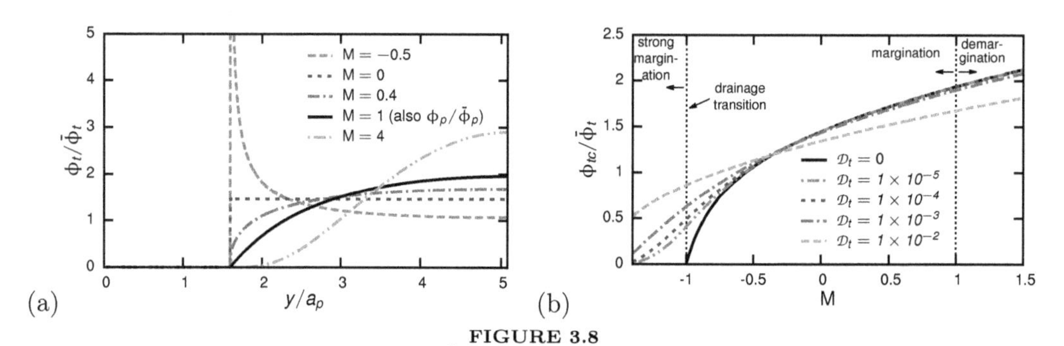

FIGURE 3.8
(a) Steady state volume fraction profiles of $\phi_p/\bar{\phi}_p$ (black solid line) and $\phi_t/\bar{\phi}_t$ for various values of M. (The curves coincide when M = 1.) Here $\bar{\phi}_p = 0.12$, $\phi_{pc} = 0.23$, C = 5.08, $\kappa_{pm} = 0.11$, $\kappa_{ppc} = 0.02$, and $\kappa_{ppd} = 0.07$, resulting in $l_d = 1.6$ (extracted from simulation results in [46]). For simplicity, $\kappa_{tpd} = \kappa_{ppd}$ and $\kappa_{tpc} = \kappa_{ppc}$. Here M is varied by changing κ_{tm}. (b) Centerline volume fraction of the trace component ϕ_{tc} scaled with the average trace volume fraction $\bar{\phi}_t$ *vs.* M for varying $\mathcal{D}_t$. Reproduced with permission from [32].

where $\phi_p(y)$ is the steady state solution for ϕ_p found above, ϕ_{tc} is the volume fraction of the trace component at the centerline, and

$$\mathrm{M} = \frac{\kappa_{ppc} + 2\kappa_{ppd}}{\kappa_{tpd}} \left(\frac{\kappa_{tm}}{\kappa_{pm}} - \frac{\kappa_{tpc} + \kappa_{tpd}}{\kappa_{ppc} + 2\kappa_{ppd}} \right). \tag{3.40}$$

Remarkably, this single quantity, which we call the *margination parameter*, determines the qualitative nature of the concentration profile as we now describe.

The sign of M is determined by the competition between the ratio of the migration velocities of the two components, $\frac{\kappa_{tm}}{\kappa_{pm}}$, and the ratio of the collisional terms, $\frac{\kappa_{tpc}+\kappa_{tpd}}{\kappa_{ppc}+2\kappa_{ppd}}$. Specifically, the migration velocity decreases as the trace particle becomes either smaller or stiffer, and stiffer cells are displaced more during a pair collision than flexible ones [43]. Thus, as a cell becomes smaller or stiffer, M moves toward negative values. The effect of shape is more complex and not completely understood [71].

Depending on M, several distinct regimes of behavior can be identified:

1. $\mathrm{M} > 1$: the trace component is displaced further from the wall than the primary component: it *demarginates*.

2. $0 < \mathrm{M} < 1$: the relative concentration of the trace component is higher near the wall than the primary component but does not display a peak: it *weakly marginates*.

3. $-1 < \mathrm{M} < 0$: the trace component displays a peak at $y = l_d$, corresponding to an integrably singular concentration profile: it *moderately marginates*.

4. $\mathrm{M} \leq -1$: here (3.39) displays a nonintegrable singularity at $y = l_d$. This steady state is physically unrealizable as it corresponds to an infinite amount of material in a finite region. In this regime collisional transport overwhelms migration, and the trace component accumulates indefinitely in the region $y \leq l_d$, indicating *strong margination*. If the trace component does not migrate (as in the case of rigid particles), then $\kappa_{tm} = 0$ and $\mathrm{M} = -(1 + \kappa_{tc}/\kappa_{td})$, which is *always* less than -1.

The black solid line in Figure 3.8b shows the ratio between the centerline concentration ϕ_{tc} and the average concentration $\bar{\phi}_t$ *vs.* M. This falls sharply to zero at $\mathrm{M} = -1$; we call this phenomenon the *drainage transition*, since for $\mathrm{M} \leq -1$ all the trace component is completely drained from the bulk. Direct simulations of binary suspensions of fluid-filled non-Brownian elastic capsules at low Reynolds number also show a drainage transition as the size or stiffness of the trace component is lowered [31, 32]. From the present results it appears that both leukocytes and platelets would satisfy the conditions for drainage in simple shear.

For pressure driven flow, which is the more generally relevant situation, a similar model can be developed. Although analytical solutions are not available, the same qualitative trends with respect to the cell-free layer and margination are found and corroborated by simulation results [32]. The primary qualitative difference observed is that when $M \gtrsim -1$, a concentration peak at the centerline is observed. This is consistent with experimental observations [69] and previous theoretical predictions [56], and arises because the shear rate, and thus the driving force for shear-induced transport, vanishes at the centerline, allowing particles to accumulate there.

For particles the size of blood cells ($> 1\mu m$) at shear rates characteristic of the microcirculation ($10^2 - 10^3\ s^{-1}$), Brownian diffusion is unimportant – a particle will be carried hundreds or thousands of times its own size in the time it takes to diffuse its size. For smaller particles, however, such as might be used for drug delivery, this may no longer be true. The impact of Brownian diffusion on trace component transport can be seen qualitatively by adding the appropriately nondimensionalized diffusion term $\mathcal{D}_t\ \partial^2\phi_t/\partial y^2$ in Eq. 3.36 for $\alpha = t$. Here $\mathcal{D}_t = D_B/a_p^2\dot{\gamma}$, where D_B is the Brownian diffusivity of the trace component. Using typical values for blood, varying $\mathcal{D}_t$ from 10^{-5} to 10^{-2} corresponds to varying a_t from $\sim 10^{-6}$ m to $\sim 10^{-9}$m. Figure 3.8b shows results for varying $\mathcal{D}_t$; the qualititative observation of a drainage transition persists but becomes less sharp and is shifted to more negative values of M. This result is consistent with experimental [76] and numerical [57] studies in which nanoparticles, which experience a stronger molecular diffusion, showed a lower margination than microparticles. It should be noted that the model for molecular diffusion used here is highly simplified, because in a suspension there is a nontrivial interaction between Brownian motion and shear-induced diffusion that is not quantitatively understood.

3.6 CONCLUSION

Many aspects of the dynamics of deformable-particle suspensions and their relevance to blood flow are well-understood, including for example, the principles underlying rheology and shear-induced transport. Many phenomena, however, are still in need of improved understanding. For examples, we cannot quantitatively model red blood cell dynamics over a wide range of Ca and λ, largely due to uncertainties in how to model RBC membrane dynamics. Margination is becoming increasingly well-understood, but we still need to better understand effects of shape and size as well as the impact of the geometric complexities of the circulation such as contractions, expansions and junctions. These issues are important not only with respect to distributions within the bloodstream but also regarding adsorption to vessel walls. More broadly, interactions of physical and biochemical effects are important to understand. These are critically coupled, for example in hemostasis [23]. There is also evidence of a biomechanical role for drugs in the process of margination of white blood cells [20]; dexamethasone and epinephrine cause softening of these cells, decreasing their propensity to marginate. Finally, it is important to understand how these phenomena are changed during pathologies such as sickle cell disease.

ACKNOWLEDGMENTS

The author's research on blood flow has been supported by the National Science Foundation, grants CBET-1436082, CBET-1132579 and CBET-0852976, and the National Institutes of Health, grant 1R21MD011590-01A1. The author thanks the students and postdocs who have worked with him in this area: Pratik Pranay, Amit Kumar, Sam Anekal, Kushal Sinha, Rafael Henriquez Rivera and Xiao Zhang, as well as the research group of Wilbur Lam.

Bibliography

[1] M. Abkarian, M. Faivre, and A. Viallat. Swinging of red blood cells under shear flow. *Phys. Rev. Lett.*, 98(18):188302, 2007.

[2] M. Abkarian and A. Viallat. On the Importance of the Deformability of Red Blood Cells in Blood Flow. In Camille Duprat and Howard A Stone, editors, *Fluid-Structure Interactions in Low-Reynolds Number Flows*, pages 347–462. Royal Society of Chemistry, 2016.

[3] J. H. Barbee and G. R. Cokelet. The Fahraeus effect. *Microvasc Res*, 3(1):6–16, January 1971.

[4] G. K. Batchelor. The stress system in a suspension of force-free particles. *J. Fluid Mech.*, 41(3):545–570, May 1970.

[5] M. Bitbol. Red blood cell orientation in orbit c = 0. *Biophys. J.*, 49(5):1055–1068, 1986.

[6] J. R. Blake. A note on the image system for a stokeslet in a no-slip boundary. *Proc Camb Philos S-M*, 70:303–310, January 1971.

[7] D. Boal. *Mechanics of the Cell.* Cambridge University Press, Cambridge, second edition, 2012.

[8] M. Brust, C. Schaefer, R. Doerr, L. Pan, M. Garcia, P. E. Arratia, and C. Wagner. Rheology of human blood plasma: viscoelastic versus Newtonian behavior. *Phys. Rev. Lett.*, 110(7):078305, February 2013.

[9] G. Bugliarello and J. W. Hayden. Detailed characteristics of the flow of blood in vitro. *Trans. Soc. Rheol.*, 7(1):209–230, 1963.

[10] P. C. H. Chan and L. G. Leal. The motion of a deformable drop in a second-order fluid. *J. Fluid Mech.*, pages 131–170, 1979.

[11] S. Chien. Shear dependence of effective cell volume as a determinant of blood viscosity. *Science*, 168(3934):977–979, May 1970.

[12] D. Cordasco, A. Yazdani, and P. Bagchi. Comparison of erythrocyte dynamics in shear flow under different stress-free configurations. *Phys. Fluids*, 26(4):041902, April 2014.

[13] R. Cortez, L. Fauci, and A. Medovikov. The method of regularized Stokeslets in three dimensions: Analysis, validation, and application to helical swimming. *Phys. Fluids*, 17(3):031504, January 2005.

[14] F. R. daCunha and E. J. Hinch. Shear-induced dispersion in a dilute suspension of rough spheres. *J. Fluid Mech.*, 309:211–223, 1996.

[15] D. Di Carlo. Inertial microfluidics. *Lab Chip*, 9(21):3038, 2009.

[16] J. Dupire, M. Abkarian, and A. Viallat. Chaotic dynamics of red blood cells in a sinusoidal flow. *Phys. Rev. Lett.*, 104(16):168101, 2010.

[17] J. Dupire, M. Socol, and A. Viallat. Full dynamics of a red blood cell in shear flow. *PNAS*, 109(51):20808–20813, 2012.

[18] E. C. Eckstein, D. G. Bailey, and A. H. Shapiro. Self-diffusion of particles in shear flow of a suspension. *J. Fluid Mech.*, 79:191–208, 1974.

[19] R. Fahraeus and T. Lindqvist. The viscosity of the blood in narrow capillary tubes. *Am J Physiol*, 96:562–568, 1931.

[20] M. E. Fay, D. R. Myers, A. Kumar, C. T. Turbyfield, R. Byler, K. Crawford, R. G. Mannino, A. Laohapant, E. A. Tyburski, Y. Sakurai, M. J. Rosenbluth, N. A. Switz, T. A. Sulchek, M. D. Graham, and W. A. Lam. Cellular softening mediates leukocyte demargination and trafficking, thereby increasing clinical blood counts. *Proc. Nat. Acad. Sci.*, 113(8):1987–1992, February 2016.

[21] T. M. Fischer, M. Stöhr-Liesen, and H. Schmid-Schönbein. The red cell as a fluid droplet: tank tread-like motion of the human erythrocyte membrane in shear flow. *Science*, 202(4370):894–896, 1978.

[22] T. M. Fischer, C. W. Haest, M. Stöhr-Liesen, H. Schmid-Schönbein, and R. Skalak. The stress-free shape of the red blood cell membrane. *Biophys. J.*, 34(3):409–422, 1981.

[23] A. L. Fogelson and K. B. Neeves. Fluid mechanics of blood clot formation. *Annu Rev Fluid Mech*, 47(1):377–403, January 2015.

[24] Y.C. Fung. *Biodynamics: Circulation.* Springer-Verlag, New York, 1983.

[25] Y. C. Fung and B. W. Zweifach. Microcirculation: mechanics of blood flow in capillaries. *Annu. Rev. Fluid Mech.*, 3(1):189–210, 1971.

[26] H. L. Goldsmith and J. Marlow. Flow behaviour of erythrocytes. I. Rotation and deformation in dilute suspensions. *Proc. R. Soc. B*, 182(1068):351–384, 1972.

[27] M. D. Graham. Fluid dynamics of dissolved polymer molecules in confined geometries. *Annu Rev Fluid Mech*, 43(1):273–298, January 2011.

[28] M. D. Graham. *Microhydrodynamics, Brownian Motion, and Complex Fluids.* Cambridge University Press, Cambridge, 2018.

[29] A. Helmy and D. Barthes-Biesel. Migration of a spherical capsule freely suspended in an unbounded parabolic flow. *Journal de Mécanique théoretique et appliquée*, 1(5):859–880, 1982.

[30] S. Henon, G. Lenormand, A. Richert, and F. Gallet. A new determination of the shear modulus of the human erythrocyte membrane using optical tweezers. *Biophys. J.*, 76(2):1145–1151, 1999.

[31] R. G. Henriquez Rivera, K. Sinha, and M. D. Graham. Margination regimes and drainage transition in confined multicomponent suspensions. *Phys. Rev. Lett.*, 114(18), 2015.

[32] R. G. Henriquez Rivera, X. Zhang, and M. D. Graham. Mechanistic theory of margination and flow-induced segregation in confined multicomponent suspensions: simple shear and Poiseuille flows. *Phys. Rev. Fluids*, 1:060501, May 2016.

[33] B. P. Ho and L. G. Leal. Inertial migration of rigid spheres in 2-dimensional unidirectional flows. *J. Fluid Mech.*, 65:365–400, 1974.

[34] R. M. Hochmuth and R. E. Waugh. Erythrocyte membrane elasticity and viscosity. *Annu. Rev. Physiol.*, 49(1):209–219, 1987.

[35] S. D. Hudson. Wall migration and shear-induced diffusion of fluid droplets in emulsions. *Phys. Fluids*, 15(5):1106–1113, January 2003.

[36] G. B. Jeffery. The motion of ellipsoidal particles immersed in a viscous fluid. *Proc. R. Soc. A*, 102(715):161–179, 1922.

[37] R. M. Jendrejack, D. C. Schwartz, J. J. de Pablo, and M. D. Graham. Shear-induced migration in flowing polymer solutions: Simulation of long-chain deoxyribose nucleic acid in microchannels. *J Chem Phys*, 120(5):2513–2529, January 2004.

[38] S. R. Keller and R. Skalak. Motion of a tank-treading ellipsoidal particle in a shear flow. *J. Fluid Mech.*, 120:27–47, 1982.

[39] D. Kim and R. L. Beissinger. Augmented mass transport of macromolecules in sheared suspensions to surfaces. *J. Colloid Interface Sci.*, 159:9–20, 1993.

[40] S. Kim, R. L. Kong, A. S. Popel, M. Intaglietta, and P. C. Johnson. Temporal and spatial variations of cell-free layer width in arterioles. *Am J Physiol-Heart C*, 293(3):H1526–H1535, January 2007.

[41] S. Kim, P. K. Ong, O. Yalcin, M. Intaglietta, and P. C. Johnson. The cell-free layer in microvascular blood flow. *Biorheology*, 46(3):181–189, 2009.

[42] D. L. Koch. Hydrodynamic diffusion near solid boundaries with applications to heat and mass transport into sheared suspensions and fixed-fibre beds. *J. Fluid Mech.*, 318:31–47, 1996.

[43] A. Kumar and M. D. Graham. Segregation by membrane rigidity in flowing binary suspensions of elastic capsules. *Phys. Rev. E*, 84(6):066316, December 2011.

[44] A. Kumar and M. D. Graham. Margination and segregation in confined flows of blood and other multicomponent suspensions. *Soft Matter*, 8(41):10536–10548, 2012.

[45] A. Kumar and M. D. Graham. Mechanism of margination in confined flows of blood and other multicomponent suspensions. *Phys. Rev. Lett.*, 109(10), 2012.

[46] A. Kumar, R. G. Henriquez Rivera, and M. D. Graham. Flow-induced segregation in confined multicomponent suspensions: Effects of particle size and rigidity. *J. Fluid Mech.*, 738:423–462, 2014.

[47] L. D. Landau and E. M. Lifschitz. *Fluid Mechanics*. Course of Theoretical Physics. Pergamon, Oxford, 2nd edition, 1984.

[48] D. T. Leighton and A. Acrivos. Measurement of shear-induced self-diffusion in concentrated suspensions of spheres. *Journal of Fluid Mechanics*, 177:109–131, 1987.

[49] D. T. Leighton and A. Acrivos. The shear-induced migration of particles in concentrated suspensions. *J. Fluid Mech.*, 181:415–439, January 1987.

[50] J. Li, M. Dao, C. T. Lim, and S. Suresh. Spectrin-level modeling of the cytoskeleton and optical tweezers stretching of the erythrocyte. *Biophys. J.*, 88(5):3707–3719, 2005.

[51] H. H. Lipowsky. In vivo studies of blood rheology in the microcirculation in an in vitro world: Past, present and future. *Biorheology*, 50:3–16, 2013.

[52] M. Lopez and M. D. Graham. Shear-induced diffusion in dilute suspensions of spherical or nonspherical particles: Effects of irreversibility and symmetry breaking. *Phys. Fluids*, 19(7):073602, July 2007.

[53] M. Lopez and M. D. Graham. Enhancement of mixing and adsorption in microfluidic devices by shear-induced diffusion and topography-induced secondary flow. *Phys. Fluids*, 20(5):053304, May 2008.

[54] H. Ma and M. D. Graham. Theory of shear-induced migration in dilute polymer solutions near solid boundaries. *Phys. Fluids*, 17(8):083103, January 2005.

[55] E. W. Merrill. Rheology of blood. *Physiological Reviews*, 49(4):863–888, 1969.

[56] R. M. Miller and J. F. Morris. Normal stress-driven migration and axial development in pressure-driven flow of concentrated suspensions. *Journal of Non-Newton. Fluid Mech.*, 135(2-3):149–165, 2006.

[57] K. Müller, D. A. Fedosov, and G. Gompper. Margination of micro- and nano-particles in blood flow and its effect on drug delivery. *Sci. Rep.*, 4, May 2014.

[58] V. Narsimhan, H. Zhao, and E. S. G. Shaqfeh. Coarse-grained theory to predict the concentration distribution of red blood cells in wall-bounded Couette flow at zero Reynolds number. *Phys. Fluids*, 25(6):061901, 2013.

[59] B. Neu and H. J. Meiselman. Depletion-mediated red blood cell aggregation in polymer solutions. *Biophysical Journal*, 83(5):2482–2490, January 2002.

[60] R. J. Phillips, R. C. Armstrong, R. A. Brown, A. L. Graham, and J. R. Abbott. A constitutive equation for concentrated suspensions that accounts for shear-induced particle migration. *Phys Fluids A-Fluid*, 4(1):30–40, January 1992.

[61] A. S. Popel and P. C. Johnson. Microcirculation and hemorheology. *Annu Rev Fluid Mech*, 37:43–69, January 2005.

[62] C. Pozrikidis. *Theoretical and Computational Fluid Dynamics*. Oxford University Press, Oxford, 1997.

[63] P. Pranay, R. G. Henriquez Rivera, and M. D. Graham. Depletion layer formation in suspensions of elastic capsules in Newtonian and viscoelastic fluids. *Phys. Fluids*, 24(6):061902, 2012.

[64] A. R. Pries, D. Neuhaus, and P. Gaehtgens. Blood-Viscosity in tube flow - dependence on diameter and hematocrit. *Am J Physiol*, 263(6):H1770–H1778, 1992.

[65] M. W. Rampling, H. J. Meiselman, B. Neu, and O. K. Baskurt. Influence of cell-specific factors on red blood cell aggregation. *Biorheology*, 41(2):91–112, January 2004.

[66] W. B. Russel, D. A. Saville, and W. R. Schowalter. *Colloidal Dispersions*. Cambridge University Press, Cambridge, 1989.

[67] H. Schmid-Schönbein and R. Wells. Fluid drop-like transition of erythrocytes under shear. *Science*, 165(3890):288–291, 1969.

[68] G. Segre and A. Silberberg. Behaviour of macroscopic rigid spheres in Poiseuille flow Part 2. Experimental results and interpretation. *J. Fluid Mech.*, 14(1):136–157, 1962.

[69] D. Semwogerere and E. R. Weeks. Shear-induced particle migration in binary colloidal suspensions. *Phys. Fluids*, 20:043306, 2008.

[70] K. Sinha and M. D. Graham. Dynamics of a single red blood cell in simple shear flow. *Phys. Rev. E*, 92:042710, October 2015.

[71] K. Sinha and M. D. Graham. Shape-mediated margination and demargination in flowing multicomponent suspensions of deformable capsules. *Soft Matter*, 12(6):1683–1700, January 2016.

[72] R. Skalak and P.I. Branemark. Deformation of red blood cells in capillaries. *Science*, 164:717–719, 1969.

[73] R. Skalak, N. Ozkaya, and T. C. Skalak. Biofluid Mechanics. *Annu Rev Fluid Mech*, 21:167–204, 1989.

[74] J.M. Skotheim and T.W. Secomb. Red blood cells and other nonspherical capsules in shear flow: oscillatory dynamics and the tank-treading-to-tumbling transition. *Phys. Rev. Lett.*, 98(7):078301, 2007.

[75] J. R. Smart and D. T. Leighton. Measurement of the drift of a droplet due to the presence of a plane. *Phys Fluids A-Fluid*, 3(1):21–28, January 1991.

[76] A. J. Thompson, E. M. Mastria, and O. Eniola-Adefeso. The margination propensity of ellipsoidal micro/nanoparticles to the endothelium in human blood flow. *Biomaterials*, 34(23):5863–5871, July 2013.

[77] G. B. Thurston. Viscoelasticity of human blood. *Biophysical Journal*, 12(9):1205–1217, September 1972.

[78] G. A. Truskey, F. Yuan, and D. F. Katz. *Transport Phenomena in Biological Systems*. Pearson Prentice Hall, 2004.

[79] A. Viallat and M. Abkarian. Red blood cell: from its mechanics to its motion in shear flow. *Int. Jnl. Lab. Hem.*, 36(3):237–243, April 2014.

[80] J. Wang, E. J. Tozzi, M. D. Graham, and D. J. Klingenberg. Flipping, scooping, and spinning: Drift of rigid curved nonchiral fibers in simple shear flow. *Phys. Fluids*, 24(12):123304, 2012.

[81] W. Yao, Z. Wen, Z. Yan, D. Sun, W. Ka, L. Xie, and Shu Chien. Low viscosity ektacytometry and its validation tested by flow chamber. *J. Biomech.*, 34(11):1501–1509, 2001.

[82] A. L. Yarin, O. Gottlieb, and I. V. Roisman. Chaotic rotation of triaxial ellipsoids in simple shear flow. *J. Fluid Mech.*, 340:83–100, June 1997.

[83] I. E. Zarraga and D. T. Leighton. Measurement of an unexpectedly large shear-induced self-diffusivity in a dilute suspension of spheres. *Phys. Fluids*, 14:2194–2201, 2002.

[84] X. Zhang and M. D. Graham. Simulations of flowing suspensions of flexible and stiff red blood cells. In preparation, 2018.

[85] M. Zurita-Gotor, J. Blawzdziewicz, and E. Wajnryb. Layering instability in a confined suspension flow. *Phys. Rev. Lett.*, 108(6):68301, 2012.

[86] A. L. Zydney and C. K. Colton. Augmented solute transport in the shear flow of a concentrated suspension. *Physicochem. Hydrodyn.*, 10:77–96, 1988.

CHAPTER 4

Microstructure and rheology of cellular blood flow, platelet margination and adhesion

Qin M. Qi
Stanford University

Eric S. G. Shaqfeh
Stanford University

CONTENTS

4.1 INTRODUCTION

A SUSPENSION of red blood cells and platelets is a special class of multi-component suspension. While red blood cells are highly deformable and occupy up to 50% volume fraction in whole blood, unactivated platelets are much smaller, more rigid, and present at far lower concentrations. In this chapter, we will discuss how red blood cells play an important role in platelet adhesion to blood vessel walls and ultimately influence hemostasis and "blood clotting".

At a vascular injury site, the sub-endothelial layer of the vessel wall is exposed, triggering interactions with platelet surface receptors. Subsequently adhered platelets form aggregates and seal the injury site. During this process, it is beneficial to have a near-wall excess concentration of platelets to reduce the cross-flow transport time. This phenomenon is called platelet margination and occurs in the presence of flow shear. In channel flow, red blood cells experience a lift force away from the wall due to their deformability and form a cell-free layer or Fahreaus-Lindqvist layer at the wall [11]. Hence rigid platelets at low concentrations are subsequently pushed closer to the wall thanks to red blood cell migration.

Red blood cell migration and platelet margination can only be discussed when a proper parameter space is specified, which may or may not be physiological. In this chapter, we focus on blood arterioles with diameters ranging from 30 μm to 100 μm and the corresponding physiological wall shear rates between 1000s^{-1} and 3000s^{-1}. The macroscopic flow properties for such length scales are typically referred to as blood microrheology.

Aside from the cross-flow movement of platelets and red blood cells, the adhesion of platelets also involves a rich biophysics and draws continuous research interest. Prior to forming any bonds, discoid-shaped platelets undergo modified Jeffery Orbit motions near the wall in channel flow [44]. The abundance of surface receptors thus leads to frequent bond formation and breaking, and causes the translocating motion of temporarily-adhered platelets. In addition, each receptor-ligand binding shows force-dependent kinetics, which, in turn, affects platelet translocation.

The fluid dynamics regulation of platelet adhesion is only partially understood. The development of microfluidic devices and experiments has helped and led to many important observations in-vitro, but many still await theoretical explanation. Red blood cells are highly aspherical and thus analytic theories can, in general, not be derived as, for example, in the cases of near-spherical drops and capsules. The pressure-driven flow condition inside the vessels leads to a varying shear rate and complicates the process of cell migration. A simple shear flow approximation needs to be justified even if the near-wall region is the focus of attention.

In recent years, whole blood simulations have served as a popular method to complement experimental studies. Common simulation techniques include boundary integral [20, 51], immersed boundary [10], dissipative particle dynamics [18, 49] and lattice-Boltzmann [8]. Regardless of the simulation technique, it is widely agreed to model the red blood cell membrane using the Skalak law [39]. Therefore, the elastic membrane is nearly incompressible and the cell deformability is represented by its resistance to shearing. The detailed descriptions of these techniques will not be repeated here, and are reviewed by Pozrikidis *et al.* [29]. In order to resolve complex and deforming shapes of each individual cell, whole blood simulations require super computing power and each simulation takes several weeks to complete. Alternatives to such expensive computations are necessary, especially for the purpose of studying platelet adhesion. Thus, it would be desirable to derive lower-order theories to capture the indirect role of red blood cells in controlling platelet adsorption.

The objective of this chapter is to describe our present understanding of the concepts of red blood cell migration, platelet margination and platelet adhesion. This understanding spans a spectrum of subjects including platelet biology and blood fluid mechanics. We will unify these concepts and outline an approach to building a lower-order multi-scale model as a direction for future research.

4.2 RHEOLOGY OF BLOOD SUSPENSIONS

4.2.1 Shear-thinning of blood

Due to the deformability of red blood cells, blood manifests rheological properties that differ from those of rigid particle suspensions. A major rheological role of red blood cells is to increase the apparent viscosity of blood, since their internal cytoplasm viscosity is around 5 cP as compared to 1.2 cP for the suspending plasma [47]. The deformability of red blood cells also leads to their non-uniform distribution in the cross-flow direction, which is the fundamental cause of the shear-thinning behavior as well as the blunted velocity profile. While the plasma can be considered Newtonian, whole blood exhibits non-Newtonian behavior for shear rates ranging from 0 to 3000 s^{-1} as shown in Figure 4.1. Recent experiments have visualized the shape distribution of red blood cells at different shear rates [21]. At low shear rates up to $O(10s^{-1})$, red blood cells aggregate and form microstructures called rouleaux due to the effect of plasma proteins. As the shear rate increases, red blood cells start to align with the flow field and undergo tank-treading motion. Thus, the apparent blood viscosity is reduced and reaches a constant value when the shear rate increases above 1000 s^{-1} (arteriole shear rate).

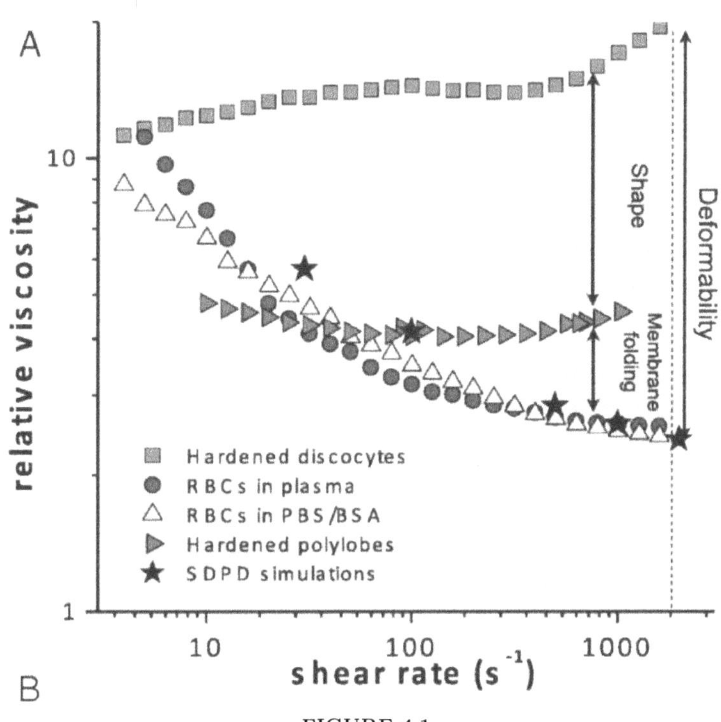

FIGURE 4.1
Relative viscosity of dense cell suspensions(Ht=45%) [21].

4.2.2 A two-phase model for blood flow

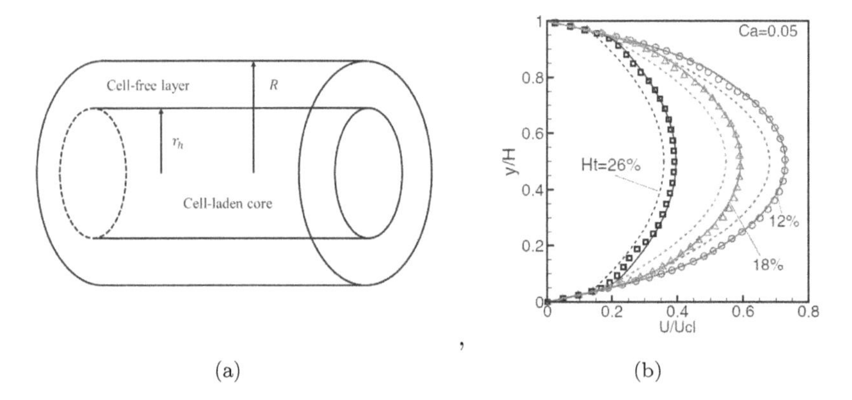

FIGURE 4.2
(a) Schematic diagram of the two-phase flow model. (b) Comparison of mean velocity obtained from simulations (symbols) and predicted by two-phase model, dotted lines, and an improved three-layer model, continuous lines. H/d=6.3, Ca= 0.05 [10].

The simplest model describing blood flow in tubes is the continuum two-phase model as shown in Figure 4.2(a). The vessel of radius R consists of a cell-laden core of radius r_h and a cell-free layer with thickness $R - r_h$. Each phase is assumed Newtonian and immiscible. Due to unmatched viscosity between the two regions, the velocity profile shows an increased wall shear rate as compared to the parabolic flow case. The two-phase model is an empirical model for the viscosity, cell-free layer and flow profile of red blood cells. It does not provide a fundamental understanding of the complex rheology of blood suspensions. In addition, the assumptions such as a homogeneous red blood cell concentration in the core region may differ greatly from the actual physics. The validity of the two-phase model is discussed in Doddi and Bagchi's paper based on particle-scale immersed boundary simulations [10] and is shown in Figure 4.2(b). They demonstrated that the two-phase model over-predicts the blunted velocity profile due to the discontinuous viscosity at the edge of the cell-free layer.

As mentioned in Section 11.1, the rheology of blood is closely related to platelet adhesion. The most direct impact is the fact that platelet surface receptors are shear-sensitive. Therefore, platelet adhesion experiments using pure plasma will be different from whole blood at matching flow rates due to the change in flow profile. The secondary effect, but perhaps, more significant, is the margination of platelets due to the formation of a cell-free layer. Therefore, the near-wall concentration of platelets will vary based on the hematocrit level even when the total platelet concentration is fixed, as discussed in Section 4.5.1.

4.3 THEORY OF RED BLOOD CELL MIGRATION

4.3.1 Overview

The mechanism of red blood cell migration can be studied in a rectangular channel as shown in Figure 4.3. In a Cartesian coordinate system, x is the flow direction, y is the vorticity direction and z is the direction of the velocity gradient. The y dimension is assumed much larger than the z dimension, and thus wall effects are ignored in the y direction. Therefore, the flow can be approximated as a two-dimensional channel flow driven by a pressure gradient. The change of geometry from a circular tube to a rectangular channel does not affect the physics of cell migration as long as the smallest dimension(z) is consistent [27, 34].

FIGURE 4.3
A schematic of red blood cells(red) and platelets(white) in wall-bound channel flow driven by a pressure gradient.

The flux of red blood cells(or other deformable suspensions) in the cross-flow direction consists of two major contributions [34, 52]: a convective flux due to the deformability-induced lift velocity u_{lift} and a flux due to shear-induced collisions F_{CC}. Therefore, the governing equation for the concentration distribution of red blood cells n_C can be written as follows:

$$\frac{\partial n_C}{\partial t} + \frac{\partial (u_{lift} n_C)}{\partial z} + \frac{\partial F_{CC}}{\partial z} - \frac{\partial}{\partial z}(D\frac{\partial n_C}{\partial z}) = 0 \tag{4.1}$$

The characteristic length scale is the cell radius a, which is typically 2.8 μm. The characteristic time scale is the inverse of a characteristic wall shear rate $\dot{\gamma}_c$, defined as the equivalent wall shear rate for a parabolic flow at matching flow rate. From now on in this chapter, we will use such dimensional scalings in our discussion unless noted otherwise.

With $Re \sim 0$ and $Pe \sim \infty$, the most important dimensionless number is the Capillary number $Ca = \frac{\mu \dot{\gamma}_c a}{E_s}$, with μ denoting the plasma viscosity and E_s denoting the cell shear modulus. It represents the flow viscous effect relative to the cell membrane elastic force, and governs both the lifting and shear-induced cell diffusion behavior. Simulations are often used to determine these individual flux contributions. By assuming that the lifting and shear-induced diffusion mechanisms are the same for individual cells and for the blood suspension, one can reduce the simulation scale and save a significant amount of computation time. However, the loss of accuracy in such approximation has yet to be fully understood.

4.3.2 Hydrodynamic lift

Deformable particles experience a lift force away from the wall even in the absence of inertia, and thus lift creates a convective flux. This lifting force is present in both the tumbling regime and the tank-treading regime of deformable particles [17]. In the physiological range of shear rates, a red blood cell tumbles in pure plasma while tank-treads in a cellular suspension [27]. To account for this difference in lifting behavior [17], $\lambda = 1$ is usually used to study a single red blood cell in theory, simulation and experiments such that the tank-treading behavior is retained. In the case of simple shear flow, the lift velocity has been well-characterized for red blood cells [27], capsules [34] and vesicles [4, 28]. In the far-field, the velocity is estimated to scale with $\frac{S_{zz}}{z^2}$, where S_{zz} is the particle stresslet and z is the particle distance to the wall [7, 34]. This argument is based on the method of images such that a particle interacts with its image stresslet opposite the wall.

A second mechanism of lifting arises from the curvature of pressure-driven flow even in the absence of the wall [7, 9]. This effect is likely to dominate when the red blood cell migrates close to the centerline. Therefore, it is not adequate to simply use a superposition of lift velocities induced by two walls to characterize the lift velocity in channels or tubes.

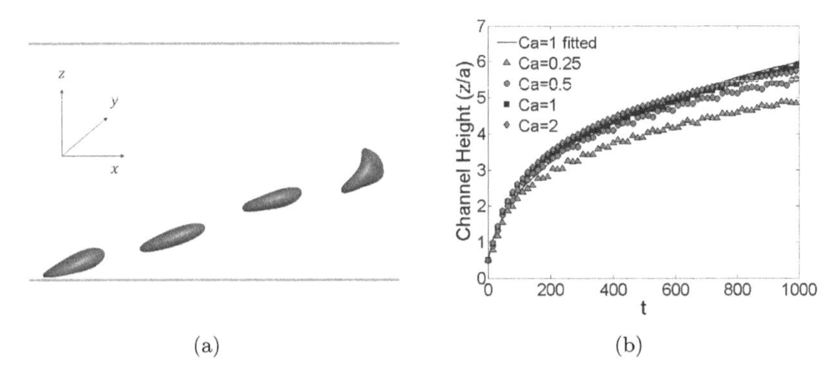

FIGURE 4.4
Simulation of a single red blood cell: (a) snapshot of a simulation (b) red blood cell lift trajectories at various Capillary numbers, channel height $H = 17.73$.

The combined effects of two walls and the flow curvature lead to the following lift velocity correlation:

$$u_{lift} = \frac{\xi\dot{\gamma}(z)}{z^{\alpha}} \tag{4.2}$$

z is the distance defined with respect to the closer wall. There are two fitting parameters in this expression: α and ξ. They can be determined from the simulation of a single red blood cell as shown in Figure 4.4.

Equation 4.2 offers a very good fit to the cell lift trajectory obtained from a boundary integral simulation [33]. The reported correlations for the dimensionless lift velocity are $u_{lift} = \frac{0.067\dot{\gamma}_0(z)}{z^{1.3}}$ for channel height $H = 12$ and $u_{lift} = \frac{0.041\dot{\gamma}_0(z)}{z^{1.2}}$ for $H = 17.73(50\ \mu\text{m})$. Unlike the $\alpha = 2$ correlation from the stresslet argument, the reported α values for red blood cells are close to those from experiments of tank-treading vesicles($\alpha = 1 \pm 0.1$) [7]. This decrease of α indicates that the enhancement of lift velocity due to flow curvature plays a more important role than the canceling effect of the stresslet interaction from the opposite wall.

4.3.3 Shear-induced collisions

In channel flow under shear, strong hydrodynamic interactions between particles lead to fluctuations in their velocities and cause a diffusive-like behavior for the suspension, as first observed by Leighton and Acrivos [22]. This shear-induced diffusion leads to low particle concentration in the high shear rate region where collisions are more frequent. In the case of red blood cells, an effective shear-induced diffusivity can be estimated [17]:

$$D = \dot{\gamma}\phi a^2 f_s \tag{4.3}$$

Where $\dot{\gamma}$ is the shear rate, ϕ is the volume fraction and f_s is a dimensionless parameter that depends on particle shape and deformability. The value of f_s has been estimated from experiments, but the accuracy of the result is limited due to experimental errors.

The primary mechanism of shear-induced diffusion is the hydrodynamic interaction between two particles. The cross-flow distance between the particle pair increases upon interaction due to symmetry breaking. As a result, the diffusional flux can be written in a Boltzmann-like collisional form as follows:

$$F_{CC} = \int_{\delta_z} \int_{\delta_y} \int_0^{\Delta_{CC}} n_C(z-b) n_C(z-b-\delta_z) \delta u \; db \; d\delta_y \; d\delta_z \tag{4.4}$$

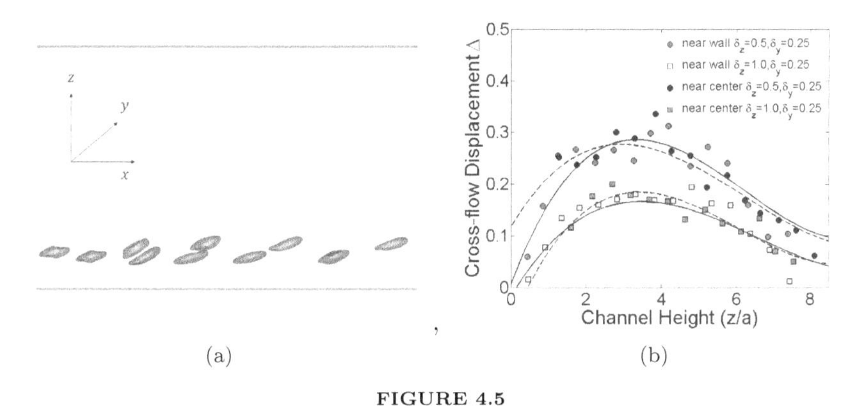

FIGURE 4.5
Simulation of a pair of red blood cells: (a) snapshot of a cell-cell collision simulation including a near-center red blood cell(red) and a near-wall red blood cell(blue) (b) cross-flow displacement Δ_{CC} values vs. z positions from cell-cell collision simulations, $H = 17.73$, Ca=1, noises are due to fluctuations in the center-of-mass position of the lifting cell.

This formulation has been used in deformable suspensions in both simple shear and pressure-driven flows [27, 34, 52]. As shown in Figure 4.5, pair-wise collisions at different separating distances δ_z and δ_y can be used to determine the post-collisional displacement Δ_{CC}. If cells also experience lifting during the collision, this effect needs to be subtracted from the cell trajectories. In the case of unbounded simple shear flow, Δ_{CC} is not a function of z and thus requires a relatively small number of simulations. However, in the case of pressure-driven channel flow, Δ_{CC} is also a function of z as shown in Figure 4.5, and the number of simulations increases with the channel height H. Two-cell simulations run much faster than whole blood simulations and therefore the number of simulations is in general not a limiting factor. The wall effect and the shear rate gradient manifest themselves in the nonlinear correlation between Δ_{CC} and z as well as the asymmetrical Δ_{CC} between two cells.

Regardless of the flow profile, the presence of a wall changes the force dipole contribution in its vicinity [27, 52]. Thus, the near-wall Δ_{CC} values deviate significantly from far-field estimates [27]. Similar observations have been made for rigid spheres in shear flow [52] and are said to enhance particle layering in polydisperse solutions.

An issue specific to pressure-driven flow is the zero shear rate at the centerline of the channel. This vanishing shear rate causes all shear-induced fluxes to decay and a singularity arises in Equation 4.1. The solution is to add a corrective flux term to remove the singularity, which is an idea borrowed from the literature on suspensions of rigid spheres [24]. At least two possible choices of corrections can be used arising from different origins. A non-local correction to the shear rate in the collisional flux expression can be added. This correction takes into account the finite size of red blood cells, and therefore the cell experiences at least $\frac{\epsilon}{2}\dot{\gamma}_c$ shear rate, where ϵ denotes the ratio of cell radius to half channel size. An alternative correction was implemented by Rivera *et al.* [34] to take into account the flow curvature. It is a shear rate correction term of $\frac{\epsilon}{2}\dot{\gamma}_c\delta_z$.

The evaluation of the integral in Equation 4.1 requires closing at a proper cutoff distance. In simple shear flow, hydrodynamic interactions are negligible for $\delta_z > 5, \delta_y = 0$ [27]. This value reduces to $\delta_z > 3, \delta_y = 0$ for the pressure-driven flow case [33], which is due to the decaying shear rate away from the wall. A smaller cutoff distance also reduces the number of pair-wise collision simulations needed.

A possible fitting parameter for shear-induced diffusion is an additional diffusivity D to account for three-or-more-body interactions. This diffusivity scales with the local shear rate $\dot{\gamma}(z)$ and thus varies with z. Grandchamp and coworkers [17] estimated that shear-induced diffusivity grows nonlinearly with hematocrit for values as low as 30%. Since most literature which ignore this higher-order diffusion correction have achieved qualitative to semi-quantitative agreement with experiments or simulations [27,34], there is evidence that two-body interactions are the dominant diffusion mechanism even at hematocrits greater than 20%.

4.3.4 Red blood cell migration at steady state

At steady state, the concentration profile of red blood cells can be solved by setting the time derivative to zero in Equation 4.1. Red blood cells exhibit a peak concentration at the center as shown in Figure 4.6, similar to the case of rigid sphere suspensions [24]. A unique feature of the red blood cell profile is the cell-free layer at the wall. In simulations, the cell-free layer thickness is calculated from time-averaging the smallest cell-to-wall distance. In lower-order modeling, it is usually defined as the distance between the closest concentration peak and the wall [27]. The nonlinear nature of red blood cell lifting and shear-induced collisions also creates additional features in the concentration profile, and the concentration profile is not monotonic. Such fine features are not solely due to temporal fluctuations and are not captured by simple analytical solutions [34]. A closer look at these features is required in the future, as they may be related to the varying shape of red blood cells in channel flow.

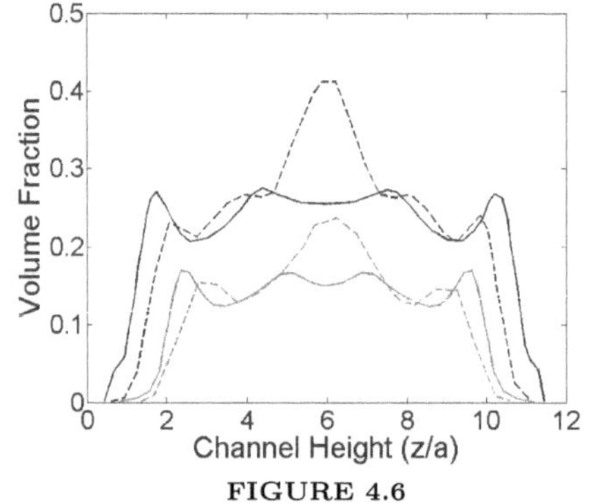

FIGURE 4.6
Red blood cell concentration distribution at $Ht = 20\%$(blue) and $Ht = 10\%$(red) for simple shear flow [27](solid)and pressure-driven flow [51](dashed), $H = 12$, $Ca = 1$.

A typical approximation in modelling red blood cells and platelets flowing in channels is for the local flow to be simple shear. It is based on the reasoning that platelet adhesion only occurs in the near-wall region where the flow is nearly linear. Thus, the entire channel can be treated as having a fixed shear rate matching the actual wall shear rate. This approximation makes the flux calculations simpler: the lift velocity is no longer influenced by the flow curvature, and the collisional displacement is uniform except near the wall due to confinement. The error created by this approximation is shown in Figure 4.6. There is a 20% reduction in the cell-free layer thickness for the simple shear flow approximation when all else is the same. This discrepancy can be explained by the stronger collisions away from the wall due to a constant shear rate. Therefore the use of the simple shear approximation should be done with some caution.

4.3.5 Migration timescales

An area often neglected by researchers is the time evolution of cell migration. Starting from a uniform distribution of red blood cells, there are two important timescales to consider: the timescale to form the cell-free layer and the timescale to reach a steady cross stream concentration profile. The entrance lengths corresponding to these two length scales are important design parameters in microfluidic devices. Katanov *et al.* reported an entrance length estimate of $25D$ (tube diameter) for the cell-free layer development based on DPD simulations [18].

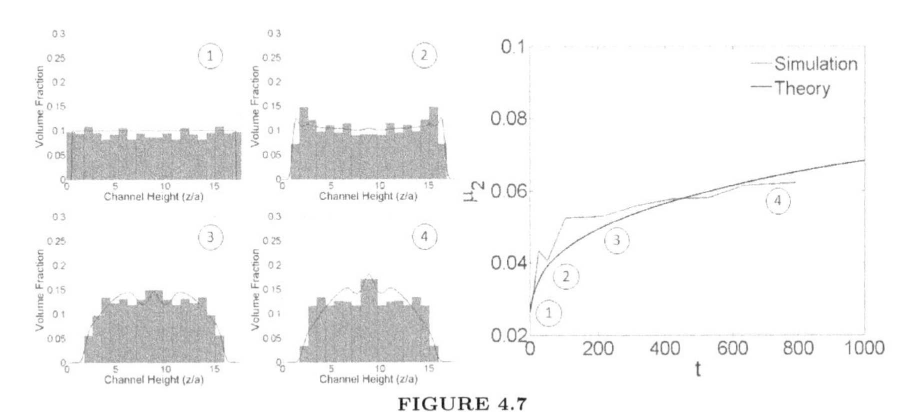

FIGURE 4.7
Comparison between simulation(pink) and theoretical(red) results for red blood cell concentration distribution $H = 17.73, Ht = 10\%, Ca = 1$. These results use methods similar to [27, 33]. (a) Snapshots at different times (b) second moment, from [32].

Alternatively, Equation 4.1 can be used to make such estimates in a simple fashion. As shown in Figure 4.7, red blood cells quickly deplete from the near-wall region and the concentration profile resembles that of simple shear flow, with a concentration peak located at the boundary of the cell-free layer. The total migration time (*i.e.* time to reach steady state), however, is set by the slow development of the center peak concentration. Due to the varying shear rate and lift velocity, it is not surprising that there is a separation of timescales and a simple shear approximation will underestimate the total migration time. The second moment of the concentration profile, defined as: $\mu_2 = \int_0^H (z - H/2)^2 n(z)dz$, can be used as a quantitative measure of reaching the steady state. A 95% quantile can be used to calculate the time to reach steady state $t_{SS,C}$ and form a cell-free layer t_{CFL}. At 20% hematocrit in an $H = 10.64(30\ \mu\text{m})$ channel, the entrance length for cell-free layer formation is estimated to be 1 mm [18] while the total entrance length is 8 mm [13].

The timescales of cell migration can also be understood from a scaling perspective. Based on the driving mechanisms in the near-wall and bulk region, the two time scales can be estimated as follows:

$$t_{CFL} = \int_0^{z_{cfl}} \frac{dz}{u_{lift}} \tag{4.5}$$

$$t_{SS,C} = \frac{(\frac{H-2z_{cfl}}{4})^2}{D_{cell}} + t_{CFL} \tag{4.6}$$

The diffusivity of red blood cells D_{cell} has been determined in various references both by using simulations and experiments. Crowl and coworkers gave an estimate of $1.5 * 10^{-6} cm^2/s$ for 20% hematocrit and wall shear rate of 1100 s^{-1} [8], which is close to a value of $1.3 * 10^{-6} cm^2/s$ at matched conditions given by Grandchamp and coworkers [17]. The validity of the above two scaling laws will be discussed in the following section.

4.3.6 Effects of hematocrit, channel height, viscosity ratio and capillary number

Variables relevant in red blood cell migration are hematocrit, channel height, viscosity and Capillary number. Below we review studies spanning this broad parameter space with an emphasis on the concentration profile and the cell-free layer thickness.

In the microcirculation, the hematocrit typically ranges between 10% and 30%, and is thus lower than whole blood hematocrit (30% to 50%). In this physiological range of hematocrit, the cell-free layer thickness decreases with hematocrit as shown in Figure 4.8. Narsimhan gave a simple argument regarding this correlation: the lift flux scales linearly with ϕ, but the collisional flux scales quadratically with ϕ. Thus, $l \sim \phi^{-1/2}$ for simple shear flow [27]. A similar argument is given by Katanov *et al.* [18] using an effective pressure argument. In pressure-driven flow, this correlation still holds, since the role of hematocrit remains the same in the corresponding flux contributions. The steady state migration profile changes with hematocrit as shown in Figure 4.6. The center peak concentration grows less linearly with hematocrit, because the cell-free layer is narrower. The steady state is also reached more quickly. For extremely low hematocrit such as 5%, experimentalists observed interesting distributions of red blood cells where the center peak is replaced by two off-center peaks [38]. In this low concentration limit, red blood cell distributions are dominated by the single cell lifting mechanism and cell-cell interactions are weak. In the other limit of high hematocrit, however, the cell-free layer thickness saturates. The shear-induced diffusional flux in Equation 4.1 needs to be revisited at very high hematocrit as three-body interactions may dominate the diffusive behavior.

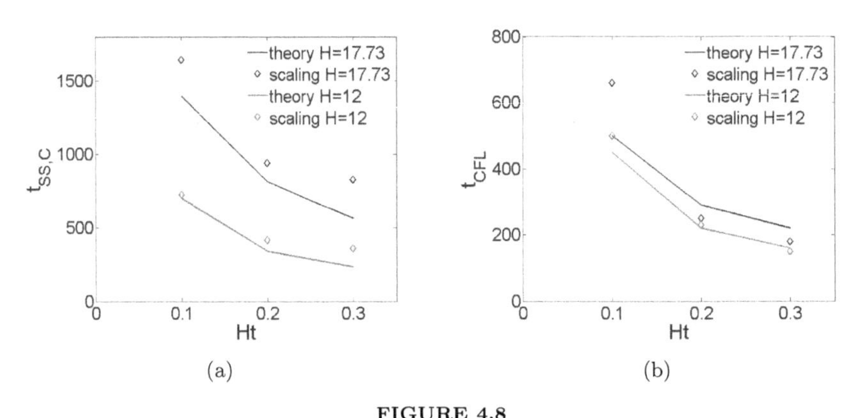

FIGURE 4.8
Comparison between theory and scaling for red blood cell migration (a) $t_{SS,C}$ (b) t_{CFL}. The theory uses methods similar to [27,33], from [32].

Common choices for the channel height in microfluidic experiments are 30 μm and 50 μm [13,14]. The red blood cell concentration distribution profiles (as measured) are similar with a slight increase in the cell-free layer thickness in the wider channel. This effect is as expected because both lift flux and collisional flux depend weakly on the channel width. The cell-free layer time scale t_{CFL} and the overall timescale $t_{SS,C}$ see an increase with channel height [18, 27]. For channels smaller than 20 μm, the center concentration peak is not observed even at high hematocrits [38]. As pointed out by Secomb, when the size of red blood cells becomes comparable to the channel size, the suspension transitions to a single-file flow. The cell shape and their interactions are different from those in disordered suspensions [37]. From a theoretical point of view, the non-local shear rate correction increases as the cells become more confined, which implies that the finite volume and shape of red blood cells become important in the channel. Therefore, rather than considering only the center-of-mass in the continuum formulation, the characteristics of individual cells must be properly considered.

The comparison of the scaling law and the time-dependent theory is shown in Figure 4.8. The convective timescale from Equation 4.5 can capture the trend of t_{CFL} nicely for both varying hematocrit and channel height. As the hematocrit increases, the cell-free layer thickness decreases and reduces the convective distance. A bigger discrepancy occurs for lower hematocrit mainly because shear-induced diffusion becomes relevant as the distance to the wall increases. Likewise, $t_{SS,C}$ decreases with hematocrit and channel height. In this case, the effect of channel height is more significant due to the larger role of shear-induced diffusion in the bulk region.

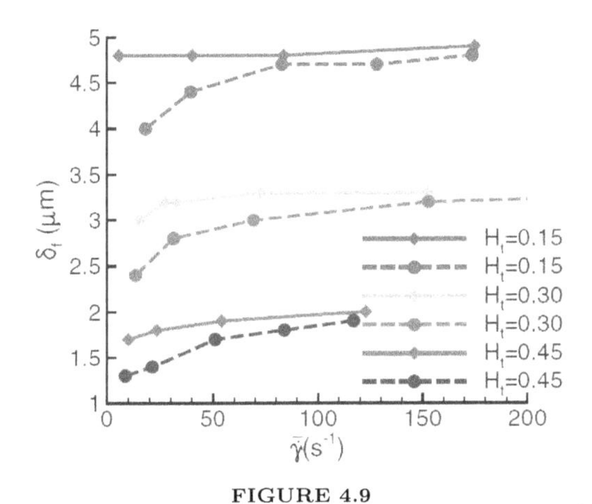

FIGURE 4.9
Cell-free layer thickness l at various hematocrits and average shear rates [18], $D = 20\,\mu m$.

As shown in Figure 4.9, the cell-free layer thickness increases at low Capillary number and saturates at high Capillary number. This finding is consistent in pressure-driven flow [18] and simple shear flow [27]. In the case of healthy red blood cells, the physiological Capillary number in arteriole flow is between 0.5 and 1.5, and the effect of Capillary number usually results from the change of shear rate. As pointed out by Zhao and coworkers [51], $Ca = 0.5$ is the onset of cell tank-treading behavior and the cell shape is relatively fixed. The lift velocity scales linearly with the shear rate. The collisional displacements Δ_{CC} do not change significantly [27], but the collision frequency is proportional to the shear rate. Thus, collisional flux also scales nearly linearly with the shear rate. Katanov and coworkers showed that the dimensionless timescales are thus nearly universal, and the entrance lengths are independent of shear rate [18].

Finally, an area recently drawing research interest is the viscosity contrast λ. As mentioned before, red blood cell suspensions exhibit tank-treading behaviors at λ=5, unlike the case of a single cell. This is confirmed by experimental research at various hematcrits and channel widths. Researchers have further demonstrated that the viscosity contrast effects become minimal for wider channels and higher hematocrits, which coincide with the physiological parameter range [38]. In the theoretical formulation, a higher viscosity ratio leads to reductions in the lift velocity as well as collisional displacements, which seems to create canceling effects that do not significantly influence the ultimate steady migration profile. From a simulation perspective, a unit viscosity ratio simplifies the boundary integral formulation and is therefore commonly assumed. In the future, it is worth investigating the role of viscosity contrast in more detail and examine whether the $\lambda = 1$ assumption is quantitative or qualitative.

4.4 MODEL OF PLATELET ADHESION

4.4.1 Receptor-ligand binding

Platelet adhesion is mediated by various receptors on its surface which interact with wall-bound Von Willebrand Factors(VWF), fibrinogen, and collagen. In the physiological range of shear rates, it is the binding between GPIb receptors and the A1 domain of VWF that dominates the initial interaction between a flowing platelet and the static vessel wall [16]. The detailed biology is not the focus of this chapter, and readers are encouraged to read the review by Fogelson and Neeves [16]. The most up-to-date measurements of association and dissociation rates at the single molecule level is done by Kim and co-workers [19]. A GPIb-VWF bond breaks under applied tensile force according to a two slip-bond model("flex-bond"). Therefore, there exist two states in which the dissociation rate increases with the applied force and can be related to the fluid mechanics of platelets in flow.

In addition to the force-dependent kinetics, the stretching of the GPIb-VWF bond also creates a force on the platelet that can be balanced by the drag force. The size of GPIb is usually neglected and the size of VWF is mediated by the applied force. VWF consists of long multimers and exists in a compact globular conformation in the absence of flow [42]. Only under high shear can it extend to a chain configuration and open up active sites for binding. The detailed biology of VWF can be found in a review by Springer [42]. A value between 200 nm and 300 nm is often used as the size of VWF under shear, which translates to a reactive distance for binding [13, 41]. Heretofore only the dynamics of a freely-flowing VWF have been examined in experiments and simulations [36]. A more relevant study would be the dynamics of a tethered VWF under shear, which could be a direction of future studies.

GPIb-VWF binding is short-lived with a lifetime of O(0.1s) [42] and therefore a stable adhesion of platelets relies on subsequent steps of binding. Most common reactions that provide stable, long-time bindings include GPIIbIIIa-fibrinogen and GPIIbIIIa-VWF interactions [35]. Such bonds form slowly, but the reverse reaction also occurs slowly, leading to prolonged bond lifetimes. Another area of future research is the change of platelet configuration upon activation. A recent model incorporates the activation mechanism and couples it with platelet deformability [50]. Much more needs to be done to complete our understanding of how activation affects platelet adhesion.

4.4.2 From single bond kinetics to platelet adhesion

Despite the transient nature of a single GPIb-VWF bond, platelets in blood flow are able to adhere to the vessel wall and finally form stable adhesion thanks to the presence of tens of thousands of adhesive receptors on their surfaces. Therefore, platelets are able to slow down from an $O(10^{-3}$ m/s) velocity in flow to an $O(10^{-6}$ m/s) rolling motion upon adhesion. Experiments reveal that adhered platelets undergo stop-go motions, commonly referred to as translocation, on the VWF-coated surfaces as shown in Figure 4.10(a). Bhatia and co-workers first modeled the platelet as a sphere to study its motion subject to simple shear flow and reversible receptor-ligand reactions [3]. They revealed that the platelet translocation motion occurs only when the receptor and ligand densities are within a proper range, as shown in Figure 4.10(b). The receptor and ligand densities govern the rate of forming new bonds, which slow down platelets, relative to the rate of breaking existing bonds, which enable platelet rolling. The border of the state diagram is governed by the single bond association rate as well as the shear rate.

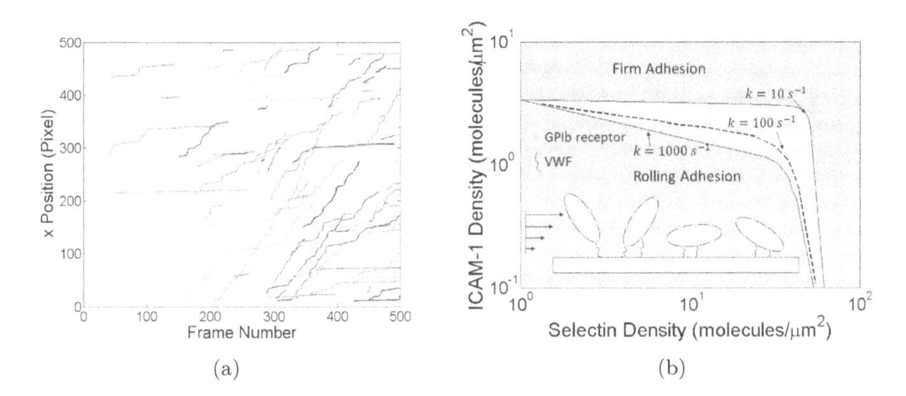

FIGURE 4.10
(a) Experimental tracks of translocating platelets on VWF-coated surfaces. Experimental setup is similar to those described in Fitzgibbon *et al.* [13], from [31]. (b) State diagram of platelet translocation(rolling) or stably adhering [3] along with the schematic of platelet rolling(translocating).

Mody and King established a model called Platelet Adhesive Dynamics, to study platelets in simple shear flow in the absence of red blood cells [25]. This model combines a Monte Carlo simulation of receptor-ligand bindings with the hydrodynamics of an ellipsoid in shear flow. Skorczewski and coworkers updated the Platelet Adhesive Dynamics model by considering platelet deformation and platelet-platelet interactions [40]. The change of geometry from a sphere to an ellipsoid leads to the discussion of the modified Jeffery orbits, because the tumbling motion of a Jeffery orbit is hindered by the existence of a wall. There is a fast flipping period when the thin edge of the platelet is closest to the wall and a slow tilting period when the long axis of the platelet is near-parallel to the wall. The fast flipping period gives the platelet the largest fractional area within the reactive distance. Thus, the first bond formation is typically observed during this period [15,25]. Later Vahidkhah and Bagchi confirmed such platelet motion in whole blood simulations without binding reactions [44].

Platelet adhesion simulations have also been performed by Fitzgibbon *et al.* using Stokesian dynamics in simple shear flow and the effect of red blood cells is considered as a fluctuation in the flow field [13]. They focused on characterizing the platelet translocating distance, which shows an exponential distribution. The origin of the translocation distance involves both the hydrodynamics and the binding kinetics. Therefore, it requires further investigation to decouple these two effects and characterize the translocating motion in more detail.

4.5 THE ROLE OF RED BLOOD CELLS IN PLATELET ADHESION

4.5.1 Platelet margination

Similar to Equation 4.1, the distribution of platelets can be written in the following form using the dominant flux terms:

$$\frac{\partial n_P}{\partial t} + \frac{\partial F_{CP}}{\partial z} + \frac{\partial F_{PP}}{\partial z} = 0 \tag{4.7}$$

Since non-activated platelets can be considered nearly rigid particles, platelets do not experience the lifting flux as red blood cells do. In addition to cell-platelet interaction F_{CP}, there also exists a platelet-platelet interaction term F_{PP}. This effect has been ignored in previous literature [23,34] due to the sparsity of platelets. However, it results in a singularity inside the cell-free layer where the near-wall platelet concentration would be ill-defined without F_{PP}. The shear-induced diffusion can be formulated in terms of two particle collisions as follows:

$$F_{PP} = \int_{\delta_z} \int_{\delta_y} \int_0^{\Delta_{PP}} n_P(z-b)n_P(z-b-\delta_z)\delta u \; db \; d\delta_y \; d\delta_z \tag{4.8}$$

$$F_{CP} = \int_{\delta_z} \int_{\delta_y} \int_0^{\Delta_{CP}} n_P(z-b)n_C(z-b-\delta_z)\delta u \; db \; d\delta_y \; d\delta_z \tag{4.9}$$

As n_C is decoupled from the platelet expression in Equation 4.1, the role of red blood cells at a given hematocrit is only to provide platelet-cell interactions whose frequency vary non-monotonically with z. The magnitude of collisional displacements for a binary capsule suspension in which the two species differ in size and deformability was discussed by Kumar and coworkers [20]. In the case of red blood cells and platelets, $\Delta_{CP} > \Delta_{CC} > \Delta_{PP} >> \Delta_{PC}$. The fact that the red blood cell trajectory is essentially unchanged by its interaction with platelets($\Delta_{PC} \sim 0$) justifies our not including such effects in Equation 4.1. Interestingly, for a near-wall pair including a red blood cell and a platelet(with the cell being closer to the wall), the platelet experiences a magnified displacement due to the lift of the red blood cell [33]. This effect may be relevant for the platelets in the bulk region above the first layer of red blood cells, making it harder for them to enter the cell-free layer.

Due to the challenge of visualizing platelet distribution in an experimental setting, platelet-sized microparticles are usually used to demonstrate the effect of margination as shown in Figure 4.11. When the hematocrit increases from 10% to 20%, the platelet concentration peak is shifted by 2.3 μm with a higher peak value. This enhanced margination can be explained by the increased cell-platelet collisional frequency due to higher hematocrit as well as a smaller cell-free layer thickness that causes further concentration of the near-wall platelets.

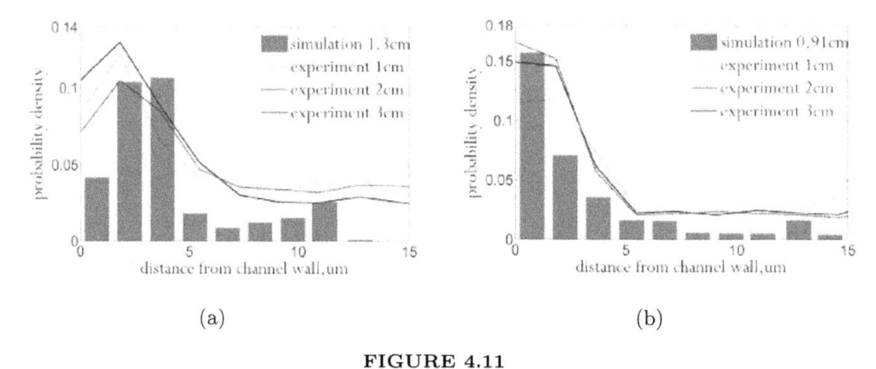

FIGURE 4.11
Comparison of platelet-sized microsphere concentration profile using experiments(histogram) and boundary integral simulations(lines) at (a) $Ht = 10\%$ (b) $Ht = 20\%$, $H = 10.63(30\ \mu\text{m})$, $Ca = 0.41$ [14].

The time-evolution of platelet margination is shown in Figure 4.12 with a uniform concentration distribution for red blood cells and platelets at $t = 0$. The margination process can be broken down to an initial formation of a near-wall peak close to the edge of the cell-free layer and a slow drainage of platelets from the cell-laden region. The first step also leads to a "slow down" of the time variation of the second moment, which is not observed if the steady state concentration profile of red blood cells is used at $t = 0$. This finding is in agreement with simulations by Crowl and co-workers [8]. The initial timescale can be defined as a fast margination timescale t_{FM}. It is measured in terms of the margination parameter PM, which is typically defined as the fraction of platelets inside the cell-free layer [5]. The overall timescale to reach full platelet margination, $t_{SS,P}$, is the longest timescale among the four timescales in the migration and margination processes: $t_{SS,C}$, $t_{SS,P}$, t_{CFL} and t_{FM}. When translated to an entrance length, it corresponds to 3.5 cm, which agrees with Fitzgibbon *et al.* [14] who gave an O(1cm) estimation in a margination experiment using microspheres. The separation of two margination timescales also agrees with findings based on tracking individual platelets from simulations. Zhao *et al.* characterized platelet margination as shear-induced diffusion in the cell-laden region and an irreversible drift towards the cell-free layer. This drift is caused by asymmetric collisions due to the gradient of red blood cell concentration [51]. Vahidkhah *et al.* [44] reported similar types of behavior in simple shear flow simulations and described platelet movement inside the cell-laden region as a slow process and the movement towards the cell-free layer as a fast process.

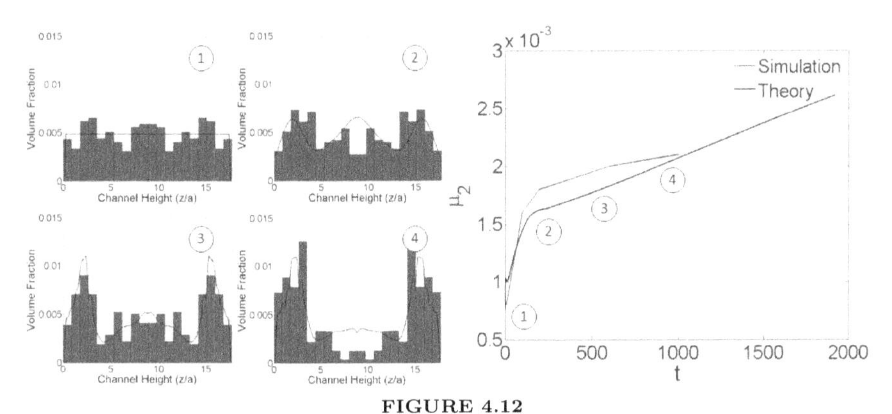

FIGURE 4.12
Comparison between simulation(purple) and theoretical(blue) results for platelet concentration distribution $H = 17.73$, $Ht = 10\%$, $Ca = 1$. These results use methods similar to [27,33]. (a) Snapshots at different times (b) second moment, from [32].

Similar to the scaling analysis for red blood cells, the two margination timescales can be approximated as follows:

$$t_{FM} = \int_0^{z_{cfl}} \frac{dz}{u_{platelet}} \tag{4.10}$$

$$t_{SS,P} = \frac{(\frac{H-2z_{cfl}}{4})^2}{D_{platelet}} + t_{FM} \tag{4.11}$$

The diffusivity of platelets $D_{platelet}$ has been estimated in the literature [5,8,44,51] and is $O(10^{-7} cm^2/s)$. The drift velocity of platelets $u_{platelet}$ has been calculated explicitly [8]. It can also be obtained based on the difference in diffusivity $D_{platelet}$ inside and outside the cell-free layer: $u_{platelet} = \Delta D_{platelet}/z_{cfl}$ [44]. Scaling arguments based on these estimated $D_{platelet}$ and $u_{platelet}$ values are compared in Figure 4.13. Estimated fast margination timescale t_{FM} based on Vahidkhah and coworkers' $D_{platelet}$ values [44] is in better agreement with the theoretical estimation. Nevertheless, the scaling law only gives a qualitative trend for the two timescales. Discrepancy between the scaling and the time-dependent theory may be explained by the sparsity of platelets, which increases the error estimation.

Both margination timescales decrease with increasing hematocrit as is also mentioned in the case of red blood cell migration, since the frequency of cell-platelet interaction is proportional to the red blood cell concentration. In addition, margination is slightly enhanced in wider channels, mainly due to the increase in cell-free layer thickness. Since cell migration is the dominating effect on platelet margination and is independent of the Capillary number, one sees little Ca effect on the margination parameter [5].

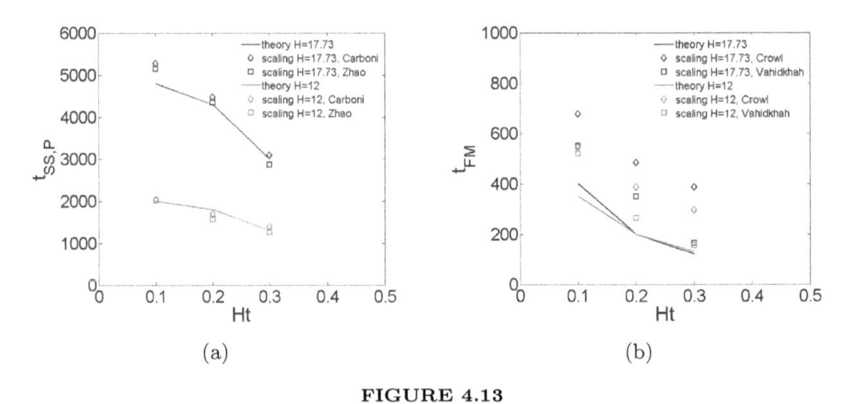

FIGURE 4.13
Comparison between theory and scaling for platelet margination (a) $t_{SS,P}$ (b) t_{FM} with drift velocity from direct [8] and indirect estimations [44]. The theory uses methods similar to [27,33], from [32].

The theory of platelet margination can be readily extended to the margination of rigid particles of various sizes and shapes. An additional Brownian flux can be added for sub-micron-particles as demonstrated by Rivera *et al.* [34]. As expected, Brownian motion weakens particle margination. In the limit when Brownian motion dominates over shear-induced diffusion, Müller *et al.* [26] pointed out that particles distribute evenly in the cell-free layer and the volume excluded by cells in the bulk region. The finding that microspheres marginate more than nanoparticles may be related to drug delivery efficiency. However, the effect of Brownian motion yet to be considered in hydrodynamic collisions, which is presumably a secondary effect.

4.5.2 Platelet adhesion

As first illustrated in Valeri's experiment [45], hematocrit plays an important role in human bleeding time. A recent experiment by Walton and co-workers confirms that elevated hematocrit increases the rate of platelet adhesion and thrombus growth in healthy mice [46]. In order to focus on the initial adhesion, anticoagulant is often added in experimental blood flows to prevent subsequent aggregation. Coated surfaces are also used to maximize platelet interaction with the wall. Therefore, platelet adhesion is observed as discrete adhesion points in images taken. For example, Chen and co-workers considered flowing blood through a 127 μm channel with VWF-tethered surfaces [6]. As shown in Figure 4.14(a), a significant increase in platelet adhesion count is observed when hematocrit is increased from 20% to 60%. A similar experiment was accomplished by Spann and co-workers using a 70 μm channel except the surface being coated with collagen [41]. Platelet adhesion was also found to increase with hematocrit as shown in Figure 4.14(b).

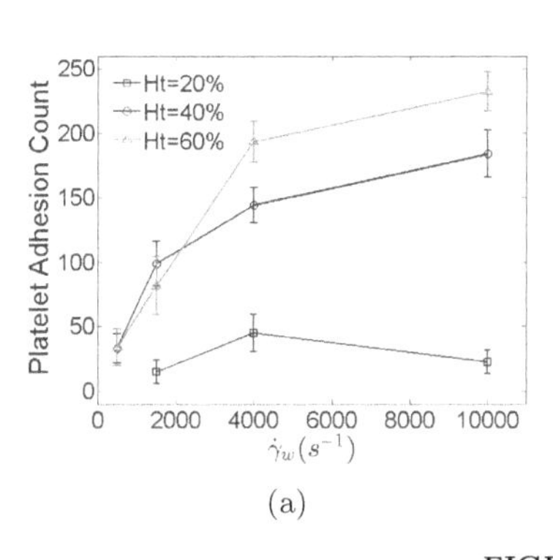

(a)

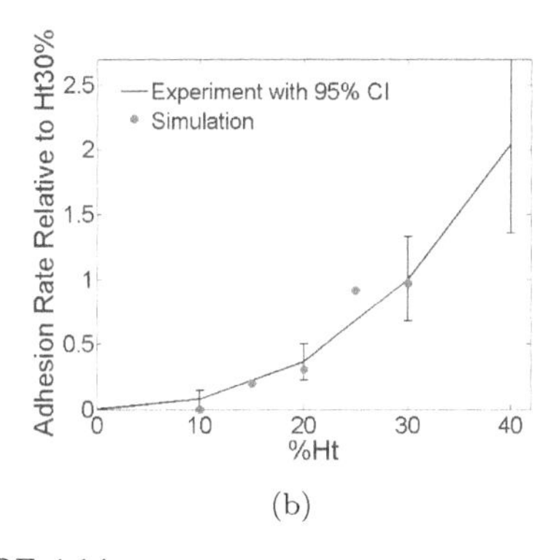

(b)

FIGURE 4.14
The effect of hematocrit on platelet adhesion using (a) VWF-coated surface [6] (b) collagen-coated surface [41].

Although we have already illustrated the role of hematocrit in terms of red blood cell migration, which in turn influences platelet margination and affects the near-wall concentration of platelets, the hematocrit has yet to be directly related to the final process of platelet adhesion on a theoretical basis. Most modeling work involves simulations, as the above two experimental papers [6, 41] include some simulation results for comparison. Fitzgibbon and co-workers categorized platelets into four states and performed a lower-order modeling of the transition between states [13]. The four states consist of platelets flowing in the bulk cell-laden region, platelets flowing inside the cell-free layer, platelets transiently adhered to VWF-coated surface due to GPIb-VWF binding and platelets stably adhered to VWF-coated surface due to GPIIbIIIa-VWF binding. Hematocrit plays a role in affecting the timescale of platelets diffusing from the cell-laden region to the cell-free layer. In addition, hematocrit affects the distribution of platelets inside the cell-free layer where the flow is approximated as simple shear. Fitzgibbon *et al.* also estimated the rate constants for state transitions as shown in Table 4.1 below. This model captures most of the physics of platelet transport across a channel.

An alternative approach of looking at platelet adhesion in a blood suspension is to include the near-wall reactions R in Equation 4.7 for platelet margination.

$$\frac{\partial n_P}{\partial t} + \frac{\partial F_{CP}}{\partial z} + \frac{\partial F_{PP}}{\partial z} + R = 0 \tag{4.12}$$

R corresponds to the net rate of platelets exiting the flow at height z. An integration of R over the entire reactive distance L yields the total change of adhered platelet concentration on the surface $n_{P,s}$:

$$\int_0^L R dz = \frac{\partial n_{P,s}}{\partial t} \tag{4.13}$$

TABLE 4.1
Rate constant estimates [13]. 1: bulk, 2: near wall, 3: transient adhesion, 4: firm adhesion.

Transition	Controlling Physics	Rate Estimate
1→2	Hematocrit, weak Brownian	0.2-0.5 s^{-1}
2→1	Hematocrit, weak Brownian	0 s^{-1}
2→3	Cell-free layer thickness, shear flow, VWF-GPIb kinetics	0.07-0.4 s^{-1}
3→2	VWF-GPIb and GPIIbIIIa kinetics	0.2 s^{-1}
3→4	VWF-GPIb and GPIIbIIIa kinetics	0.2 s^{-1}

The density of adhered platelets can be further divided into transiently adhered platelets $n_{P,s1}$ and stably adhered platelets $n_{P,s2}$.

$$R(z) = \alpha k_{on}(z) A_{\alpha}(z) n_P(z) - \kappa_{off} \frac{\xi(z)}{H} n_{P,s1} \tag{4.14a}$$

$$\frac{\partial n_{P,s1}}{\partial t} = \int_0^L R dz - \beta \tilde{k}_{on}^2 \tilde{A}_{\beta} n_{P,s1} \tag{4.14b}$$

The rate of forming transient adhesion depends on the rate of GPIb-VWF binding k_{on}, the GPIb count per platelet α, the area fraction within the reactive distance A_{α} and the local flowing platelet density n_P. Similarly, the rate of forming stable adhesion depends on the rate of GPIIbIIIa-VWF binding k_{on}^2, the GPIIbIIIa count per platelet β, the area fraction within the reactive distance A_{β} and the density of transiently adhered platelets $n_{P,s1}$. The rate at which transiently adhered platelets re-enter the flow can be modeled as having an average value κ_{off} with a probability density function of ξ at various z positions. This detachment rate requires breaking all GPIb-VWF bonds, each with rate k_{off} and is the result of competition between k_{on} and k_{off}. κ_{off} can be obtained from Monte Carlo simulations or can be estimated analytically under appropriate conditions.

While α and β values have known average values of 25000 and 50000, respectively, for healthy individuals, A_{α} and A_{β} have not yet been reported. These values should be available in current simulation results of near-wall platelet dynamics and are functions of the platelet center-of-mass z. Values of k_{on}, k_{off} and k_{on}^2 as determined from single molecule experiments, can be used to calculate various transition rates of platelet adhesion [19]. The reverse can also be done to compare bond-level kinetics with literature. While Fitzgibbon *et al.* used a mean field theory to represent platelet-cell interactions [13], the term F_{CP} in Equation 4.12 provides more complete information about their interactions and accounts for changes in platelet concentration in flow due to near-wall adhesion reactions.

4.6 RED BLOOD CELLS AND PLATELETS IN COMPLEX GEOMETRIES

While a straight channel serves as a good starting point for modeling blood flow in the microcirculation, the actual blood vessels have a complex geometry and the relevant research is in its nascent stage. As pointed out by Fitzgibbon *et al.*, the lengthscale of margination is longer than the lengthscale of branching [14], which supports the need to investigate such complex geometries. In this section, we consider two complex geometries as shown in Figure 4.15 which are often used in experiments and simulations: channels with constrictions and channels with branched geometries.

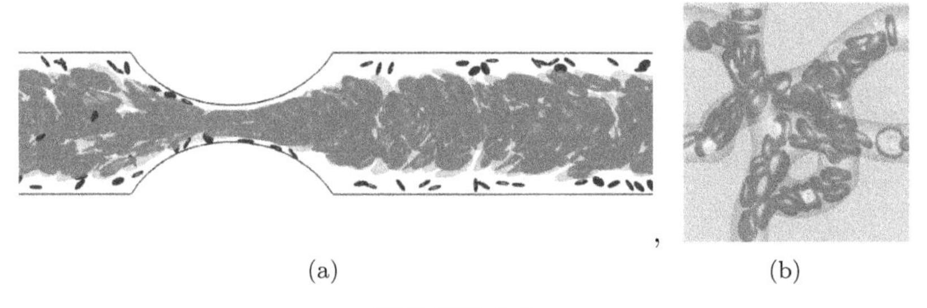

(a) (b)

FIGURE 4.15
Simulating blood in complex geometries (a) a constricted channel [49] (b) branched vessels [43].

Microfluidic channels with constrictions are often used as models for studying blood flow in the stenotic region. Caused by the formation of a clot or a thrombus, these segments of narrow pathways usually have elevated wall shear rates up to $10000s^{-1}$. Favire and coworkers conducted microfluidic experiments of blood at low hematocrit through constrictions and compared the concentration distribution of red blood cells upstream and downstream of the constriction [12]. Their major finding is an increase of cell-free layer thickness downstream, which eventually relaxes to the upstream value over a long distance. Therefore, the constriction enhances red blood cell migration. However, the margination of platelets was not considered.

As shown in Figure 4.15(a), Yazdani *et al.* simulated stenosed 30 μm channels for a range of hematocrit, shear rate and degree of constriction. They confirmed the previous experimental findings for an enhancement of red blood cell migration downstream. Furthermore, they observed that the focusing of platelets in the constriction is hindered by the focused red blood cells. As a result, platelet margination is also enhanced and increases with hematocrit. Bacher and co-workers simulated a suspension of red blood cells and rigid spheres. They found an increase of sphere concentration close to the entrance of the constriction [1], and explained it on the basis that concentrated red blood cell suspension in the constriction creates a barrier for spheres to enter. Although this may be relevant for micron-sized drug delivery carriers, it may not apply to platelets as their aspherical shape is beneficial for traveling in the reduced cell-free layer. Moreover, the sphere concentration used in the simulations is higher than the normal platelet concentration(150000~400000/μl).

Vahidikhah and coworkers [43] considered the case of asymmetric constrictions, which is a more realistic model of actual plugs. As expected, the asymmetric geometry leads to asymmetric distributions of red blood cells and cell-free layer thickness. Another important finding is the significant increase in the root-mean-squared velocity of red blood cells in the constriction as they now distribute in an almost single-file fashion. When the constriction is only 10~20% of the original vessel size and is comparable to the cell size, cell migration and platelet margination should be related to the high confinement regime discussed in Section 4.3.6. The transition and the associated timescales will also depend on the slope of channel height variation and require further investigation.

As mentioned earlier, it is necessary to consider blood flow in a branched geometry instead of a straight channel in in-vivo situations. It is well known from experimental results that when blood flows through bifurcations, the Zweifach-Fung(Z-F) effect occurs [30]: the daughter branch with a lower flow rate receives a smaller fraction of red blood cells. The Z-F effect is closely related to the formation of the cell-free layer, because the red blood cell trajectories do not sample the whole range of flow streamlines. Platelets follow the flow streamlines upon branching. Since they are concentrated inside the cell-free layer, we expect the partitioning to be an inverse Z-F effect. In addition to the proportionality of platelets flowing to two daughter branches, the allocation of red blood cells will result in changes in the cell-free layer thickness and thus influence downstream platelet margination.

So far most studies involving blood flow in branches are experimental [48]. Very few simulations, especially three-dimensional simulations are available. An important issue when comparing this research is the various geometries used to model the branching. Thus, the simulations may be hard to verify if the geometry is not carefully chosen. Among the qualitative observations, Podgorski [38] reported an inversion of Z-F effect at low hematocrit. They attributed it to the microstructure of such dilute suspensions where cell-cell interactions are weak and red blood cells exhibit a five-layer structure in their distribution. It can be related to the low hematocrit effect we previously discussed for straight channels.

The Z-F effect is also determined by the viscosity ratio and the confinement ratio; both can again be traced back to our discussion of these effects in straight channels in Section 4.3.6. Balough and Bagchi published a pioneering work on using the immersed-boundary method to perform a three-dimensional simulation of blood flowing through an arbitrarily complex geometry [2] as shown in Figure 4.15(b). This methodology may lead to more studies regarding branching in the near future.

4.7 SUMMARY AND OUTLOOK

The microrheology of blood is governed by red blood cells, the most abundant species in a blood suspension. Due to their highly deformable shapes, red blood cells migrate away from the wall, leaving a cell-free layer in which platelets are concentrated. The concentration profile of red blood cells shows non-monotonic features due to the competition between hydrodynamic lift and shear-induced diffusion. Both processes are nonlinear in shear rate and can be separately studied using small-scale simulations. So far there has not been a complete theory for blood in the microcirculation that does not require any simulation outputs but limiting the number of outputs is a worthy research effort.

The cell-free layer thickness depends on the hematocrit and has a direct influence on platelet margination. The rate of platelet adhesion not only depends on the binding kinetics of individual receptor-ligand pairs, but also on the near-wall concentration of platelets. Platelets experience wall-hindered tumbling motion under shear, which may be related to the ongoing research of modified Jeffery orbits of ellipsoids near the wall. With the development of near-wall theories, the fractional reactive area of a platelet should be better characterized in the future. The flow shear not only regulates the dynamics of flowing platelets, but also mediates binding kinetics and the size of the Von Willebrand Factor. The change of VWF configuration due to shear can be related to the theory of wall-tethered polymers [42]. Thus, more simulations and experiments are needed to accurately determine the reactive lengthscale of VWF at proper wall shear rates.

Finally, the multi-scale model we propose in this chapter can provide solutions when all parameters are specified. Thus, the problem of platelet adhesion in a straight channel can be determined at a relatively low computational cost. Simulation efforts may be most effectively directed at the study of complex geometries which are more appropriate models for the actual vascular network.

Bibliography

[1] Christian Bächer, Lukas Schrack, and Stephan Gekle. Clustering of microscopic particles in constricted blood flow. *Physical Review Fluids*, 2(1):013102, 2017.

[2] Peter Balogh and Prosenjit Bagchi. A computational approach to modeling cellular-scale blood flow in complex geometry. *Journal of Computational Physics*, 334:280–307, 2017.

[3] Sujata K Bhatia, Michael R King, and Daniel A Hammer. The state diagram for cell adhesion mediated by two receptors. *Biophysical Journal*, 84(4):2671–2690, 2003.

[4] Natacha Callens, Christophe Minetti, G Coupier, M-A Mader, Frank Dubois, Chaouqi Misbah, and Thomas Podgorski. Hydrodynamic lift of vesicles under shear flow in microgravity. *EPL (Europhysics Letters)*, 83(2):24002, 2008.

[5] Erik J Carboni, Brice H Bognet, Grant M Bouchillon, Andrea L Kadilak, Leslie M Shor, Michael D Ward, and Anson WK Ma. Direct tracking of particles and quantification of margination in blood flow. *Biophysical Journal*, 111(7):1487–1495, 2016.

[6] Hsieh Chen, Jennifer I Angerer, Marina Napoleone, Armin J Reininger, Stefan W Schneider, Achim Wixforth, Matthias F Schneider, and Alfredo Alexander-Katz. Hematocrit and flow rate regulate the adhesion of platelets to von willebrand factor. *Biomicrofluidics*, 7(6):064113, 2013.

[7] Gwennou Coupier, Badr Kaoui, Thomas Podgorski, and Chaouqi Misbah. Noninertial lateral migration of vesicles in bounded poiseuille flow. *Physics of Fluids (1994-present)*, 20(11):111702, 2008.

[8] L Crowl and Aaron L Fogelson. Analysis of mechanisms for platelet near-wall excess under arterial blood flow conditions. *Journal of Fluid Mechanics*, 676:348–375, 2011.

[9] Gerrit Danker, Petia M Vlahovska, and Chaouqi Misbah. Vesicles in poiseuille flow. *Physical Review Letters*, 102(14):148102, 2009.

[10] Sai K Doddi and Prosenjit Bagchi. Three-dimensional computational modeling of multiple deformable cells flowing in microvessels. *Physical Review E*, 79(4):046318, 2009.

[11] Robin Fahraeus and Torsten Lindqvist. The viscosity of the blood in narrow capillary tubes. *American Journal of Physiology–Legacy Content*, 96(3):562–568, 1931.

[12] Magalie Faivre, Manouk Abkarian, Kimberly Bickraj, and Howard A Stone. Geometrical focusing of cells in a microfluidic device: an approach to separate blood plasma. *Biorheology*, 43(2):147–159, 2006.

[13] Sean Fitzgibbon, Jonathan Cowman, Antonio J Ricco, Dermot Kenny, and Eric SG Shaqfeh. Examining platelet adhesion via stokes flow simulations and microfluidic experiments. *Soft Matter*, 11(2):355–367, 2015.

[14] Sean Fitzgibbon, Andrew P Spann, Qin M Qi, and Eric SG Shaqfeh. In vitro measurement of particle margination in the microchannel flow: Effect of varying hematocrit. *Biophysical Journal*, 108(10):2601–2608, 2015.

[15] Sean Richard Fitzgibbon. *Platelet Dynamics in Whole Blood and Interpreting the Interfacial Stress Rheometer*. PhD thesis, Stanford University, 2014.

[16] Aaron L Fogelson and Keith B Neeves. Fluid mechanics of blood clot formation. *Annual Review of Fluid Mechanics*, 47:377–403, 2015.

[17] Xavier Grandchamp, Gwennou Coupier, Aparna Srivastav, Christophe Minetti, and Thomas Podgorski. Lift and down-gradient shear-induced diffusion in red blood cell suspensions. *Physical Review Letters*, 110(10):108101, 2013.

[18] Dinar Katanov, Gerhard Gompper, and Dmitry A Fedosov. Microvascular blood flow resistance: Role of red blood cell migration and dispersion. *Microvascular Research*, 99:57–66, 2015.

[19] Jongseong Kim, Nathan E Hudson, and Timothy A Springer. Force-induced on-rate switching and modulation by mutations in gain-of-function von willebrand diseases. *Proceedings of the National Academy of Sciences*, 112(15):4648–4653, 2015.

[20] Amit Kumar, Rafael G Henríquez Rivera, and Michael D Graham. Flow-induced segregation in confined multicomponent suspensions: effects of particle size and rigidity. *Journal of Fluid Mechanics*, 738:423–462, 2014.

[21] Luca Lanotte, Johannes Mauer, Simon Mendez, Dmitry A Fedosov, Jean-Marc Fromental, Viviana Claveria, Franck Nicoud, Gerhard Gompper, and Manouk Abkarian. Red cells' dynamic morphologies govern blood shear thinning under microcirculatory flow conditions. *Proceedings of the National Academy of Sciences*, page 201608074, 2016.

[22] David Leighton and Andreas Acrivos. The shear-induced migration of particles in concentrated suspensions. *Journal of Fluid Mechanics*, 181:415–439, 1987.

[23] Marmar Mehrabadi, David N Ku, and Cyrus K Aidun. A continuum model for platelet transport in flowing blood based on direct numerical simulations of cellular blood flow. *Annals of Biomedical Engineering*, 43(6):1410–1421, 2015.

[24] Ryan M Miller and Jeffrey F Morris. Normal stress-driven migration and axial development in pressure-driven flow of concentrated suspensions. *Journal of Non-Newtonian Fluid Mechanics*, 135(2):149–165, 2006.

[25] Nipa A Mody and Michael R King. Platelet adhesive dynamics. Part i: Characterization of platelet hydrodynamic collisions and wall effects. *Biophysical Journal*, 95(5):2539–2555, 2008.

[26] Kathrin Müller, Dmitry A Fedosov, and Gerhard Gompper. Margination of micro-and nano-particles in blood flow and its effect on drug delivery. *Scientific Reports*, 4, 2014.

[27] Vivek Narsimhan, Hong Zhao, and Eric SG Shaqfeh. Coarse-grained theory to predict the concentration distribution of red blood cells in wall-bounded couette flow at zero reynolds number. *Physics of Fluids (1994-present)*, 25(6):061901, 2013.

[28] Piero Olla. The lift on a tank-treading ellipsoidal cell in a shear flow. *Journal de Physique II*, 7(10):1533–1540, 1997.

[29] Constantine Pozrikidis. *Computational Hydrodynamics of Capsules and Biological Cells*. CRC Press, 2010.

[30] AR Pries, K Ley, M Claassen, and P Gaehtgens. Red cell distribution at microvascular bifurcations. *Microvascular Research*, 38(1):81–101, 1989.

[31] Qin M Qi, Eimear Dunne, Irene Oglesby, Ingmar Schoen, Antonio J Ricco, Dermot Kenny, and Eric SG Shaqfeh. In vitro measurement and modeling of platelet adhesion on vwf-coated surfaces in channel flow. *Biophysical Journal*, 116(6):1136–1151, 2019.

[32] Qin M Qi and Eric SG Shaqfeh. Time-dependent particle migration and margination in the pressure-driven channel flow of blood. *Physical Review Fluids*, 3(3):034302, 2018.

[33] Q.M. Qi and E.S.G. Shaqfeh. Theory to predict particle migration and margination in the pressure-driven channel flow of blood. *Physical Review Fluids*, 2(9):093102, 2017.

[34] Rafael G Henríquez Rivera, Xiao Zhang, and Michael D Graham. Mechanistic theory of margination and flow-induced segregation in confined multicomponent suspensions: Simple shear and poiseuille flows. *Physical Review Fluids*, 1(6):060501, 2016.

[35] Brian Savage, SJ Shattil, and ZM Ruggeri. Modulation of platelet function through adhesion receptors. a dual role for glycoprotein iib-iiia (integrin alpha iib beta 3) mediated by fibrinogen and glycoprotein ib-von willebrand factor. *Journal of Biological Chemistry*, 267(16):11300–11306, 1992.

[36] SW Schneider, S Nuschele, A Wixforth, C Gorzelanny, A Alexander-Katz, RR Netz, and MF Schneider. Shear-induced unfolding triggers adhesion of von willebrand factor fibers. *Proceedings of the National Academy of Sciences*, 104(19):7899–7903, 2007.

[37] Timothy W Secomb. Blood flow in the microcirculation. *Annual Review of Fluid Mechanics*, 49:443–461, 2017.

[38] Zaiyi Shen, Gwennou Coupier, Badr Kaoui, Benoît Polack, Jens Harting, Chaouqi Misbah, and Thomas Podgorski. Inversion of hematocrit partition at microfluidic bifurcations. *Microvascular Research*, 105:40–46, 2016.

[39] R Skalak, A Tozeren, RP Zarda, and S Chien. Strain energy function of red blood cell membranes. *Biophysical Journal*, 13(3):245–264, 1973.

[40] Tyler Skorczewski, Boyce E Griffith, and Aaron L Fogelson. Multi-bond models of platelet adhesion and cohesion. *AMS Contemporary Mathematics Series on Biofluids*, 628:149–72, 2014.

[41] Andrew P Spann, James E Campbell, Sean R Fitzgibbon, Armando Rodriguez, Andrew P Cap, Lorne H Blackbourne, and Eric SG Shaqfeh. The effect of hematocrit on platelet adhesion: experiments and simulations. *Biophysical Journal*, 111(3):577–588, 2016.

[42] Timothy A Springer. Von willebrand factor, jedi knight of the bloodstream. *Blood*, 124(9):1412–1425, 2014.

[43] Koohyar Vahidkhah, Peter Balogh, and Prosenjit Bagchi. Flow of red blood cells in stenosed microvessels. *Scientific Reports*, 6, 2016.

[44] Koohyar Vahidkhah, Scott L Diamond, and Prosenjit Bagchi. Platelet dynamics in three-dimensional simulation of whole blood. *Biophysical Journal*, 106(11):2529–2540, 2014.

[45] C Robert Valeri, George Cassidy, Linda E Pivacek, Gina Ragno, Wilfred Lieberthal, James P Crowley, Shukri F Khuri, and Joseph Loscalzo. Anemia-induced increase in the bleeding time: implications for treatment of nonsurgical blood loss. *Transfusion*, 41(8):977–983, 2001.

[46] Bethany L Walton, Marcus Lehmann, Tyler Skorczewski, Lori A Holle, Joan D Beckman, Jeremy A Cribb, Micah J Mooberry, Adam R Wufsus, Brian C Cooley, Jonathan W Homeister, et al. Elevated hematocrit enhances platelet accumulation following vascular injury. *Blood*, 129(18):2537–2546, 2017.

[47] R Wells and H Schmid-Schönbein. Red cell deformation and fluidity of concentrated cell suspensions. *Journal of Applied Physiology*, 27(2):213–217, 1969.

[48] Sung Yang, Akif Ündar, and Jeffrey D Zahn. A microfluidic device for continuous, real time blood plasma separation. *Lab on a Chip*, 6(7):871–880, 2006.

[49] Alireza Yazdani and George Em Karniadakis. Sub-cellular modeling of platelet transport in blood flow through microchannels with constriction. *Soft Matter*, 12(19):4339–4351, 2016.

[50] Peng Zhang, Li Zhang, Marvin J Slepian, Yuefan Deng, and Danny Bluestein. A multiscale biomechanical model of platelets: Correlating with in-vitro results. *Journal of Biomechanics*, 50:26–33, 2017.

[51] Hong Zhao, Eric S. G. Shaqfeh, and Vivek Narsimhan. Shear-induced particle migration and margination in a cellular suspension. *Physics of Fluids*, 24:011902, 2012.

[52] M Zurita-Gotor, J Bławzdziewicz, and E Wajnryb. Layering instability in a confined suspension flow. *Physical Review Letters*, 108(6):068301, 2012.

CHAPTER 5

Single red blood cell dynamics in shear flow and its role in hemorheology

Simon Mendez
IMAG, CNRS, Univ Montpellier, Montpellier, France

Manouk Abkarian
CBS, CNRS, INSERM, Univ Montpellier, Montpellier, France

CONTENTS

5.1 INTRODUCTION

The red blood cells (RBCs) are some of the most abundant cells in the body and all together they form a fluid organ responsible for the transport of oxygen to tissues and the removal of carbon dioxide from the tissues to the lungs. Unlike all other cells in the body (except the platelets), the RBC does not have a nucleus and can be seen as made of a deformable envelope containing a Newtonian solution of hemoglobin proteins, the moieties in charge of carrying the oxygen in the cells. The excess surface area of their membrane relative to the enclosed volume allows the RBCs to sneak into capillaries of a few microns in diameter within the deepest tissues, in order to fulfill their carrier function.

The membrane of an RBC has a complex composite structure that provides it with a remarkable resilience, knowing that an RBC spends an average of 120 days in the circulation and travels during its lifetime more than 250 km of the complex microcirculatory network of vessels more than 40,000 km long put end to end. The membrane consists of a lipid bilayer, which can be considered as an incompressible two-dimensional fluid, under which is anchored a two-dimensional cross-linked network of long spectrin filaments. This lamellar structure allows RBCs to develop large deformations while keeping a constant surface area. The complexity of the mechanical behavior of the membrane associated with the interaction with the flow greatly influences the transport of blood in physiological, pathological or artificial situations. Understanding such a behavior requires a multi-scale approach capable of integrating a complete vision of the behavior of a single RBC and transposing it to the case of a concentrated suspension. In such problems, the object affects the flow which in turn exerts a force on the object and deforms it.

In this chapter, we will discuss the dynamics associated with the single RBC in shear flow and its role in hemorheology. RBCs' apparent structural simplicity contrasts with the complexity of predicting their behavior in any flow conditions. Understanding the RBC flow dynamics comes down to two main issues: how its shape adapts to the viscous stress by local bending and in-plane deformations of its membrane and how the internal flow produced impacts cell behavior. Despite decades of research, the simplest configuration of pure shear flow is not fully understood, as demonstrated by the recent discovery of new dynamics and shapes in physiological conditions of outer fluid viscosity and high shear rates [90]. This chapter starts by providing the reader with basic elements about RBCs' membrane mechanics before discussing their rich dynamics and their role in hemorheology.

5.2 THE HUMAN RED BLOOD CELL AND ITS MECHANICAL MODELING

5.2.1 The structure and the geometry of the red blood cell

A healthy RBC can be considered for mechanical purposes as a deformable micron-scale bag of the shape of a biconcave disk at rest, with an average diameter of about 7.5 μm, containing a Newtonian solution of hemoglobin of viscosity η_i, typically nine times the viscosity of water at 37°C and five times higher than the outer viscosity η_o of the suspending plasma in physiological conditions. This solution is enclosed by a thin composite membrane made of an incompressible fluid lipid bilayer, scaffolded inside by two-dimensional highly folded elastic filaments of spectrin proteins anchored by two protein complexes, namely the Ankyrin and the 4.1 complexes, and forming a hexagonal-like lattice as represented in Figure 5.1.

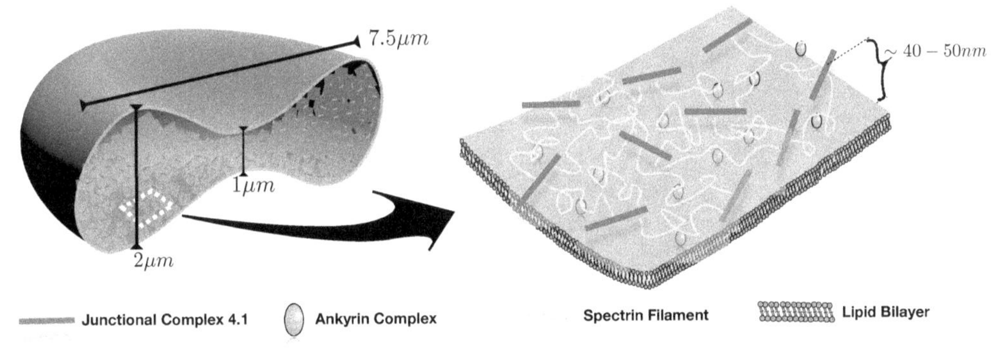

FIGURE 5.1
A schematic representation of the red blood cell membrane structure.

Human RBCs have an average surface area S of 135 μm^2 encapsulating a mean cellular volume V of 94 μm^3 [47, 67]. This volume would be coverable by only 100 μm^2 showing that the cells are actually in a deflated state. The corresponding reduced volume, which represents the ratio of the volume of the cell V normalized by the volume of the sphere with the measured surface area S, can be calculated to be around 0.6, allowing the cell to undergo extensive deformations. However, during its lifetime, the cell loses about 30% of its volume and 20% of its hemoglobin content and surface area, whereas the hemoglobin concentration increases by 14%, indicating that relatively more water than hemoglobin is lost [103, 163], through the shedding of small membrane vesicles of 100–200 nm in diameter devoid of cytoskeleton. These sizes are similar to the 70 nm mesh-size of the underlying spectrin network [163]. The oldest cells have therefore a smaller average surface area of approximately 117 μm^2 enclosing a smaller volume of 78 μm^3 compared to the average area of 148 μm^2 of the youngest cells enclosing 98 μm^3. The reduced volume of RBCs thus increases in time, from 0.58 to 0.64. Moreover, this relative dehydration results in a marked increase of hemoglobin concentration and therefore in cytoplasm viscosity which reduces cells' deformability. As a consequence, old cells have more difficulty to fit in the smallest capillaries of the body, which is an important determinant of their removal from the circulation by the spleen [98].

5.2.2 Viscoelastic properties of the RBC membrane

5.2.2.1 Modes of deformation of the membrane

Three independent modes of deformation only are necessary to describe any static elastic state of the membrane since it is a thin structure (see Figure 5.2): isotropic stretching or compression, simple shear at constant surface area, and out of plane bending. Only three elastic moduli are therefore necessary to describe any instantaneous state of stress in the membrane: the stretching modulus K_α, the shear modulus G_s and the bending modulus B, respectively. However, unlike a uniform elastic thin sheet, the various elastic moduli are carried by different parts of the composite membrane of the cell. It is the lipid bilayer that mainly resists both bending and stretching deformations, while the underlying spectrin cytoskeleton bears the resistance to shear deformation (Figure 5.2).

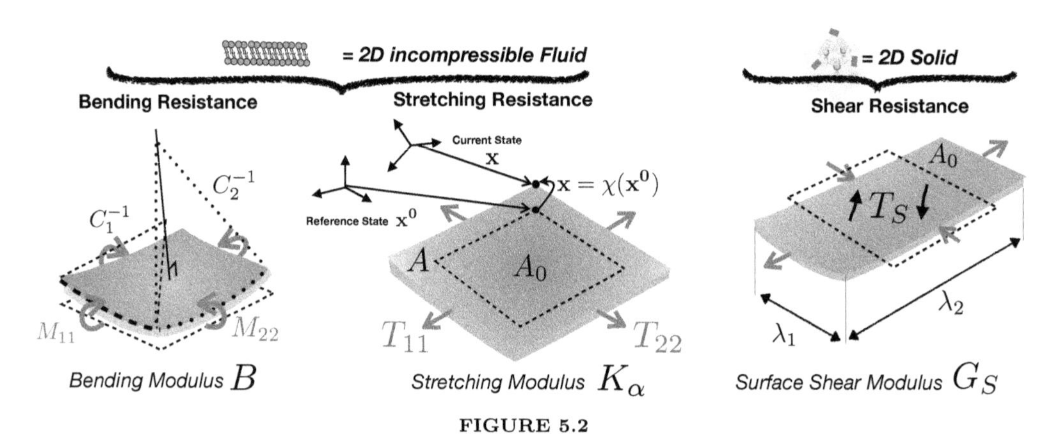

FIGURE 5.2
A schematic representation of the three modes of deformations carried by the lipid bilayer (left) and the cytoskeleton (right).

5.2.2.2 Simple definitions to quantify deformations in the membrane

For any deformation, one can find three mutually orthogonal directions along which the membrane undergoes a pure stretch. These directions are called principal axes of deformation and the associated stretches are called principal stretches. They are respectively eigenvectors and eigenvalues of the Lagrangian strain tensor e_{ij} (see Chapter 1). In these principal axes of deformation for which $\frac{\partial x_i}{\partial x_i^0} \equiv \lambda_i$, we have $e_{ii} = \frac{1}{2}(\lambda_i^2 - 1)$, where the λ_i physically describe the fractional elongations or contractions along the two local principal axes of strain of the infinitesimal neighborhood of a point over the membrane (see Figure 5.2).

In general, in-plane deformations are completely described by two strain invariants which are proper combinations of λ_i, denoted by α and β, and defined by:

$$\begin{aligned} \alpha &= \lambda_1\lambda_2 - 1 \\ \beta &= \frac{1}{2}\left[\left(\frac{\lambda_1}{\lambda_2} - 1\right) + \left(\frac{\lambda_2}{\lambda_1} - 1\right)\right] = \frac{\lambda_1^2 + \lambda_2^2}{2\lambda_1\lambda_2} - 1. \end{aligned} \tag{5.1}$$

One can immediately see that for isotropic stretching where $\lambda_1 = \lambda_2 = \hat{\lambda}$,

$$\alpha = \hat{\lambda}^2 - 1 \equiv \frac{A - A_0}{A_0} = \frac{\Delta A}{A_0} \quad \text{and} \quad \beta = 0.$$

α represents the relative variation of surface area without shear. For pure shear with a constant area, $\lambda_1 = \hat{\lambda} = 1/\lambda_2$,

$$\alpha = 0 \quad \text{and} \quad \beta = \frac{1}{2}(\hat{\lambda}^2 + \frac{1}{\hat{\lambda}^2}) - 1. \tag{5.2}$$

5.2.2.3 Mechanical constitutive law: in-plane tensions

These deformations and their rates of variation are induced by traction forces T_{ij} acting along the edge of a deformed membrane element and tangentially to it, as well as bending moments M_{ij} per unit length, which act around the element edges as illustrated in Figure 5.2 and Figure 5.3A. This traction tensor is a two-dimensional analogue of the stress tensor describing the stress acting on any surface element in the bulk of three-dimensional materials. Noteworthy, T_{ij} are forces per unit length and are also called in the literature tension forces and are a tensorial generalization of the concept of surface tension for liquid-liquid interfaces.

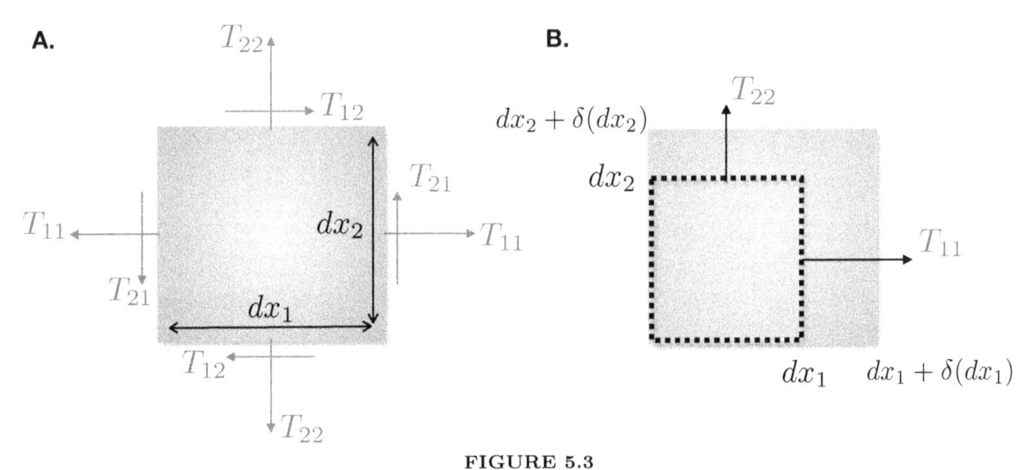

FIGURE 5.3
A. Schematic of the traction forces acting on a membrane element. B. Deformation of a small membrane element in the principal axes.

Following Evans and Skalak, one can use the laws of thermodynamics to provide the relationship between the internal elastic energy U stored in the material of the membrane to its total free energy F and its heat content Q [48]. In fact, $dF = dU - TdS$, where $dQ = TdS$, T is the temperature of the membrane supposed constant and S the entropy of the system.

In the frame of reference of the principal directions of deformation (which we will note by the subscript 1 and 2), if one considers an elastic membrane locally isotropic and homogeneous, then one can write the infinitesimal work dW associated to the adiabatic deformation of a membrane element of size $dx_1 \times dx_2$ by the traction forces T_{11} and T_{22}, as represented in Figure 5.3B:

$$dF = dW = \underbrace{(T_{11}dx_2)}_{\text{force}} \underbrace{\delta(dx_1)}_{\text{displacement}} + (T_{22}dx_1)\delta(dx_2). \tag{5.3}$$

Since, by definition, $\lambda_i \equiv dx_i/dx_i^0$, then $\delta(dx_i) = dx_i^0\delta(\lambda_i)$. Therefore, the infinitesimal free energy density $d\tilde{F} = dF/dx_1^0 dx_2^0$ normalized by the undeformed surface area $dx_1^0 dx_2^0$ associated to this local deformation can be written as:

$$d\tilde{F} = (T_{11}\lambda_2\delta(\lambda_1)) + (T_{22}\lambda_1\delta(\lambda_2)). \tag{5.4}$$

This relation allows the definition of the T_{ij} as a function of the derivatives of $\tilde{F}$:

$$T_{11} = \frac{1}{\lambda_2}\frac{\partial \tilde{F}}{\partial \lambda_1}, T_{22} = \frac{1}{\lambda_1}\frac{\partial \tilde{F}}{\partial \lambda_2}. \tag{5.5}$$

In fact, $\tilde{F}$ is also called the strain energy functional in the literature. The T_{ij} can then be expressed more generally, by a change in variable as a function of the derivative of $\tilde{F}(\alpha, \beta)$ with respect to the two invariants α and β we described in the former section:

$$T_{11} = \left.\frac{\partial \tilde{F}}{\partial \alpha}\right|_\beta + \frac{\lambda_1^2 - \lambda_2^2}{2\lambda_1^2\lambda_2^2}\left.\frac{\partial \tilde{F}}{\partial \beta}\right|_\alpha \tag{5.6}$$

$$T_{22} = \left.\frac{\partial \tilde{F}}{\partial \alpha}\right|_\beta + \frac{\lambda_2^2 - \lambda_1^2}{2\lambda_1^2\lambda_2^2}\left.\frac{\partial \tilde{F}}{\partial \beta}\right|_\alpha \tag{5.7}$$

The strain energy functional $\tilde{F}$ of an isotropic membrane should be a symmetric function of λ_1 and λ_2. A simple expression for $\tilde{F}$ is given by a first order development of $\tilde{F}(\alpha, \beta)$:

$$\tilde{F} = \frac{1}{2}K_\alpha\alpha^2 + G_s\beta, \tag{5.8}$$

that has been used widely in the literature, especially in simulations [96]. It represents the sum of the stretching energy and the shear energy. The associated elastic stresses can then be separated into an isotropic term describing a change in surface area without a local change of the shape and a deviatoric part which deals with the tensions produced by a change of shape for a fixed surface area. The expression of the tension tensor elements becomes:

$$\begin{aligned} T_{11}^e &= (\bar{T}_0 + K_\alpha\alpha) + \frac{1}{2}G_s\left(\frac{1}{\lambda_2^2} - \frac{1}{\lambda_1^2}\right) \\ T_{22}^e &= (\bar{T}_0 + K_\alpha\alpha) + \frac{1}{2}G_s\left(\frac{1}{\lambda_1^2} - \frac{1}{\lambda_2^2}\right), \end{aligned} \tag{5.9}$$

and $\bar{T}_0$ is the residual initial isotropic average tension (also called the pre-stress). In this case, the formal definition in a large deformation framework of K_α and G_s becomes [72,73]:

$$K_\alpha = \left.\frac{\partial^2 \tilde{F}}{\partial^2 \alpha}\right|_\beta \quad \text{and } G_s = \left.\frac{\partial \tilde{F}}{\partial \beta}\right|_\alpha . \tag{5.10}$$

Equivalently, one defines the maximum shear stress T_S, with the associated maximum shear strain e_S, as well as the isotropic tension $\bar{T}$, as:

$$T_S \equiv \frac{|T_{11} - T_{22}|}{2} = \frac{|\lambda_1^2 - \lambda_2^2|}{2\lambda_1^2\lambda_2^2} \left.\frac{\partial \tilde{F}}{\partial \beta}\right|_\alpha , \bar{T} \equiv \frac{T_{11} + T_{22}}{2} = \left.\frac{\partial \tilde{F}}{\partial \alpha}\right|_\beta , \tag{5.11}$$

then, formally for any $\tilde{F}$, we have:

$$G_s = \frac{T_S}{\lambda_S}, \qquad \text{with } \lambda_S = \frac{|\lambda_1^2 - \lambda_2^2|}{2\lambda_1^2\lambda_2^2}, \qquad \text{and } K_\alpha = \left.\frac{\partial \bar{T}}{\partial \alpha}\right|_\beta \tag{5.12}$$

For small strains, these definitions reduce to the one of Evans and Skalak [48]:

$$G_s = \frac{T_S}{2e_S} \qquad \text{where } e_S = \frac{|\lambda_1^2 - \lambda_2^2|}{4}, \qquad \text{and } K_\alpha = \frac{\bar{T} - \bar{T}_0}{\alpha} \tag{5.13}$$

Measurements under shear flow either in rheoscopes or ektacytometry systems have shown that RBCs' deformation increases logarithmically with the shear stress in both moderate and large deformation regimes [74, 109] suggesting that the cell membrane has a strain-hardening nature not correctly accounted for in the first approximation of the strain energy function given in Eq. (5.8). In fact, the linear approach is even strain-softening [34]. However, as we will see, a linear approach allows analytical modeling of RBCs' dynamics and helps in deciphering the physical mechanisms at play (see Sec. 5.4).

One successful constitutive law used in simulations to describe both small and moderate deformation regimes for RBCs' membrane has been proposed by Skalak and coworkers [145]. In their work, the authors take into consideration the elastic response in shear deformation, with an elastic modulus G_s and the intrinsic resistance to area dilation, with a corresponding modulus K_α, without imposing local incompressibility, which would imply $\lambda_1\lambda_2 \equiv 1$. Rather, the authors proposed a strain energy functional which adds non linearly these two effects in the following manner:

$$\tilde{F} = \frac{G_s}{4}\left(\left(\lambda_1^2 + \lambda_2^2 - 2\right)^2 + 2\left(\lambda_1^2 + \lambda_2^2 - \lambda_1^2\lambda_2^2 - 1\right) + C\left(\lambda_1^2\lambda_2^2 - 1\right)^2\right), \tag{5.14}$$

where $C \equiv K_\alpha/G_s \gg 1$, indicating a strong area-dilation stiffness in comparison to shear resistance. The principal tensions that can be deduced from this functional are given by:

$$\begin{aligned} T_{11} &= \frac{G_s}{\lambda_1\lambda_2}\left(\lambda_1^2(\lambda_1^2 - 1) + C(\lambda_1\lambda_2)^2[(\lambda_1\lambda_2)^2 - 1]\right) \\ T_{22} &= \frac{G_s}{\lambda_1\lambda_2}\left(\lambda_2^2(\lambda_2^2 - 1) + C(\lambda_1\lambda_2)^2[(\lambda_1\lambda_2)^2 - 1]\right) \end{aligned} \tag{5.15}$$

Noteworthy, this law remains intrinsically strain-hardening even when imposing local incompressibility [34].

Continuous modeling of the RBC membrane is not the only option that has been used in the literature. Another popular choice is to calculate the effect of in-plane resistance using a flexible network of springs of fixed connectivity forming a triangulation of the membrane [53, 55, 126, 135]. The model is defined through the energy associated with the stretching of each spring. The linear mechanical properties of such a model can be calculated [31], relating the parameters of the spring model to the mechanical moduli of the membrane.

5.2.2.4 Stress-free shape of the membrane

Calculations of the in-plane tensions rely on the difference between a current configuration and a reference configuration, for which the tensions in the membrane elements are zero. This reference configuration is thus referred to as unstressed or stress-free. The question of the stress-free shape of the RBC membrane has long been recognized as an essential element to understand the RBC shape and behavior under flow. Early papers use either a sphere or the equilibrium biconcave shape as the stress-free shape, and tend to conclude in favor of the biconcave shape [56, 168]. In 2004, the shape memory of RBCs was demonstrated [58]. Fischer followed the movement of a micron-size bead stuck to the membrane of an RBC in flow. After several revolutions, the bead always comes back to the same relative position on the cell when the flow is stopped. For instance, if initially placed on the rim, the bead is found back on the rim of the cell. This definitely rules out the possibility of the sphere as the RBC stress-free shape, an isotropic shape being unable to yield shape memory. Lim and coworkers [95, 96] showed that a model for the mechanics of the RBC membrane could reproduce complex shape changes (from discocyte RBCs to stomatocytes or echinocytes) obtained when treating the membrane with specific chemical agents. However, the ability of the model to reproduce the various shape of RBCs depends on the stress-free shape of the RBC membrane, and the best stress-free shape seemed to be an ellipsoid of reduced volume $\vartheta_{ref} = 0.95$. Since then, numerical simulations have considered either the biconcave shape or a quasi-spherical ellipsoid as a stress-free shape [29, 104, 123, 124, 159]. Note that the modeling of the RBC determines how the stress-free shape is imposed: if in-plane elasticity and bending resistance are considered separately, it is possible to impose reference (unstressed) configurations that differ for the two elements of the membrane model.

5.2.2.5 Mechanical constitutive law: bending moments

We also need to consider the deformations out of the plane of the membrane which are related to the bending moments M_{ij} acting on the edge of a membrane element. Considering the curvature of the membrane elements C_1 and C_2 in the principal directions of curvature, since the membrane is locally isotropic, two principle moments M_{11} and M_{22} acting along the edges of the bent elements as indicated in Figure 5.2 are simply given by:

$$\begin{aligned} M_{11} &= B(C_1 - C_0) \\ M_{22} &= B(C_2 - C_0), \end{aligned} \tag{5.16}$$

where B is the bending modulus, C_0 is the reference initial curvature, also called the natural or spontaneous curvature, for which the bending moments vanish. One can define the isotropic bending moment:

$$\bar{M} = \frac{M_{11} + M_{22}}{2} = B(H - C_0) \tag{5.17}$$

where H is the mean local curvature of the membrane.

A common way of accounting for bending resistance is to endow the membrane with a curvature energy, from which bending forces are calculated. Assuming that the energy density along the membrane surface only depends on the local shape of the membrane, it is possible, by considering the lowest order terms, to obtain a bending energy formulation that depends on the mean curvature H and Gaussian curvature K:

$$E_b = \frac{B}{2} \int_S (H - C_0)^2 dS + \kappa_g \int_S K dS, \tag{5.18}$$

κ_g being the Gaussian modulus. This form of bending energy is compatible with Euclidean invariance, and membrane isotropy [95]. The Gaussian term of Eq. (5.18) does not contribute to the shape problem in the case of closed surfaces. This term is thus generally omitted in the expression of E_b, which is generally referred to as the Helfrich bending energy [77]. The bending moment associated with this bending energy is indeed that of Eq. (5.17).

The first variation of the Helfrich bending energy yields the bending force density acting on the fluid [173]

$$\mathbf{F}_b = B \left[(2H - C_0)(2H^2 - 2K + C_0 H) + 2\Delta_{LB} H \right] \mathbf{n}. \tag{5.19}$$

Δ_{LB} denotes the surface Laplacian operator (also called the Laplace-Beltrami operator). $\mathbf{n}$ is the outward normal vector. This force can be calculated and accounted for in simulations to model bending resistance. Another technique consists in using the principle of virtual work to calculate a local force density from the expression of the bending energy [53, 172].

5.2.2.6 Simplest approach to membrane viscosity

Evans and Hochmuth [50] proposed the simplest visco-elastic relationship for the membrane, the so-called Kelvin-Voigt model, in which T_{ij} is the sum of the elastic term T^e_{ij} described in Sec. 5.2.2.3 and depending on the instantaneous strain and a viscous term T^v_{ij} depending on the instantaneous rate of strain.

The linear elastic term can be written in the Eulerian representation as[i]:

$$T^e_{ij} = (\bar{T}_0 + K\alpha)\delta_{ij} + 2G_s E_{ij}, \tag{5.20}$$

where δ_{ij} is the identity matrix.

The viscous term is then simply the classical viscous stresses associated to shear and dilatational viscosities:

$$T^v_{ij} = 2\eta_m \left[\mathcal{D}_{ij} - \frac{1}{2} Tr(\mathcal{D})\delta_{ij} \right] + \eta_\alpha Tr\,(\mathcal{D})\,\delta_{ij}. \tag{5.21}$$

We note that the trace of the strain-rate tensor represents the rate of variation of the volume or equivalently here in two dimensions, the rate of variation of the surface area. We will see soon that the membrane can be considered quasi-incompressible for most of the situations encountered in this chapter. In this case, incompressibility is expressed simply by $Tr\,(\mathcal{D}) \equiv 0$, and in most cases η_α is not needed. The viscous tensions become:

$$\begin{aligned} T^v_{11} &= 2\eta_m \frac{1}{\lambda_1}\frac{D\lambda_1}{Dt} \\ T^v_{22} &= 2\eta_m \frac{1}{\lambda_2}\frac{D\lambda_2}{Dt} \end{aligned} \tag{5.22}$$

[i] Noteworthy, to add this tension term to the viscous one we properly express the tension with the Eulerian strain tensor and not the Lagrangian one (see Chapter 1).

where η_m represents the membrane shear viscosity and D/Dt is the material derivative. In the Eulerian representation, this derivative is written [117] (see Chapter 1):

$$\frac{D}{Dt} \equiv \left.\frac{\partial}{\partial t}\right|_{\mathbf{x}^0} \equiv \left.\frac{\partial}{\partial t}\right|_{\mathbf{x}} + \mathbf{v} \cdot \nabla. \tag{5.23}$$

When rapidly released from a micropipette with a diameter smaller than the cells, RBCs return to a discocyte shape in a characteristic time of the order of $t_c = 2\eta_m/G_s \sim 0.1-0.3$ s. Given the values of the shear modulus of the order of 2.5-9 μN/m, η_m appears to be of the order of 0.1-1 μN.s/m [80]. This surface viscosity has units of a bulk viscosity multiplied by a typical length over which the viscous dissipation occurs. This value can be converted into an equivalent bulk viscosity dividing η_m by the thickness of the membrane $\sim$ 40-50 nm. The bulk viscosity is predicted to be of the order from 2.5-6 Pa.s, a few thousand times more viscous than the cytosol ($\sim 10^{-3}$ Pa.s). Noteworthy, both temperature and inner hemoglobin concentration influence η_m, while the cholesterol content does not seem to produce any effect [80]. This high value of viscosity indicates that membrane viscosity could drive many aspects of fast deformations of RBCs in flow in the same manner as inner cytoplasm viscosity.

5.2.2.7 Membrane mechanical properties

A large toolbox of single-cell probing techniques has been introduced in the literature to measure the different material constants of RBCs' membranes [3, 87, 154]. Techniques of deformation can be divided into two categories: local and non-local probing. The first type of approaches imply localized deformations of the membrane, as during the aspiration with micropipettes [19, 21, 46, 49, 50, 79–81, 161], microneedles [30] and AFM indentation [20, 75, 83], optical [32, 78, 91, 92, 94, 109] and magnetic tweezers [131] manipulations with beads bound to the membrane, for example. The second type of approaches are nonlocal and produce a distributed stress field over the entire cell body that depends on the current deforming cell shape. Examples are optical stretchers [71,97,134], dielectric forces [45] as well as flow cells [2, 39, 151, 155–158], ektacytometry[13, 35, 113] or filtration through cylindrical micro-sieves [40]. Finally, recent techniques use thermal fluctuations as a mechanical probing either at one point [15] or over the entire surface of the membrane with quantitative phase imaging [118–120, 128] and reflection interference contrast spectroscopy [125, 149, 169, 174, 175].

We report in Table 5.1 the average and the dispersion of values of the different viscoelastic moduli measured over almost fifty years with this toolbox.

We would like also to attract the readers' attention to several fundamental review papers on membrane mechanics written by the pioneers in the field [46, 79, 80, 114], as well as to two more general reviews on deformability and biomechanics properties of RBCs [87, 154].

5.2.3 The fluids inside and outside the red blood cell

5.2.3.1 The red blood cell cytosol and its viscosity

The cytosol of RBCs is an aqueous suspension of salts (Na^+,Cl^-,K^+,HCO_3^-...) and different proteins of the metabolic machinery maintaining homeostasis of the cells and their membrane integrity during their 120 days lifetime passed in circulation [114]. The most abundant components are the oxygen-carrying proteins called hemoglobins.

The rheology of this hemoglobin-rich solution has been measured at a physiological concentration of 33 g/dl by several groups [23, 24, 162] at the end of the 1960s. Their

TABLE 5.1
Table of RBCs membrane mechanical properties and geometrical parameters. From [3] inspired from [154].

Moduli	Average value and dispersion
Shear modulus G_s (μN/m)	5.5 ± 3.3
Stretching modulus K_α (mN/m)	399 ± 110
Bending modulus B (in k_BT units=4 $\times 10^{-21}$ J)	28.75 ± 22.5
Surface shear viscosity η_m (μN.s/m)	0.515 ± 0.39

Geometric Parameters	Average value and dispersion
Volume V	94 μm^3
Surface S	135 μm^2
Reduced Volume ϑ	0.6
Whole Membrane Thickness [76]	40-50 nm

measurements show that the RBC cytoplasm is a Newtonian fluid of viscosity η_i around 6.5 cP at 37°C which increases for decreasing temperatures, though thorough and systematic studies are still missing. Estimates based on semi-empirical hard particle models describe the dependence of the viscosity over a wide range of weight concentration as shown in Figure 5.4 for 37°C. They have been used to interpret rheoscope studies of RBCs in shear flow leading to possible values of η_i of about 10 cP at 25°C[157]. Noteworthy, the variations of η_i with temperature have not been yet correctly established. The law indicated in the inset of Figure 5.4 suggests a slow temperature variation following the one of the solvent viscosity $\eta_{sol}(T)$. However, rare measurements exist at lower temperature then 37°C. One series of measurements from Artmann and coworkers [5] at the end of the 1990s plotted in the same figure seem to indicate for instance a faster increase with temperature then the one expected for the solvent viscosity as predicted in the semi-empirical hard particles models, in disagreement with estimates obtained in the rheoscope studies also done at 25°C [157]. Inner viscosity could be very sensitive to the salts content and type present in the cytosol. Ions such as calcium could induce strong modifications to these values [85] and further studies are still necessary to conclude on the topic.

Around the physiological values of hemoglobin concentration of 33 g/dL, the cytoplasm viscosity varies significantly for small changes in concentration, as depicted in Figure 5.4. 10% increase for instance can double the viscosity. Such an increase is physiologically relevant, since the mean cell hemoglobin concentration increases due to dehydration during RBCs aging *in vivo*. This leads to the drastic increase of the viscous time constant for shape recovery of these cells and therefore a global deformability reduction. Similarly, in certain pathological conditions such as sickle cell anemia or hemoglobin C diseases, the presence of an abnormal hemoglobin increases RBC inner viscosity with severe clinical consequences on blood flow in the microcirculation.

5.2.3.2 Rheology of the surrounding plasma

RBCs are suspended in what is called the plasma, a fluid composed of a large variety of high molecular weight proteins which impact whole blood rheology. In shear flow conditions, the plasma is a Newtonian fluid, with a viscosity about 1.2 mPa.s at 37°C. In comparison with the viscosity of water at the same temperature which is about 0.7 mPa.s, there is a 40% variation primarily due to the presence of the macromolecules such as fibrinogen [133].

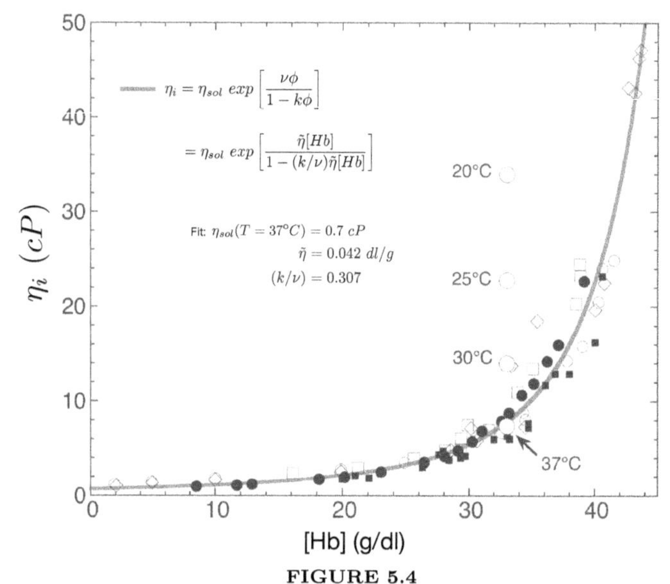

FIGURE 5.4
Human hemoglobin viscosity as a function of its concentration at 37°C. Data from (•) [162], (■) [24], (small open symbols) three different samples [23]. (Large circles) Data for 33 g/dL at different temperatures from [5]. The solid line represents a fit of the modified Mooney law described in [139] with the equation given in the inset. (Inset): $\tilde{\eta}$ represents the intrinsic viscosity of the protein suspension and can be related to the volume fraction ϕ by: $\lim_{\phi\to 0} \frac{\eta_i-\eta_{sol}}{\eta_{sol}} = \nu\phi = [Hb]\tilde{\eta}$, where [Hb] is the weight concentration of the protein, ν is a dimensionless parameter which is Einstein's coefficient of 2.5 for spheres and greater for non spherical proteins, while η_{sol} is the solvent viscosity that depends on temperature.

Noteworthy, large proteins like albumin and especially fibrinogen[ii] impact also the rheology of blood since they are at the origin of the aggregation process of RBCs into a network of so-called rouleaux, as discussed in Chapter 6. This stacking is suspected to come from depletion forces that macromolecules can induce when cells are sufficiently close to each other [11, 148]. These loose large aggregates lead to a very high zero-shear viscosity of blood, of the order of several hundreds of mPa.s and to strong shear-thinning observed when shear rates are increased from a few s^{-1} to tens of s^{-1}, due to disassembling the rouleaux into single flowing RBCs [24].

Recent experiments of filament extension in droplet breaking and flow in microfluidic constrictions [22] have revealed an unusual non-Newtonian behavior of plasma in these extensional flows whose implications remain to be understood in the conditions of the microcirculation.

5.2.4 Fluid-structure interaction for RBC dynamics

Computing the dynamics of an RBC under flow consists of coupling the equations governing the flow inside and outside the RBC (typically the Navier-Stokes equations) to the membrane mechanics. Note that the Reynolds number for flows of RBCs is generally small at the scale of the RBC. In shear flow, for instance, $Re_p = \frac{\rho_o \dot{\gamma} a_0^2}{\eta_o}$, where a_0 is a characteristic length of the RBC, $\dot{\gamma}$ the shear rate and ρ_o and η_o are the density and the dynamic viscosity of the outer fluid. a_0 being of the order of a few microns, inertia is generally negligible, except in extreme and generally non-physiological situations. The Stokes equations may thus generally be used.

[ii]present at concentrations of 3.5g/l [133].

The coupling conditions are both kinematic and dynamic:

$$\mathbf{v}_o(\mathbf{x}) = \mathbf{v}_i(\mathbf{x}) = \mathbf{v}_m(\mathbf{x}) = \frac{\partial \mathbf{x}}{\partial t}, \quad \mathbf{x} \in \text{Membrane} \tag{5.24}$$

where $\mathbf{v}_o$ denotes the velocity vector of the outer fluid, $\mathbf{v}_i$ the velocity vector of the inner fluid and $\mathbf{v}_m$ the membrane velocity.

In addition, the traction jump across the membrane is related to the membrane forces (unlike fluids with only surface tension as described in Chapter 1):

$$(\mathbf{\Sigma}_o(\mathbf{x}) - \mathbf{\Sigma}_i(\mathbf{x}))\mathbf{n} = -\mathbf{F}_m(\mathbf{x}), \quad \mathbf{x} \in \text{Membrane} \tag{5.25}$$

where the left-hand side denotes a surface traction jump between the outer and the inner fluid and $\mathbf{F}_m$ membrane forces. Such a definition is general and the expression of $\mathbf{F}_m$ may be adapted depending on the model and the approach used to treat the problem.

In the framework of thin-shell theory, Pozrikidis shows that

$$\mathbf{F}_m = \nabla_\mathbf{S}.(\mathbf{T} + \mathbf{q}\,\mathbf{n}), \tag{5.26}$$

where $\mathbf{T}$ is the in-plane tension tensor, $\mathbf{q}$ the transverse shear tension, related to the bending moment [129], and $\nabla_\mathbf{S}$ the surface divergence operator.

Another option, very common in numerical simulations, is to derive the membrane forces from the energy equation of the membrane, using the principle of virtual work [129]. In general, membrane stress and bending moment are recast in terms of membrane forces, which are the sum of an elastic contribution generally coming from in-plane elasticity only and a bending contribution often derived from the Helfrich elastic energy.

Many methods have been developed during the past years to solve the fluid-structure interaction problem governing the RBC dynamics. Full numerical simulations consist in predicting both the membrane deformation and the fluid motions along time, generally accounting for large displacements and large deformations of the membrane. In general, the membrane is discretized using a triangular mesh over which the membrane forces are calculated, but more sophisticated methods have also been used, using for example the representation of the membrane with spherical harmonics [65, 170]. The fluid flow may be solved using the Boundary Element Method (restricted to Stokes equations [36, 37, 65, 66, 121, 123, 124, 130, 144, 159, 170, 171]), finite-element [88, 99], finite-difference [7, 28, 29, 165, 166] or finite-volume methods [90, 104, 107, 108, 143], the lattice-Boltzmann method [6, 70, 89, 100, 115, 135, 150, 152] or particle methods such as Multiple-Particle Collision dynamics or Dissipative Particle Dynamics [53, 54, 90, 104–106, 122, 127, 167], for instance. In some methods, the membrane vertices are provided with a fictitious mass and their position advanced using Newton's second law, fed with forces coming from the fluid and the neighboring membrane elements [53, 54, 100, 135, 159]. In others, the membrane is massless and transported at the velocity of the fluid [7, 28, 29, 66, 90, 104, 107, 150, 152, 170, 171]. Whatever the details of the methods, they take as input data the set of fluid and membrane parameters characterizing their rheology, a fluid domain with boundary conditions, a stress-free shape for the membrane and initial conditions for the fluid and membrane. They predict along time the position of the membrane elements and thus its deformation, together with the fluid flow in the computational domain.

Note also that some of the motions of RBCs are experienced by particles with simpler rheology, notably polymeric capsules (whose membrane is a viscoelastic solid with very low bending resistance) or vesicles (whose membrane is constituted by a lipid layer or bilayer, in which shear resistance does not exist). The dynamics of such simpler particles are sometimes studied to mimic RBC movements [4, 8, 9, 43, 52, 112].

Low-order approaches also exist as an alternative to full numerical simulations. Quasi-spherical capsules or vesicles sometimes behave similarly to RBCs, tank-treading [62, 140] under flow, for instance (see Sec. 5.8). In the framework of small perturbations of a membrane immersed in fluids governed by the Stokes equations, the equation of the dynamics of such objects may be solved [10, 18, 111, 141, 160]. Alternatively, some low-order models exist for particles with aspect ratios similar to that of the RBC [1,3,41,86,107,146]. In this family of models, the RBC is modeled as a fixed-shape ellipsoid (out-of-plane membrane deformations are not authorized), but in-plane deformations may occur as the membrane circulates along the ellipsoid surface. This assumption is relevant to moderate stresses, where observations show that RBCs almost preserve their resting shape under flow [1, 42]. In such models, the membrane dynamics are restricted to its position, orientation, and membrane circulation. The RBC dynamics can thus be solved with a small number of degrees of freedom.

In the remainder of this chapter, we will describe specifically the case of RBCs suspended in a pure shear flow. Physiological flows being shear-dominated, this configuration has been the most popular in terms of RBC dynamics of the last decades. In order to discuss the behavior of RBCs, we will rely on existing experiment and full numerical simulations data published in the literature, and also discuss the predictions of low-order models assuming a fixed shape under flow [1, 3, 41, 86, 107, 146] to enlighten the dynamics of an RBC under moderate shear flow.

5.3 THE MOVEMENTS OF AN ISOLATED RED BLOOD CELL IN PURE SHEAR FLOW

In this section, we report how an RBC behaves in pure shear flow, and in particular how the dynamics change as a function of two control parameters: the imposed shear rate and the viscosity of the suspending fluid. Tens of papers have been published on the motions of a single RBC in shear flow. Due to the richness of the problem, they often study a specific motion or a particular region of the phase diagram. The motions have been referred to using a wealth of metaphoric terms, which are now classical and will not be detailed in the core of the chapter. However, to clarify the discussion and possibly help the non-specialist, it seemed important to propose a dictionary of the RBC dynamics, which is presented in the Appendix (Sec. 5.8).

5.3.1 Isolated red blood cells under pure shear flow: governing parameters

In the remainder of this chapter, we focus on the dynamics of isolated RBCs and blood under pure shear flow. The purpose of this section is to review the configuration and the governing parameters of the problem of the isolated RBC in shear flow. We consider the configuration represented in Figure 5.5.

The RBC is subjected to the pure shear flow $\hat{u}_1 = \dot{\gamma}\hat{x}_2$, where $\dot{\gamma}$ is the shear rate and the fixed laboratory frame is denoted by the coordinates ($\hat{x}_1$, $\hat{x}_2$, $\hat{x}_3$). We will restrict our discussion to pure shear flow. Note that in experiments, RBCs are sometimes studied in large pipes or channels, where the curvature of the velocity profile remains moderate enough and the distance to the wall sufficient to be neglected. In an experiment, different external parameters can be controlled: the shear rate $\dot{\gamma}$, the external temperature, the osmolarity of the medium, the external fluid density ρ_o, the external fluid rheology and in particular its viscosity η_o for Newtonian fluids, the degree of confinement, *i.e.* the distance from the RBCs to the walls. The effect of the shear rate and of the viscosity of the external fluid

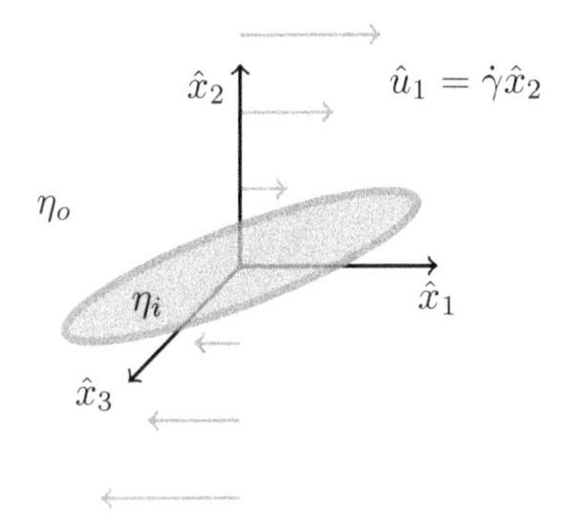

FIGURE 5.5
Isolated red blood cell in shear flow. Schematic of the configuration.

have been extensively studied, and we will concentrate on these two parameters. We are not aware of detailed studies of isolated RBC dynamics as a function of the temperature, but all the rheological parameters depend on it, as discussed in Sec. 5.2.3.1. Finally, note that the effect of confinement has been studied mainly for vesicles [84, 153] and has been shown to favor tank-treading.

The dynamics of the cell depend both on its geometry and its mechanics. The geometry is characterized by the volume V of the cytosol and by the surface area of the membrane, S. A typical length scale a_0 may be built as the radius of the sphere with the same volume V as the cell ($a_0 \approx 2.82$ μm). On the other hand, the cytosol is characterized by the internal fluid density ρ_i and viscosity η_i and the membrane by the viscosity η_m, the shear modulus G_s and area modulus K_α (the area modulus being much larger than the shear modulus), the bending modulus B and the stress-free shape of the cytoskeleton (see Sec. 5.2).

We now introduce classical non-dimensional numbers that are useful to understand the dynamics of RBCs in shear flow.

$\vartheta = 3\sqrt{4\pi}V/S^{3/2}$: the reduced volume ϑ compares the volume of the cell to the volume of the sphere having the same surface area as the cell, S. The typical value for an RBC is $\vartheta = 0.6$. The osmolarity of the medium may change this reduced volume [136], as well as cell aging (see Sec. 5.2.1).

ρ_i/ρ_o: the internal density of RBCs is slightly higher than that of water or plasma (typically, $\rho_i/\rho_o \approx 1.1$). RBCs thus sediment, which has led experimentalists to use dextran solutions as external fluids to limit sedimentation during the experiments. Recently, Optiprep (iodixanol solution, Axis-Shield), has also been used to match the density of RBCs [110, 137]. Apart from sedimentation itself, no effect is expected on the dynamics of the isolated RBC.

$\lambda = \eta_i/\eta_o$: the internal-to-external viscosity ratio λ is one of the most important non-dimensional parameters controlling the dynamics of RBCs in shear flow. As already discussed (see Sec. 5.2.3), the typical value for the viscosity ratio at 37°C is $\lambda = 5.0$. An effective viscosity ratio is sometimes introduced to account for membrane viscosity [1].

$Ca = \dfrac{\eta_o\dot{\gamma}a_0}{G_s}$: Ca is the capillary number and compares the shear stress $\eta_o\dot{\gamma}$ applied by external flow with the shear elastic resistance of the cell. It measures the ability of the shear flow to deform the whole cell. Note that the competition between the shear stress and the mechanics of the cell is sometimes presented in terms of bending resistance: a similar non-dimensional number based on the bending modulus may also be built as $\eta_o\dot{\gamma}a_0^3/B$. Note that in experiments, Ca cannot be used as the shear modulus of the tested RBCs is generally unknown: experimental results are generally presented with respect to the shear stress $\eta_o\dot{\gamma}$ or the shear rate $\dot{\gamma}$ in the literature.

$Re_p = \dfrac{\rho_o\dot{\gamma}a_0^2}{\eta_o}$: Re_p is the particle Reynolds number, comparing inertial to viscous effects at the scale of the particle. The size of the RBC makes Re_p small in physiological flows, so that inertial effects may be neglected in general. Inertial effects should be accounted for when the shear rate is higher than $\dot{\gamma} \approx 10^4$ s^{-1}. Such shear rates may be encountered in some cardiovascular devices [33].

$\mathcal{F} = G_s a_0^2/B$: $\mathcal{F}$ is the Föppl-von Karman number. It is a non-dimensional number built from the mechanical properties of the membrane. It quantifies the relative importance in the free energy of the non-flat membrane shear elasticity relative to its bending resistance. Typically, for thin elastic shells, since G_s and B are related to the material two-dimensional Young modulus, this number is a purely geometric ratio: $\mathcal{F} \sim (R/e)^2$, where e is the shell thickness and R the typical size of the shell or over which deformation is happening (it could be the radius of the shell for spherical shells). For the RBC membrane however, shear and bending are carried by two different material components (Sec. 5.2.2.1) and $\mathcal{F}$ does not reduce to a simple geometric ratio. In fact, one can define a new length in the problem $\Lambda_e = \sqrt{B/G_s}$ and rewrite $\mathcal{F} = (a_0/\Lambda_e)^2$. For RBCs membrane, $\Lambda_e \sim 250\ nm \ll a_0$ and therefore $\mathcal{F} \gg 1$. Shear elasticity is thus essential at the scale of an RBC, and bending dominates only for very small scales, when the membrane strongly folds in flow, as we will see later in the chapter or for strong shape modifications induced by physico-chemical changes in the surrounding of the cell leading to a discocyte-to-echinocyte shape transition, where spicules appear over the entire cell [116].

ϑ_{ref}: the stress-free shape cannot be directly measured from an RBC. On the contrary, in numerical simulations, it has to be specified. It is generally considered to be either the biconcave equilibrium shape or, more recently, an oblate ellipsoid with the same surface area as the RBC (see Sec. 5.2.2.4). In this case, the stress-free reference shape is defined by its reduced volume ϑ_{ref}.

In general, experiments have tried to characterize the dynamics of RBCs varying the shear rate and the viscosity of the external medium [1,17,42,56,57,59–62,90,110,121,122,124,164]. Simulations have often considered fixed properties of RBCs (the effect of the Föppl-von Karman number being rarely investigated, for instance) and predicted the dynamics as a function of λ and Ca, to mimic experiments [36,37,44,90,104,107,144,150,165,166]. The dependence to the reduced volume of the reference shape has also been studied [29,123,159], as the stress-free shape of an RBC is not known. Some rare studies also considered the effect of RBC swelling or inertia [38,102].

5.3.2 The emblematic dynamics: flipping and tank-treading

In the 1960s and 1970s, two emblematic motions of RBCs were studied in detail: tumbling and tank-treading. Tumbling, or flipping, is the motion of the RBC that resembles that of a solid disk. It has been studied mainly at low shear rates and high values of the viscosity ratio (physiological conditions) [69]. On the contrary, tank-treading [62,140], in which the

RBC membrane rotates around the cytosol with a constant inclination angle (which is a motion similar to that of droplets), was obtained when subjecting an RBC to high shear rates in viscous external media (low viscosity ratio). The RBC was then thought to present a dual "solid and liquid behavior" [79] depending on the conditions. These motions and the transition between them received considerable attention, in particular because a mimetic system of the RBC, the vesicle, presented the same behaviors and a transition between them with the viscosity ratio [4, 12, 16, 86].

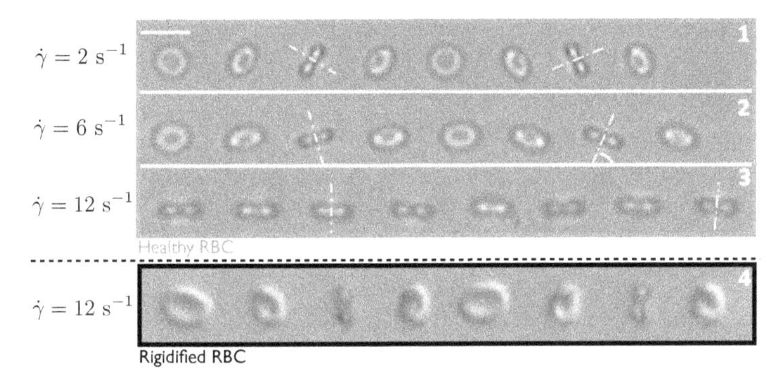

FIGURE 5.6
Shear rate dependent orbital drift of healthy RBCs toward rolling in a dextran solution of viscosity $\eta_o = 7.15$ mPa.s (Top view, parallel to the vorticity direction, flow from left to right). Noteworthy, a rigidified cell maintains a flipping movement (from [42]).

However, several works demonstrated that RBCs' dynamics is much more complex than initially thought, and recent studies have separated the effects of shear rate and external viscosity. Even the flipping movement was found to be different from that of a solid: Bitbol [17] showed that contrary to a perfectly rigid object, whose flipping orbit depends on the initial orientation only at zero Reynolds number, the RBC has a flipping orbit that drifts with the shear rate until rolling is reached (Figure 5.6). Rolling is the motion where the small axis of the cell is aligned with the vorticity axis and the cell rotates like a wheel. In addition, experiments have shown that at very low shear rates, the orbit is not stable [42, 110]. Orbit selection and drift only occur above $\eta_o\dot{\gamma} \approx 0.05$ Pa. This is completely different from the vesicle tumbling, which is a flipping around the vorticity axis only [4].

5.3.3 Motions in a viscous external medium

Detailed studies of the effect of the shear rate on the RBC dynamics in a viscous medium [1, 42, 110] now allow us to propose the following summary of the behavior at low values of the viscosity ratio. At high shear rates (high Ca), the RBC is elongated and tank-treads. When the shear rate is decreased, the elongation of the RBC is reduced and its inclination angle relative to the direction of flow increases. When the deformation is small enough, tank-treading is perturbed by an additional oscillation of the inclination angle: this motion is called swinging and is well visible for external stresses of the order of 0.2 Pa [1, 110] (Figure 5.7). When the shear rate is further decreased, the angle oscillations increase until they are so large that the RBC actually flips: at low shear rates, flipping is obtained. However, this transition was shown to be hysteretic: when the shear rate is increased from the flipping motion, an orbital drift of the flipping is observed until rolling is reached (for $\eta_o\dot{\gamma}$ between 0.05 and 0.1, typically). Further increase in the shear rate makes the RBC go to swinging/tank-treading, possibly via a hovering/frisbee motion [42].

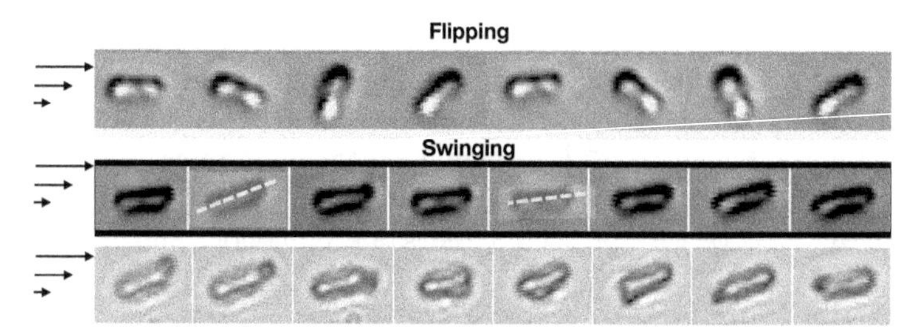

FIGURE 5.7
Flipping and swinging movement under shear flow (from [1]).

Note also that the hysteretic transition region is particularly interesting, as illustrated by the large deformations sometimes obtained in this region during the transitions [1,93]. Unpublished numerical simulations by the authors have shown a very strong effect of the reference stress-free shape of the membrane on the dynamics in this region.

5.3.4 Motions in a low-viscosity external medium

The motions in physiological conditions of viscosity (RBCs typically suspended in plasma or phosphate-buffered saline, PBS, solutions, for example) have only been studied in detail recently [90,104] (see Figure 5.8). At low shear rates, flipping is obtained, with an orbital drift to rolling with increasing Ca, as for low values of λ. However, orbital drift is seen for lower values of $\eta_o\dot{\gamma}$ and rolling is reached for lower shear stresses than in viscous media (typically 0.02 Pa). Then, rolling is stable for a much wider range of Ca, over which deformation increases [164]. When Ca is increased, the RBC deforms until compression forces make it buckle: a rolling stomatocyte is obtained dynamically (if the shear flow is stopped, the RBC goes back to its discocyte shape). Around $\eta_0\dot{\gamma} = 0.1$ Pa, stomatocytes are mainly observed. The elongation of this stomatocyte increases with Ca until the dynamics is unstable ($\eta_0\dot{\gamma} \approx 0.3$ Pa). The RBC changes its orientation and a tumbling stomatocyte is observed. Increasing Ca further makes the RBC transition to other multilobe shapes, like trilobes and pyramids/hexalobes [90]. Trilobes are then mainly observed up to shear stresses of 1.0 Pa, approximately.

5.3.5 Summary: the phase diagram

A qualitative phase diagram ($\eta_o\dot{\gamma}$,λ) is now presented in Figure 5.9 to summarize the previous statements. In addition to the dynamics and transitions already commented upon, we draw attention on some specific points. First, for low values of λ, the transition between flipping and swinging is quite complex, with hysteretic behaviors [1,110] and possibly large transient deformations [1,93]. Depending on the stress-free shape, simulations either predict the transition to dynamic stomatocytes flipping/rolling, or the existence of a stable hovering/frisbee motion. Such a stable frisbee motion has not been reported in experiments. However, Dupire and coworkers [42] observed a frisbee-like motion during a transient increase in shear rate.

The viscosity ratio mainly controls one transition between swinging/tank-treading at low viscosity ratio and rolling-stomatocytes-polylobes at high viscosity ratio. Simulations have shown that this transition is around $\lambda = 3$, although a systematic study of this limit has not been undertaken. Of course, λ also modulates the dynamics, for instance by inhibiting deformations and membrane circulation (tank-treading), in general. The viscosity

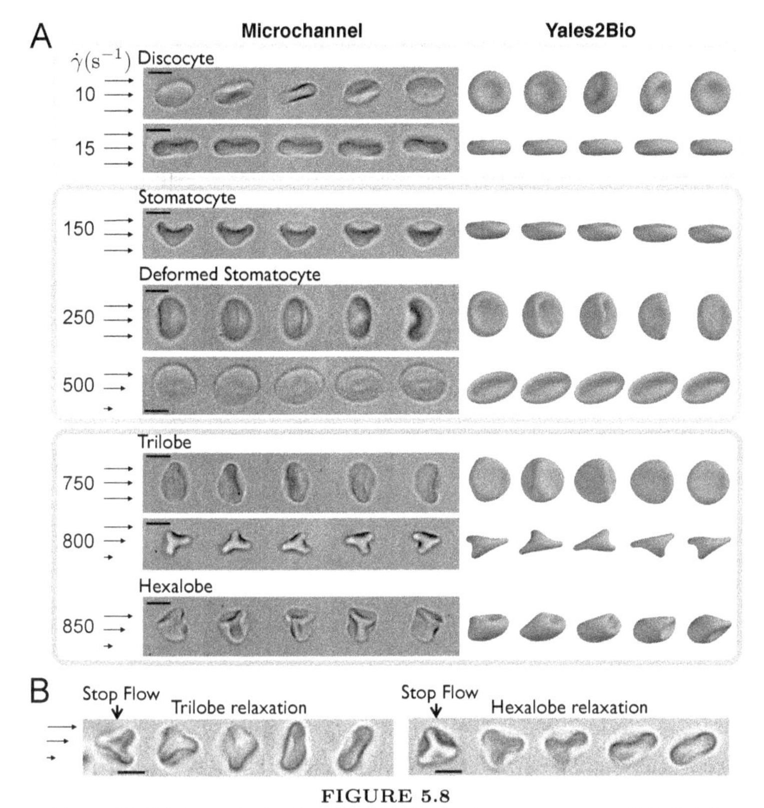

FIGURE 5.8
Microfluidic observations of RBC dynamics in shear flow. (A) Timelapse sequences of RBC deformations at various shear rates. The right side of the figure shows analogous time sequences of RBCs obtained with YALES2BIO simulations. (B) Stop-flow sequences of (Left) a trilobe with a relaxation time of 1 s and intermediate images separated by 0.23 and 0.56 s and (Right) a hexalobe with relaxation of 1.2 s and successive images with a time interval of 0.32 and 0.71 s. (Scale bars, 5 μm.). From [90].

ratio also influences the critical shear stress at which shear-controlled transitions occur (rolling is for instance reached for lower shear stresses when increasing the viscosity ratio). Finally, note that many dynamics reported in Figure 5.9 (flipping, rolling, frisbee, swinging) occur without significant deformations of the cell. This opens the way to an explanation of those dynamics using low-order modeling of the RBC movement where out-of-plane RBC deformations are neglected. This is the object of the following section.

5.4 DYNAMICS AT LOW SHEAR RATES: FROM LOW-ORDER MODELING TO PHYSICAL UNDERSTANDING

In 1922, Jeffery published a famous paper detailing the dynamics of a solid ellipsoid in a general linear Stokes flow [82]. It provides the rotation rates of an ellipsoid along time, thus the complete characterization of its movement. In 1982, Keller and Skalak (hereafter referred to as KS) proposed an extension of this model to the case of a fixed-shape ellipsoid containing an internal fluid [86] in a pure shear flow, albeit with an inelastic membrane. The extension to account for membrane elasticity was published in 2007 by Abkarian, Faivre and Viallat [1] and Skotheim and Secomb [146] (the elastic model will be referred to as the AFV-SS model). However, the KS and the AFV-SS models assume that the symmetry axis of the fluid ellipsoid is in the shear plane, precluding the description of off-plane motion

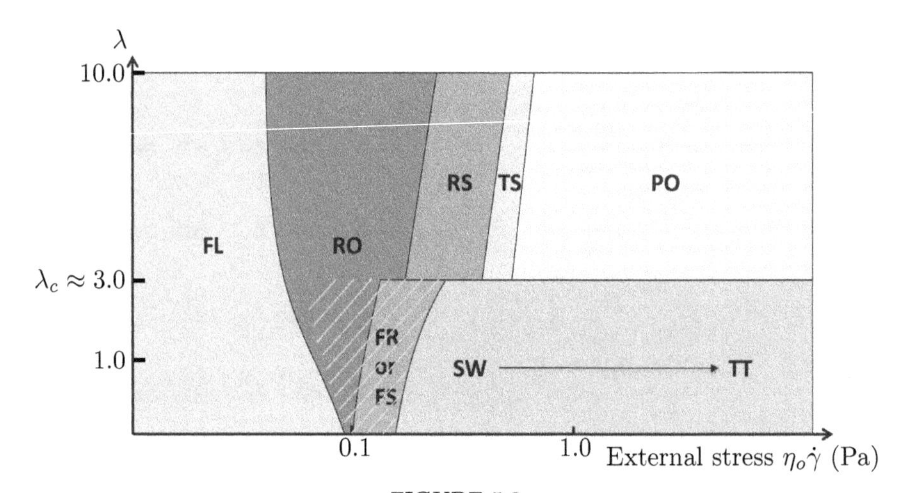

FIGURE 5.9
Approximate phase diagram inferred from existing simulations and experiments. The diagram is separated into a low-viscosity-ratio region and a high-viscosity-ratio region. The critical value for the ratio of viscosity is $\lambda_c \approx 3.0$. Regions of flipping (FL), frisbee or flipping stomatocyte (FR or FS), swinging (SW) to tank-treading (TT), rolling discocyte (RO), rolling stomatocyte (RS), tumbling stomatocyte (TS) and tumbling polylobes (PO) are displayed. The hatched zone corresponds to a region of the phase diagram where hysteresis has been reported experimentally, depending on the history of shear stress applied. The exact frontiers are unknown and depend on the RBC characteristics. See Appendix 5.8 for details on the different dynamics.

(orbital drift, rolling, frisbee motions). The extension of the AFV-SS model to 3-D was recently made by Mendez and Abkarian [107]. The purpose of this section is to highlight the physical understanding gained thanks to these shape-preserving models, which are relevant to moderate stresses (small out-of-plane deformations).

5.4.1 Shape-preserving models for the shear-plane dynamics of red blood cells in shear flow

As already stated, early studies of RBCs had clearly identified two different behaviors in shear flow: flipping (tumbling) and tank-treading. These experimental observations motivated the work of KS [86] in the early 1980s. KS considered an ellipsoid of prescribed geometry, whose membrane, an interface without elasticity, can circulate along the shape. The ellipsoid is suspended in an infinite shear flow at zero Reynolds number. One axis of the ellipsoid is aligned with the vorticity axis, so that the two other axes are in the shear plane. The ellipsoid contains fluid. The dynamics of the ellipsoid may be flipping or tank-treading, and the parameters are the geometry of the ellipsoid and the ratio of viscosity between the internal fluid and the external fluid, λ. Writing that the moment about the axis of vorticity exerted by the external fluid on the particle is zero yields an equation for the angular speed of the ellipsoid, which depends on the tank-treading velocity. A second equation is obtained by specifying that the rate of work done by the fluid on the particle W_p has to be equal to the dissipation rate D' produced by the flow inside the particle and the membrane: $W_p = D'$. This yields an equation for the tank-treading rate, which closes the system. The KS model [86] predicts that under a threshold value for λ, the particle tank-treads with a fixed orientation under flow, while it flips for high values of λ.

However, the KS model [86] does not reproduce two experimental results obtained by Abkarian and coworkers [1]: 1) the tumbling-tank-treading transition with shear rate; 2) the swinging motion. These findings could only be explained by accounting for the elasticity of the membrane: the KS model was thus extended [1, 146] so that the membrane stores and restores energy during circulation. One way of accounting for membrane elasticity is to revisit the energy balance written by KS [86] to state that the rate of work on the particle W_p can either be dissipated or stored in the membrane under the form of elastic energy. The balance becomes $W_p = D' + P_{el}$, where P_{el} is the elastic power stored in the membrane. This yields a couple of differential equations for the orientation of the particle and the circulation of the membrane (details of the calculation can be found in [3]). The AFV-SS model involves the capillary number, so that results depend on the shear stress applied, which was not the case for the KS model. It predicts that the RBC tumbles at low shear stress and swings above a critical shear stress, with the amplitude of swinging oscillations decreasing with shear stress [1,146] .

The shape-preserving models with membrane elasticity [1,146] demonstrated that the dynamics of RBCs in the shear plane may be understood if one accounts for their shape memory. The relevant mechanism here is the fact that when the cytoskeleton circulates around the center of mass during tank-treading, its elements move away from their preferred state, in which the shape is at equilibrium. This has an elastic energy cost that modifies the dynamics in several ways: first, it means that the full circulation of the membrane can only be realized if an energy barrier is overcome, the energy associated with the displacement of rim elements to the dimple and dimple elements to the rim. If the flow strength is not large enough, the energy barrier is not overcome, and circulation is prevented by membrane elasticity. Second, even when the membrane circulates, the elastic torque may alternatively favor or hamper membrane circulation. This directly impacts the angular speed of the ellipsoid, which explains why RBCs swing. Further understanding of membrane shape memory and the importance of the stress-free shape of the cytoskeleton has been provided by Dupire and coworkers [41], who managed to account for different stress-free shapes for the membrane. They demonstrated that the closer the stress-free shape to a sphere, the lower the energy barrier.

Despite the merits of these models, their restriction to the shear plane is problematic: indeed, in their experiments, Abkarian and coworkers [1] discarded RBCs showing out-of-plane dynamics from their analysis, in order to obtain clean transitions between tumbling and tank-treading. However, we know now that out-of-plane dynamics are characteristic of the RBC movement at low shear stress. An extension of the existing model to 3-D is thus indispensable to understand the mechanism controlling the full dynamics of RBCs at low shear stresses.

5.4.2 A 3-D shape-preserving model for the dynamics of red blood cells in shear flow

For the purpose of our discussion, we will restrict ourselves to the case of a pure shear flow, but the model may be extended to general linear flows. The schematic of the problem is provided in Figure 5.10. An ellipsoid is suspended in a pure shear flow. Two frames of reference of the same origin are used, with two systems of Cartesian coordinates: the fixed reference frame $\hat{\mathcal{R}}$, of axes $\hat{x}_i$ (of unit vectors $\hat{e}_i$), and the so-called body frame $\mathcal{R}$, of axes x_i (of unit vectors e_i). The axes of the body frame are aligned with the principal axes of the ellipsoid. In order to simplify the discussion, we will consider an axisymmetric oblate ellipsoid, whose small axis of symmetry is aligned with e_3. In the fixed frame, $\hat{x}_1$, $\hat{x}_2$ and $\hat{x}_3$ are the flow, the velocity gradient and the vorticity directions, respectively. The external flow reads $\hat{u}_1 = \dot{\gamma}\hat{x}_2$, with $\dot{\gamma}$ the shear rate. It is assumed that the shape of the ellipsoid is always preserved, so that the state of the ellipsoid with respect to the fixed frame can be defined by a simple rotation. R is the rotation matrix between the bases e_i and $\hat{e}_i$:

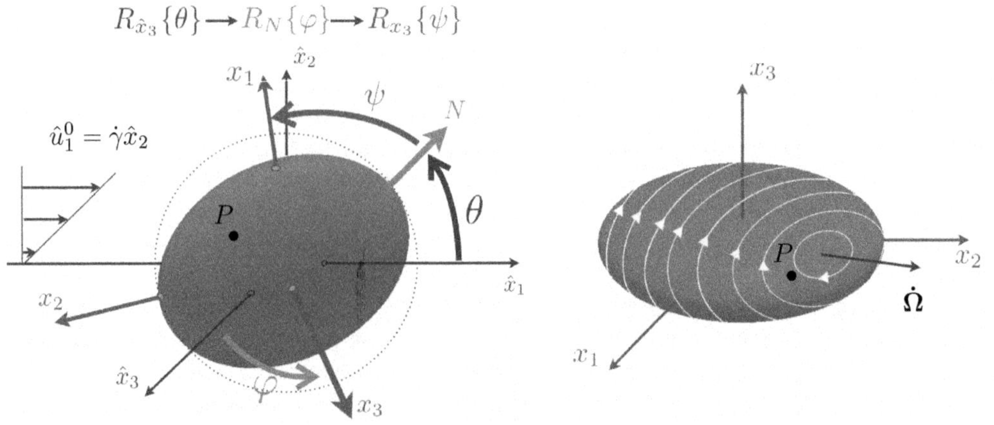

FIGURE 5.10
Schematic of the configuration for the fixed-shape model. An ellipsoid envelope of semi-axes $a_1 = a_2$, a_3, containing fluid of internal viscosity η_i, is suspended in a fluid of viscosity η_o and subjected to pure shear flow. Left: definition of the orientation of the ellipsoid with respect to the fixed frame ($\hat{x}_1$, $\hat{x}_2$, $\hat{x}_3$). The body frame (x_1, x_2, x_3) is obtained by rotation of the fixed frame. The rotation is parametrized by the Euler angles and reads $R_{\hat{x}_3}\{\theta\} \circ R_N\{\varphi\} \circ R_{x_3}\{\psi\}$, where $R_{\hat{x}_3}\{\theta\}$ is a rotation of angle θ around the axis $\hat{x}_3$. The element of the membrane located at rest at $x_3 = a_3$, denoted by P, is tracked along the dynamics as a measure of the membrane circulation. Right: definition of the membrane velocity field (materialized by streamlines over the membrane, depicted using white lines), defined using the tank-treading rates vector $\dot{\boldsymbol{\Omega}}$.

$\hat{x}_i = R_{ij}x_j$. The orientation of the ellipsoid is parametrized using the Euler angles θ, φ, ψ, as sketched in Figure 5.10:

$$R = \begin{pmatrix} \cos\theta & -\sin\theta & 0 \\ \sin\theta & \cos\theta & 0 \\ 0 & 0 & 1 \end{pmatrix} \begin{pmatrix} 1 & 0 & 0 \\ 0 & \cos\varphi & -\sin\varphi \\ 0 & \sin\varphi & \cos\varphi \end{pmatrix} \begin{pmatrix} \cos\psi & -\sin\psi & 0 \\ \sin\psi & \cos\psi & 0 \\ 0 & 0 & 1 \end{pmatrix}. \tag{5.27}$$

The ellipsoid has a membrane that is allowed to circulate (tank-tread) around the fixed ellipsoid. The tank-treading velocity v_i^m is defined as in the Keller and Skalak model [86]:

$$v_i^m = a_i\, \epsilon_{ijk}\, \dot{\Omega}_j\, x_k/a_k, \tag{5.28}$$

where ϵ_{ijk} is the Levi-Civita symbol in three dimensions and the $\dot{\Omega}_j$ are the tank-treading rates of the membrane circulation around the axes x_j, which have the dimension of a frequency. We will see that the dynamics of the fluid ellipsoid depends on the position of the membrane with respect to its position at rest: more specifically, this position can be defined using the location of one point over the membrane. To do so, we track in a Lagrangian way the membrane point which is at rest at $x_1 = x_2 = 0$ and $x_3 = a_3$, along the small axis of the ellipsoid. During the movement, this point circulates over the ellipsoid at the circulation velocity defined by Eq. (5.28). We denote P this Lagrangian point, which is the small axis of the membrane. During the dynamics, it will be defined by its coordinates projected on a sphere of radius unity: $\xi_i = x_i/a_i$ (no summation). Solving the fluid-structure interaction problem for this fluid ellipsoid in shear flow means determining the values of the Euler angles θ, φ, ψ and the coordinates of point P, ξ_i along time, given an initial value for the unknowns and the parameters of the problem. In the particular case of the KS and the AFV-SS models, ξ and e_3 are restricted to the shear plane ($\hat{x}_1$, $\hat{x}_2$), $\varphi = \pi/2$ and the rotation is defined with a unique angle θ.

The ellipsoid is neutrally buoyant and inertia is not considered. The condition that the total moment on such a particle must vanish yields a system of equations relating the spins around the axes of the particles to the tank-treading rates, which is a generalization of the result by KS [86] to 3-D. First, the components of the resultant moment in the body frame are $M_i = \int_{\partial E} \epsilon_{ijk} x_j c_{kl} n_l dS$, where E is the ellipsoid surface, n_l the coordinates of the normal to E and c_{kl} the components of the external stress tensor in the body frame. KS provide the general expression for c_{kl}, so that it can be shown that:

$$M_1 = \frac{C_1}{(a_1 a_2 a_3)^{2/3}} \left[(a_2^2 - a_3^2)(e_{32}^0 - e_{32}^m) + (a_2^2 + a_3^2)((\zeta_{32}^0 - \zeta_{32}^m) - \dot{\omega}_1) \right], \tag{5.29}$$

where C_1 is a constant depending on the shape of the RBC [86]. Similar expressions can be obtained for M_2 and M_3 by cyclic permutation of the indices. e_{ij}^0, e_{ij}^m are the components of the symmetric part of velocity gradient tensors (the strain-rate tensors) built from the pure shear flow (superscript 0) and the membrane flow (superscript m) defined by Eq. (5.28), respectively. ζ_{ij}^0 and ζ_{ij}^m are the components of the skew-symmetric part of the same velocity gradient tensors. The total moment acting on the particle is thus the sum of three contributions: a contribution from the external shear flow (superscript 0), a contribution from the tank-treading of the membrane (superscript m) and the one from the rotation of the particle around the axis (the $\dot{\omega}$ term).

e_{ij}^0, e_{ij}^m, ζ_{ij}^0 and ζ_{ij}^m are all expressed in the body frame.

e_{ij}^m and ζ_{ij}^m are directly obtained from Eq. (5.28). The e_{ij}^0 and the ζ_{ij}^0 read:

$$\begin{cases} e_{32}^0 = \frac{\dot{\gamma}}{2}\left(\frac{1}{2}\cos\psi \sin 2\theta \sin 2\varphi + \cos 2\theta \sin\psi \sin\varphi\right), \\ e_{13}^0 = \frac{\dot{\gamma}}{2}\left(\frac{1}{2}\sin\psi \sin 2\theta \sin 2\varphi - \cos 2\theta \cos\psi \sin\varphi\right), \\ e_{21}^0 = \frac{\dot{\gamma}}{2}\left(\cos\varphi \cos 2\theta \cos 2\psi - \frac{1+\cos^2\varphi}{2} \sin 2\theta \sin 2\psi\right). \end{cases} \tag{5.30}$$

$$\begin{cases} \zeta_{32}^0 = -\frac{\dot{\gamma}}{2} \sin\varphi \sin\psi, \\ \zeta_{13}^0 = -\frac{\dot{\gamma}}{2} \sin\varphi \cos\psi, \\ \zeta_{21}^0 = -\frac{\dot{\gamma}}{2} \cos\varphi. \end{cases} \tag{5.31}$$

From the expressions of the moments and imposing $M_i = 0$, the expressions of the spins of the ellipsoid around its axes are obtained:

$$\begin{cases} \dot{\omega}_1 = \zeta_{32}^0 + \frac{a_2^2 - a_3^2}{a_2^2 + a_3^2} e_{32}^0 - \frac{2a_2 a_3}{a_2^2 + a_3^2} \dot{\Omega}_1, \\ \dot{\omega}_2 = \zeta_{13}^0 + \frac{a_3^2 - a_1^2}{a_3^2 + a_1^2} e_{13}^0 - \frac{2a_3 a_1}{a_3^2 + a_1^2} \dot{\Omega}_2, \\ \dot{\omega}_3 + \dot{\Omega}_3 = \zeta_{21}^0. \end{cases} \tag{5.32}$$

with a_1, a_2 and a_3 the semi-axes of the ellipsoid ($a_1 = a_2$ for the axisymmetric ellipsoid considered). $\dot{\omega}_i$ is the spin around axis x_i, which can be expressed as a function of the Euler angles: $\dot{\omega}_1 = \dot{\theta} \sin\varphi \sin\psi + \dot{\varphi} \cos\psi$, $\dot{\omega}_2 = \dot{\theta} \sin\varphi \cos\psi - \dot{\varphi} \sin\psi$, $\dot{\omega}_3 = \dot{\theta} \cos\varphi + \dot{\psi}$. In other words, the spin of the fluid ellipsoid around its axes is due to the rotational part of the external flow (ζ_{ij}^0 terms), the strain part of the external flow (e_{ij}^0 terms) and the circulation of the membrane ($\dot{\Omega}_i$ terms), the latter two depending on the particle aspect ratios. Note that $\dot{\Omega}_3$ cannot be distinguished from $\dot{\omega}_3$ (see Eq. 5.32) because x_3 is an axis of symmetry: tank-treading and spinning around an axis of symmetry are defined by the same expression. For the sake of simplicity, it is thus chosen that $\dot{\Omega}_3 = 0$.

In the KS and AFV-SS models, the kinetic energy balance equation provides an additional equation to determine the tank-treading rate [1,3,41,86,146]: the total rate at which work is exerted by the external flow on the particle W_p is equal to the sum of 1) the viscous dissipation rate inside the membrane and particle, D' and 2) the elastic power on the membrane P_{el}. As an example, in the AFV model, the membrane is viewed as a 3-D Kelvin-Voigt viscoelastic material of volume V_m, of bulk shear modulus G and bulk membrane viscosity η_m. The stress in the membrane then reads:

$$\sigma_{ij} = 2GE_{ij} + 2\eta_m \mathcal{D}_{ij}, \tag{5.33}$$

where E_{ij} is the Euler-Almansi strain tensor (see Sec. 5.2.2) and $\mathcal{D}_{ij}$ is the Eulerian strain-rate tensor. This can be viewed as the 3-D generalization of the membrane viscoelastic model presented in Sec. 5.2.2.6. As discussed in Sec. 5.2.2, such a model does not reproduce the shear-hardening behavior of the RBC, but it has allowed an explicit calculation of the power in the membrane using $P_m = \int_{V_m} Tr(\sigma : \boldsymbol{\mathcal{D}})dV$. This elastic part of σ yields the elastic power P_{el} while the viscous part of σ contributes to the total dissipation D' rate [3].

Under the assumption of a 3-D Kelvin-Voigt viscoelastic material for the membrane, P_{el} had been calculated for the particular case where the particle lies in the shear plane [1,3,41] and the kinetic energy balance equation reads:

$$\underbrace{\eta_o V\dot{\Omega}\left(f_2\dot{\Omega} + 2f_3 e^0\right)}_{W_p} = \underbrace{V\eta_{eff} f_1 \dot{\Omega}^2}_{D'} + \underbrace{\frac{1}{2} f_1 \dot{\Omega} V_m G \sin(2\Omega)}_{P_{el}}. \tag{5.34}$$

In this equation, e^0 is the strain rate in the body frame, $\dot{\Omega}$ is the tank-treading rate and $\Omega = \int \dot{\omega} dt$ is a measure of how much the membrane has circulated. f_i are geometric factors, defined further (see Eq. 5.37), V_m is the volume of the membrane and V the volume of the cell. η_{eff} is an effective particle viscosity combining the internal viscosity η_i and the membrane viscosity η_m:

$$\eta_{eff} = \eta_i + \eta_m \frac{V_m}{V}. \tag{5.35}$$

In the KS and AFV-SS models, the strain rate e^0 is only in the shear plane and the tank-treading rate $\dot{\Omega}$ is only around the vorticity axis [86]. The expression of the tank-treading rate can be obtained from Eq. (5.34) and we refer the reader to the papers by KS, AFV and SS for the discussion of the model and the results when the movement is restricted to the shear plane [1,3,41,86,146]. Eq. (5.34) can actually be generalized to 3-D and projected in the different directions to yield separate equations for the tank-treading rates. The tank-treading rates read:

$$\begin{cases} \dot{\Omega}_1 = \dfrac{-2f_3}{f_2 - \lambda_{eff} f_1}\left[e^0_{32} - \dfrac{\dot{\gamma}}{Ca^*}\xi_2\xi_3\right], \\[2ex] \dot{\Omega}_2 = \dfrac{2f_3}{f_2 - \lambda_{eff} f_1}\left[e^0_{13} - \dfrac{\dot{\gamma}}{Ca^*}\xi_1\xi_3\right]. \end{cases} \tag{5.36}$$

They depend on the strain-rate tensor of the unperturbed flow e^0_{ij} and on the non-dimensionalized coordinates of point P, ξ_i. The ξ_i in Eq. (5.36) allow us to quantify how much the membrane has circulated, as Ω in Eq. (5.34). The expression of the membrane velocity (Eq. 5.28) allows us to advance the position of point P ($\dot{\xi}_i = \epsilon_{ijk}\,\dot{\Omega}_j\,\xi_k$) and to close the equations for the spins. The f_i are geometric factors [86] which are a function of the axes of the ellipsoid and are given by:

$$\begin{aligned} f_1 &\equiv 4z_1^2, \qquad f_2 \equiv 4z_1^2(1 - \frac{2}{z_2}), \qquad f_3 \equiv -4\frac{z_1}{z_2}, \quad \text{with} \\ z_1 &\equiv \frac{1}{2}(\frac{a_2}{a_3} - \frac{a_3}{a_2}), \qquad z_2 \equiv g_1'(\alpha_2^2 + \alpha_3^2), \quad \text{with } \alpha_i = \frac{a_i}{(a_1 a_2 a_3)^{1/3}}, \\ g_1' &\equiv \int_0^\infty \frac{ds}{(\alpha_2^2 + s)(\alpha_3^2 + s)\Delta} \quad \text{and} \quad \Delta^2 \equiv (\alpha_1^2 + s)(\alpha_2^2 + s)(\alpha_3^2 + s). \end{aligned} \tag{5.37}$$

Ca^* is a non-dimensional number which compares the stress exerted by the external shear flow to the elastic resistance of the cell. It is thus similar to a capillary number. As out-of-plane deformations are proscribed in the model, it can be viewed as an in-plane capillary number, associated with the membrane elasticity when it deforms in the plane of the membrane only. This signification of Ca^* will be discussed in detail in Sec. 5.4.5. Finally, $\lambda_{eff} = \eta_{eff}/\eta_o$ is an effective viscosity ratio accounting for membrane viscosity. The dynamics as a function of the shear rate and the external medium viscosity can be thus explored by changing the two non-dimensional parameters λ_{eff} and Ca^*.

5.4.3 Predictions of the theoretical shape-preserving model

An axisymmetric ellipsoid of small axes $a_1 = a_2 = 4.2375$ μm and $a_3 = 1.2511$ μm is considered. Such an ellipsoid has a volume of 94.1 μm^3 and a surface area of 132.9 μm^2 [107]. For such a geometry, the parameters in the model are $f_1 \approx 9.5590$, $f_2 \approx -6.5645$, $f_3 \approx -5.2150$. The initial orientation of the ellipsoid is given by $\theta(t = 0) = \psi(t = 0) = 0$ and $\varphi(t = 0) = -\pi/4$. The dynamics is solved by advancing the differential equations of the model, with the numerical method described by Mendez and Abkarian [107]. Figures 5.11 and 5.12 display typical dynamics predicted by the shape-preserving model for $\lambda_{eff} = 1.0$ and $\lambda_{eff} = 5.0$, respectively, and varying Ca^*, the non-dimensional shear stress applied on the particle. In order to visualize the dynamics, the direction of the small axis of the ellipsoid is followed in the fixed frame and projected in the shear plane ($\hat{x}_1$,$\hat{x}_2$). When the small axis is on the circle $\hat{x}_1^2 + \hat{x}_2^2 = 1.0$, the ellipsoid has its small axis in the shear plane (this is the case for swinging and tank-treading, for instance).

Figure 5.11 shows the sequence of dynamics predicted by the model typical of low viscosity ratios (here $\lambda_{eff} = 1.0$), and compared to the rigid case ($Ca^* = 0$). For the rigid case, the particle remains on the initial orbit. This is not the case when Ca^* is finite, even if small. For small values of Ca^* ($Ca^* = 0.1$ for instance), the ellipsoid slowly drifts towards a large limit orbit. When Ca^* increases, the orbit shrinks towards the vorticity axis ($\hat{x}_3$, $\hat{x}_1 = \hat{x}_2 = 0$) and gets close to rolling when Ca^* is close to 1.0. For $Ca^* \approx 1.09$, the dynamics transitions to a frisbee dynamics, with a fixed inclination angle reached ($Ca^* = 1.20$ and 1.80 in Figure 5.11). This inclination angle slowly drifts from the vorticity axis towards the shear plane (typically up to $Ca^* \approx 2.0$). For $Ca^* > 2.0$, the ellipsoid reorients and its small axis goes to the shear plane, where it oscillates (swinging, $Ca^* = 2.10$). This movement may be preceded by transient oscillations around the shear plane (a motion sometimes called kayaking). The higher Ca^*, the smaller the oscillations around and in the shear plane: for $Ca^* \to \infty$, the ellipsoid directly reorients to the shear plane and tank-treads without oscillating, as illustrated by the case $Ca^* = 20.0$.

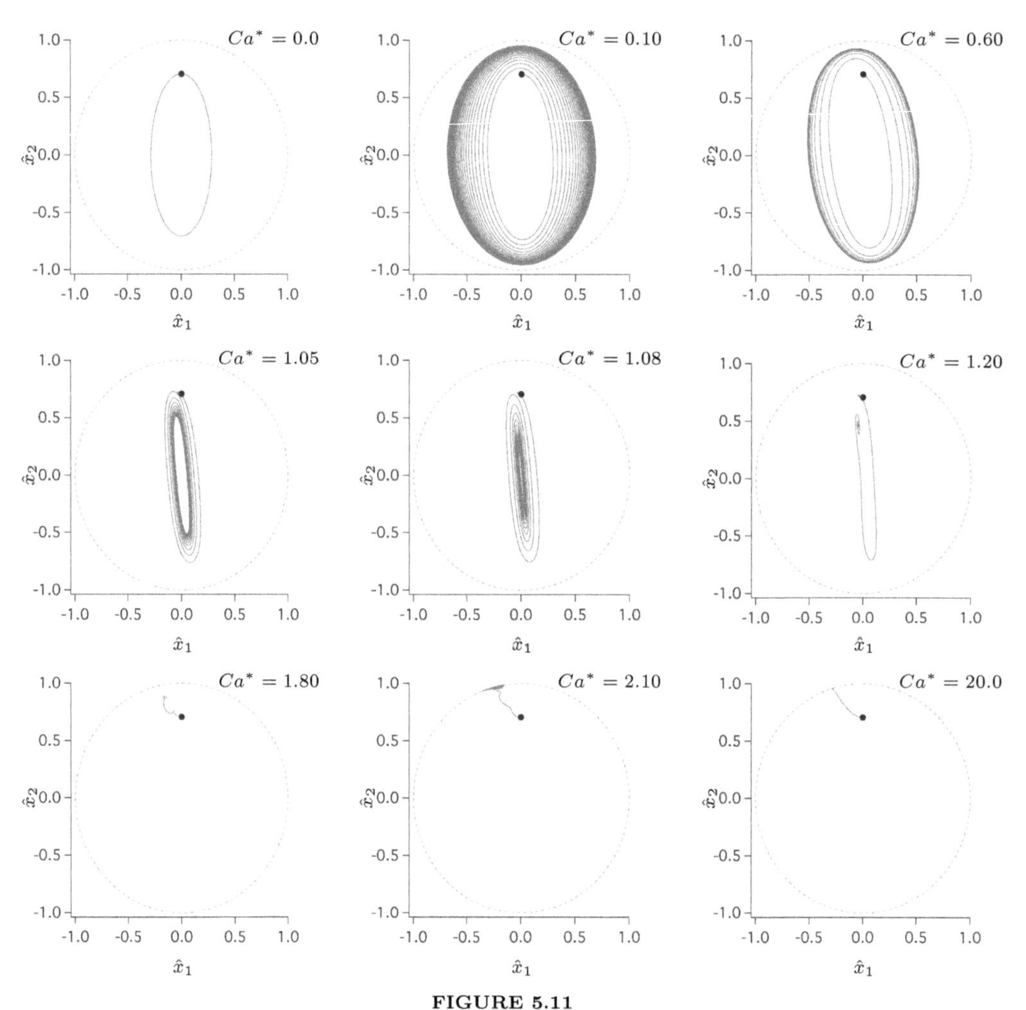

FIGURE 5.11
Predictions of the shape-preserving model for a fluid ellipsoid with $a_1 = a_2 = 4.2375$ μm, $a_3 = 1.2511$ μm and $\lambda_{eff} = 1.0$, for varying in-plane capillary number Ca^*. The direction of the small axis of the ellipsoid in the fixed frame, $(e_3)_{\hat{\mathcal{R}}}$, projected in the $(\hat{x}_1, \hat{x}_2)$ plane, is plotted along time.

Figure 5.12 shows the sequence of dynamics predicted by the model at $\lambda_{eff} = 5.0$ and typical of high viscosity ratios. The rigid case ($Ca^* = 0$) is identical to the one at $\lambda_{eff} = 1.0$, as the internal fluid does not play any role. For small Ca^*, a limit orbit is reached, similar to the one obtained at $\lambda_{eff} = 1.0$. Then, drifting occurs when increasing Ca^*. However, two differences may be seen compared to the $\lambda_{eff} = 1.0$ cases: first, rolling is reached earlier (typically at $Ca^* \approx 0.5$). Another difference is the aspect ratio of the orbits, which is larger than for $\lambda_{eff} = 1.0$. In addition, for $Ca^* > 0.5$, the model only predicts rolling, the small axis of the ellipsoid going to the vorticity axis. Interestingly, the time to reach rolling depends on Ca^*. When Ca^* is very large, the time needed to leave the initial orbit to reach rolling increases.

Figure 5.13 displays the phase diagram from the shape-preserving model for $a_1 = a_2 = 4.2375$ μm and $a_3 = 1.2511$ μm and varying λ_{eff} and Ca^*. The typical dynamics detailed earlier are gathered in this phase diagram. We also remark a clear separation between 'low' and 'high' viscosity ratios, when Ca^* is high enough. For low values of Ca^*, flipping over orbits with orbital drift is obtained whatever the value of λ_{eff}.

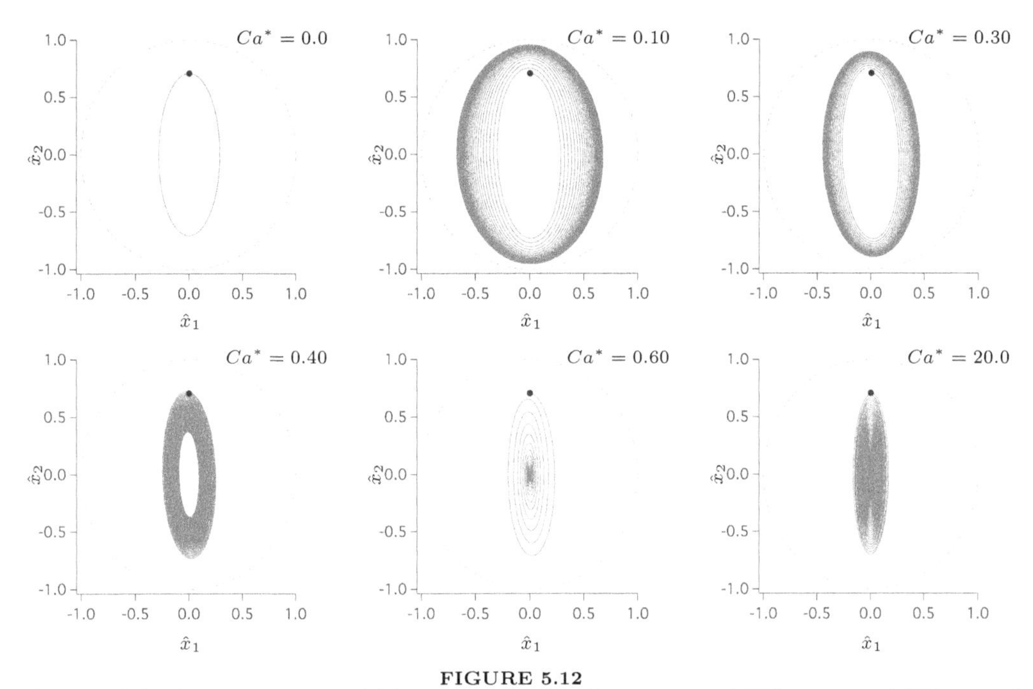

FIGURE 5.12
Predictions of the shape-preserving model for a fluid ellipsoid with $a_1 = a_2 = 4.2375$ μm, $a_3 = 1.2511$ μm and $\lambda_{eff} = 5.0$, for varying in-plane capillary number Ca^*. The direction of the small axis of the ellipsoid in the fixed frame, $(e_3)_{\hat{\mathcal{R}}}$, projected in the $(\hat{x}_1,\hat{x}_2)$ plane, is plotted along time.

To be complete, the dependence of such a diagram on the initial orientation is discussed: when the initial orientation is in the shear plane, it remains in the shear plane, even if the position may be unstable (this is also true for the rolling position, which is an equilibrium for all cases, stable or not). When forced to remain in the shear plane, tumbling in the shear plane is observed at low shear rates or for high viscosity ratios. In addition, the intermittent behavior predicted by the 2-D shape preserving AFV-SS model [1,146] is actually unstable dynamics (the ellipsoid leaves the shear plane if allowed to). Apart from the particular cases of the small axis lying in the shear plane or aligned with the vorticity axis, the long-term dynamics reached almost never depends on the initial orientation. The only exception is a small part of the phase digram for low values of λ_{eff}, essentially in the frisbee regime: if the initial orientation is close enough to the shear plane (see Figure 5.13), swinging is obtained. Interestingly, experiments also report hysteresis in the transition between flipping and swinging at low values of λ [42,110], thus in the same region of the phase diagram.

Overall, the qualitative agreement between experiments and simulations is excellent, for a model which has only 5 degrees of freedom. The predictions of the model indeed resemble the phase diagram gathered from experiments and simulations presented in Figure 5.9, at least at low shear stress and low Ca^*, when the RBC deformations are small. The fixed-shape model is of course unable to reproduce the dynamics at large shear stresses. However, two questions remain: 1) while the model predicts the correct dynamics, can we go further and use it to explain the different motions and transitions between them? 2) The model introduces a non-dimensional number which is similar to the classical capillary number (see Sec. 5.3.1), but different from it: what is the relation between them and what is their relative meaning? This is the purpose of the next sections to answer these two questions.

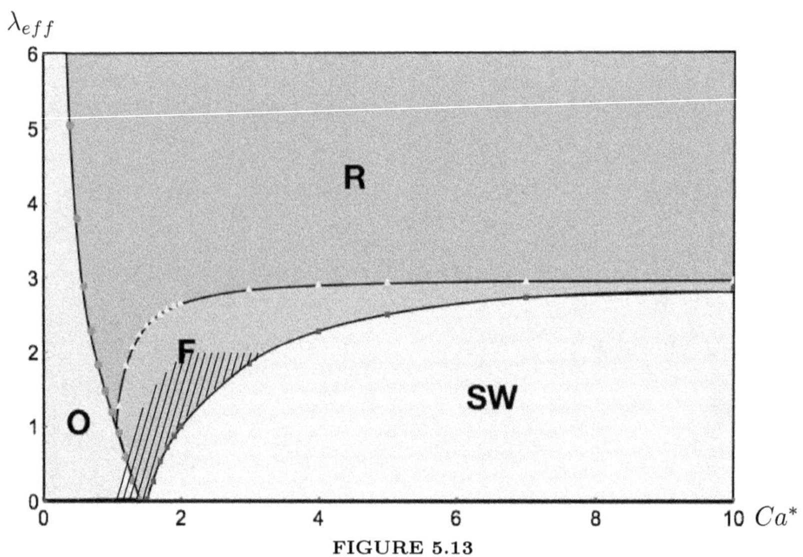

FIGURE 5.13
Phase diagram from the 3-D shape-preserving model in the (Ca^*, λ_{eff}) plane. The hatched zone denotes a hysteresis region where the swinging motion can be obtained if the initial orientation of the particle is close enough to the shear plane.

5.4.4 Mechanism of orbital change

Compared to previous existing results in 3-D, which were limited to the rigid case, the present model introduces two elements: the internal medium and the membrane viscoelasticity, *i.e.* the energy dissipation inside the ellipsoid and in the membrane and elastic energy storage in the membrane. The associated non-dimensional parameters are the viscosity ratio λ_{eff} and the capillary number Ca^*. In order to analyze the results of the model, it is interesting to separate the effects coming from these two physical mechanisms, to clarify their respective contribution in the modification of the behaviors with respect to the rigid case (Jeffery model). For a rigid ellipsoid, Jeffery [82] showed that:

$$\dot{\theta} = -\frac{\dot{\gamma}}{a_2^2 + a_3^2}\left(a_3^2 \cos^2(\theta) + a_2^2 \sin^2(\theta)\right) \quad \text{and} \quad \dot{\varphi} = \frac{\dot{\gamma}}{4}\frac{a_2^2 - a_3^2}{a_2^2 + a_3^2} \sin(2\theta)\sin(2\varphi). \tag{5.38}$$

The result is that trajectories are closed orbits only depending on the initial orientation and the aspect ratio of the ellipsoid.

5.4.4.1 3-D dynamics without membrane elasticity

We first discuss how internal dissipation modifies the behavior of a fluid ellipsoid. The model proposed above is simplified by neglecting the membrane elasticity ($Ca \to \infty$). The aim is to determine how much of the 3-D phase diagram can be explained without membrane elasticity. The result is a 3-D extension of the KS model, where the expression of the ellipsoid spins is unchanged, but those of the tank-treading rates simply read:

$$\dot{\Omega}_1 = -\Lambda\, e_{32}^0, \quad \dot{\Omega}_2 = \Lambda\, e_{13}^0, \quad \text{with} \quad \Lambda = \frac{2 f_3}{f_2 - \lambda_{eff} f_1}, \tag{5.39}$$

The expression of the tank-treading rates can be substituted in those of the spins, which can be combined to obtain the equations for $\dot{\theta}$ and $\dot{\varphi}$:

$$\dot{\theta} = -\frac{\dot{\gamma}}{a_2^2 + a_3^2}\left((a_3^2 - \Lambda a_2 a_3)\cos^2(\theta) + (a_2^2 + \Lambda a_2 a_3)\sin^2(\theta)\right). \tag{5.40}$$

$$\dot{\varphi} = \frac{\dot{\gamma}}{4}\frac{a_2^2 - a_3^2 + 2\Lambda a_2 a_3}{a_2^2 + a_3^2}\sin(2\theta)\sin(2\varphi). \tag{5.41}$$

By analogy with the rigid ellipsoid formulae, we can analyze Eqs. (5.40) and (5.41) as follows: if $a_3 > \Lambda a_2$, these equations are the same as Eq. (5.38), but with an apparent aspect ratio of the ellipsoid modified by the effect of λ_{eff}, and equal to the geometric aspect ratio only if $\lambda_{eff} \to \infty$ (rigid limit).

The former equation for $\dot{\theta}$ can also be written with the following form:

$$\dot{\theta} = -\frac{\dot{\gamma}}{2} + \frac{a_2^2 - a_3^2 + 2\Lambda a_2 a_3}{a_2^2 + a_3^2}\frac{\dot{\gamma}}{2}\cos(2\theta). \tag{5.42}$$

This equation is actually identical to the one found by KS [86]. In their work, θ is the orientation angle in the shear plane, while in 3-D, θ is the first Euler angle. As the equation for θ is identical, the whole discussion by [86] about the transition from tumbling to tank-treading as a function of the viscosity ratio holds here: the critical value for the transition is the same whatever the initial orientation of the cell: the critical viscosity ratio λ_c is obtained by writing the condition $a_3 = \Lambda a_2$, yielding:

$$\lambda_c = \frac{f_2 a_3 - 2 f_3 a_2}{f_1 a_3}. \tag{5.43}$$

If $\lambda_{eff} < \lambda_c$, the RBC tank-treads, so that $\dot{\theta} = 0$ [86]. As θ is constant, Eq. (5.41) can be written $\dot{\varphi} = A\sin(2\varphi)$, with A a positive constant. This differential equation can be integrated as

$$\varphi = \arctan(\tan(\varphi_0)e^{2At}), \tag{5.44}$$

with $\varphi(t = 0) = \varphi_0$. This shows that once θ is constant, $\tan(\varphi)$ tends to infinity exponentially ($\varphi \to \pi/2$), so that the ellipsoid is rapidly attracted by the shear plane whatever its initial orientation.

In summary, for a fluid ellipsoid without membrane elasticity, two behaviors are possible: tank-treading in the shear plane and tumbling along modified Jeffery orbits. The parameter that controls these different dynamics is the viscosity ratio, with the same critical λ_c value as in the analysis of Keller and Skalak [86], which is of the order of 3.0 for geometries relevant to RBCs. When $\lambda \leq \lambda_c$, the ellipsoid goes tank-treading in the shear plane. It is the only stable dynamics. When $\lambda > \lambda_c$, the ellipsoid tumbles on an orbit that is fixed by the initial orientation, as for rigid ellipsoids. The orbits are Jeffery orbits, but their exact characteristics do not depend only on the geometry of the ellipsoid, but also on the viscosity ratio. The effective aspect ratio of the ellipsoid is actually larger than its geometrical aspect ratio.

The results are summarized in Figure 5.14, in which the dynamics of the fluid ellipsoid without membrane elasticity is displayed, by tracking the trajectory of the symmetry axis of the cell along time, over a sphere of radius unity. First, the influence of the initial orientation is shown in Figure 5.14(a). As for rigid ellipsoids, when flipping is predicted, the orbit depends on the initial orientation. On the contrary, for low viscosity ratios, the tank-treading solution is unique and reached whatever the initial orientation. Figure 5.14(b) then shows the influence of the viscosity ratio for a given initial orientation. In the flipping regime (cases a-c), the aspect ratio of the orbit decreases with λ_{eff}, as demonstrated by the analysis of

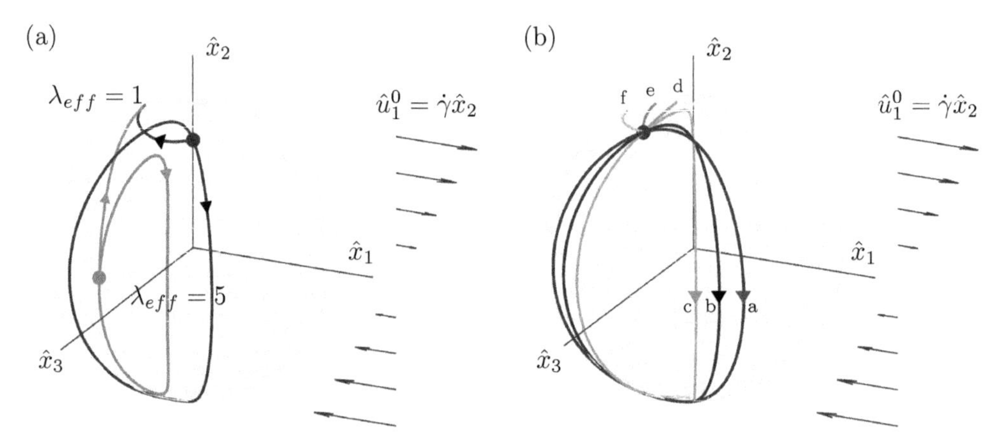

FIGURE 5.14
Trajectory of the symmetry vector of an oblate ellipsoid of small axes 4, 4 and 1.5 during its motion in shear flow for different values of the viscosity ratio. The initial orientation is labeled with a thick dot. (a) Trajectories for velocity ratios 1 and 5, for two different initial orientations. (b) Evolutions for a unique initial orientation and for different values of the viscosity ratio labeled from a to f in decreasing order: $\lambda_{eff} = 1000$, 6.25 and 3.57 (tumbling with closed orbits), 3.0, 1.67 and 0.1 (open trajectories to tank treading in the shear plane $\hat{x}_3 = 0$).

the equations. In the tank-treading regime, when $\lambda_{eff} < \lambda_c$ (cases d-f), the orientation of the ellipsoid during tank-treading is a function of λ_{eff}, as shown by KS. The lower the viscosity ratio, the higher the angle between the ellipsoid and the flow direction.

This clarifies the role of viscous dissipation in the model, with respect to the rigid case: when internal viscosity is small enough, the dynamics is completely changed, the ellipsoid directly goes tank-treading. For high viscosity, internal dissipation acts by changing the aspect ratio of the orbit. However, the type of dynamics is qualitatively unchanged: the ellipsoid stays in its initial orbit, which does not depend on the shear rate. The shear rate only determines the frequency at which the ellipsoid flips, as for the rigid case.

5.4.4.2 Role of membrane elasticity

In order to analyze the behavior of the full model, we separate the different contributions to the spin of the ellipsoid: $\dot{\omega} = \dot{\omega}_{Ext.Flow} + \dot{\omega}_{Visc.Circ.} + \dot{\omega}_{Elast.}$, with

$$\dot{\omega}_{Ext.Flow} = \begin{cases} \zeta_{32}^0 + \dfrac{a_2^2 - a_3^2}{a_2^2 + a_3^2} e_{32}^0 \\[2ex] \zeta_{13}^0 + \dfrac{a_3^2 - a_1^2}{a_3^2 + a_1^2} e_{13}^0 \\[2ex] \zeta_{21}^0, \end{cases} \tag{5.45}$$

$$\dot{\omega}_{Visc.Circ.} = \begin{cases} -\dfrac{2a_2a_3}{a_2^2 + a_3^2} \dfrac{-2f_3}{f_2 - \lambda_{eff} f_1} e_{32}^0 \\[2ex] -\dfrac{2a_3a_1}{a_3^2 + a_1^2} \dfrac{2f_3}{f_2 - \lambda_{eff} f_1} e_{13}^0 \\[2ex] 0, \end{cases} \tag{5.46}$$

$$\dot{\omega}_{Elast.} = \begin{cases} \dfrac{2a_2a_3}{a_2^2+a_3^2}\dfrac{-2f_3}{f_2-\lambda_{eff}f_1}\dfrac{\dot{\gamma}}{Ca^*}\xi_2\xi_3 \\ \dfrac{2a_3a_1}{a_3^2+a_1^2}\dfrac{2f_3}{f_2-\lambda_{eff}f_1}\dfrac{\dot{\gamma}}{Ca^*}\xi_1\xi_3 \\ 0. \end{cases} \tag{5.47}$$

$\dot{\omega}_{Ext.Flow}$ corresponds to the spin generated by the external shear flow on a rigid ellipsoid. $\dot{\omega}_{Visc.Circ.}$ is the change in the spin due to the circulation of the membrane. $\dot{\omega}_{Elast.}$ is the change in the spins due to membrane elasticity. We now study a particular case to show how $\omega_{Visc.Circ.}$ and $\dot{\omega}_{Elast.}$ modify Jeffery's orbit. Note that the vectors defined by Eqs. (5.45-5.47) are defined in the body frame and have been rotated to the fixed frame to be displayed in Figure 5.15.

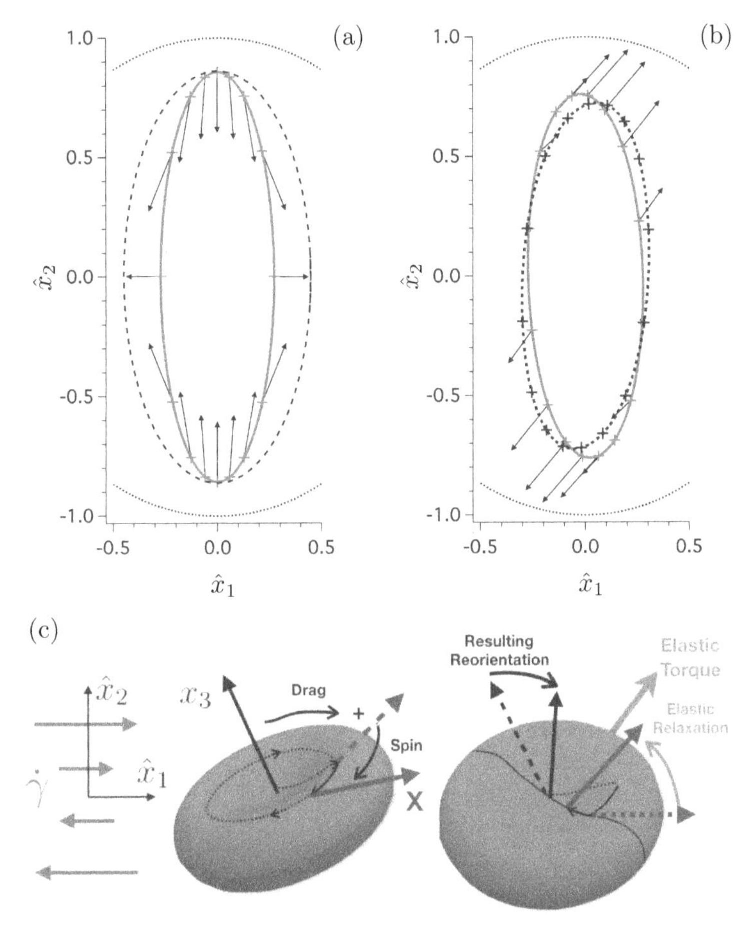

FIGURE 5.15

Example of closed orbits (followed in the clockwise direction) in shear flow ($a_1 = a_2 = 4.2375$ μm, $a_3 = 1.2511$ μm, $\lambda_{eff} = 5$). The circle of radius 1 is shown in dotted lines. (a) Cases $Ca^* = 0$ (i.e. solid ellipsoid, black dashed line) and $Ca^* = +\infty$ (without membrane elasticity, solid magenta line). Vectors $\dot{\omega}_{Visc.Circ.}$ are displayed at 16 evenly spaced instants over the period. (b) Case $Ca^* = 0.364$: trajectory of $(e_3)_{\hat{\mathcal{R}}}$ (solid red line) and $(\xi)_{\hat{\mathcal{R}}}$ (blue dashed line). Vectors $\dot{\omega}_{Elast.}$ are displayed. (c) Schematic of the drift due to the elastic torque.

Figure 5.15(a) first shows two trajectories of the small axis of the ellipsoid in the solid ($Ca^* = 0$, Jeffery orbits) and purely viscous ($Ca^* = +\infty$) cases. Vectors $\dot{\omega}_{Visc.Circ.}$ are displayed to analyze how circulation of the membrane modifies Jeffery's trajectory. Note that $\dot{\omega}$ vectors pointing inwards slow down the particle without changing its trajectory. On the other hand, torques in the same (resp. opposite) direction as the trajectory deviate the trajectory towards the inside (resp. outside). Figure 5.15(a) thus shows how the additional moment associated with the circulation of the membrane makes the orbit thinner for $Ca^* = +\infty$. Note also that $\dot{\omega}_{Visc.Circ.}$ has the same symmetries as $\dot{\omega}_{Ext.Flow}$, which explains why the orbit is similar in the rigid and purely viscous cases.

Regarding membrane elasticity, it is first useful to note that $\dot{\omega}_{Elast.}$ can be written $\dot{\omega}_{Elast.} = \dfrac{2a_3a_1}{a_3^2+a_1^2}\dfrac{2f_3}{f_2-\lambda_{eff}f_1}\dfrac{\dot{\gamma}}{Ca^*}(e_3\cdot\xi)e_3\times\xi$: its direction is normal to e_3 and ξ, while its norm is determined by the angle between e_3 and ξ. Thus, a critical aspect to understand the effect of membrane elasticity is the relative locations of e_3 and ξ, that is to say the locations of the small axis of the ellipsoid and the small axis of the membrane (point P).

Figure 5.15(b) thus displays the trajectories of $(e_3)_{\hat{\mathcal{R}}}$ and $(\xi)_{\hat{\mathcal{R}}}$ in a case with membrane elasticity. Contrary to $\dot{\omega}_{Visc.Circ.}$, the elastic torque associated with $\dot{\omega}_{Elast.}$ is not symmetric with respect to $\hat{x}_1$ and $\hat{x}_2$: the orbit thus becomes tilted. The elastic torque is of course generated by the circulation of the membrane, which is responsible for the storage of in-plane elastic energy. However, a second mechanism determines the direction of the elastic torque: the ellipsoid spins around its symmetry axis (see Eq. 5.32: $\dot{\omega}_3 = \zeta_{21}^0 = -\frac{\dot{\gamma}}{2}\cos\varphi$). This spinning is responsible for a rotation of ξ around e_3, and consequently of a reorientation of the elastic torque. This is seen in Figure 5.15 by the difference in the orientation of $\dot{\omega}_{Visc.Circ.}$ and $\dot{\omega}_{Elast.}$. The spinning of the ellipsoid thus makes the membrane restore its elastic energy with a torque oriented differently from the viscous torque which made the membrane circulate. This element is essential for the understanding of how elasticity controls the dynamics of RBCs, in particular by modifying their orbits, which are not Jeffery's orbits.

This mechanism is sketched in Figure 5.15(c): a fluid ellipsoid may have a circulating membrane due to drag (external shear stress). This displaces ξ with respect to e_3 and stores elastic energy. At the same time, the evolution of the ellipsoid orientation is modified, as part of the external stress is transferred to membrane circulation. When ξ and e_3 lie in the shear plane (classical case of AFV-SS model), the elastic energy is then restored, but $\dot{\omega}_{Elast.}$ is directed in the $\hat{x}_3$ direction, so that the orbit is not changed: the ellipsoid remains in the shear plane, and membrane elasticity makes it swing instead of tank-treading. On the contrary, if the ellipsoid is out of the shear plane ($\cos\varphi \neq 0$), it spins around its small axis at an angular velocity $\dot{\omega}_3 = -\frac{\dot{\gamma}}{2}\cos\varphi$. In that case, the orientation of the elastic torque associated with membrane in-plane elasticity changes with time: membrane elastic energy is restored, but not in the same direction as the one in which it was stored, hence the orbital drift.

What is explained here is the role of membrane elasticity in the change of orbit. However, we still do not have a simple argument regarding why the orbital drift is from large orbits to rolling with increasing shear stress. This would need to be elucidated in the future. Nevertheless, this mechanism of orbital change by the coupling of membrane circulation and ellipsoid spinning is helpful to understand some aspects of the dynamics: the transient motion before swinging (the kayaking) can for instance easily be explained. When the particle reaches the shear plane, the membrane has circulated and an elastic torque is still present to push the particle out of the shear plane. However, the spinning around the small axis being small near the shear plane, the mechanism changing the orientation of the elastic torque progressively disappears and the particle converges to the shear plane.

5.4.5 Discussion on the shape-preserving model: comparison with experiments

We have demonstrated that a low-order model for a fixed-shape ellipsoid is able to predict many features of the dynamics of an RBC at low shear stresses, where out-of-plane deformations of the cell are small. In addition to the geometry of the ellipsoid, the model is a function of two non-dimensional parameters, the viscosity ratio and an in-plane capillary number denoted by Ca^*. However, the in-plane capillary number is not the classical capillary number and cannot be calculated in an experiment. The aim of this section is to define the in-plane capillary number and discuss how it may be related to more classical quantities.

The capillary number can usually be defined as:

$$Ca = \frac{\eta_o \dot{\gamma} a_0}{G_s}, \tag{5.48}$$

where $a_0 = (a_1 a_2 a_3)^{1/3}$ and G_s is the surface shear modulus. This capillary number is a ratio of the viscous shear stress over the elastic resistance of the cell and characterizes the out-of-plane deformations of the cell. Cell deformations are typically small when $Ca \leq 0.1$.

Regarding the in-plane capillary number, different authors have come to different definitions, which are interesting to compare. Skotheim and Secomb (2007) write Ca^* as a function of E_0, the maximum of the increase of elastic energy stored in the membrane during a full tank-treading (with V the volume of the ellipsoid):

$$Ca^* = -\frac{f_3 \eta_o \dot{\gamma} V}{E_0}. \tag{5.49}$$

It measures the ability of the external shear stress to overcome the energy barrier E_0 for the membrane to tank-tread. This definition is interesting as it involves directly the energy barrier for the tank-treading, the difference between the maximum energy during tank-treading (when dimple elements are on the rim and vice versa) and the energy at equilibrium. The value is not explicit but has the advantage of providing a generic definition, whatever the membrane material and prestress. However, E_0 cannot be directly measured.

Abkarian, Faivre and Viallat (2007) calculated explicitly this value when the membrane is a 3-D Kelvin-Voigt viscoelastic material. Later, Dupire, Abkarian and Viallat (2015) extended this calculation to the case of a prestressed 3-D Kelvin-Voigt viscoelastic material. Mendez and Abkarian (2018) extended the formula to 3-D. They found that:

$$Ca^* = -\frac{2 f_3}{f_1} \frac{\eta_o \dot{\gamma} V}{V_m G C}. \tag{5.50}$$

Note the minus sign in the expression of Ca^*. This is due to the fact that f_1 and f_3 are of opposite signs: Ca^* is thus positive. G is the bulk shear modulus of the membrane, assumed to be a Kelvin-Voigt viscoelastic material. The membrane volume is V_m and the membrane surface is S. It is a thin material, so that $V_m = Se$, where e is the thickness of the membrane. C is a non-dimensional constant accounting for membrane prestress [41]. Dupire and coworkers [41] computed the stress associated with the circulation of the membrane assuming that the stress-free shape of the ellipsoid is not the ellipsoid itself but another ellipsoid close to a sphere, as in numerical simulations. They showed that the membrane behaves as if its elastic modulus was decreased by a coefficient which depends on the stress-free shape and the equilibrium shape [41]. As for E_0, C is not a quantity which can be directly measured, as the stress-free shape of the membrane is unknown.

From the two expressions of Ca^* above, the value of E_0 can be explicitly written, when the membrane is a prestressed 3-D Kelvin-Voigt material, whose reference stress-free shape is an ellipsoid:

$$E_0 = \frac{1}{2} f_1 V_m G C, \qquad \text{or} \qquad E_0 = \frac{1}{2} f_1 S G_s C. \tag{5.51}$$

The energy barrier E_0 is thus found to be proportional to the surface area of the membrane S, the surface shear modulus G_s and C, the prestress constant. When the stress-free shape of the membrane gets close to a sphere, C decreases. E_0 is all the smaller as the stress-free shape is close to a sphere. Typical values of $C = 0.01$ have been found when the stress-free shape is an ellipsoid with aspect ratio around 0.95 [41]. Such a value of $C = 0.01$ yields reasonable comparisons between the model and tumbling-to-tank-treading transitions experiments [1,3,41], which is another indication that the quasi-sphere is a relevant stress-free shape for the RBC.

It is also interesting to rewrite Ca^* as a function of Ca:

$$Ca^* = -\frac{2f_3}{f_1} \frac{V}{Sa_0} \frac{1}{C} Ca = \frac{Ca}{\beta C}, \tag{5.52}$$

with

$$\beta = -\frac{f_1}{2f_3} \frac{Sa_0}{V}. \tag{5.53}$$

β is thus a purely geometric quantity. In the case considered, $\beta \approx 3.64$. This result shows that for a given particle, Ca^* and Ca are proportional. However, there may be a large ratio between the two values: Ca^* is typically one or two orders of magnitude larger than Ca: indeed, transitions from flipping to rolling or swinging are predicted for $Ca^* \approx 1.0$, while they are observed experimentally for $Ca \approx 0.03 - 0.1$ [1,42].

In an experiment, it is virtually possible to measure the shear modulus of an RBC, then subject it to shear flow. In that case, the capillary number would be known. However, the stress-free shape (or the energy barrier during tank-treading are unknown). The identification of the dynamics varying the shear rate should enable us to infer these characteristics of the cell. If two cells with identical size, shape, shear modulus and internal viscosity transition to rolling at different Ca, it would mean that they have different stress-free shapes: the cell that transitions at lower Ca has a lower energy barrier, a lower C, thus a stress-free shape closer to a sphere.

We now understand the RBC dynamics as a competition between the external stress imposed by the shear flow and different aspects of the membrane mechanics. If Ca is close to unity, the flow strength is sufficient to deform the RBC and different shapes (stomatocytes, trilobes, elongated cells, see Sec. 5.5) may be obtained depending on the conditions. However, $Ca << 1.0$ does not mean that the cell behaves as a solid: even when the Ca is small, the in-plane capillary number Ca^* may be close to unity. In that case, the RBC experiences different types of movements without cell deformations, depending on the conditions: orbital flipping and drift, rolling, frisbee, swinging. All these movements result from a coupling between the flow and the in-plane elasticity of the membrane, out-of-plane deformations being unnecessary to predict them. The possibility for an RBC to have $Ca << 1.0$ and $Ca^* \approx 1.0$ depends on its reference stress-free shape, which is thus a critical parameter of the problem: the closer it is to a sphere, the higher the ratio between Ca^* and Ca. This notably explains why first numerical simulations of RBC dynamics predicted the transition from flipping to tank-treading with large deformations of the cells (contrary to the experimental evidence), as they imposed a biconcave stress-free shape, which yields similar values of Ca and Ca^*: in that case, membrane circulation cannot occur without cell deformation.

5.5 DYNAMICS AT HIGH SHEAR RATES: COMPRESSIVE INSTABILITIES CONTROLLED BY IN-PLANE ELASTICITY

In the physiological regimes of high shear rates and high viscosity ratio ($\lambda \sim 5$ at 37°C), as we described in the former sections, two remarkable dynamical transitions occur. The first one is around tens of s^{-1} where the rolling discocytes transform into rolling cup-like shapes called stomatocytes. For higher shear rates, of the order of 900 s^{-1}, a second transition occurs where shapes with large lobes appear. In the following section, we will discuss these dynamical morphologies in the context of shear flows and show how in-plane elasticity and stress-free shapes are still the key ingredients explaining RBCs' behavior.

5.5.1 Rolling discocyte-to-stomatocyte transition

The rolling discocyte to rolling stomatocyte transition is first discussed. Let us first consider the decomposition of the shear flow into its rotational and extensional components. The first flow is at the origin of the rolling movement of the cell. The extensional part however submits the cell to both extension forces in the positive quadrant of 45° and to a compressive stress in the negative quadrant of the flow (see Figure 5.16A). Which one induces the loss of the dimple? Mauer and coworkers [104] answered this question thanks to a simulation of the effect of forces acting on a resting discocyte to see which component can induce the buckling transition, by considering separately cell stretching and compression.

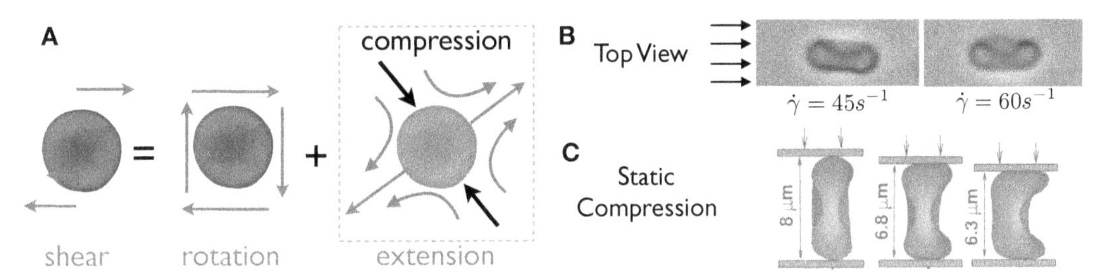

FIGURE 5.16
Buckling of a compressed RBC. Figure C is from [104].

To mimic the extensional component of the flow, an RBC is stretched at two diametrically opposed points similarly to the RBC deformation obtained by optical tweezers [32, 53, 143]. Even for very strong stretching deformations, an RBC remains symmetric and no transition to a stomatocyte-like shape occurs. In a second step, Mauer and coworkers placed an RBC with its largest diameter of about 8 μm between two parallel walls, as shown in the Figure 5.16C, and compressed the cell by moving the upper wall down. When the distance between the walls became approximately 6.3μm, the RBC exhibited a buckling transition and became a stomatocyte. Note that this buckling transition under compressive stress occurs for their model with a nearly spherical stress-free shape (reduced volume of 0.96), while for a biconcave stress-free shape (reduced volume of 0.64), the cell remains biconcave confirming that the stress-free shape of the spectrin network of an RBC is likely to be close to a sphere, consistent with previous studies [29, 123]. Noteworthy, this transition occurs for all considered viscosity contrasts.

In conclusion, buckling occurs because the cell is maintained in the rolling state by the in-plane elasticity, and the compressive stress of the extensional part of the shear flow transforms the cell in a rolling stomatocyte thanks to its quasi-spherical stress-free shape.

5.5.2 Swinging-to-trilobe transition

Following the rolling discocyte-to-stomatocyte transition, the other important change of dynamics occurs at high shear rates around a few hundreds inverse seconds and at viscosity contrasts larger then 3.0. As we already described, RBCs adopt folded three-lobe shapes which are referred to as trilobes. At such high shear rates, RBCs elongate in the vorticity direction and display three lobes that rotate around the center of mass. An even more solid-like rotation is observed for shear rates of around a thousand inverse seconds, where tetrahedral shapes called hexalobes appear.

The exact mechanism leading to these intricate dynamical shapes is still not understood, but recent simulations highlight the importance of in-plane shear elasticity. We now illustrate the transition to trilobes for a fixed value of $Ca = 1.35$ and increasing viscosity contrast λ. Numerical simulations performed with the in-house YALES2BIO solver [90,104,107,108,142,143] and modeling the membrane with the Skalak law (see Sec. 5.2.2.1) are commented upon.

Figure 5.17(a) shows the inclination angle and the aspect ratio of an RBC with a fixed orientation as a function of the viscosity contrast, in the tank-treading regime. As λ increases, both the cell's extension and its inclination angle decrease. At $\lambda \approx 3.2$, a transition to a trilobe shape occurs, even though the inclination angle is still nonzero (equal to about 5°). Figure 5.17(b) illustrates the time evolution of the shape at this transition; the membrane first forms small bumps at the top and the bottom, then very rapidly forms only three rotating lobes.

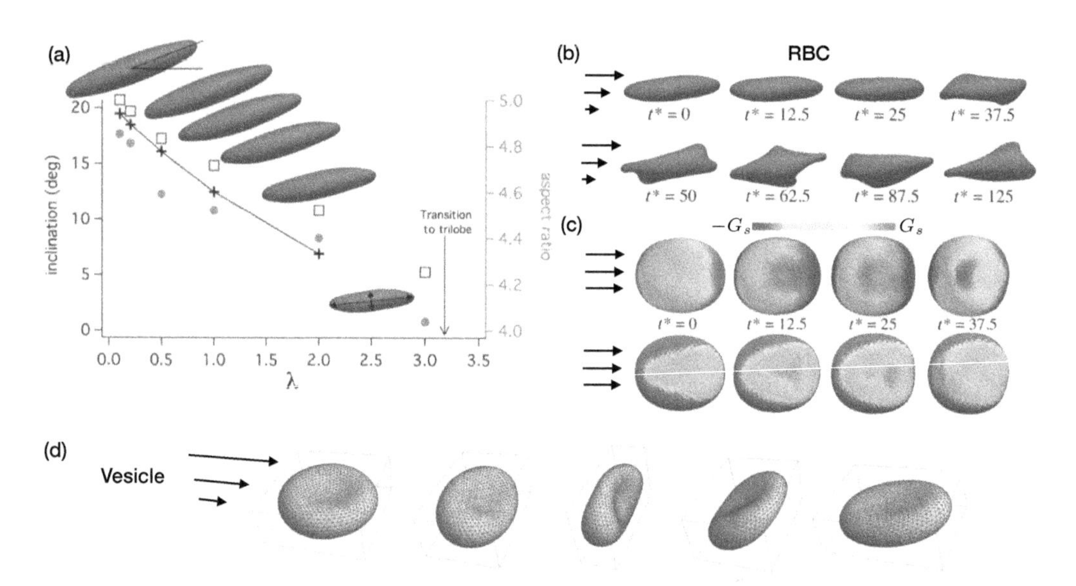

FIGURE 5.17
Trilobe formation. (a) Inclination angle (square and plus symbols, left axis) and aspect ratio (maximum to minimum size ratio, red bullets, right axis) as a function of λ, at $Ca = 1.35$, from simulations [104]. The square and bullet symbols correspond to simulation measurements, while the plus symbols are obtained from the KS theory for fluid vesicles [104] using cell dimensions measured in simulations. (b) Time evolution ($t^* = \dot{\gamma}t$) of shapes at $\lambda = 3.5$ (starting from a simulation with $\lambda = 3.1$), when tank-treading becomes unstable and a trilobe is formed. (c) Membrane's elastic tension in the direction of the smallest (top) and largest (bottom) local strains. (d) Simulation of a vesicle of reduced volume=0.6, $\lambda = 3.7$, and bending capillary number of 6. From [147].

In fact, this is another example of a buckling transition mediated by the elasticity of the membrane, as shown in Figure 5.17(c), which displays membrane's elastic tension at four time instances with the corresponding shapes in Figure 5.17(b). In particular, one can see in Figure 5.17(c) that large parts of the membrane experience negative tension in both principal directions, which is most pronounced in regions where the membrane buckles out-of-plane. A similar appearance of local negative tension has been also reported for elastic capsules in shear flow [63]. A comparison of the inclination angle for simulated RBCs to the Keller and Skalak theory for fluid ellipsoids [86] in Figure 5.17(a) shows a good agreement for $\lambda \lesssim 0.5$, but strong discrepancies for $\lambda \gtrsim 0.5$ due to both the non-ellipsoidal shape of cells and the strong compressions appearing in the membrane, which lead to membrane buckling and negative inclination. When comparing to the dynamics of simple lipid vesicles [147] presenting a viscosity contrast at the same critical λ of 3.2, the regime of high capillary number shows indeed that the vesicles deform globally presenting a kind of flipping transition of their cross-section. The membrane is never locally in compression before the transition induced by the viscosity contrast, as in the case of RBCs.

5.6 ON THE INFLUENCE OF RED BLOOD CELLS' DYNAMICAL SHAPES ON BLOOD RHEOLOGY

5.6.1 Influence of the hematocrit on the shape of red blood cells under shear

To evaluate the prevalence of polylobed shapes at physiologically relevant shear rates and hematocrits (Hts), Lanotte and coworkers realized hardening experiments in rheometers of suspensions of RBCs in PBS solution with different Ht values of 5, 15, 22, 35, and 45%. The goal of the experiment was to observe the typical shapes taken by the cells in shear experiments of concentrated suspensions. $\dot{\gamma} = 900$ s^{-1} (at 37°C) has been chosen since it was shown that a maximum of the number of trilobes appear in a given population of cells at this shear rate [90]. Figure 5.18 shows a significant decrease in the probability to obtain polylobed RBCs for increasing Ht. The percentage of trilobes and hexalobes, which amounts to about 55% of the entire population at Ht = 5%, is more than halved when the Ht reaches 45%. In addition, at high enough Ht, many RBCs display a deformed shape with multiple irregular lobes, making it difficult to precisely classify their morphology. These cells were named 'multilobes' and are illustrated in Figure 5.18, Lower Inset. The fraction of these multilobed RBCs seems to be rather stable, amounting to 25%-35% of the total population. However, the onset of a new cell morphology is detected for increasing Ht. Flattened discocytes characterized by one or several grooves or creases on their membrane have been found and are illustrated in Figure 5.18, Upper Inset, by optical and confocal microscopy. The fraction of creased discocytes increases significantly at high Ht values, amounting to almost 50% of the sample at Ht = 45%. In contrast to the vast majority of polylobed RBCs in dilute suspensions, the formation of flat cells with creases is a clear effect of mutual hydrodynamic interactions between cells in such dense suspensions. Indeed, the total percentage of lobed cells with increasing Ht has the exact inverse evolution of that for creased discocytes, as shown in Figure 5.18. Interestingly, physiological Hts in the microcirculation lie in the range 15-30% (light yellow region in Figure 5.18), in which different cell morphologies coexist, with a majority of lobed shapes.

5.6.2 Revisiting shear-thinning

At end of the 1960s, Chien and collaborators showed in their seminal work that blood is a shear-thinning fluid [25]. In fact, blood is one of the most emblematic complex fluids of the literature. Chien and his coworkers demonstrated that there are in fact two regimes of shear-thinning. At shear rates lower than a few inverse seconds, the primarily decrease in viscosity is related to the breakage of the large aggregates that blood cells form in these conditions,

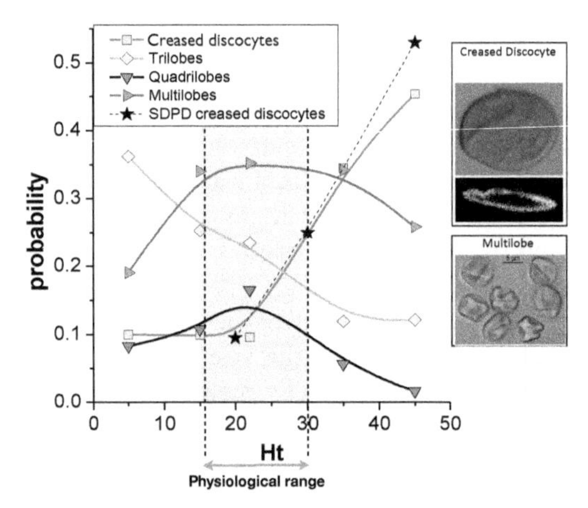

FIGURE 5.18
Shape distributions of RBCs at 900 s^{-1} in suspensions with different Ht. RBCs are hardened under flow to count the shapes after the experiment. For clarity, discocyte and stomatocyte number densities have been omitted because they are negligible. (Inset) Black frames on the right, a top view and a cross-section of a creased discocyte acquired by confocal microscopy as well as an image of multilobes in bright field microscopy. Physiological Hts are indicated by the light yellow region. From [90].

due to depletion forces induced by the presence of macromolecules such as fibrinogen in the plasma (see Chapter 6). The viscosity drops from 100 cP down to about 10 cP. However, for shear rates higher than 5-10 s^{-1}, the viscosity continues to drop down to 2-3 cP, notably due to the deformability of the cells: when RBCs were rigidified at rest in their discocyte shape, their viscosity remained constant around 10 cP as the shear rate was increased (example of such rheograms can be found in Figure 5.19).

In an attempt to better understand the behavior of RBCs in such a dense suspension and at high enough stress of 0.1-1 Pa, Fischer and his collaborators introduced their seminal study of the movement of single RBCs in viscous fluids using a rheoscope [62]. The high viscosity was used to both reach high enough stresses at lower shear rates for observational purposes and to mimic the presence of neighboring RBCs. However, it concomitantly reduced the viscosity ratio that exists physiologically between the inner hemoglobin cytoplasm and the plasma. Doing so, the studies showed the unique ability that had the RBCs to present a droplet-like behavior for high enough shear stresses. The cells maintain a stationary orientation relative to the direction of the flow thanks to the circulation of their membrane which produces a counter-viscous torque that resists the rotational torque applied by the shear flow. Therefore, the shear-thinning paradigm was rooted in RBCs' ability to tank-tread whatever the Ht in response to high shear stresses [54].

However, plasma viscosity under physiological conditions is about five times smaller than that of RBCs' cytosol. New studies have therefore investigated RBC local dynamics at such low physiological viscosity conditions and high shear rates [90]. It was observed that the solid-like tumbling motion is not replaced by tank-treading for increasing shear rates, but rather by a typical solid rolling motion, where the axis of symmetry of the discocyte lies in the direction of vorticity [42]. For stresses up to 0.5 Pa no fluidization of the membrane was observed, demonstrating the important role the inner-to-outer viscosity ratio plays in local dynamics. For further increase in shear stress, Lanotte and coworkers demonstrated that RBCs, rather than tank-treading, change their cross-section against the flow by folding into trilobes and hexalobes [90] (see sequence of shapes in Figure 5.8).

These observations change the interpretation of classical shear-thinning rheograms for human blood. In fact, in a simple experiment, Lanotte and coworkers could show how

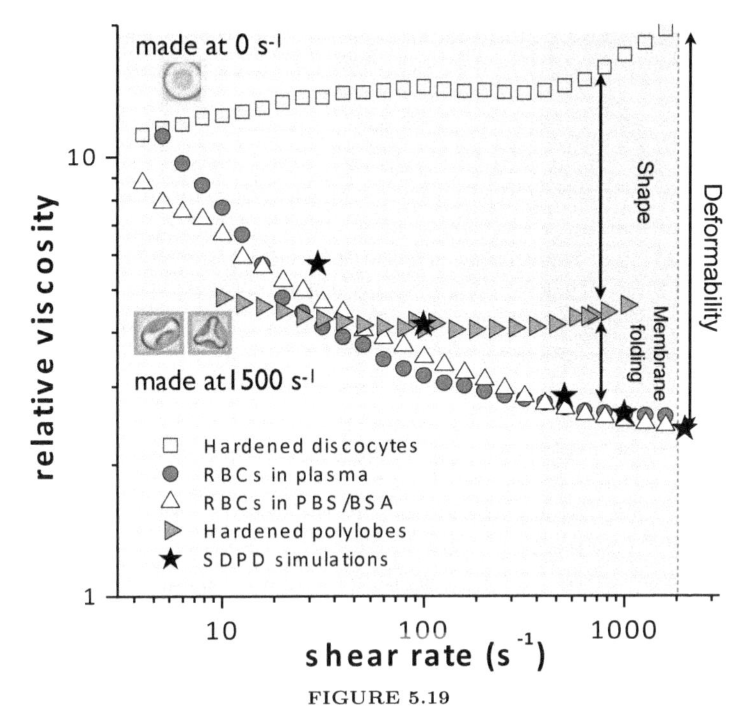

FIGURE 5.19
Rheology of dense suspensions of RBCs (Ht = 45%). (A) Relative viscosity of the suspensions of deformable RBCs in plasma (○) and in PBS/BSA (△) as a function of the shear rate in comparison with the suspensions of washed cells hardened at rest (□) and at 1500 s^{-1} (▷). Suspensions of deformable RBCs show a typical shear thinning for increasing shear rate, whereas hardened samples have a nearly Newtonian behavior. SDPD simulation data (black stars) are also shown for deformable cells and agree well with the experimental results for RBCs suspended in plasma or in PBS/BSA. From [90].

much these shape transitions contribute to the drop of viscosity. They performed rheology measurements at Ht=45% of blood cells suspension dispersed in saline (what they call washed blood) compared with that of two hardened RBC suspensions at same Ht: one with stiff discocytes hardened at rest and the other one with polylobed cells suspension obtained by hardening dilute suspensions of cells with glutaraldehyde in the rheometer at 1500 s^{-1} and resuspending them in PBS/BSA at the desired 45% Ht. Both suspensions showed nearly Newtonian behavior values with viscosity values larger than those for blood in saline. However, the sample with hardened polylobes yields a 70% lower viscosity than that for stiffened discocytes, as seen in Figure 5.19. These results indicate that from 10 s^{-1} up to a few hundred s^{-1} shear-thinning is largely determined by the change in RBCs' dynamics from tumbling to rolling and by the deformation of RBCs into elongated stomatocytes. Beyond 400 s^{-1}, however, a further viscosity drop is related to the formation of polylobed and flattened shapes discussed above.

In summary, blood shear-thinning is related to a rich behavior of RBCs in shear flow convoluted with a large distribution of cell shapes for any given flow condition. The lack of membrane fluidity for high viscosity contrast is the key feature that controls RBCs' behavior. Besides shear-thinning, several fundamental physiological phenomena have been analyzed under the assumption of membrane tank-treading, such as vasoregulatory ATP release by RBCs in strong shear flows [64] or the formation of a few-micron-thick cell-free layer adjacent to the vessel walls that is responsible for the apparent viscosity drop with decreasing vessel diameter, the so-called Fåhraeus-Lindqvist effect [51]. Future researchers should head into re-exploring both experimentally and theoretically blood rheology for physiologically relevant viscosity and stress conditions both in health and diseases.

5.7 CONCLUSION

The dynamics of an isolated RBC in pure shear flow has motivated a large number of experimental, theoretical and numerical studies over the last 5 decades. This configuration has aroused the interest of researchers not only to understand blood rheology, but also for itself, as a complex fluid-structure interaction problem between two fluids and a composite biological membrane. In this chapter, we have reviewed the rheological properties of the different elements of the problem: the internal and external fluids and the RBC membrane, notably in order to provide the reader with the information necessary to understand the modeling efforts that have been made to predict and shed light on the different motions of an RBC in shear flow. We have also reviewed these motions and the conditions under which they appear.

One of the most striking aspects about the research on the RBC dynamics in shear flow is the progressive construction of the knowledge along time. The dynamics of an RBC had long been interpreted as a dual solid-liquid behavior [79], in the light of the most emblematic movements of the single RBC: the solid-like flipping and the droplet-like tank-treading. However, after a clarification of the conditions under which those movements occur, researchers have found that the RBC dynamics is far richer, with off-shear plane motions and complex convoluted shapes identified, in particular over the last decade. The discovery of these motions has little by little shown the specificities of the movement of an RBC with respect to other simplified models (solid particles, droplet, vesicles, capsules).

Theoretical models have been developed to explain the different motions and the transitions between them, in particular at low shear stresses, where the RBC deformations remain small. Full numerical simulations solving the fluid-structure interaction problem have also shown their ability to reproduce the experimental results and have proven useful to interpreting them by giving access to quantities that are not possible to extract experimentally. Both theoretical models and simulations have underlined the importance of the cytoskeleton elasticity in the motions of the RBC. However, if the full phase diagram of an RBC under shear flow has become clearer as a function of the shear stress and the viscosity of the suspending fluid, it remains to be explained in detail: some particular dynamics and above all the transitions between them are still not understood, especially at high shear rates.

Finally, the fine knowledge of the dynamics of one RBC in shear flow has not been fully included in a more integrated view of the behavior of blood, where RBCs are generally concentrated at high volume fractions. The relation between the individual dynamics of RBCs, their interactions over short and large distances and the consequences on the effective behavior of the suspension remains to be explored.

5.8 APPENDIX: A DICTIONARY OF THE DYNAMICS OF AN RBC IN SHEAR FLOW

Tens, maybe hundreds of papers have been published on the motions of a single RBC in shear flow. Due to the richness of the problem, they often study a specific motion or a particular region of the phase diagram. The motion has been referred to using a wealth of metaphoric terms. In order to clarify the discussion and gather the results, it seemed important to propose a dictionary of the RBC dynamics. The order in which they are presented follows more or less the chronological order in which they were identified. The reader is referred to Figure 5.9 for the phase diagram specifying under which conditions such motions are observable. Citations of some classical papers have been reported in italics.

All images used in this section of time sequences of numerical simulations solving the fluid-membrane interaction problem use an in-house finite-volume software based on the immersed boundary method. The membrane is modeled as infinitely thin, its in-plane resistance is defined by the Skalak model and its bending resistance using the Helfrich model (see Sec. 5.2). The stress-free shape of the cytoskeleton is a quasi-spherical ellipsoid. Details about the numerical methods and validation are provided in existing publications [90, 104, 107, 108, 142, 143].

5.8.1 Flipping or Tumbling

One of the most studied motions of a single RBC under shear flow is called 'flipping' or 'tumbling'. During flipping, the cell behaves like a solid disk in a pure shear flow. It rotates indefinitely, with a rotation speed depending on its orientation with respect to the flow. This motion, identified by Goldsmith and Marlow [69], has been studied by analogy with an oblate ellipsoid in pure shear flow at zero Reynolds number, for which the theoretical results from Jeffery [82] hold. The axis of symmetry of the RBC precesses around the vorticity direction, describing a closed trajectory named 'orbit' by Jeffery. If φ denotes the angle between the vorticity direction and the symmetry axis of the RBC, φ varies between two extrema during a Jeffery orbit: $tan(\varphi_1) = Cr_p$ and $tan(\varphi_2) = C$, where r_p is the aspect ratio of the ellipsoid in Jeffery's theory and C is a constant characterizing the orbit [3, 69]. Goldsmith and Marlow identified two limiting cases: *(i) when C=0 the axis of revolution lines up with the vorticity axis, and the disk then spins about its own axis in the median plane without change in orientation of its upper face; (ii) when $C = \infty$, $\varphi = \pi/2$ throughout the orbit and the axis of revolution describes a circle lying in the XY-plane and the particle is then seen rotating edge on.* The two limiting cases are generally called 'rolling' (see specific section further) and 'tumbling'. Examples of orbits are displayed in Figure 5.20 and in Figure 5.21. Goldsmith and Marlow [69] observed that at low shear rates, the RBC follows the dynamics of a solid oblate ellipsoid having a aspect ratio of $r_p \approx 0.38$, while the geometric aspect ratio of an RBC (thickness over diameter) is closer to 0.24 [69].

However, even at low shear rates, RBCs do not behave exactly as solid objects. Jeffery [82] demonstrated that the orbit of an ellipsoid only depends on its initial orientation, the shear rate only controlling the speed at which this orbit is followed. In other terms, Jeffery's orbits are degenerate. This is not the case for RBCs. Bitbol [17] and Dupire and coworkers [42] identified that the orbit of an RBC depends on the shear rate. More precisely, the RBC drifts from orbits close to the tumbling orbit (orbit $C = \infty$) to the rolling motion (orbit $C = 0$) with shear rate. During this drift, the RBC remains a discocyte and no major deformation of the RBC shape is observed [42].

The flipping motion of RBCs is a characteristic dynamics of low shear stresses and has been reported both for low and high-viscosity suspending medium [17, 42, 69].

Note that the particular case where the symmetry axis lies in the shear plane has received much attention, notably because it is simpler to study theoretically [1, 41, 86] and numerically and this tumbling is also characteristic of vesicles. The transition between this motion and tank-treading (see Sec. 5.8.2) has also been extensively studied [1, 61, 166].

Remarks on flipping:

- The orbital drift takes place for a range of applied shear stress which weakly depends on the external viscosity [42, 107, 110].

- The orbits at low shear stress are close to the shear plane, but are not stable in time [42,110].
- Small displacements of the membrane are predicted by models even without shape deformations [41,107].
- This regime is sometimes referred to as kayaking [144] or wobbling [3,26,43].

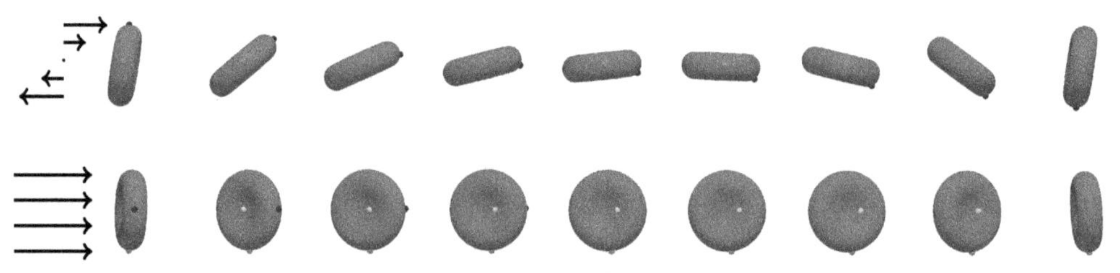

FIGURE 5.20
Time series of an RBC flipping in shear flow with an orbit very close to the shear plane (often called tumbling): views from the vorticity direction (first row) and shear gradient direction (second row). Case $v_{ref} = 0.997$, $Ca = 0.018$, $\lambda = 0.2$. The time between two images is $1.0\,\dot{\gamma}^{-1}$. The white bead marks the short axis of the reference shape and is located in the dimple at rest. The blue and the green beads are located at the rim at rest.

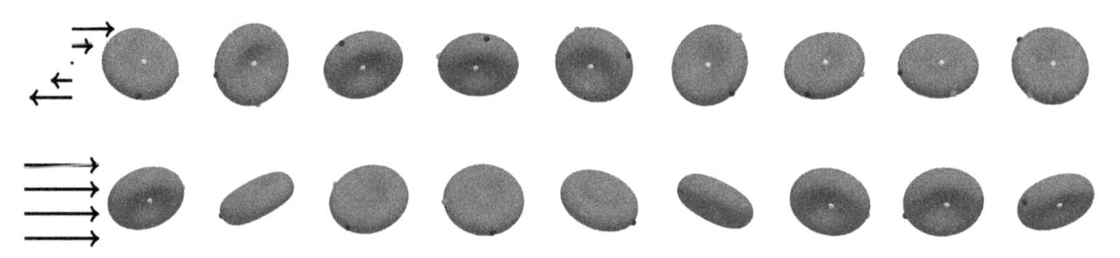

FIGURE 5.21
Time series of an RBC flipping in shear flow: views from the vorticity direction (first row) and shear gradient direction (second row). Case $v_{ref} = 0.96$, $Ca = 0.015$, $\lambda = 1$. The time between two images is $2.5\,\dot{\gamma}^{-1}$. The white bead marks the short axis of the reference shape and is located in the dimple at rest. The blue and the green beads are located at the rim at rest.

5.8.2 Tank-treading

At the end of the 1960s, Goldsmith [68] observed the behavior of ghost RBCs in suspension and made the analogy with fluid droplets. To test the hypothesis that an RBC, when subjected to high shear, *assumes the flow properties of a fluid drop*, Schmid-Schönbein and Wells [140] observed RBCs in dextran solutions flowing in a rheoscope. The viscosity of the external fluid was thus much higher (62 cP) than that of plasma, which has the advantages of slowing down sedimentation and enabling the application of high stresses at low shear rates. Except for very low shear rates, they observed that RBCs align with the flow direction and deform into prolate ellipsoids. They also reported a circulation of the cell membrane, later beautifully displayed by Fischer and coworkers [62], in particular. This rotational motion of the cell membrane was said to *resemble the rotation of the tread of a tank around its wheels* [140], hence the term tank-treading. An example of a tank-treading cell is shown in Figure 5.22. This observation, together with the fact that similar tank-treading was observed in a suspension of RBCs by the same authors, led to the paradigm that tank-treading was responsible for blood shear-thinning at high shear rates [140], which is now challenged by the recent discoveries about the motions of RBCs at high values of the shear rate and of the viscosity ratio [90].

Tank-treading was known to be a characteristic motion of single RBCs suspended in a viscous medium (low viscosity ratio) and high shear rates. Keller and Skalak [86] predicted in a theoretical work that the viscosity ratio has to be lower than a threshold for a liquid-filled capsule to tank-tread. For an RBC, this threshold was about 2.8-3.0, which has been confirmed in numerical simulations, in particular.

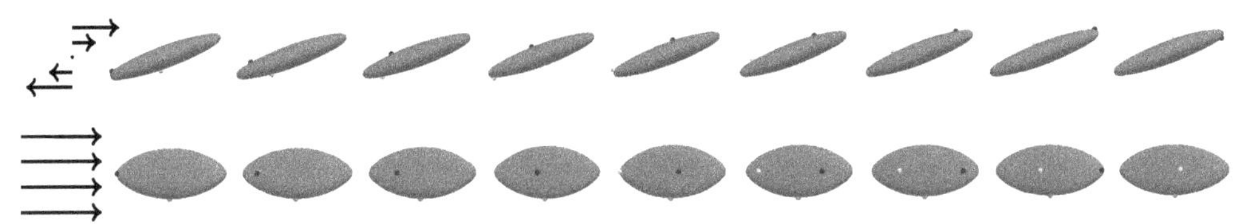

FIGURE 5.22
Time series of an RBC tank-treading under shear flow: views from the vorticity direction (first row) and shear gradient direction (second row). Case $v_{ref} = 0.997$, $Ca = 1.85$, $\lambda = 0.2$. The time between two images is $5\,\dot{\gamma}^{-1}$. The white bead marks the short axis of the reference shape and is located in the dimple at rest. The blue and the green beads are located at the rim at rest.

Tank-treading has been extensively studied over the past fifty years, in particular to explain the transition between flipping and tank-treading with shear rate. However, numerous results have been obtained using vesicles, which exhibit both tumbling and tank-treading motions and the transition with viscosity ratio.

Remarks on tank-treading:

- While RBCs keep their discocyte shape while tank-treading at low shear rates, they then elongate in the flow direction and become prolate ellipsoids of increasing aspect ratios. Their orientation is also more and more aligned with the flow.

- The tank-treading frequency is almost linear with shear rate [59], with a slope slightly increasing with the external viscosity. [122] found that confinement makes this dependence quasi-linear, while a non-linear dependence is predicted in less confined situation, with the frequency varying like $\dot{\gamma}^{\beta}$, with $\beta \approx 0.9$.

5.8.3 Swinging

In 2007, Abkarian, Faivre and Viallat presented side views of RBCs tank-treading in a viscous medium [1]. They observed that RBCs' tank-treading is not identical to that of a droplet: during the circulation of the membrane, the inclination of the RBCs is not constant, but oscillates with twice the frequency of the tank-treading. They refer to this motion as swinging (Figure 5.23). Swinging is all the more visible when the shear rate is low, the amplitude of the angle oscillations decreasing with the applied shear rate. This motion was understood in the light of low-order modeling of the RBC dynamics, by adding the contribution of membrane elasticity [1, 146] to Keller and Skalak's model [86]. Fischer [58] has demonstrated that RBCs have shape memory: this means that all elements of the cytoskeleton are not identical. Therefore, while the membrane circulates around the cytoplasm during tank-treading, the cytoskeleton stores then releases mechanical energy. This perturbs the motion of the membrane and the orientation of the RBC.

Swinging can really be seen as a perturbation of tank-treading due to membrane elasticity. Swinging occurs only when the membrane circulates around the RBC shape, which means that it is necessary to suspend the cell in a viscous medium to observe swinging. Swinging is barely visible when cells are deformed (high shear stresses). Therefore, it is characteristic of intermediate shear stresses, between flipping and deformed tank-treading.

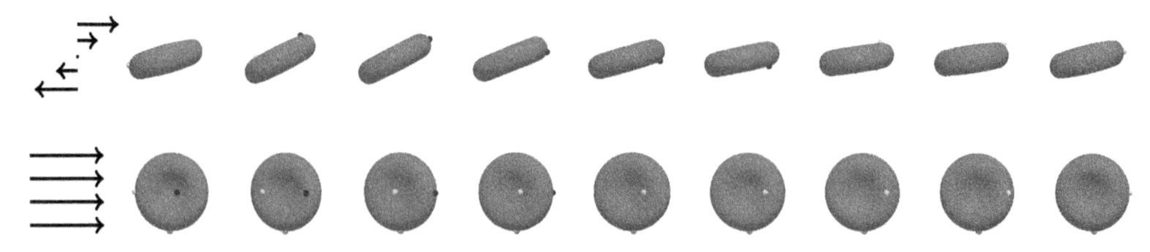

FIGURE 5.23
Time series of an RBC swinging under shear flow: views from the vorticity direction (first row) and shear gradient direction (second row). Case $v_{ref} = 0.997$, $Ca = 0.04$, $\lambda = 0.2$. The time between two images is $5\,\dot{\gamma}^{-1}$. The white bead marks the short axis of the reference shape and is located in the dimple at rest. The blue and the green beads are located at the rim at rest.

Remarks on swinging:

- Vesicles do not have membrane elasticity. They do not swing in shear flow. Note however that they can present angle oscillations and shape breathing during tank-treading, a motion often referred to as vacillating-breathing [52, 111].

- Non-spherical elastic capsules may also present angle oscillations [132], but they are associated with large deformations. It is now clear that the particular swinging motion of RBCs with angle oscillations without shape changes is due to the particular stress-free shape of the RBC, close to a sphere rather than to its resting shape. Simulations of RBCs' dynamics initially failed in reproducing the swinging motion without deformation as they considered a biconcave reference shape [166]. See the detailed discussion in Sec. 5.4.5.

- For low values of shear stress, swinging and flipping may alternate: an intermittent regime is found [1, 27, 41, 146]. Low-order models however predict that such a motion is unstable and only obtained when the RBC is forced to remain in the shear plane.

5.8.4 Rolling

Rolling motion is a particular case of RBC flipping, in which the small axis of the cell is aligned with the vorticity axis (often referred to as orbit C=0). In that case, the RBC rotates around its axis of symmetry at the same angular velocity as a solid disk, if undeformed. The RBC thus behaves as a wheel rolling in the flow (Figure 5.24). Rolling is the final state of the orbital drift of flipping cells with increasing shear rate [17, 42, 43, 107]. It has been reported for all viscosity ratios. However, the range of stability for rolling is larger when the viscosity ratio is high. In a high-viscosity medium, a rolling RBC transitions to swinging when the shear rate is further increased [42]. In a low-viscosity medium, as the shear stress increases, the cell can be deformed in the rolling motion. This has been proposed in low-viscosity ektacytometry as a way of measuring the membrane shear modulus [164].

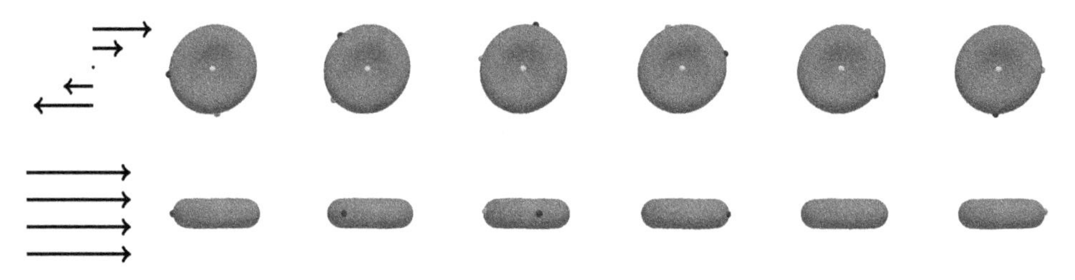

FIGURE 5.24
Time series of an RBC rolling under shear flow: views from the vorticity direction (first row) and shear gradient direction (second row). Case $v_{ref} = 0.96$, $Ca = 0.024$, $\lambda = 5.0$. The time between two images is $2\,\dot{\gamma}^{-1}$. The white bead marks the short axis of the reference shape and is located in the dimple at rest. The blue and the green beads are located at the rim at rest.

Remarks on rolling:

- No rolling has been reported for vesicles. Non-spherical capsules may roll in shear flow, yet with a deformed shape [43].

- When the deformation remains small, the angular velocity of rolling cells is close to that of a solid disk, $\dot{\gamma}/2$.

5.8.5 Kayaking

The word 'kayaking' is often used as a synonym of flipping along Jeffery orbits, in particular for prolate ellipsoids [101, 138]. In the context of RBC dynamics it has thus been used to describe the flipping motion at low shear stresses (Sec. 5.8.1), but also a motion where the RBC small axis oscillates about the shear plane [26, 43]. This motion, however, has been shown to be unstable, and characteristic of an RBC initially tilted with respect to the shear plane and reaching the swinging motion with transient oscillations about the shear plane due to membrane circulation [43]. We will not use the term 'kayaking' in this chapter as its definition is not consistent over the RBC literature.

5.8.6 Frisbee/Hovering

Several papers mention a type of Frisbee motion, also called hovering [29, 42, 107, 144] in which the small axis of the cell has a constant orientation, tilted with respect to the shear plane and the vorticity direction and spins around a fixed axis. It could be viewed as an inclined rolling (Figure 5.25). This motion has been found to be stable in numerical simulations [29,107,144], but in a limited range of shear rate, between those when rolling and swinging are obtained, and at low viscosity ratio. Frisbee has been observed experimentally as a transient state between rolling and swinging when the shear rate is increased.

Depending on the initial orientation of an RBC in simulations, swinging or hovering may be obtained. Interestingly, the region of the phase diagram where frisbee is seen has been found to present hysteresis in experiments [42, 110] and models (see Sec. 5.4.3).

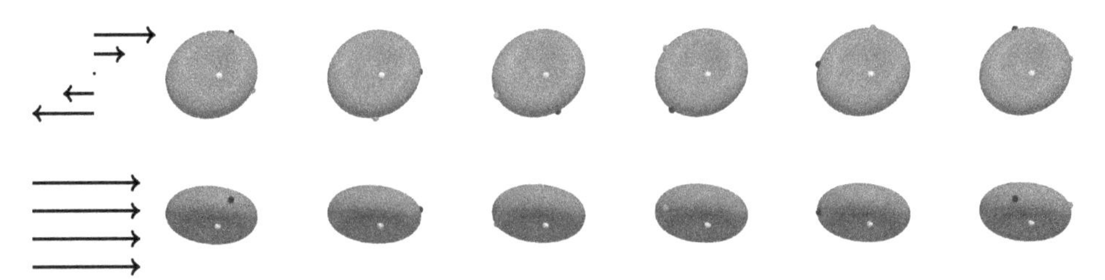

FIGURE 5.25
Time series of an RBC hovering under shear flow: views from the vorticity direction (first row) and shear gradient direction (second row). Case $v_{ref} = 0.997$, $Ca = 0.031$, $\lambda = 0.2$. The time between two images is $5\,\dot{\gamma}^{-1}$. The white bead marks the short axis of the reference shape and is located in the dimple at rest. The blue and the green beads are located at the rim at rest.

5.8.7 Dynamic stomatocyte

The word stomatocyte refers to a particular shape of RBCs, which may look like a cup. The term comes from the word *stoma* in Greek, mouth. Stomatocytes present only one deep concavity instead of the two concavities of the discocyte. Stomatocytes may be characteristic of some hereditary or acquired hemolytic states or obtained by changing the pH of the ambient medium, for instance [14].

However, it has been recently discovered that RBCs can become stomatocytes as a response to shear stress, in a stable yet reversible way. At high viscosity ratio, when increasing the shear rate, the cell flips over different orbits drifting to rolling, then rolls. As the shear rate is further increased, the cell extends in an intermediate direction between the velocity gradient and the flow directions. However, the cell does not stay discocyte when the deformation is too large and one of the dimples buckles [90, 104].

The RBC then performs a kind of rolling stomatocyte motion, which is similar with the classical rolling motion with higher deformation and one dimple buckled (Figure 5.26). For higher shear rates, rolling stomatocyte is not stable and the stomatocyte reorients its small axis to the shear plane. The cell then tumbles in its stomatocyte shape (Figure 5.27).

For some combinations of stress-free shape and low viscosity ratio, some simulations have displayed flipping stomatocytes, the buckling of the discocytes intervening for lower shear stress than that where orbital drift is complete.

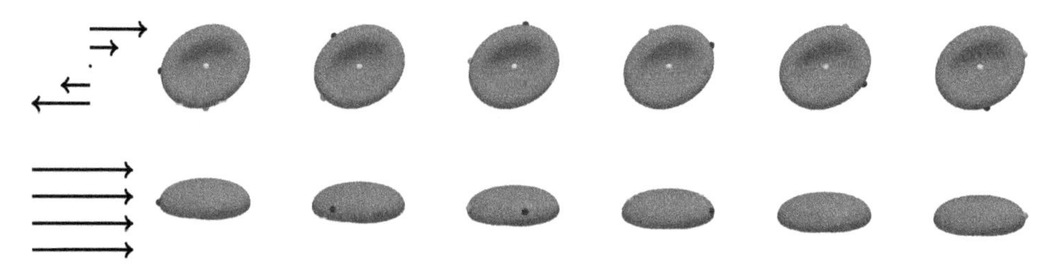

FIGURE 5.26
Time series of an RBC rolling under shear flow in a stomatocyte shape. Views from the vorticity direction (first row) and shear gradient direction (second row). Case $v_{ref} = 0.96$, $Ca = 0.07$, $\lambda = 5.0$. The time between two images is $2\,\dot{\gamma}^{-1}$. The white bead marks the short axis of the reference shape and is located in the dimple at rest. The blue and the green beads are located at the rim at rest.

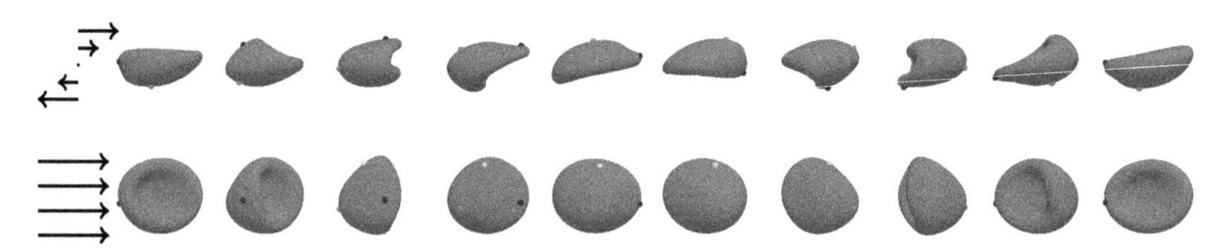

FIGURE 5.27
Time series of an RBC tumbling under shear flow in a stomatocyte shape. Views from the vorticity direction (first row) and shear gradient direction (second row). Case $v_{ref} = 0.96$, $Ca = 0.58$, $\lambda = 5.0$. The time between two images is $2.5\,\dot{\gamma}^{-1}$. The white bead marks the short axis of the reference shape and is located in the dimple at rest. The blue and the green beads are located at the rim at rest.

Remarks on dynamically obtained stomatocytes:

- Depending on the reference stress-free shape, stomatocytes may also be obtained in numerical simulations (unpublished author data) at low viscosity ratio.

- During the tumbling stomatocyte dynamics, simulations have shown that the membrane barely circulates. The higher the internal viscosity the less circulation is observed.

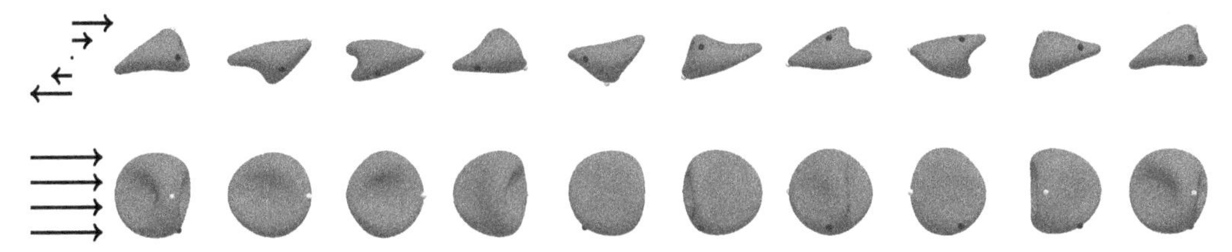

FIGURE 5.28
Time series of a trilobe tumbling under shear flow. Views from the vorticity direction (first row) and shear gradient direction (second row). Case $v_{ref} = 0.96$, $Ca = 0.81$, $\lambda = 5.0$. The time between two images is $2.5\,\dot{\gamma}^{-1}$. The white bead marks the short axis of the reference shape and is located in the dimple at rest. The blue and the green beads are located at the rim at rest.

5.8.8 Polylobes

A detailed study of the dynamics of isolated RBCs at high shear rates and physiological viscosity ratios have shown that the characteristic shapes displayed by RBCs in this regime are polylobed shapes [90, 104]. For shear rates above a few hundreds inverse of seconds, RBCs become trilobes that tumble in the shear flow, as shown in Figure 5.28. During the trilobe dynamics, simulations have shown that the membrane barely circulates [90]. The higher the internal viscosity, the less circulation is observed. Experiments at shear rates above 1000 s^{-1} have also reported pyramid/hexalobe shapes that have not been found stable in reported simulations [90].

Bibliography

[1] M. Abkarian, M. Faivre, and A. Viallat. Swinging of red blood cells under shear flow. *Phys. Rev. Lett.*, 98(188302), 2007.

[2] M. Abkarian and A. Viallat. Dynamics of vesicles in a wall-bounded shear flow. *Biophys. J.*, 89(02):1055–1066, 2005.

[3] M. Abkarian and A. Viallat. *Fluid-Structure Interactions in Low-Reynolds-Number Flows*, chapter: On the importance of red blood cells deformability in blood flow, pages 347–462. Royal Society of Chemistry, 2016.

[4] D. Abreu, M. Levant, V. Steinberg, and U. Seifert. Fluid vesicles in flow. *Adv. Coll. & Interf. Sc.*, 208:129–141, 2014.

[5] G. M. Artmann, C. Kelemen, D. Porst, G. Buldt, and S. Chien. Temperature transitions of protein properties in human red blood cells. *Biophys. J*, 75:3179–3183, 1998.

[6] C. Baecher, A. Kihm, L. Schrak, L. Kaestner, M. W. Laschke, C. Wagner, and S. Gekle. Antimargination of microparticles and platelets in the vicinity of branching vessels. *Biophys. J.*, 115:411–425, 2018.

[7] P. Balogh and P. Bagchi. Direct numerical simulation of cellular-scale blood flow in 3d microvascular networks. *Biophys. J.*, 113:2815–2826, 2017.

[8] B. Barthès-Biesel. Motion and deformation of elastic capsules and vesicles in flow. *Ann. Rev. Fluid Mech.*, 48:25–52, 2016.

[9] D. Barthès-Biesel. Capsule motion in flow: Deformation and membrane buckling. *Comp. Rend. Phys.*, 10(8):764–774, 2009.

[10] D. Barthès-Biesel and J. M. Rallison. The time-dependent deformation of a capsule freely suspended in a linear shear flow. *J. Fluid Mech.*, 113:251–267, 1981.

[11] O. Baskurt, B. Neu, and H.J. Meiselman. *Red Blood Cell Aggregation.* CRC Press, Boca Raton, New York, 2012.

[12] J. Beaucourt, F. Rioual, T. Séon, T. Biben, and C. Misbah. Steady to unsteady dynamics of a vesicle in a flow. *Phys. Rev. E*, 69(011906), 2004.

[13] M. Bessis, N. Mohandas, and C. Feo. Automated ektacytometry: a new method of measuring red cell deformability and red cell indices. *Blood Cells*, 6:315–327, 1979.

[14] M. Bessis, R. I. Weed, and P. F. Leblond, editors. *RED CELL SHAPE. Physiology, Pathology, Ultrastructure.* Springer Verlag, 1973.

[15] T. Betz, M. Lenz, J.-F. Joanny, and C. Sykes. Atp-dependent mechanics of red blood cells. *Proc. Natl. Acad. Sci. USA*, 106:15312–15317, 2009.

[16] T. Biben and C. Misbah. Tumbling of vesicles under shear flow within an advected-field approach. *Phys. Rev. E*, 67(031908), 2003.

[17] M. Bitbol. Red blood cell orientation in orbit C=0. *Biophys. J.*, 49:1055–1068, 1986.

[18] G. Boedec, M. Jaeger, and M. Leonetti. Settling of a vesicle in the limit of quasi-spherical shapes. *J. Fluid Mech.*, 690:227–261, 2012.

[19] N. Borghi and F. Brochard-Wyart. Tether extrusion from red blood cells: Integral proteins unbinding from cytoskeleton. *Biophys. J.*, 93, 2007.

[20] K.E. Bremmell, A. Evans, and C.A. Prestidge. Deformation and nano-rheology of red blood cells: An afm investigation. *Colloids and Surf. B*, 50:43–48, 2006.

[21] F. Brochard-Wyart, N. Borghi, D. Cuvelier, and P. Nassoy. Hydrodynamic narrowing of tubes extruded from cells. *Proc. Natl. Acad. Sci. USA*, 103:7660–7663, 2006.

[22] M. Brust, C. Schaefer, R. Doerr, L. Pan, M. Garcia, P.E. Arratia, and C. Wagner. Rheology of human blood plasma: Viscoelastic versus newtonian behavior. *Phys. Rev. Lett.*, 110:078305, 2013.

[23] S. Charache, C. Lockard Conley, D. F. Waugh, R. J. Ugoretz, and E. Gayle J. R. Spurrell. Pathogenesis of hemolytic anemia in homozygous hemoglobin c disease. *J. Clin. Invest.*, 46:1795–1811, 1967.

[24] S. Chien, S. Usami, and J.F. Bertles. Abnormal rheology of oxygenated blood in sickle cell anemia. *J. Clin. Invest.*, 49:623–634, 1970.

[25] S. Chien, S. Usami, R.J. Dellenback, and M.I. Gregersen. Blood viscosity: Influence of erythrocyte deformation. *Science*, 157:827–831, 1967.

[26] D. Cordasco and P. Bagchi. Orbital drift of capsules and red blood cells in shear flow. *Phys. Fluids*, 25(091902), 2013.

[27] D. Cordasco and P. Bagchi. Intermittency and synchronized motion of red blood cell dynamics in shear flow. *J. Fluid Mech.*, 759:472–488, 2014.

[28] D. Cordasco and P. Bagchi. On the shape memory of red blood cells. *Phys. Fluids*, 29(041901), 2017.

[29] D. Cordasco, Yazdani, and P. Bagchi. Comparison of erythrocyte dynamics in shear flow under different stress-free configurations. *Phys. Fluids*, 26(041902), 2014.

[30] B. Daily, E.L. Elson, and G.I. Zahalak. Cell poking : Determination of the elastic area compressibility modulus of the erythrocyte membrane. *Biophys. J.*, 45:671–682, 1984.

[31] M. Dao, J. Li, and S. Suresh. Molecularly based analysis of deformation of spectrin network and human erythrocyte. *Mat. Sci. Eng. C*, 26:1232–1244, 2006.

[32] M.C. Dao, C.T. Lim, and S. Suresh. Mechanics of the human red blood cell deformed by optical tweezers. *J. Mech. and Phys. Solids*, 51:2259–2280, 2003.

[33] S. Deutsch, J. M. Tarbell, K. B. Manning, G. Rosenberg, and A. A. Fontaines. Experimental fluid mechanics of pulsatile artificial blood pumps. *Ann. Rev. Fluid Mech.*, 38:65–86, 2006.

[34] P. Dimitrakopoulos. Analysis of the variation in the determination of the shear modulus of the erythrocyte membrane: Effects of the constitutive law and membrane modeling. *Phys. Rev. E*, 85:041918, 2012.

[35] J.G.G. Dobbe, M.R. Hardeman, G.J. Streekstra, J. Strackee, C. Ince, and C.A. Grimbergen. Analyzing red blood cell-deformability distributions. *Blood Cells, Molecules, and Diseases.*, 28:373–384, 2002.

[36] W. R. Dodson III and P. Dimitrakopoulos. Tank-treading of erythrocytes in strong shear flows via a nonstiff cytoskeleton-based continuum computational modeling. *Biophys. J.*, 99:2906–2916, 2010.

[37] W. R. Dodson III and P. Dimitrakopoulos. Oscillatory tank-treading motion of erythrocytes in shear flows. *Phys. Rev. E*, 84(011913), 2011.

[38] W. R. Dodson III and P. Dimitrakopoulos. Tank-treading of swollen erythrocytes in shear flows. *Phys. Rev. E*, 85(021922), 2012.

[39] A. Drochon. Rheology of dilute suspensions of red blood cells: experimental and theoretical approaches. *Eur. Phys. J. Appl. Phys.*, 22:155–162, 2003.

[40] A. Drochon. Use of cell transit analyser pulse height to study the deformation of erythrocytes in microchannels. *Med. Eng. Phys.*, 27:157–165, 2005.

[41] J. Dupire, M. Abkarian, and A. Viallat. A simple model to understand the effect of membrane shear elasticity and stress-free shape on the motion of red blood cells in shear flow. *Soft Mat.*, 11:8372–8382, 2015.

[42] J. Dupire, M. Socol, and A. Viallat. Full dynamics of a red blood cell in shear flow. *Proc. Natl Acad. Sc. USA*, 109(51):20808–20813, 2012.

[43] C. Dupont, F. Delahaye, B. Barthès-Biesel, and A.-V. Salsac. Stable equilibrium configurations of an oblate capsule in simple shear flow. *J. Fluid Mech.*, 791:738–757, 2016.

[44] C. D. Eggleton and A. S. Popel. Large deformation of red blood cell ghosts in a simple shear flow. *Phys. Fluids*, 10(8):1834–1845, 1998.

[45] H. Engelhardt and E. Sackmann. On the measurement of shear elastic-moduli and viscosities of erythrocyte plasma-membranes by transient deformation in high-frequency electric-fields. *Biophys. J.*, 54:495–508, 1988.

[46] E. Evans. Structure and deformation properties of red blood cells: concepts and quantitative methods. *Method Enzymol.*, 173:3–35, 1989.

[47] E. Evans and Y.C. Fung. Improved measurements of the erythrocyte geometry. *Microvasc. Res.*, 4:335–347, 1972.

[48] E. Evans and R. Skalak, editors. *Mechanics and Thermodynamics of Biomembranes.* CRC Press, 1980.

[49] E.A. Evans. Constitutive relation for red cell membrane. *Biophys. J.*, 16:597–600, 1976.

[50] E.A. Evans and R.M. Hochmuth. Membrane viscoelasticity. *Biophys. J.*, 16:1–11, 1976.

[51] R. Fahraeus and T. Lindqvist. The viscosity of the blood in narrow capillary tubes. *Am. J. Physiol.*, 96:562–568, 1931.

[52] A. Farutin and C. Misbah. Squaring, parity breaking, and S tumbling of vesicles under shear flow. *Phys. Rev. Lett.*, 109(248106), 2012.

[53] D. A. Fedosov, B. Caswell, and G. E. Karniadakis. A multiscale red blood cell model with accurate mechanics, rheology, and dynamics. *Biophys. J.*, 98:2215–2225, 2010.

[54] D. A. Fedosov, W. Pan, B. B. Caswell, and G. E. Karniadakis. Predicting human blood viscosity in silico. *Proc. Natl. Acad. Sci. USA*, 108:11772–11777, 2011.

[55] D. A. Fedosov, M. Peltomäki, and G. Gompper. Deformation and dynamics of red blood cells in flow through cylindrical microchannels. *Soft Mat.*, 10:4258–4267, 2014.

[56] T. M. Fischer. On the energy dissipation in a tank-treading human red blood cell. *Biophys. J.*, 32:863–868, 1980.

[57] T. M. Fischer. Is the surface area of the red cell membrane skeleton locally conserved? *Biophys. J.*, 61:298–305, 1992.

[58] T. M. Fischer. Shape memory of human red blood cells. *Biophys. J.*, 86:3304–3313, 2004.

[59] T. M. Fischer. Tank-tread frequency of the red cell membrane: Dependence on the viscosity of the suspending medium. *Biophys. J.*, 93:2553–2561, 2007.

[60] T. M. Fischer, C. W. M. Haest, M. Stöhr-Liesen, H. Schmid-Schönbein, and R. Skalak. The stress-free shape of the red blood cell membrane. *Biophys. J.*, 34:409–422, 1981.

[61] T. M. Fischer and R. Korzeniewski. Threshold shear stress for the transition between tumbling and tank-treading of red blood cells in shear flow: dependence on the viscosity of the suspending medium. *J. Fluid Mech.*, 736:351–365, 2013.

[62] T. M. Fischer, M. Stöhr-Liesen, and H. Schmid-Schönbein. The red cell as a fluid droplet: Tank tread-like motion of the human erythrocyte membrane in shear flow. *Science*, 202:894–896, 1978.

[63] E. Foessel, J. Walter, A.-V. Salsac, and D. Barthès-Biesel. Influence of internal viscosity on the large deformation and buckling of a spherical capsule in a simple shear flow. *J. Fluid Mech.*, 672:477–486, 2011.

[64] A. M. Forsyth, J. Wan, P. D. Owrutsky, M. Abkarian, and Stone H. A. Multiscale approach to link red blood cell dynamics, shear viscosity, and ATP release. *Proc. Natl Acad. Sc. USA*, 108(27):10986–10991, 2011.

[65] J. B. Freund. Numerical simulation of flowing blood cells. *Ann. Rev. Fluid Mech.*, 46:67–95, 2014.

[66] J. B. Freund and M. M. Orescanin. Cellular flow in a small blood vessel. *J. Fluid Mech.*, 671:466–490, 2011.

[67] Y.C. Fung. *Biomechanics: Mechanical Properties of Living Tissues.* Springer, New York, 2 edition, 1993.

[68] H. L. Goldsmith. The microrheology of red blood cell suspensions. *J. Gen. Physiol.*, 52(1):5–28, 1968.

[69] H. L. Goldsmith and J. Marlow. Flow behaviour of erythrocytes. I. Rotation and deformation in dilute suspensions. *Proc. R. Soc. Lond. B*, 182:351–384, 1972.

[70] M. Gross, T. Krüger, and F. Varnik. Rheology of dense suspensions of elastic capsules: normal stresses, yield stress, jamming and confinement effects. *Soft Mat.*, 10:4360–4372, 2014.

[71] J. Guck, R. Ananthakrishnan, H. Mahmood, T.J. Moon, C.C. Cunningham, and J. Kas. The optical stretcher: A novel laser tool to micromanipulate cells. *Biophys. J.*, 81:767–784, 2001.

[72] J.C. Hansen, R. Skalak, S. Chien, and A. Hoger. An elastic network model based on the structure of the red blood cell membrane skeleton. *Biophys. J.*, 70:146–166, 1996.

[73] J.C. Hansen, R. Skalak, S. Chien, and A. Hoger. Influence of network topology on the elasticity of the red blood cell membrane skeleton. *Biophys. J.*, 72:2369–2381, 1997.

[74] M.R. Hardeman, P.T. Goedhart, Dobbe J.G.G., and K.P. Lettinga. Laser-assisted optical rotational cell analyser (l.o.r.c.a.); I. A new instrument for measurement of various structural hemorheological parameters. *Clin. Hemorheol.*, 14:605–618, 1994.

[75] A. Hategan, R. Law, S. Kahn, and D.E. Discher. Adhesively-tensed cell membranes: Lysis kinetics and atomic force microscopy probing. *Biophys. J.*, 85:2746–2759, 2003.

[76] V. Heinrich, K. Ritchie, N. Mohandas, and E. Evans. Elastic thickness compressibilty of the red cell membrane. *Biophys. J.*, 81:1452–1463, 2001.

[77] W. Helfrich. Elastic properties of lipid bilayers: Theory and possible experiments. *Z. Naturforsch*, 28 c:693–703, 1973.

[78] S. Henon, G. Lenormand, A. Richert, and F. Gallet. A new determination of the shear modulus of the human erythrocyte membrane using optical tweezers. *Biophys. J.*, 76(2):1145–1151, 1999.

[79] R.M. Hochmuth. Solid and liquid behavior of red cell membrane. *Ann. Rev. Biophys. Bioeng.*, 11:43–55, 1982.

[80] R.M. Hochmuth and R.E. Waugh. Erythrocyte membrane elasticity and viscosity. *Ann. Rev. Physiol.*, 49:209–219, 1987.

[81] W.C. Hwang and R.E. Waugh. Energy of dissociation of lipid bilayer from the membrane skeleton of red blood cells. *Biophys. J.*, 72(6):2669–2678, 1997.

[82] G. B. Jeffery. The motion of ellipsoidal particles immersed in a viscous fluid. *Proc. R. Soc. Lond. A*, 102(715):161–179, 1922.

[83] A.S.M. Kamruzzahan, F. Kienberger, C.M. Stroh, J. Berg, R. Huss, A. Ebner, R. Zhu, C. Rankl, H.J. Gruber, and P. Hinterdorfer. Imaging morphological details and pathological differences of red blood cells using tapping-mode afm. *Biol. Chem.*, 385:955–960, 2004.

[84] B. Kaoui, T. Krüger, and J. Harting. How does confinement affect the dynamics of viscous vesicles and red blood cells? *Soft Mat.*, 8:9246–9252, 2012.

[85] C. Kelemen, S. Chien, and G.M. Artmann. Temperature transition of human hemoglobin at body temperature: effects of calcium. *Biophys. J.*, 80:2622–2630, 2001.

[86] S. R. Keller and R. Skalak. Motion of a tank-treading ellipsoidal particle in a shear flow. *J. Fluid Mech.*, 120:27–47, 1982.

[87] Y. Kim, K. Kim, and Y. Park. *Blood Cell - An Overview of Studies in Hematology*, chapter: Measurement techniques for red blood cell deformability: Recent advances. InTech, ebook, 2012.

[88] T. Klöppel and W. A. Wall. A novel two-layer, coupled finite element approach for modeling the nonlinear elastic and viscoelastic behavior of human erythrocytes. *Biomech. Model. Mechanobiol.*, 10:445–459, 2011.

[89] T. Krüger, M. Gross, D. Raabe, and F. Varnik. Crossover from tumbling to tank-treading-like motion in dense simulated suspensions of red blood cells. *Soft Mat.*, 9:9008–90015, 2013.

[90] L. Lanotte, J. Mauer, S. Mendez, D. A. Fedosov, J.-M. Fromental, V. Clavería, F. Nicoud, G. Gompper, and M. Abkarian. Red cells' dynamic morphologies govern blood shear thinning under microcirculatory flow conditions. *Proc. Natl Acad. Sc. USA*, 113(47):13289–13294, 2016.

[91] G. Lenormand, S. Henon, A. Richert, J. Simeon, and F. Gallet. Direct measurement of the area expansion and shear moduli of the human red blood cell membrane skeleton. *Biophys. J.*, 81:43–56, 2001.

[92] G. Lenormand, S. Henon, A. Richert, J. Simeon, and F. Gallet. Elasticity of the human red blood cell skeleton. *Biorheology*, 40:247–251, 2003.

[93] M. Levant and V. Steinberg. Intermediate regime and a phase diagram of red blood cell dynamics in a linear flow. *Phys. Rev. E*, 94:062412, 2016.

[94] C.T. Lim, M. Dao, S. Suresh, C.H. Sow, and K.T. Chew. Large deformation of living cells using laser traps. *Acta Materialia*, 52:1837–1845, 2004.

[95] G. H. W. Lim, M. Wortiz, and R. Mukhopadhyay. *Red Blood Cell Shapes and Shape Transformations: Newtonian Mechanics of a Composite Membrane*, volume Lipid Bilayers and Red Blood Cells of *Soft Matter*, chapter 2, pages 94–269. WILEY-VCH Verlag GmbH & Co. KGaA, 2008.

[96] G.H.W. Lim, M. Wortis, and R. Mukhopadhyay. Stomatocyte-discocyte-echinocyte sequence of the human red blood cell: evidence for the bilayer-couple hypothesis from membrane mechanics. *Proc. Natl. Acad. Sci. USA*, 99, 2002.

[97] B. Lincoln, H.M. Erickson, S. Schinkinger, F. Wottawah, D. Mitchell, S. Ulvick, C. Bilby, and J. Guck. Deformability-based flow cytometry. *Cytometry A*, 59:203–209, 2004.

[98] O. Linderkamp and H.J. Meiselman. Geometric, osmotic, and membrane mechanical properties of density-separated human red cells. *Blood*, 59:1121–1127, 1982.

[99] Y. Liu and W. K. Liu. Rheology of red blood cell aggregation by computer simulation. *J. Comput. Phys.*, 220:139–154, 2006.

[100] Z. Liu, J. R. Clausen, R. R. Rao, and C. K. Aidun. Nanoparticle diffusion in sheared cellular blood flow. *J. Fluid Mech.*, 871:636–667, 2019.

[101] F. Lundell. The effect of particle inertia on triaxial ellipsoids in creeping shear: From drift toward chaos to a single periodic solution. *Phys. Fluids*, 23(011704), 2011.

[102] Z. Y. Luo, S. Q. Wang, L. He, F. Xu, and B. F. Bai. Inertia-dependent dynamics of three-dimensional vesicles and red blood cells in shear flow. *Soft Mat.*, 9:9651–9660, 2013.

[103] H.U. Lutz, S-C. Liu, and J. Palek. Release of spectrin-free vesicles from human erythrocytes during atp depletion. I. Characterization of spectrin-free vesicles. *J. Cell Biol.*, 73:548–560, 1977.

[104] J. Mauer, S. Mendez, L. Lanotte, F. Nicoud, M. Abkarian, G. Gompper, and D. A. Fedosov. Flow-induced transitions of red blood cell shapes under shear. *Phys. Rev. Lett.*, 121(118103), 2018.

[105] J. L. McWhirter, H. Noguchi, and G. Gompper. Flow-induced clustering and alignment of vesicles and red blood cells in microcapillaries. *Proc. Natl Acad. Sc. USA*, 106(15):6039–6043, 2009.

[106] J. L. McWhirter, H. Noguchi, and G. Gompper. Deformation and clustering of red blood cells in microcapillary flows. *Soft Mat.*, 7:10967–10977, 2011.

[107] S. Mendez and M. Abkarian. In-plane elasticity controls the full dynamics of red blood cells in shear flow. *Phys. Rev. Fluids*, 3(101101(R)), 2018.

[108] S. Mendez, E. Gibaud, and F. Nicoud. An unstructured solver for simulations of deformable particles in flows at arbitrary Reynolds numbers. *J. Comput. Phys.*, 256(1):465–483, 2014.

[109] J.P. Mills, L. Qie, M. Dao, C.T. Lim, and S. Suresh. Nonlinear elastic and viscoelastic deformation of the human red blood cell with optical tweezers. *Mech. and Chem. Biosystems*, 1:169–180, 2004.

[110] C. Minetti, V. Audemar, T. Podgorski, and G. Coupier. Dynamics of a large population of red blood cells under shear flow. *J. Fluid Mech.*, 864:408–448, 2019.

[111] C. Misbah. Vacillating breathing and tumbling of vesicles under shear flow. *Phys. Rev. Lett.*, 96(028104), 2006.

[112] C. Misbah. Vesicles, capsules and red blood cells under flow. *J. Phys.: Conf. Series*, 392:012005, 2012.

[113] N. Mohandas, M.R. Clark, M.S. Jacobs, and S.B. Shohet. Analysis of factors regulating erythrocyte deformability. *J. Clin. Invest.*, 66:563–573, 1980.

[114] N. Mohandas and E. Evans. Mechanical properties of the red cell membrane in relation to molecular structure and genetic defects. *Annu. Rev. Biophys. Struct.*, 23:787–818, 1994.

[115] L. Mountrakis, E. Lorenz, and A. G. Hoekstra. Scaling of shear-induced diffusion and clustering in a blood-like suspension. *Europhys. Lett.*, 114(14002):1–6, 2016.

[116] R. Mukhopadhyay, G. H. W. Lim, and M. Wortis. Echinocyte shapes: bending, stretching, and shear determine spicule shape and spacing. *Biophys. J.*, 82:1756–1772, 2002.

[117] R.W. Ogden. *Non-Linear Elastic Deformations.* Dover Publications, 1997.

[118] Y. Park, C.A. Best, T. Auth, N.S. Gov, S.A. Safran, G. Popescu, S. Suresh, and M.S. Feld. Metabolic remodeling of the human red blood cell membrane. *Proc. Natl. Acad. Sci. USA*, 107:1289–1294, 2010.

[119] Y. Park, C.A. Best, K. Badizadegan, R.R. Dasari, M. S. Feld, T. Kuriabova, M.L. Henle, A.J. Levine, and G. Popescu. Measurement of red blood cell mechanics during morphological changes. *Proc. Natl. Acad. Sci. USA*, 107:6731–6736, 2010.

[120] Y. Park, C.A. Best, T. Kuriabova, M.L. Henle, M.S. Feld, A.J. Levine, and G. Popescu. Measurement of the nonlinear elasticity of red blood cell membranes. *Phys. Rev. E*, 83:051925, 2011.

[121] Z. Peng, R. J. Asaro, and Q. Zhu. Multiscale modelling of erythrocytes in Stokes flow. *J. Fluid Mech.*, 686:299–337, 2011.

[122] Z. Peng, X. Li, I. V. Pivkin, M. Dao, G. E. Karniadakis, and S. Suresh. Lipid bilayer and cytoskeletal interactions in a red blood cell. *Proc. Natl Acad. Sc. USA*, 110(33):13356–13361, 2013.

[123] Z. Peng, A. Mashayekh, and Q. Zhu. Erythrocyte responses in low-shear-rate flows: effects of non-biconcave stress-free state in the cytoskeleton. *J. Fluid Mech.*, 742:96–118, 2014.

[124] Z. Peng, S. Salehyar, and Q. Zhu. Stability of the tank treading modes of erythrocytes and its dependence on cytoskeleton reference states. *J. Fluid Mech.*, 771:449–467, 2015.

[125] M.A. Peterson, H. Strey, and E. Sackmann. Theoretical and phase-contrast microscopic eigenmode analysis of erythrocyte flicker - amplitudes. *J. Phys. II*, 2:1273–1285, 1992.

[126] I. V. Pivkin and G. E. Karniadakis. Accurate coarse-grained modeling of red blood cells. *Phys. Rev. Lett.*, 101(118105), 2008.

[127] I. V. Pivkin, Z. Peng, G. E. Karniadakis, P. Buffet, M. Dao, and S. Suresh. Biomechanics of red blood cells in human spleen and consequences for physiology and disease. *Proc. Natl Acad. Sc. USA*, 113(28):7804–7809, 2016.

[128] G. Popescu, T. Ikeda, K. Goda, C.A. Best-Popescu, M. Laposata, S. Manley, R.R. Dasari, K. Badizadegan, and M.S. Feld. Optical measurement of cell membrane tension. *Phys. Rev. Lett.*, 97:218101, 2006.

[129] C. Pozrikidis. *Modeling and Simulation of Capsules and Biological Cells.* Boca Raton: Chapman & Hall/CRC, 2003.

[130] C. Pozrikidis. Axisymmetric motion of a file of red blood cells through capillaries. *Phys. Fluids*, 17(031503), 2005.

[131] M. Puig-De-Morales-Marinkovic, K.T. Turner, J.P. Butler, J.J. Fredberg, and S. Suresh. Viscoelasticity of the human red blood cell. *American J. Physiol.- Cell Physiol.*, 293:C597–C605, 2007.

[132] S. Ramanujan and C. Pozrikidis. Deformation of liquid capsules enclosed by elastic membranes in simple shear flow: large deformations and the effect of fluid viscosities. *J. Fluid Mech.*, 361:117–143, 1998.

[133] M.W. Rampling. *Handbook of Hemorheology and Hemodynamics*, volume 69 of *Biomedical and Health Research*, chapter: Compositional properties of blood. IOS Press, Amsterdam, Netherlands, 2007.

[134] S. Rancourt-Grenier, M.T. Wei, J.J. Bai, A. Chiou, P.P. Bareil, P.L. Duval, and Y.L. Sheng. Dynamic deformation of red blood cell in dual-trap optical tweezers. *Optics Express*, 18:10462–10472, 2014.

[135] D. A. Reasor, J. R. Clausen, and C. K. Aidun. Rheological characterization of cellular blood in shear. *J. Fluid Mech.*, 726:497–516, 2013.

[136] C. Renoux, M. Faivre, A. Bessaa, L. Da Costa, P. Joly, A. Gauthier, and P. Connes. Impact of surface-area-to-volume ratio, internal viscosity and membrane viscoelasticity on red blood cell deformability measured in isotonic condition. *Sc. Rep.*, 9(6771):1–7, 2019.

[137] S. Roman, A. Merlo, P. Duru, F. Risso, and S. Lorthois. Going beyond 20 μm-sized channels for studying red blood cell phase separation in microfluidic bifurcations. *Biomicrofluid.*, 10(034103), 2016.

[138] T. Rosen, F. Lundell, and C. K. Aidun. Effect of fluid inertia on the dynamics and scaling of neutrally buoyant particles in shear flow. *J. Fluid Mech.*, 738:563–590, 2014.

[139] P.D. Ross and A.P. Minton. Hard quasispherical model for the viscosity of hemoglobin solutions. *Biochemical and Biophysical Research Communications*, 76:971–976, 1977.

[140] H. Schmid-Schönbein and R. Wells. Fluid drop-like transition of erythrocytes under shear. *Science*, 165:288–291, 1969.

[141] U. Seifert. Fluid membranes in hydrodynamic flow fields: Formalism and an application to fluctuating quasispherical vesicles in shear flow. *Eur. Phys. J. B*, 8:405–415, 1999.

[142] J. Sigüenza, S. Mendez, D. Ambard, F. Dubois, F. Jourdan, R. Mozul, and F. Nicoud. Validation of an immersed thick boundary method for simulating fluid-structure interactions of deformable membranes. *J. Comput. Phys.*, 322:723–746, 2016.

[143] J. Sigüenza, S. Mendez, and F. Nicoud. How should the optical tweezers experiment be used to characterize the red blood cell membrane mechanics? *Biomech. Model. Mechanobiol.*, 16:1645, 2017.

[144] K. Sinha and M. D. Graham. Dynamics of a single red blood cell in simple shear flow. *Phys. Rev. E*, 92(042710), 2015.

[145] R. Skalak, A. Tozeren, R.P. Zarda, and S. Chien. Strain energy function of red blood cell membranes. *Biophys. J.*, 13:245–264, 1973.

[146] J. M. Skotheim and T. W. Secomb. Red blood cells and other nonspherical capsules in shear flow: Oscillatory dynamics and the tank-treading-to-tumbling transition. *Phys. Rev. Lett.*, 98(078301), 2007.

[147] A.P. Spann, H. Zhao, and E.S.G. Shaqfeh. Loop subdivision surface boundary integral method simulations of vesicles at low reduced volume ratio in shear and extensional flow. *Phys. Fluids*, 26:031902, 2014.

[148] P. Steffen, C. Verdier, and C. Wagner. Quantification of depletion-induced adhesion of red blood cells. *Phys. Rev. Lett.*, 110:018102, 2013.

[149] H. Strey and M. Peterson. Measurement of erythrocyte-membrane elasticity by flicker eigenmode decomposition. *Biophys. J.*, 69:478–488, 1995.

[150] Y. Sui, Y. T. Chew, P. Roy, Y. P. Cheng, and H. T. Low. Dynamic motion of red blood cells in simple shear flow. *Phys. Fluids*, 20(112106), 2008.

[151] S.P. Sutera, P.R. Pierre, and G.I. Zahalak. Deduction of intrinsic mechanical properties of the erythrocyte membrane from observations of tank-treading in the rheoscope. *Biorheology*, 26:177–197, 1989.

[152] N. Takeishi, M. E. Rosti, Y. Imai, S. Wada, and Brand. Haemorheology in dilute, semi-dilute and dense suspensions of red blood cells. *J. Fluid Mech.*, 872:818–848, 2019.

[153] M. Thiébaud, Z. Shen, J. Harting, and C. Misbah. Prediction of anomalous blood viscosity in confined shear flow. *Phys. Rev. Lett.*, 112(238304):1–5, 2014.

[154] G. Tomaiuolo. Biomechanical properties of red blood cells in health and disease towards microfluidics. *Biomicrofluidics*, 8:051501, 2014.

[155] G. Tomaiuolo, M. Barra, V. Preziosi, A. Cassinese, B. Rotoli, and S. Guido. Microfluidics analysis of red blood cell membrane viscoelasticity. *Lab. Chip*, 11:449–454, 2011.

[156] M. Toner and D. Irimia. Blood-on-a-chip. *Annu. Rev. Biomed. Eng.*, 7:77–103, 2005.

[157] R. Tran-Son-Tay, S.P. Sutera, and P.R. Rao. Determination of red blood cell membrane viscosity from rheoscopic observations of tank-treading motion. *Biophys. J.*, 46:65–72, 1984.

[158] R. Tran-Son-Tay, S.P. Sutera, G.I. Zahalak, and P.R. Rao. Membrane stress and internal pressure in a red blood cell freely suspended in a shear flow. *Biophys. J.*, 51:915–924, 1987.

[159] K.-I. Tsubota, S. Wada, and H. Liu. Elastic behavior of a red blood cell with the membrane's nonuniform natural state: equilibrium shape, motion transition under shear flow, and elongation during tank-treading motion. *Biomech. Model. Mechanobiol.*, 13(4):735–746, 2014.

[160] P. M. Vlahovska, Y.-N. Young, G. Danker, and C. Misbah. Dynamics of a non-spherical microcapsule with incompressible interface in shear flow. *J. Fluid Mech.*, 678:221–247, 2011.

[161] R.E. Waugh and R.G. Bauserman. Physical measurements of bilayer-skeletal separation forces. *Ann. Biomed. Eng.*, 23:308–321, 1995.

[162] R. Wells and H. Schmid-Schönbein. Red cell deformation and fluidity of concentrated cell suspensions. *J. App. Physiol.*, 27:213–217, 1969.

[163] J.M. Were, F.L.A. Willekens, F.H. Bosch, L.D. De Haan, S.G.L. Van der Vegt, A.G. Van den Bos, and G.J.C.G.M. Bosman. The red cell revisited matters of life and death. *Cell. Mol. Biol.*, 50:139–145, 2004.

[164] W. Yao, Z. Wen, Z. Yan, D. Sun, W. Ka, L. Xie, and S. Chien. Low viscosity Ektacytometry and its validation tested by flow chamber. *J. Biomech.*, 34:1501–1509, 2001.

[165] A. Z. K. Yazdani and P. Bagchi. Phase diagram and breathing dynamics of a single red blood cell and a biconcave capsule in dilute shear flow. *Phys. Rev. E*, 84(026314), 2011.

[166] A. Z. K. Yazdani, R. M. Kalluri, and P. Bagchi. Tank-treading and tumbling frequencies of capsules and red blood cells. *Phys. Rev. E*, 83(046305), 2011.

[167] T. Ye, N. Phan-Tien, C. T. Lim, and Y. Li. Red blood cell motion and deformation in a curved microvessel. *J. Biomech.*, 65:12–22, 2017.

[168] R. P. Zarda, S. Chien, and R. Skalak. Elastic deformations of red blood cells. *J. Biomech.*, 10:211–221, 1977.

[169] K. Zeman, H. Engelhard, and E. Sackmann. Bending undulations and elasticity of the erythrocyte-membrane - effects of cell-shape and membrane organization. *Eur. Biophys. J.*, 18:203–219, 1990.

[170] H. Zhao, A. H. G. Isfahani, L. N. Olson, and J. B. Freund. A spectral boundary integral method for flowing blood cells. *J. Comput. Phys.*, 229:3726–3744, 2010.

[171] H. Zhao, E. S. G. Shaqfeh, and V. Narsimhan. Shear-induced particle migration and margination in a cellular suspension. *Phys. Fluids*, 24(011902):1–21, 2012.

[172] H. Zhao, A. Spann, and E. S. G. Shaqfeh. The dynamics of a vesicle in a wall-bound shear flow. *Phys. Fluids*, 23(121901), 2011.

[173] O. Y. Zhong-can and W. Helfrich. Bending energy of vesicle membranes: General expressions for the first, second, and third variation of the shape energy and applications to spheres and cylinders. *Phys. Rev. A*, 39(10):5280–5288, 1989.

[174] A. Zilker, H. Engelhardt, and E. Sackmann. Dynamic reflection interference contrast (ric-) microscopy - a new method to study surface excitations of cells and to measure membrane bending elastic-moduli. *J. Phys.*, 48:2139–2151, 1987.

[175] A. Zilker, M. Ziegler, and E. Sackmann. Spectral-analysis of erythrocyte flickering in the 0.3-4 μ m^{-1} regime by microinterferometry combined with fast image-processing. *Phys. Rev. A*, 46:7998–8002, 1992.

CHAPTER 6

Aggregation and blood flow in health and disease

Viviana Clavería

CBS, CNRS, INSERM, Univ Montpellier, Montpellier, France

Christian Wagner

Experimental Physics, Saarland Univ., Germany and Physics and Materials Science Research Unit, Univ. of Luxembourg, Luxembourg

Philippe Connes

Laboratoire Interuniversitaire, de Biologie de la Motricité (LIBM) EA7424, Equipe "Biologie vasculaire et du globule rouge", UCBL1, Lyon, France and Laboratoire d'Excellence (Labex) GR-Ex, Paris, France

CONTENTS

THE MECHANISMS OF AGGREGATION or clustering of red blood cells in flow are discussed in this chapter. Besides the physiological case of plasma protein induced aggregation we focus on diseases for which red blood cell (RBC) aggregation and deformability are supposed to play an important role, such as sickle cell disease.

6.1 INTRODUCTION

If one looks under the microscope at a sample of blood that is freshly drawn from a healthy donor, one will always see that the majority of the blood cells, the RBCs, form lengthy aggregates that look like stack of coins and are therefore called rouleaux (see Figure 1, right panel). The aggregates can be broken up by shear forces in the flow and at stasis the cells aggregate again. The first studies on the effect of this aggregation mechanism were presented by Fåhraeus in 1958 [53] and since the seminal work by Merril in 1963 [108], it is known that plasma fibrinogen is the main protein responsible for the aggregation. Many of the findings on aggregation of RBCs are reviewed in the book "Red Blood Cell Aggregation" by Baskurt, Neu and Meiselmann from 2011 [21]. This chapter will concentrate on the most recent findings in respect to RBC aggregation and set a focus on recent in-vitro experiments that model capillary flow.

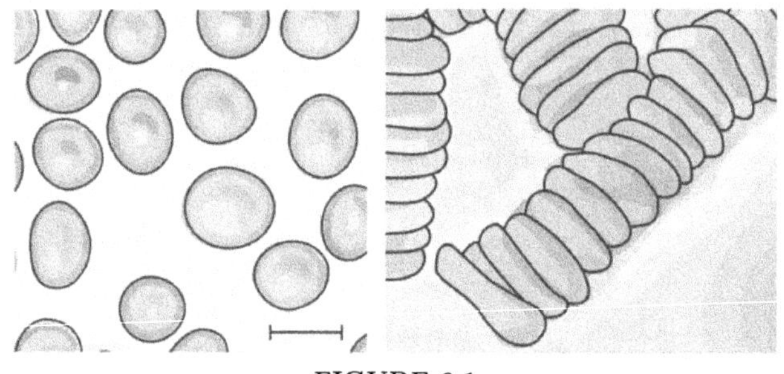

FIGURE 6.1
RBCs suspended in buffer solution (left panel) compared to RBCs suspended in autologous plasma forming a rouleau (right panel). The scale on the left panel is equivalent to 10 μm.

Studies on RBC aggregation have been intensively carried out during the last few decades and showed that RBC aggregation is a significant diagnostic parameter that allows monitoring the state of blood microcirculation. For example, the unambiguous strong correlation of fibrinogen concentration with the amount of aggregates is used in clinical tests, in the so-called erythrocyte sedimentation rate (ESR). ESR is a very common unspecific pathological test for inflammatory diseases in the human body. In ESR, blood is kept in an upright capillary and the RBCs sediment faster if they form aggregates. By measuring the sedimentation front one gets a measure of the fibrinogen level which is increased in inflammatory states. Often the ESR is directly related to the level of C-reactive protein (CRP), another acute phase indicator that is tested in antibody tests, but recently it was shown that CRP has no direct aggregating effect [63].

Of course an obvious question is how the aggregation of RBCs affects vascular flow. Macroscopically, the reversible aggregation process is the dominant mechanism for the shear thinning of blood. At low shear rates the aggregates form a gel-like network that is broken up with increasing shear rate and thus the viscosity decreases. The change in blood microstructure due to aggregation disruption is accompanied by a strong viscosity drop from approximately 100 to 10 cP [39]. However, for shear rates common in microcirculation ranging from 10 to 1500 1/s, only deformability and dynamics of RBCs can account for a further decrease in blood viscosity, reaching values down to 2 cP [39]; value strongly depends on the number of RBCs and the RBC hemoglobin viscosity, values of which have been reported at around 5 cP at 37°C [155]. When vessel diameter decreased from about 300 to 10 μm the apparent (also called effective) viscosity of the blood decreases. Additionally, the hematocrit ratio between the tube hematocrit (Ht) and discharged tube hematocrit (Hd), Ht/Hd decreases. These two effects are known as the Fåhraeus-Lindqvist and Fåhraeus effects, respectively [52,54]. Both of these effects are understood as a result of lift forces that drive the RBCs away from the capillary wall towards the capillary center [6], producing a plasma layer free of RBCs next to the wall vessel, typically called cell free layer or CFL. As the diameter of the capillary decreases, the relative ratio of the cross sectional area of this cell-free layer increases in respect to the tube diameter. At tube diameters of approximately 10 μm, cells flow one after the other. We will see that cells typically flow in the form of clusters and that there are at least two mechanisms that lead to clusters: hydrodynamic interaction and rouleaux formation. Also, the hematocrit Ht observed at these small capillaries (of around 10 to 20%) is much lower than the average hematocrit for whole blood (of around 45%), as a consequence of the Fåhraeus effect and the partitioning of RBCs and plasma at vessel bifurcations [66]. The RBC volume fraction in micro-vessels is strongly fluctuating. However, if the vessel diameter is decreased below 10 μm, the apparent/effective viscosity begins to increase, a phenomenon known as the inverse Fåhraeus-Lindqvist effect.

In order to develop quantitative predictive theoretical and numerical models on blood flow one must know the physical properties of the cells and the interaction mechanisms and strength. Quantifying the interaction potential between the cells is therefore crucial to obtain the correct input parameters for the theoretical and numerical models.

Through this chapter, we will start discussing the possible molecular origin of red blood cell aggregation and the models proposed to explain the mechanism of aggregation. In the next sections, the methods typically used to measure RBC aggregation and to quantify single-cell adhesion strength are discussed. As a next step, we discuss the impact of aggregation on the bulk blood viscosity and the pathophysiological factors involved in RBC aggregation, including the case of sickle cell anemia. Finally, we want the reader to have an overview of the behavior of blood flow through capillaries and the impact of the aggregation of RBC in this behavior. Finally an outlook on the current perspectives and challenges is given.

6.2 POSSIBLE MOLECULAR ORIGIN OF PHYSIOLOGICAL RBC AGGREGATION

Aggregation is caused by the presence of macromolecules. The strength and the number of cells forming aggregates, as well as their structure depend on the hematocrit level, macromolecule concentration, molecular weight, as well as other factors such as osmolarity and pH. For example, aggregation strength increases when fibrinogen concentration increases and in the case of the polysaccharide dextran, the aggregation strength increases when the concentration of dextran increases in the suspending medium until it reaches a maximum, where the aggregation strength starts to decrease again.

Until today, the mechanisms that explain RBC aggregation have not been fully understood and two coexisting models have been proposed to explain it: the so-called bridging and depletion models. In the following subsections, each of the mechanisms will be explained in more extended detail.

6.2.1 Bridging model

This model assumes that large macromolecules adsorb onto the RBCs' surface and bridge two cells when they approach close enough. The hypothesis is based on studies that focused on the inter-cellular distance between two adjacent cells [17,38,39]. It was observed that the size of the hydrated molecules is longer than the inter-cellular distance between adjacent RBCs, leading to the assumption that the terminal portions of the flexible polymers should be adsorbed onto the surfaces resulting in a cell-cell adhesion as represented in Figure 6.2. In this process, a specific binding (link to a specific component of the membrane) or unspecific binding (link to group of components of the membrane) of the flexible polymers to the RBC membrane could occur.

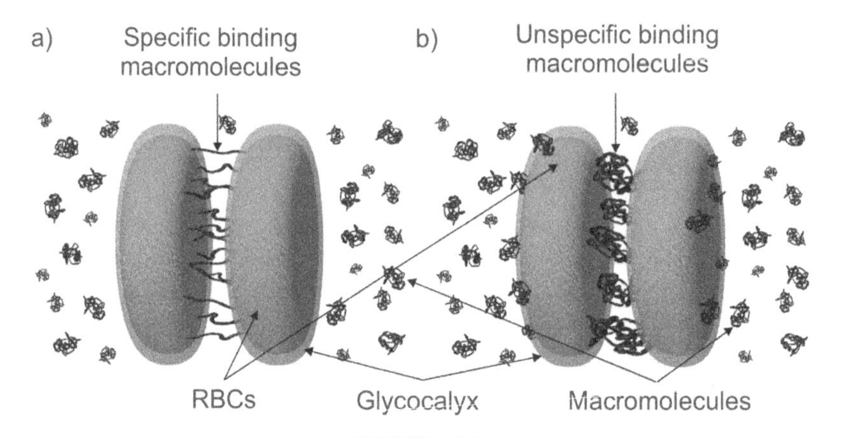

FIGURE 6.2
A specific binding to proteins of macromolecules (black) between two red blood cells (red) (a) could be unspecific if no membrane proteins are involved (b). The glycocalyx is marked slightly red. Different sizes of macromolecules indicating the presence of imperfect macromolecules.

Interestingly, the inter-cellular distance has been observed to increase when polymer molecular weight increases, and independently of the polymer size, the inter-cellular distance has been observed to be smaller than the hydrated polymer [37]. Other data that support the bridging model results from adsorption measurements [98] and from measurements of interaction energies as a function of the size of the interaction area [97].

To estimate the aggregation force $\mathbf{f}_{agg}$ per unit length of the cell membrane, it is assumed that the bonds of the cell's cross-linked polymers behave like stretched springs, leading to the following equation:

$$\mathbf{f}_{agg} = k_b(l - l_0)n_b\frac{\mathbf{x}}{l} \tag{6.1}$$

where k_b is the spring-constant (force per stretch length), l and l_0 are the stretch and unstretched bond length, respectively, x the inter-cellular distance and n_b the bond density defined by the following reaction equation:

$$\frac{\partial n_b}{\partial t} = 2\left[k_+\left(n - \frac{n_b}{2}\right)^2 - k_- n_b^2\right] \tag{6.2}$$

where n is the density of cross-linked polymers on each cell (it is assumed that each cell has the same number of cross-linked polymers). k_+ and k_- are the forward and reverse reaction rate coefficients given by:

$$k_+ = k_+^0 exp\left[-\frac{k_{ts}\left(l - l_0\right)^2}{2k_B T}\right], |\mathbf{x}| = l < l_t \tag{6.3}$$

$$k_- = k_-^0 exp\left[-\frac{\left(k_b - k_{ts}\right)\left(l - l_0\right)^2}{2k_B T}\right] \tag{6.4}$$

where k_+^0 and k_-^0 are the rate coefficients in equilibrium, k_B the Boltzmann constant, T the temperature, k_{ts} the transition spring constant and l_t the distance below which the bond formation is initiated. If it is assumed that the polymers cannot interact with themselves, then the maximum interaction force should be reached when 50% of the cell membrane is covered by polymers.

6.2.2 Depletion model

Depletion force is an attractive force that arises when small particles (polymers or macromolecules) are present within the cell suspending medium. A first explanation was given by Asakura and Oosawa [10], who discovered that the presence of small spheres (i.e., macromolecules) can induce effective forces between two larger particles if the distance between them is small enough. The origin of these forces is purely entropic and can be used to explain the attraction of several systems like plates immersed in a solution of rigid spherical macromolecules or oil droplets surrounded by micelles. Basically, if two large bodies approach one another, the macromolecules around are excluded from the region in between, usually called forbidden zone or area (see Figure 6.3), leading to an uncompensated osmotic pressure within the depleted layer. In consequence, the depletion interaction scales with the osmotic pressure of the macromolecules as well as with the depleted volume in between the interacting bodies. More precisely, if we consider that the thickness of the depletion layer on each body equals the radius of the smaller macromolecules, when they overlap, an additional free volume is available for the macromolecules causing an increase in the entropy of the system. The increase in entropy leads to an increase in Helmholtz's free energy leading to an effective osmotic pressure that causes an attractive force between the bodies. In diluted solutions below the overlap concentration [69], polymers can be treated as rigid spheres considering their radius of gyration [125].

Depletion model of RBCs

The osmotic force Π experienced by polymer chains on a surface can be calculated by a power series in concentration. The coefficients higher than second virial coefficient (B_2) can be neglected since the concentrations relevant for RBC aggregation are comparably small [114]:

$$\Pi = \frac{RT}{M_2}c_2 + B_2\left(c_2\right)^2 = -\frac{\left(\mu_1 - \mu_1^0\right)}{\nu_1} \tag{6.5}$$

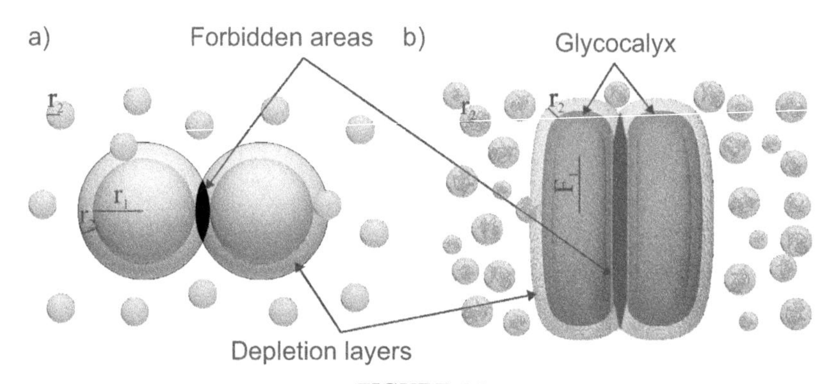

FIGURE 6.3
(a) Depletion of large spheres by small ones (b) RBC depletion by approximately spherical macromolecules. The depletion layer is illustrated in blue and the forbidden areas in black.

where R, T, ν_1, M_2 and c_2 are the gas constant, absolute temperature, molecular volume of the solvent, the molecular mass of the polymer and the bulk polymer concentration, respectively. μ_1 and μ_1^0 are the chemical potentials of the solvent in the polymer solution and in a polymer-free solution.

In the case of RBCs it is important to take into consideration that their plasma lipid bilayer membrane is covered by the glycocalyx, a complex layer of approximately $5nm$ thick composed of proteins (~52%), lipids (~41%) and carbohydrates (~7%) [77]. The presence of the carboxyl group of sialic acids in the cell membrane creates a negative charge and a repulsive electric zeta potential between cells [117]. Due to this electrostatic repulsion, the inter-cellular distance at which the minimal interaction energy occurs or in other words, where the maximum adhesion strength occurs, can be considered greater than twice the thickness of the glycocalyx [114]. Therefore, steric interactions between glycocalyxes can be neglected from a possible RBCs adhesion model. Van der Waals forces are also neglected.

On the other hand, the glycocalyx can be considered as a soft and hairy structure where macromolecules can penetrate in part or entirely [114, 153]. If we consider extreme cases of a perfectly hard surface or a perfectly soft surface, the penetration depth p of the polymers would be zero or as large as the glycocalyx thickness δ respectively. However, the glycocalyx can be considered as a non-perfect soft surface and in consequence the penetration depth p should have a value between zero and the glycocalyx thickness δ. The depletion layer thickness Δ is calculated based on the equilibrium between the elastic free energy and the osmotic force Π as a function of the bulk polymer concentration c_2 [152].

In general, the depletion interaction energy is obtained by multiplying the osmotic force Π produced by the polymers by the excluded volume generated by the colloidal objects. In RBCs, this will be equivalent to calculating the depletion interaction energy per unit area as a function of the cell-cell separation d, the glycocalyx thickness δ and the penetration depth p. Due to the lack of knowledge about the physiochemical properties of the glycocalyx, we can consider the penetration depth p to approach asymptotically to δ at high concentrations of c_2. In a first approximation, c_2 can be assumed to be constant. However, for a more accurate description it would be necessary to express c_2 as a function of the used polymer type and polymer concentration [21].

Additionally, because the surfaces of RBCs are negatively charged, an electrostatic repulsion has to be taken into account to calculate the total interaction energy between two RBCs assuming an evenly distributed constant surface charge of both cells within the glycocalyx. The electrostatic interaction energy is calculated considering the free energy of the two cells at a separation distance d, followed by the assumption of the free energy of single cells ($d \rightarrow \infty$). The electrostatic potential can be approximated as a superposition of the potential of two single cells since the Debye length is small compared to both the glycocalyx thickness δ and the cell-cell distance d. To calculate the electrostatic potential, the Poisson-Boltzmann equation has to be solved and considering a moderate electrical potential, a linear approximation can be used. The details of the calculation can be found in Baskurt et al. [21].

6.3 QUANTIFYING INTERACTION FORCES AMONG RBC

6.3.1 Methods to measure RBC aggregation

Different methods have been developed to measure RBC aggregation [87, 114]. The most widely used is the optical aggregometry Laser-assisted Optical Rotational Cell Analyzer, LORCA, where the backscattered light intensity of a laser beam going through a blood suspension is detected and analyzed [23, 121]. The Myrenne aggregometer is based on the same principle, except that this is the light intensity of a laser beam transmitted through a suspension, and not backscattered, that is analyzed [19]. A second very common method uses light microscopy coupled with image analyses [26, 40]. Finally, RBC aggregation has been also characterized by ultrasound techniques, a non-invasive imaging technique that seems to be promising due to its potential to be applied *in vivo* and *in situ* [64, 157]. From these methods, several and different parameters are extracted and related to RBC aggregate size, formation time and aggregate strength [23, 133]. However, they do not give information regarding adhesion strength among individual RBCs. For this last case, single-cell methods are needed. In the next section, we will briefly describe the most relevant methods used to quantify adhesion strength among individual RBCs.

6.3.2 Methods to quantify single-cell adhesion strength

Evans in 1980 was the first one to introduce an analytical method considering minimum free energy that provides a way to quantify the affinity of RBCs for other surfaces [50]. Evans considered the elastic properties of RBCs and their inner characteristics like inner viscosity. Later on, studies using scanning electron microscopy (SEM) yielded the first estimates of RBC interaction energy and speculated about the mechanism of RBCs interaction [38, 137]. In more recent time, single-cell force spectroscopy (SCFS) approaches using different techniques have been developed, where optical microscopes are used to observe the process of cell adhesion while force measurements are made. The mechanism of cell manipulation includes methods like micropipette aspiration [33, 50], atomic force microscopy (AFM) and optical tweezers (OT) adhesion strength measurements techniques. These last methods allow a deeper understanding of the RBC aggregation process and the estimation of adhesion forces between cells under well controlled conditions [30, 96, 120, 141]. Another less direct approach was more recently presented, where the buckling of the membrane of two adherent cells was compared with theoretical predictions and in this way the interaction energy could be estimated, at least for an ensemble of measurements [62]. One of the famous experiments used to measure aggregation energy is the micropipette based approach performed by Buxbaum in 1982 [33]. Buxbaum used various concentrations of the polysaccharide dextran to induce aggregation among RBCs and measured the extent of encapsulation of a spherical induced RBC by an intact/flaccid RBC as shown in Figure 6.4.

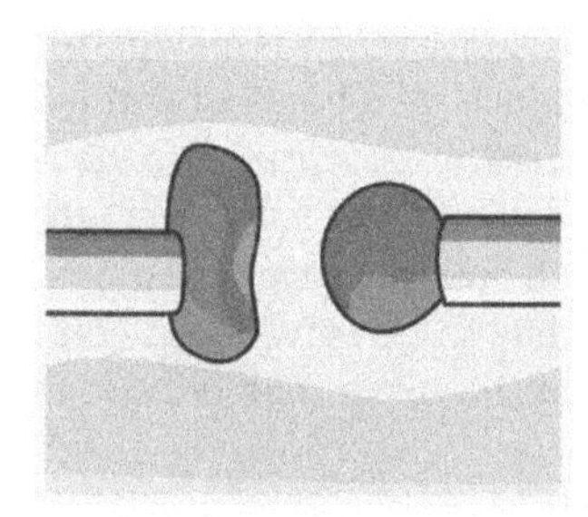

FIGURE 6.4
Spherical RBC (right) and an intact RBC (left) before encapsulating one inside the other to measure adhesion strength among them [33].

Based on the depletion model for the interaction energy among individual RBCs (see previous sections for more details), Neu et al.[114] calculated the interaction energy curves to fit with the experimentally obtained interaction energy curves reported by Buxbaum for dextran 70 and 150 kDa in molecular weights. The model seems to fit well with the experimental data reported. However, it should be taken into account that the data points reported by Buxbaum remained limited and with large error bars due to aberrations and inaccurate determination of aspiration pressure values. A second technique used to measure interaction energy among aggregated RBCs is the AFM-based technique. This technique was introduced in 1986 by Binnig and Gerber [67]. The basic idea of this technique is to use a reflecting cantilever in combination with a laser beam and a photodiode as a laser detector. The photodiode can indirectly measure the deflection of the cantilever when it enters in contact with the sample due to a position shift of the laser in the detector. The result is a force-distance curve. The distance is given by the position of the piezo motors that controls the cantilever positioning while the force is deduced mainly by the mechanical properties of the cantilever. AFM can be used in air and liquid sample environments. This last characteristic offers several advantages for biological applications and cell physics. In the case of RBCs, Steffen et al. [141] have use AFM-based technique attaching one RBC to a cantilever and another RBC to a petri dish surface. Both cells were brought into close contact and during the withdrawing the cantilever positioning was registered. The adhesion force between both cells was obtained after interpreting the cantilever positioning curves considering the mechanical properties of the cantilever, the contact mode, and the suspending medium characteristics such as viscosity and temperature. Figure 6.5 shows the dependence of the interaction energy on the concentration of the dextran used, measured with a cantilever velocity of 5 $\mu m/s$. For high interaction energies arising in the dextran 150 measurements the measured values were corrected to smaller interaction energies (marked zone between lines in Figure 6.5). The measured interaction energies are still in good agreement with the predicted interaction energies given by Neu et al. [114]. In their study, they used the depletion model previously discussed to calculate the interaction energy between two RBCs. Neu's model combines electrostatic repulsion due to RBC surface charge and osmotic attractive forces due to polymer depletion near the RBC surface. The theory is developed for soft surfaces like RBCs and the penetration depth p of polymers into the surface or glycocalyx. The penetration depth p depends mainly on the polymer concentration and molecular size (as well as molecule type) and is expected to be larger for small molecules and to increase with increasing polymer concentration due to increasing osmotic pressure. Penetration of macromolecules into the RBCs' surface deepens as osmotic pressure increases, impeding the depletion of macromolecules between both cells and hence reducing the interaction energy. Above a threshold in concentration, this effect becomes dominant causing a decrease of the interaction energy, producing a bell-shaped dependence of the interaction energy on the dextran concentration as shown in the solid line in Figure 6.5.

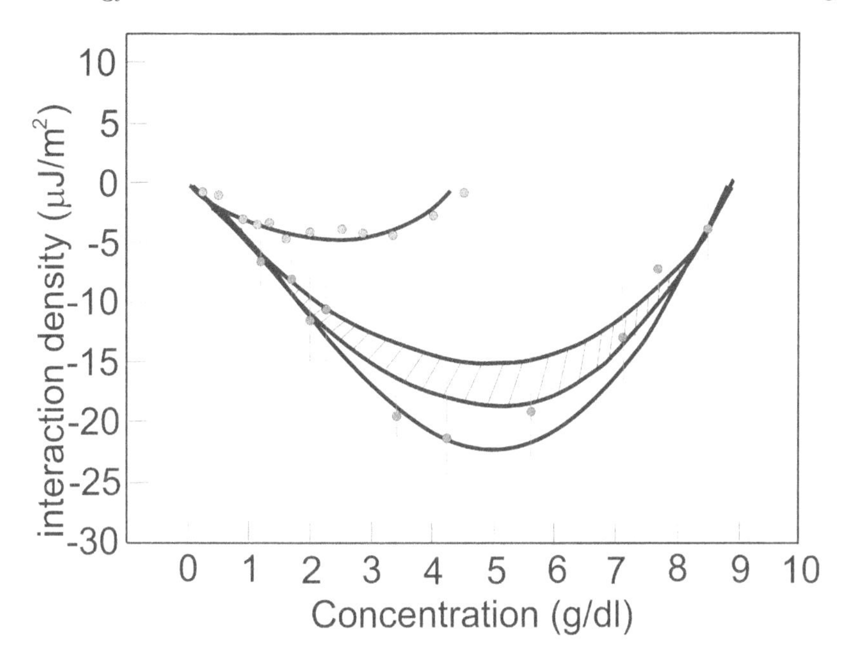

FIGURE 6.5
Interaction energy of two red blood cells according to the concentration of two different dextran types measured with (5 $\mu m/s$) (green spots: dextran 70 kDa, red spots: dextran 150 kDa). The solid line represents the theoretical curve predicted by Neu et al [114]. The marked zone gives the range of interaction energy extrapolated to smaller velocities (2 $\mu m/s$). Graph adapted from [141]

Finally, optical tweezers became most recently a very versatile tool to measure interaction energy among aggregate RBCs. Arthur Ashkin was the first to develop optical traps able to confine single particles due to radiation pressure against the gravitational force[11,13,15]. The invention of high focusing microscope objectives facilitated the realization of a special form of optical traps that are referred to as optical tweezers (OT) [12,14]. OT turned out to be a versatile tool for several applications in physics, chemistry and foremost biology [28,92,144]. With the help of OT, microscopic objects like dielectric particles and living cells can be trapped and manipulated via a high focused laser beam with a minimal invasion to the manipulation of living cells measuring forces in the range of pN [11]. Regarding RBC, the processes of aggregation and disaggregation in plasma and other suspending media like a buffer containing fibrinogen, albumin or fibrinogen and albumin have been studied using OT [97].

Results have shown that aggregation and disaggregation processes behave differently [30,97]. In effect, the force required to stop aggregation of RBCs is about a few pN, while the force required to separate the RBCs equals to about a few dozen pN [84,97,104]. On the other hand, the aggregation force has been found to decrease with the interaction area while the disaggregation decreases with the same parameter. The first effect has been attributed to depletion forces. On the other hand, the disaggregation process has been speculated to be described as a manifestation of the surface cross-bridges that stabilize the aggregates and can be formed as cells come close to each other [97]. The interpretation is that more and more polymers are dragged through the liquid membrane when cells are laterally separated from each other and in this way more adhesive contact points are formed. Therefore, higher forces are required to disaggregate RBCs.

Consequences

While all three methods (micropipettes, AFM and OT) have their advantages and disadvantages, optical tweezers seem to be the most versatile tool because first they manipulate the cells in the bulk of the liquid and not adhered at a surface and second they can be combined with a microfluidic device which allows the exchange of polymers. All methods indicate that the adhesion force increases linearly with the concentration of fibrinogen within the physiological range and at a concentration of 4 mgl/ml forces in the range of 5 pN for OT [97] and 40 pN for AFM measurements [141]. However, one must keep in mind that in these measurements denaturated fibrinogen in powder form was re-dissolved in the red blood cell solution and might have been quite different from its natural conformation and the separation mode (lateral vs. vertical) is quite different. Adhesion strengths in plasma measured with OT were around 20 pN [97]. Dextran 70 kDa at a concentration of 30 mgl/ml showed comparable adhesion forces as fibrinogen at 4 mg/ml in AFM measurements [31] and was used as a model in both in vitro experiments and in numerical simulations with good agreement as we will see in Section 6.7.3. In hard-sphere colloidal systems, it has recently been shown that both bridging and depletion can simultaneously affect the process of crystallization and self-assembly in flow [90]. However, there is quantitative disagreement between numerical predictions and experiments on the kinetics of phase transitions, and it is predicted that hydrodynamic interactions must be considered to resolve this discrepancy [124]. Although there are now quantitative data on the adhesion strength between red blood cells available and quite sophisticated measurement methods are used to distinguish between aggregating and separating forces that seem to allow a clear distinction between the two microscopic models that try to explain the aggregation, bridging and depletion; the situation is still not fully clear. At least, the most recent OT measurements indicate that bridging might always play a role. Still, the effect of depletion is also always present and one must answer the question of which effect predominates under which conditions, but this is still an open task.

6.4 IMPACT OF AGGREGATION ON THE BULK RHEOLOGY OF BLOOD AND VASCULAR REACTIVITY

6.4.1 Bulk blood viscosity

Blood viscosity is an important determinant of local flow characteristics. Blood exhibits shear thinning behavior: i.e., its viscosity decreases exponentially with increasing shear rates. In addition, blood has visco-elastic and thixotropic properties, which also affect local hemodynamics. A thixotropic fluid is a fluid whose viscosity is a function not only of the shearing condition, but also of the previous history of motion within the fluid [151]. Indeed, for a given flow and shear rate ($\dot{\gamma}$), blood viscosity usually decreases with the length of time the fluid is in motion. The relative contribution of RBCs is represented by the hematocrit (Hct) value. A rise in Hct increases blood viscosity at all $\dot{\gamma}$ and thixotropy, more particularly at low $\dot{\gamma}$, like in veins and venules [20,151]. The shear thinning and non-Newtonian behavior of blood is determined primarily by the mechanical properties of circulating RBCs. There are two unique RBC characteristics that are primarily responsible for this non-Newtonian behavior: RBC deformability and RBC aggregation (see Figure 6.6, [35,93]).

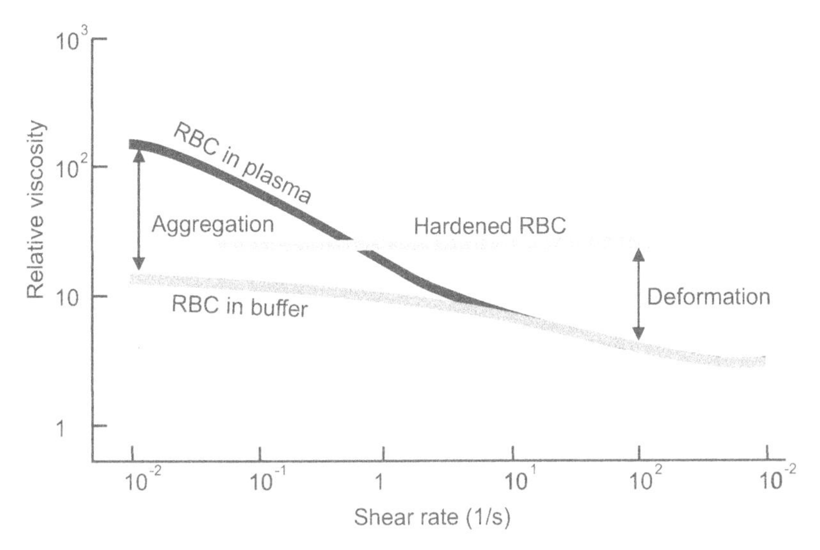

FIGURE 6.6
Rheology of RBC suspensions at 45% hematocrit [35]. Curves are the logarithmic relation between viscosity and shear rate in three types of suspensions: deformable (normal) RBC suspended in autologous plasma (black curve), deformable (normal) and hardened RBC suspended in a buffer solution containing 11% of albumin (gray and light gray curves respectively).

At high $\dot{\gamma}$ (characterized by high shear forces), RBCs undergo an extensive passive shape change (RBC deformability) forcing them to align parallel with laminar flow streamlines [20]. As a consequence, blood viscosity and the internal resistance of blood to flow decreases [20]. Of note and in contrast to the current paradigm, which assumes that RBCs align steadily around the flow direction while their membranes and cytoplasm circulate [60], it was recently demonstrated that RBCs might also successively tumble, roll, deform into rolling stomatocytes, and, finally, adopt highly deformed polylobed shapes for increasing shear rates [93]. RBC deformability is determined by cell geometry, cytoplasmic composition, internal viscosity and membrane characteristics [36].

Under low $\dot{\gamma}$ or static conditions (mainly in veins), blood viscosity increases compared to high $\dot{\gamma}$ conditions. This exponential increase in blood viscosity is due to the tendency of RBCs to the rouleaux formation [56, 75, 112, 136]. Moreover, given sufficient time and a non-confining space, individual rouleaux can cluster, thereby forming three-dimensional structures (see right panel in Figure 6.1). As a consequence, any increase in RBC aggregation would tend to increase vascular resistance, mainly in veins, and reduce blood flow [27]. This unique process requires low energy and is reversible: RBC aggregates may disaggregate, at least partly, in vascular regions where $\dot{\gamma}$ is high, such as in arteries and arterioles. The disaggregation of RBC aggregates is accompanied by a decrease in blood viscosity.
Of course, the determination of blood viscosity in simulations that are based on physical models of the cells are a first benchmark test in numerics. A shear thinning that quantitatively compares to the experimental data can be only reproduced if aggregation is taken into account and there is some quite good agreement [56].

6.4.2 Vascular reactivity

In addition to its effects on blood viscosity, RBC aggregation has been shown to directly modulate vascular reactivity. A study performed by Baskurt et al. [24] demonstrated that enhanced RBC aggregation resulted in suppressed expression of nitric oxide (NO) synthesizing mechanisms, thereby leading to altered vasomotor tonus; the mechanisms involved most likely relate to decreased wall shear stresses due to decreased blood flow and/or increased axial accumulation of RBCs in blood vessels. Indeed, increased RBC aggregation is able to modulate blood flow resistance in the venous compartment but also in arterial/arterioles compartments where $\dot{\gamma}$ is higher. However, it is highly possible that the effects of RBC aggregation on NO metabolism and vascular reactivity would be dependent on the orientation of vessels. Previous experiments compared the hemodynamic behavior of RBCs flowing in a vertical versus horizontal tube [127], [43]. These studies showed that, at low flow rates, RBC aggregation promotes RBC sedimentation in a horizontal tube, resulting in asymmetric distribution of flowing RBCs. In contrast, no marked asymmetry was observed in vertical tubes, although irregularities in the axial and marginal flow areas were noted. As a consequence, flow resistance and apparent viscosity were higher in horizontal than in vertical tubes [34]. However, vessels are not tubes and endothelial NO-synthase activation is sensitive to wall shear stress. Whether NO production and vasodilation are higher in horizontal than in vertical tubes remains an open question. Moreover, Cokelet and Goldsmith compared the hydrodynamic resistance in a vertical tube for RBC suspension flowing upward or downward. As expected, at low flow rates (i.e., when RBC aggregates are present),flow resistance was higher in upward than in downward condition because RBCs had to flow in the opposite direction of the sedimentation effect. Every vessel is oriented differently in the human body and the orientation can vary when humans change their position (for example, lying position versus sitting position). As a consequence, the impact of RBC aggregation on global and systemic flow resistance is probably changing over time, depending on the physiological situation and vessels orientation.

Finally, the group of Nash et al. [113] also demonstrated that enhanced RBC aggregation in low $\dot{\gamma}$ regions results in increased white blood cells margination and adhesion to the vessel wall [2, 113]. Given the importance of circulating-endothelial adhesive phenomenon in the setting of atherosclerosis and in various inflammatory syndromes, one may expect a key role of RBC aggregation in the control of endothelial function/biology. Increased white blood cells adhesion has also been demonstrated to participate to the onset of painful vaso-occlusive crises in sickle cell anemia [83].

6.5 PATHOLOGICAL RED BLOOD CELL AGGREGATION

6.5.1 Pathophysiological factors involved in RBC aggregation modulation

RBC aggregation has been investigated in various acute and chronic disorders [21]. Usually, disorders where inflammation is enhanced are characterized by increased RBC aggregation and increased RBC aggregates strength, which would impair blood flow in both the micro- and macrocirculation [20]. For instance, RBC aggregation has been reported to be increased in Crohn's disease [115], unstable angina [8], obesity [129], Gaucher disease [65, 159], deep venous thrombosis [88], obstructive sleep apnea [116,135], bowel disease [103] and sepsis [22]. The increased RBC aggregation is attributed to the rise in fibrinogen level accompanying inflammation but not exclusively. Oxidative stress also plays a key role in RBC aggregation properties. Baskurt et al. [22] and Hierso et al. [80] demonstrated that oxidative agents are able to decrease RBC deformability and aggregation but reinforce the strength of RBC aggregates in normal and sickle RBCs. Any rise in RBC aggregates' stickiness is expected to impair blood flow in microcirculation, notably at the entry of small capillaries where most of the RBCs need to flow as single cells and not as aggregates. We previously mentioned (see above) the extent of RBC aggregation may affect vascular control mechanisms through its effects on NO production [24]. But NO also affects RBC aggregation [134]. Bor-Kucukatay et al. [29] found that sodium nitroprusside, a NO-donor, reduced RBC aggregation in hypertensive rats. The underlying mechanisms are unknown but the direct effect of NO on RBC deformability could indirectly affect RBC aggregation. This result is of particular importance since several disorders, notably those where hemolysis is enhanced, are characterized by decreased NO bioavailability [82].

Several studies also reported increased RBC aggregation in metabolic disorders such as insulin resistance, obesity, diabetes and others [7, 76]. The chronic oxidative stress and inflammation are probably involved in the increase in RBC aggregation but modifications of the RBC membrane properties (cellular factors), such as membrane cholesterol content, phosphatidylserine externalization or sialic acid content, also play an important role (i.e., increased RBC aggregability). Finally, one may mention the effects of vascular geometry and local $\dot{\gamma}$ on the formation of RBC aggregates in various diseases where changes in blood flow structure and vascular remodeling are frequent, such as in the case of sickle cell disease [101, 126]. In summary, the following factors may affect RBC aggregation:

Oxidative stress

Nitric oxide

Inflammation

Metabolic disorders (cholesterol, glucose, insulin)

Vascular geometry because it may affect $\dot{\gamma}$

6.5.2 Focus on sickle cell disease

RBC aggregation properties have been poorly studied in sickle cell disease but several studies reported that RBC aggregation is rather decreased in sickle cell patients compared to healthy individuals. In contrast, once formed, sickle RBC aggregates are 2 to 3 fold more robust than healthy RBC aggregates [44, 149]. Recently, we demonstrated that fibrinogen, thrombospondin and von-Willebrand factor could re-enforce the RBC aggregates in sickle cell patients [111]. The lower RBC aggregation seems to be due to the low ability of the rigid irreversible sickle RBCs to form aggregates [41, 44]. Deoxygenating sickle cell blood in vitro results in a further decrease of RBC deformability and RBC aggregation [32]. To date, very few studies measured the changes in RBC aggregation during a vaso-occlusive event in sickle cell disease. Recently, Lapouméroulie et al. [94] reported increased RBC aggregation and a trend through increased RBC aggregates' robustness in sickle cell patients during vaso-occlusive crisis compared to steady-state. Limited data is also available on erythrocyte sedimentation rate (ESR), a parameter that is affected by both Hct and RBC aggregation [16, 95], showing increased ESR during crisis. The increase of fibrinogen during vaso-occlusive events could promote RBC aggregation [16]. The findings of Loiseau et al. [101] suggest that these sticky RBC aggregates would preferentially deposit at bifurcation level in vascular networks, which could participate to the pathophysiology of vaso-occlusive events. Indeed, Lamarre et al. [91] found an association between RBC aggregates' strength and the occurrence of acute chest syndrome (ACS) in SCA.

6.6 BLOOD FLOW STRUCTURING IN BIG TUBES, VISCOSITY BEHAVIOR AND EFFECTS OF RBC DEFORMABILITY AND AGGREGATION

In the middle of 19th century Poiseuille, in his effort to understand the pressure distribution on the circulatory system, developed several experiments until he derived a relationship between flow rate, the driving pressure and the tube length and diameter of water flowing through glass capillaries [142]. In the following years, many efforts were done to know whether blood obeys Poiseuille law or not. In 1915 Hess [79], summarized and re-analyzed the previous studies and concluded that blood follows Poiseuille law only for high flow rates and high $\dot{\gamma}$ [72, 79]. He also pointed out that at small pressure drops or in small capillaries, the nature of blood as a fluid formed by colloidal particles was an important fact to be considered to understand its behavior. Following this line, Fåhraeus studied in more detail blood behavior at small pressure drops and in small capillaries [52, 72, 142].

The two most important findings of Fåhraeus are known as the Fåhraeus and Fåhraeus-Lindqvist effect [52, 54]. At high flow rates and in capillaries of diameters less than 300 μm, RBCs tend to migrate to the center and due to the parabolic flow profile, the mean velocity of RBCs is higher than the surrounding fluid resulting in a decrease of Hct compared to the feeding hematocrit Hcf. This is known as Fåhraeus effect. The ratio between Hct and Hcf decreases when capillary diameter decreases.

An important observation is related to the tube viscosity. In 1931, Fåhraeus and Lindqvist observed that in capillaries of diameters of less than 300 μm, blood viscosity decreases when capillary diameter decreases without keeping a constant value. This phenomenon is today known as Fåhraeus-Lindqvist effect [54]. At diameters below 10 μm, the feature of blood as a colloidal suspension is evident and the tube viscosity increases instead with decreasing diameter [48, 72]. Even though the Fåhraeus-Lindqvist effect is generally accepted [48], the physics behind the RBC lateral migration remained unclear for some decades and a model for blood flow behavior is still not available. In the next lines, we will briefly discuss the state of art of the lateral migration involving RBCs.

6.6.1 Lift force of deformable objects

The lateral migration of rigid spheres and deformable objects were first studied by Segré and Silberberg [132] and Goldsmith and Mason [73] respectively. In the later years, inertial effects were considered to develop a theoretical description of the lateral migration of solid spheres [128] and later a theory for the non-inertial lift on deformable objects was developed. The rate of migration of deformable objects at low Reynolds number was found to increase with the object size, the fluid flow rate and the distance of the object from the axis (lateral displacement) [73]. The theory was based on the variation of the velocity gradient along the cross-section considering the curvature of the flow profile [74].

RBCs as deformable objects migrate to the channel center by generating a lift force acting on capillary walls [4]. However, the collision of cells counteracts this migration: when two deformable cells collide, the cells experience an irreversible cross-stream displacement. Over time, collisions lead to a diffusive behavior called shear induced diffusion [99]. To understand this behavior, one has to understand how a particle disturbs the flow around it, the effect of the presence of a wall close to the particle, and the effect of the collisions between particles when there is more than one particle involved. A detailed discussion on the microstructure and rheology of cellular blood flow can be found in Chapter 6. Nevertheless and for further discussion, it is necessary for us to know that the balance between the deformability-induced lift force and the shear-induced diffusion created by hydrodynamic interactions in the suspension results in both a peak concentration of RBCs at the center of the channel center and a depleted cell layer or cell-free layer (CFL) (see Figure 6.7) or Fåhraeus-Lindqvist layer near the walls [49,71,99,100]. In this dynamics, the role of hematocrit has a strong impact on RBC distribution and the determination of the CFL thickness. Interestingly, in diseases like sickle cell anemia or malaria, cells were found to have a weaker lifting force and thus a reduced CFL thickness as already mentioned, in the last case, in previous sections [55,143,146].

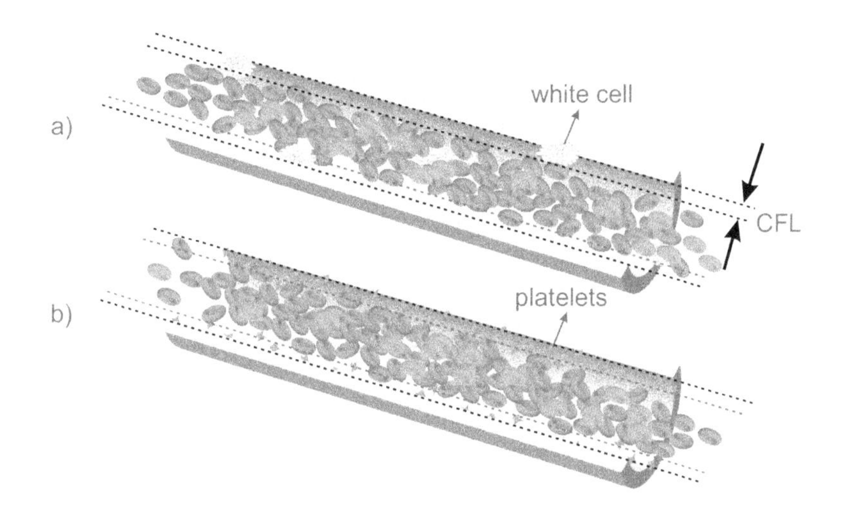

FIGURE 6.7
Schematic of margination. a) cell free layer and margination of white blood cells, b) margination of platelets

At the arteriole level, RBC axial migration influences the placement of the other circulating cells as white cells and platelets, with respect to the vascular wall [57,59,145], a phenomenon known as margination (see Figure 6.7). Margination of white cells and platelets has a physiological importance as cells need to be near the vessel walls to deal with inflammation and hemostasis [59,145]. As has been discussed by numerous researchers [61,139,158], studying cell migration, particle margination, and their relationship has a great significance in the understanding of the human biology of blood.

6.6.2 The importance of the cell-free layer, CFL

The CFL contributes importantly to microvascular function since effective blood viscosity and wall shear rate determine wall shear stresses, for release of the vaso-dilators, NO, and prostaglandins by the endothelium [85]. Its thickness lubricates blood flow by reducing the friction between RBCs and the endothelial cells, hence reducing the flow resistance [55,112], and influences the gas exchange processes like oxygen delivery and carbon dioxide removal by adding an extra barrier between RBCs and wall cells [112]. It has been found that RBC aggregation influences the migration of RBCs, increasing the relative thickness of the lower-viscosity marginal fluid [108], impacting the microvascular function related to CFL characteristics. However, RBC aggregation is also well known to enhance RBC sedimentation [51]. Then, the influence of sedimentation on axial accumulation and phase separation needs to be considered, especially in relation to the orientation of the tube with respect to gravity. In the following subsection, we will present some studies that have shown the impact of RBC aggregation on the formation of the CFL.

6.6.3 Impact of RBC aggregation on CFL formation

In vitro experiments on healthy blood have shown that RBC aggregation increases the thickness of the CFL [102] while *in vivo* experiments have reported that the thickness remains the same with or without aggregating agents in the suspending media. For example, Kim et al. [85] performed *in vivo* experiments in the rat cremaster muscle. The authors used dextran of 500 kDa to induce RBC aggregation at the levels seen in blood of healthy humans. They found no significant difference in the thickness of the CFL that remains around 3 μm in arterioles with diameters of around 50 μm. *In vitro* experiments performed by Maeda et al. [102] on hardened (H) and elastic (E) micro-vessels isolated from rabbit mesentery, showed that the thickness of the CFL decreased by increasing hematocrit for both H and E microvessels. Maeda et al. also explored the effect of cell rigidity on CFL thickness. Results showed that CFL thickness decreases when cells become more rigid, a condition that could be comparable to what happens with sickle RBCs, that in average are more rigid than healthy RBCs. In the same study, Maeda et al. explored the role of RBC aggregation in the CFL formation adding different concentrations of dextran of 70 kDa to the suspending media. Increasing dextran concentration produced an increase of the CFL thickness. It should be noticed that the range of velocities explored by Maeda et al. would produce shear rates near the wall too low to break the aggregates, meaning that at these low shear rate values, aggregation among RBCs would be the main determinant of the CFL thickness.

6.7 BLOOD FLOW IN SMALL TUBES

The biggest vessels are connected to the heart, decreasing in size, bifurcating, and creating a network of vessels as they are spread out into the tissues. Capillaries are the smallest and most numerous blood vessels in the circulatory system, with diameters ranging from about 2 to 10 μm. The typical distance between bifurcations is hundreds of micrometers to a few millimeters [118]. Since the diameter of a RBC at rest is bigger than the diameter of many capillaries, RBCs are subjected to high deformation in order to pass through them. As previously mentioned, rouleaux structures are reversibly and continuously breaking down to single flowing discocytes for increasing shear rates above 10 s^{-1} (see Figure 6.6). This observation has been done in bulk flow and classical shear rheology using bulk viscometers (e.g., cone-plate rheometers). However, in a very confined capillaries, shear stresses cannot fully act on the breaking process due to the confinement of the cells.

6.7.1 Flow of RBCs through small capillaries

Soft objects such as RBC, liquid droplets, lipid vesicles show complex behavior in flow mainly due to their long structural relaxation times and large deformation as a shear forces reaction. Droplets break-up and elastic capsules or lipid vesicles wrinkle [1, 58, 81, 150] in shear flows. RBCs and lipid vesicles have shown parachute-like shape when flowing through capillaries [68, 136, 143]. RBCs at rest have a biconcave discocytic shape, of about 8 μm of longer diameter and 2 μm in height. Typically it is considered to maintain a constant area S and volume V [66]. The RBC membrane consisting of a lipid bilayer supported by an attached cytoeskeleton, produces in the cells a resistance to bend and to shear strain, respectively. Bending is controlled by curvature elasticity with a bending rigidity, k and the cytoeskeleton resistance to shear strain is characterized by a shear modulus, μ. Measurements by micropipette aspiration [109] and stretching by optical tweezers [47] have given as a result $k/k_BT \sim 50$ and $\mu R_0^2/k_BT \sim 10^4$ under physiological conditions [47, 66, 109]

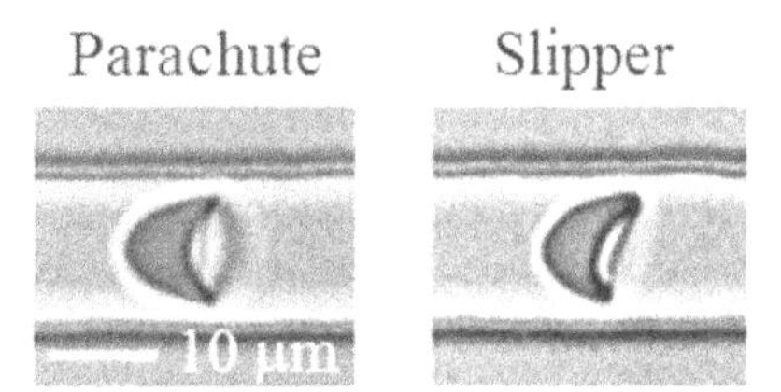

FIGURE 6.8
Parachute (left) and *slipper-like* (right) shapes of RBCs from 2D images observed when cells flow through micro-capillaries of approximately 10μm diameter.

where $R_0 = 3.3\mu$m have been considered as a typical RBC radius. The small bending resistance and the large surface area-to-volume ratio in a RBC produces an easily deformable soft body. This deformability has an important physiological impact since RBCs are able to change their shape and cross through extremely narrow capillaries with diameters as small as 2 μm, much less than the maximum diameter of the RBC. RBC deformation in microvessels seems to be an important parameter to regulate oxygen delivery since RBC deformation induces ATP release from RBCs, which induces nitric oxide (NO) synthesis and results in an increase of the vascular caliber [140].

As mentioned, when RBCs flow in tubes with diameters above but close to 10 μm, the typical shapes observed are *slipper-like* and centered *parachute* as is shown in Figure 6.8. *Slipper-like* shape is non axisymmetrical while the *parachute* shape is considered to be fully axisymmetrical.

While these shapes were first observed in 1963 and 1969 [75, 136], several studies have consecutively showed the existence of both for different experimental conditions [3, 18, 68, 89, 147], although all these studies considered 2D images. Recent 3D reconstruction of single RBCs flowing through confined channels showed that a third shape named *croissant* can be considered (see Figure 6.9). A *croissant* shape observed by one of its sides (in a 2D image) may be confused to be a perfect axisymmetrical *parachute* shape.

The important elastic properties of the RBCs to model their behavior in the capillary blood flow are the shear modulus and the bending rigidity of the membrane as mentioned, considering a constant RBC surface area and volume. The RBC hemoglobin can be considered as an isotropic and incompressible flow. As a consequence, when a RBC flows through a liquid, the interaction of the membrane with the interior and exterior fluid behaves as a dissipative system with a defined viscosity contrast. If the vessel diameter is decreased below 10 μm, the apparent (or effective) viscosity begins to increase. The axial-train or stacked-coin model [156] which employed the lubrication approximation for a regular array of cylindrically shaped cells and piecewise-parabolic flow profiles, reproduced this experimentally

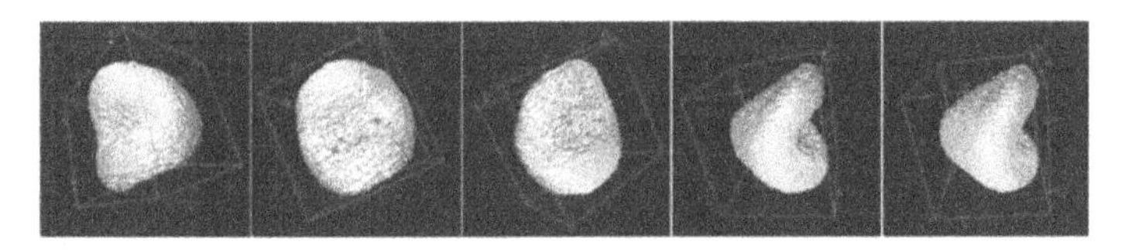

FIGURE 6.9
Sequence of images of a reconstructed 3D *croissant* shape of a RBC flowing through a highly confined microchannel of dimensions of approximately 10 x 10 μ m in width and height. The 3D reconstructions were made by A. Christ from AG Wagner using the technique described by Quint et al. [123].

observed effect qualitatively, being the first theoretical model of RBCs in narrow capillary flow. The cells were modeled as rigid, non-deformable spheres, spheroids, cylinders, and discocytes. The Stokes equation was used to compute the fluid flow around a periodic (equally spaced) array of RBCs. The results were qualitatively independent of cell shape. The relative apparent viscosity (relative to a fluid free of cells) was found to increase with increasing hematocrit and with decreasing capillary diameter [154, 156]. The additional pressure drop that each RBC contributes under this condition becomes independent of the cell-to-cell distance once this distance becomes larger than the capillary diameter, becoming ' hydrodynamically independent' or ' hydrodynamically isolated', meaning that confinement screens long-range hydrodynamic interactions between immersed RBCs or colloids, effectively reducing the range of these interactions to the diameter of the confining vessel [46].

6.7.2 Hydrodynamic interaction

In vivo and *in vitro* observations, show that RBCs tend to form organized clusters [42,68,70,148] in microcapillaries even if the tube hematocrit Hct is as low as 1% [31,42,86]. The RBCs can be attracted by hydrodynamical interactions and due to this attraction, we can find clusters formed exclusively by RBCs that have a clear distance in between their surfaces (see a) in Figure 6.10). However, after the addition of dextran 70 kDa clusters can be formed by a combination of aggregated RBCs (RBCs whose surfaces are in contact) and hydrodynamical attracted RBCs (see b) in Figure 6.10). Finally, aggregation of RBCs can happen in flow when two cells are first hydrodynamically attracted (t1 and t2 in Figure 6.10) and then their surfaces touch (t3 to t6 in Figure 6.10) [42].

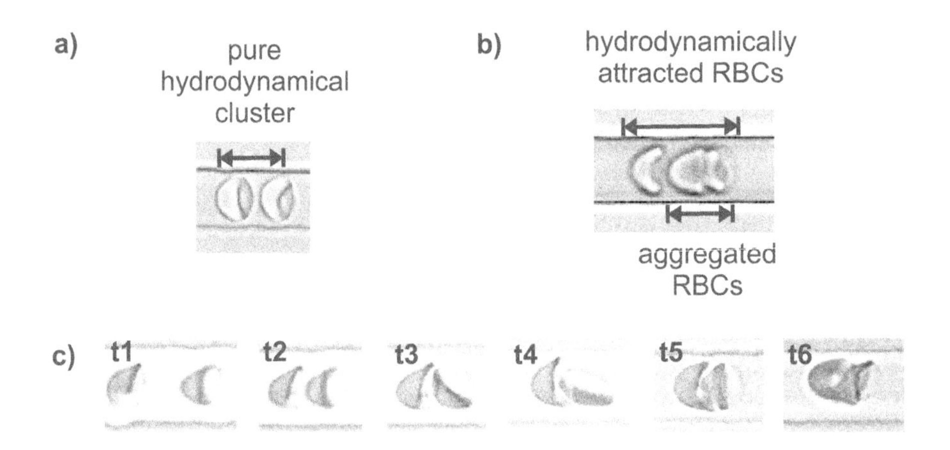

FIGURE 6.10
a) pure hydrodynamical cluster. Notice that RBCs have a clear distance in between their surfaces. b) cluster formed by a combination of two aggregated RBCs (right side of the cluster) and a single cell (left side of the cluster) hydrodynamically attracted. c) sequence of images in time from t1 to t6 showing how RBCs can aggregate in flow. At times t1 and t2 two single RBCs are at first hydrodynamically attracted and later at times from t3 to t6, their surfaces touch forming an aggregate. In all cases, cells are flowing from left to right at a velocity of ~1 mm/s.

On the other hand, *parachute* (or *croissant*) and *slipper-like* shapes may be present in the same cluster, and in the case of clusters formed by two RBCs (see Figure 6.11 or a) in Figure 6.10), a pronounced bimodal distribution of the cell-to-cell distances in the hydrodynamic clusters has been reported by Claveria et al. [42].

Recent theoretical efforts have been devoted to understand the shape changes of RBCs in narrow capillaries applying the lubrication approximation [131, 138] and by employing several numerical simulation techniques [107, 119, 122] One important finding is that the discocyte-to-parachute transition reduces the flow resistance. Hydrodynamic interactions coupled with RBC deformations induce clustering where parachute shapes prevail around a given velocity value. Two critical distances have been found to mark the boundaries between different levels of hydrodynamic interaction in simulations and also in experiments [42, 106]. For cell-to-cell distance larger than the capillary diameter RBCs are hydrodynamically isolated and do not interact. While for short distances, the shapes of two neighboring cells depend on each other producing or dissolving a bolus-flow structure between them. Therefore, changes in flow structure are strongly correlated with changes in RBC shape.

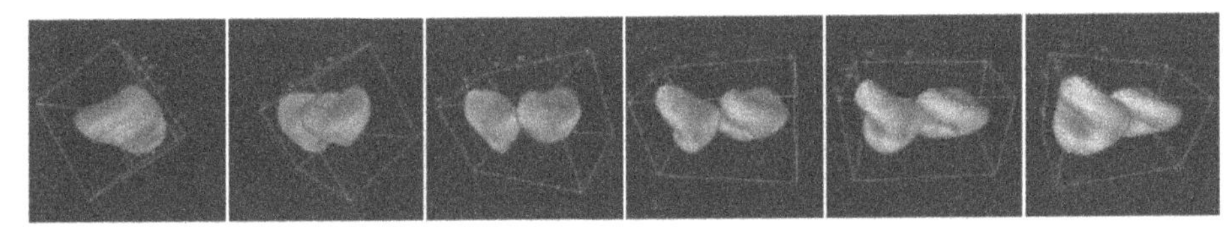

FIGURE 6.11
Sequence of images rotated counter-clockwise of a reconstructed 3D cluster formed by 2 RBCs. Cells were flowing through a highly confined microchannel at an approximate velocity of 0.5 mm/s. The 3D reconstructions were made by A. Christ from AG Wagner group, Saarland University, using the technique described by Quint et al. [123].

The physical origin of the pronounced bimodal distribution of the cell-to-cell distances in the hydrodynamic clusters can first result from changes in the cell shapes. However, even for the case of identical cells numerical simulations indicate a bi-stable situation of cell distances due to the confinement [9]. The physical origin of cluster formation can be explained to be either by long-range hydrodynamic interaction [68, 70, 106, 130, 148] or at distances of the order of the particle size [160]. In this last case, numerical simulations have shown how the streamlines disturbed around spherical particles due to the presence of the wall can overlap, causing particle interaction.

Experiments on hard spheres [45] and droplet suspensions [25] have given some knowledge of the mechanisms of hydrodynamic interaction in a confined flow where phonon-like excitations have been observed. However, much less is known about the case of soft deformable objects since hydrodynamic forces in colloidal systems are simpler because they remain at least roughly spherical and the flow field is easier to determine. Recent detailed numerical predictions show that soft objects should have a much stronger tendency to form clusters and display richer dynamics than hard systems [105]. RBCs can be considered as model objects to study the flow of soft objects in confinement [5], where the physical parameters, such as deformability are as important as the surrounding hydrodynamics.

6.7.3 Hydrodynamic versus macromolecule induced interaction

The effects of macromolecule induced clusters were tested in channels with diameters smaller than a RBC diameter, with the effect of hydrodynamic attraction minimized [31].Depending on the concentration of either dextran 70 kDa or fibrinogen clusters with two and up to five cells where observed. 2-D numerical simulations based on a vesicle model and a heuristic

Morse interaction potential comparable to the interaction energies from the AFM data showed a good qualitative agreement.

Another set of experiments was performed in channels with diameters in the order of the diameter of RBCs allowing both hydrodynamic and short-range aggregation mechanism [42] in the cluster formation. Macromolecule-induced interactions are not fully overcome by shear stresses within the physiological range (velocities around 1 mm/s), and they contribute to cluster stability. Yet, it was shown by two-dimensional numerical simulations [42], that cluster stabilization by hydrodynamics becomes predominant for mean velocities higher than 1 mm/s.

6.7.4 Consequences of clusters formation on flow resistance

Even though the mechanism of RBC interactions are still not fully understood, the RBC hematocrit and blood cell interactions in the microcirculation are known to impact the overall pressure drop in an organ [78]. Abkarian et al. [3] proposed an interesting method to quantify the impact of RBCs clusters passing through microcapillaries measuring the pressure drop produced. As a main observation, the pressure drop systematically increases as the number of cells increase in the channel or bigger clusters pass through. Nevertheless, the pressure drop did not increase proportionally to the number of cells and a cluster formed by five cells produces a pressure drop smaller than 5 times the pressure drop produced by a single RBC. We could conclude from these results that even though an increased number of cells increases flow resistance, the formation of clusters seems to be favorable in terms of flow resistance at the microcapillary level.

The clustering of RBCs observed *in vivo* in narrow arterioles of a rabbit [110] have been explained based on the polydispersity of a blood sample, where RBCs have a variation in size and flexibility. The membrane of more flexible RBCs or the lower RBC reduced volume should produce a RBC that can more easily deform and presumable migrate closer to the capillary axis, moving faster than the rest of the RBCs and crowding behind the more slow moving, less deformable RBCs. But numerical simulations have shown that clustering can occur even in a monodisperse suspension of RBCs that are identical in their size and flexibility [9,106].

6.8 CONCLUSION AND PERSPECTIVES

Most of the rheological works have focused on the impact of blood viscosity and/or RBC deformability in diseases and blood flow. A growing number of studies now pay attention to RBC aggregation. RBC aggregation may affect blood flow resistance in small and big vessels, cell adhesion phenomena and endothelium biology. This chapter provides important information regarding the mechanisms of RBC aggregation and its impact on blood viscosity and blood flow structuring in both the micro- and macrocirculation.

There are many remaining questions that should be answered by future investigations, and from our perspective the most important ones are:

- How big is the respective contribution of the bridging and the depletion effect in physiological conditions? A key issue will be the determination of the adsorption constants of the proteins or macromolecules on the cell surface to quantify the bridging effect on the level of a single molecule.

- What is the effect of aggregation on flow instabilities in bulk flow? Flow instabilities can cause many cardiovascular pathologies and at the same time flow instabilities are known to be very sensitive to the microstructure of the fluid.

- How much does RBC aggregation contribute to thrombus formation? Especially in so called deep vein thrombosis many red blood cells are found and it is not clear how the rouleaux formation contributes here.

Bibliography

[1] A. Walter, H. Rehage and H. Leonhard. Shear induced deformation of microcapsules: shape oscillations and membrane folding. *Colloids and Surfaces A: Physicochemical and Engineering Aspects*, 183-185:123 – 132, 2001.

[2] K. Abbitt and G. Nash. Rheological properties of the blood influencing selectin-mediated adhesion of flowing leukocytes. *American Journal of Physiology-Heart and Circulatory Physiology*, 2003.

[3] M. Abkarian, M. Faivre, and H. Stone. High-speed microfluidic differential manometer for cellular-scale hydrodynamics. *Proceedings of the National Academy of Sciences*, 2006.

[4] M. Abkarian and A. Viallat. Vesicles and red blood cells in shear flow. *Soft Matter*, 2008.

[5] M. Abkarian and A. Viallat. Chapter 10: On the importance of the deformability of red blood cells in blood flow. *RSC Soft Matter*, 2016.

[6] Manouk Abkarian, Colette Lartigue, and Annie Viallat. Tank treading and unbinding of deformable vesicles in shear flow: Determination of the lift force. *Phys. Rev. Lett.*, 2002.

[7] I. Aloulou, E. Varlet-Marie, J. Mercier, and J-F. Brun. *Hemorheological Disturbances Correlate with the Lipid Profile but not with the NCEP-ATPIII Score of the Metabolic Syndrome.* IOS Press, 2006.

[8] R. Ami, G. Barshtein, D. Zeltser, Y. Goldberg, I. Shapira, A. Roth, G. Keren, H. Miller, V. Prochorov, A. Eldor, S. Berliner, and S. Yedgar. Parameters of red blood cell aggregation as correlates of the inflammatory state. *American Journal of Physiology-Heart and Circulatory Physiology*, 2000.

[9] O. Aouane, A. Farutin, M. Thiébaud, A. Benyoussef, C. Wagner, and C. Misbah. Hydrodynamic pairing of soft particles in a confined flow. *Phys. Rev. Fluids*, 2017.

[10] S. Asakura and F. Oosawa. Interaction between particles suspended in solutions of macromolecules. *Journal of Polymer Science*, 1958.

[11] A. Ashkin. Acceleration and trapping of particles by radiation pressure. *Physical Review Letters*, 1970.

[12] A. Ashkin. Trapping of atoms by resonance radiation pressure. *Physical Review Letters*, 1978.

[13] A. Ashkin and J. Dziedzic. Optical levitation of liquid drops by radiation pressure. *Science*, 1975.

[14] A. Ashkin, J. Dziedzic, and S. Chu. Observation of a single-beam gradient force optical trap for dielectric particles. *Optics Letters*, 1986.

[15] A. Ashkin, J. Dziedzic, Bjorkholm J. E., and S. Chu. Applications of laser radiation pressure. *Science*, 1980.

[16] O. Awodu, A. Famodu, O. Ajayi, M. Enosolease, O. Olufemi, and E. Olayemi. Using serial haemorheological parameters to assess clinical status in sickle cell anaemia patients in vaso-occlusive crisis. *Clinical Hemorheology and Microcirculation*, 2009.

[17] P. Bagchi, P. Johnson, and A. Popel. Computational fluid dynamic simulation of aggregation of deformable cells in a shear flow. *Journal of Biomechanical Engineering*, 2005.

[18] U. Bagge, P. I. Branemark, R. Karlsson, and R. Skalak. Three-dimensional observations of red blood cell deformation in capillaries. *Blood Cells*, 1980.

[19] O. Baskurt, M. Boynard, G. Cokelet, P. Connes, B. Cooke, S. Forconi, F. Liao, M. Hardeman, F. Jung, H. Meiselman, G. Nash, N. Nemeth, B. Sandhagen, S. Shin, G. Thurston, and J. Wautier. New guidelines for hemorheological laboratory techniques. *Clinical Hemorheology and Microcirculation*, 2009.

[20] O. Baskurt and H. Meiselman. Blood Rheology and Hemodynamics. *Seminars in Thrombosis and Hemostasis*, 2003.

[21] O. Baskurt, B. Neu, and H. Meiselman. *Red Blood Cell Aggregation*. CRC Press, 2011.

[22] O. Baskurt, A. Temiz, and H. Meiselman. Effect of superoxide anions on red blood cell rheologic properties. *Free Radical Biology and Medicine*, 1997.

[23] O. Baskurt, M. Uyuklu, P. Ulker, M. Cengiz, N. Nemeth, T. Alexy, S. Shin, M. Hardeman, and H. Meiselman. Comparison of three instruments for measuring red blood cell aggregation. *Clinical Hemorheology and Microcirculation*, 2009.

[24] O. Baskurt, O. Yalcin, S. Ozdem, J. Armstrong, and H. Meiselman. Modulation of endothelial nitric oxide synthase expression by red blood cell aggregation. *American Journal of Physiology-Heart and Circulatory Physiology*, 2004.

[25] T. Beatus, R. Bar-Ziv, and T. Tlusty. Anomalous microfluidic phonons induced by the interplay of hydrodynamic screening and incompressibility. *Physical Review Letters*, 2007.

[26] S. Berliner, T. Mardi, G. Barshtein, O. Elkayam, R. Ben-Ami, S. Yedgar, and V. Deutch. A synergistic effect of albumin and fibrinogen on immunoglobulin-induced red blood cell aggregation. *American Journal of Physiology-Heart and Circulatory Physiology*, 2015.

[27] J. Bishop, P. Nance, Al. Popel, M. Intaglietta, and P. Johnson. Effect of erythrocyte aggregation on velocity profiles in venules. *American Journal of Physiology-Heart and Circulatory Physiology*, 2001.

[28] S. Block. Making light work with optical tweezers. *Nature*, 1992.

[29] M. Bor-Kucukatay, R. Wenby, H. Meiselman, and O. Baskurt. Effects of nitric oxide on red blood cell deformability. *American Journal of Physiology-Heart and Circulatory Physiology*, 2003.

[30] G. Brakenhoff, J. Brimbergen, J. Sixma, E. Nijhof, and P. Bronkhorst. The mechanism of red blood cell aggregation investigated by means of direct cell manipulation using multiple optical trapping. *Biorheology*, 2002.

[31] M. Brust, O. Aouane, M. Thiébaud, D. Flormann, C. Verdier, L. Kaestner, M. W. Laschke, H. Selmi, A. Benyoussef, T. Podgorski, G. Coupier, C. Misbah, and C. Wagner. The plasma protein fibrinogen stabilizes clusters of red blood cells in microcapillary flows. *Scientific Reports*, 2014.

[32] C. Bucherer, J. Ladjouzi, C. Lacombe, J.C. Leliévre, H. Vandewalle, Y. Beuzard, and F. Galacteros. Effect of deoxygenation on rheological behavior of density separated sickle cell suspensions. *Clinical Hemorheology and Microcirculation*, 1992.

[33] K. Buxbaum, E. Evans, and D. Brooks. Quantitation of surface affinities of red blood cells in dextran solutions and plasma. *Biochemistry*, 1982.

[34] Alonso C., Pries A., Kießlich O., Lerche D., and Gaehtgens P. Transient rheological behavior of blood in low-shear tube flow: velocity profiles and effective viscosity. *The American Journal of Physiology*, 1995.

[35] S. Chien. Shear dependence of effective cell volume as a determinant of blood viscosity. *Science*, 168:977–979, 1970.

[36] S. Chien. Red cell deformability and its relevance to blood flow. *Annual Review of Physiology*, 1987.

[37] S. Chien. Biophysical Behavior of Red Cells in Suspensions. In *The Red Blood Cell*. Academic Press, 2013.

[38] S. Chien and K. Jan. Ultrastructural basis of the mechanism of rouleaux formation. *Microvascular Research*, 1973.

[39] S. Chien, S. Simchon, R. Abbott, and K-M. Jan. Surface adsorption of dextrans on human red cell membrane. *Journal of Colloid and Interface Science*, 1977.

[40] S. Chien, L. Sung, S. Simchon, M. Lee, K.-m. Jan, and R. Skalak. Energy balance in red cell interactions. *Annals of the New York Academy of Sciences*, 1983.

[41] S. Chien, S. Usami, and J. Bertles. Abnormal rheology of oxygenated blood in sickle cell anemia. *The Journal of Clinical Investigation*, 1970.

[42] V. Claverıa, O. Aouane, M. Thiebaud, M. Abkarian, G. Coupier, Ch. Misbah, T. John, and Ch. Wagner. Clusters of red blood cells in microcapillary flow : hydrodynamic versus macromolecule induced interaction. *Soft Matter*, 2016.

[43] Giles R. Cokelet and Harry L. Goldsmith. Decreased hydrodynamic resistance in the two-phase flow of blood through small vertical tubes at low flow rates. *Circulation Research*, 1991.

[44] P. Connes, Y. Lamarre, X. Waltz, S. Ballas, N. Lemonne, M. Etienne-Julan, O. Hue, M-D. Hardy-Dessources, and M. Romana. Haemolysis and abnormal haemorheology in sickle cell anaemia. *British Journal of Haematology*, 2014.

[45] B. Cui, H. Diamant, B. Lin, and S. A. Rice. Anomalous hydrodynamic interaction in a quasi-two-dimensional suspension. *Physical Review Letters*, 2004.

[46] Bianxiao Cui, Haim Diamant, and Binhua Lin. Screened hydrodynamic interaction in a narrow channel. *Phys. Rev. Lett.*, 2002.

[47] M. Dao, C.T. Lim, and S. Suresh. Mechanics of the human red blood cell deformed by optical tweezers. *Journal of the Mechanics and Physics of Solids*, 2003.

[48] L. Dintenfass. Inversion of the Fahreaus-Lindqvist phenomenon in blood flow through capillaries of dimishing radius. *Nature*, 1967.

[49] E. Eckstein, A. Tilles, and F. Millero. Conditions for the occurrence of large near-wall excesses of small particles during blood flow. *Microvascular Research*, 1988.

[50] E. Evans. Minimum energy analysis of membrane deformation applied to pipet aspiration and surface adhesion of red blood cells. *Biophysical Journal*, 1980.

[51] T. Fabry. Mechanism of erythrocyte aggregation and sedimentation. *Blood*, 1987.

[52] R. Fahraeus. The suspension stability of blood. *Physiological Reviews*, 1929.

[53] R. Fahraeus. The Influence of the rouleau formation of the erythrocytes on the rheology of the blood. *Acta Medica Scandinavica*, 1958.

[54] R. Fahraeus and Lindqvist T. The viscosity of the blood in narrow capillary tubes. *The American Journal of Physiology*, 1931.

[55] D. Fedosov, B. Caswell, and G. Karniadakis. Systematic coarse-graining of spectrin-level red blood cell models. *Computer Methods in Applied Mechanics and Engineering*, 2010.

[56] D. Fedosov, B. Caswell, A. Popel, and G. Karniadakis. Blood flow and cell-free layer in Microvessels. *Microcirculation*, 2010.

[57] D. Fedosov and G. Gompper. White blood cell margination in microcirculation. *Soft Matter*, 2014.

[58] R Finken and U Seifert. Wrinkling of microcapsules in shear flow. *Journal of Physics: Condensed Matter*, 18(15):L185–L191, 2006.

[59] J. C. Firrell and H. H. Lipowsky. Leukocyte margination and deformation in mesenteric venules of rat. *American Journal of Physiology-Heart and Circulatory Physiology*, 1989.

[60] T. Fischer. On the energy dissipation in a tank-treading human red blood cell. *Biophysical journal*, 1980.

[61] S. Fitzgibbon, A. P. Spann, Q. M. Qi, and E. S. G. Shaqfeh. In vitro measurement of particle margination in the microchannel flow: Effect of varying hematocrit. *Biophysical Journal*, 2015.

[62] D. Flormann, O. Aouane, L. Kaestner, C. Ruloff, C. Misbah, C. Wagner, and T. Podgorski. The buckling instability of aggregating red blood cells. *Scientific Reports*, 2017.

[63] D. Flormann, E. Kuder, P. Lipp, C. Wagner, and L. Kaestner. Is there a role of C-reactive protein in red blood cell aggregation? *International Journal of Laboratory Hematology*, 2015.

[64] É. Franceschini, F. Yu, and G. Cloutier. Simultaneous estimation of attenuation and structure parameters of aggregated red blood cells from backscatter measurements. *JASA Express Letters*, 2008.

[65] M. Franco, E. Collec, P. Connes, E. Van Den Akker, T. Billette de Villemeur, N. Belmatoug, M. Von Lindern, N. Ameziane, O. Hermine, Y. Colin, C. Le Van Kim, and C. Mignot. Abnormal properties of red blood cells suggest a role in the pathophysiology of Gaucher disease. *Blood*, 2012.

[66] Y.C. Fung. Biomechanics. In *Biomechanics*. Springer, New York, NY, 1997.

[67] G. G Binnig, C.F. Quate, and Ch. Gerber. Atomic force microscope. *Physical Review Letters*, 1986.

[68] P. Gaehtgens, C. Dührssen, and K. H. Albrecht. Motion, deformation, and interaction of blood cells and plasma during flow through narrow capillary tubes. *Blood Cells*, 1980.

[69] P. Gennes. *Scaling Concepts in Polymer Physics*. Cornell University Press, 1979.

[70] G. Ghigliotti, H. Selmi, L. El Asmi, and Ch. Misbah. Why and how does collective red blood cells motion occur in the blood microcirculation? *Physics of Fluids*, 2012.

[71] H. Goldsmith. The microrheology of red blood cell suspensions. *The Journal of General Physiology*, 1968.

[72] H. Goldsmith, G. Cokelet, and P. Gaehtgens. Robin Fahraeus: evolution of his concepts in cardiovascular physiology. *American Journal of Physiology-Heart and Circulatory Physiology*, 1989.

[73] H. Goldsmith and S. Mason. Axial migration of particles in poiseuille flow. *Nature*, 1961.

[74] H. Goldsmith and S. Mason. The flow of suspensions through tubes. I. Single spheres, rods, and discs. *Journal of Colloid Science*, 1962.

[75] M. Guest, T. Bond, R. Coopert, and J. Derrick. Red blood cells: change in shape in capillaries. *Science*, 1963.

[76] P. Gyawali, R. Richards, P. Tinley, and E. Nwose. Hemorheology, ankle brachial pressure index (ABPI) and toe brachial pressure index (TBPI) in metabolic syndrome. *Microvascular Research*, 2014.

[77] C. Haest. Distribution and Movement of Membrane Lipids. In *Red Cell Membrane Transport in Health and Disease*. Springer Berlin Heidelberg, 2013.

[78] B. Helmke, S. Bremner, B. Zweifach, R. Skalak, and G. Schmid-Schönbein. Mechanisms for increased blood flow resistance due to leukocytes. *American Journal of Physiology-Heart and Circulatory Physiology*, 1997.

[79] W. Hess. Gehorcht das Blut dem allgemeinen Strömungsgesetz der Flüssigkeiten? *Pflüger's Archiv für die Gesamte Physiologie des Menschen und der Thiere*, 1915.

[80] R. Hierso, X. Waltz, P. Mora, M. Romana, N. Lemonne, P. Connes, and M-D. Hardy-Dessources. Effects of oxidative stress on red blood cell rheology in sickle cell patients. *British Journal of Haematology*, 2014.

[81] Vasiliy Kantsler, Enrico Segre, and Victor Steinberg. Vesicle dynamics in time-dependent elongation flow: Wrinkling instability. *Phys. Rev. Lett.*, 2007.

[82] G. Kato, F. Piel, C. Reid, M. Gaston, K. Ohene-Frempong, L. Krishnamurti, W. Smith, J. Panepinto, D. Weatherall, F. Costa, and E. Vichinsky. Sickle-Cell Disease. In *Nature Reviews. Disease Primers.* Macmillan Publishers Limited, 2018.

[83] D. Kaul, R. Nagel, D. Chen, and H. M. Tsai. Sickle erythrocyte-endothelial interactions in microcirculation: the role of von willebrand factor and implications for vasoocclusion. *Blood*, 1993.

[84] M. Khokhlova, E. Lyubin, A. Zhdanov, S. Rykova, I. Sokolova, and A. Fedyanin. Normal and system lupus erythematosus red blood cell interactions studied by double trap optical tweezers: direct measurements of aggregation forces. *Journal of Biomedical Optics*, 2012.

[85] S. Kim, R. Kong, A. Popel, M. Intaglietta, and P. Johnson. Temporal and spatial variations of cell-free layer width in arterioles. *American Journal of Physiology-Heart and Circulatory Physiology*, 2007.

[86] B. Klitzman and B. Duling. Microvascular hematocrit and red cell flow in resting and contracting striated muscle. *American Journal of Physiology-Heart and Circulatory Physiology*, 1979.

[87] A. Korolevich and I. Meglinsky. Experimental study of the potential use of diffusing wave spectroscopy to investigate the structural characteristics of blood under multiple scattering. *Bioelectrochemistry*, 2000.

[88] E. Krieger, B. van Der Loo, B. Amann-Vesti, V. Rousson, and R. Koppensteiner. C-reactive protein and red cell aggregation correlate with late venous function after acute deep venous thrombosis. *Journal of Vascular Surgery*, 2004.

[89] K. Kubota, J. Tamura, T. Shirakura, M. Kimura, K. Yamanaka, T. Isozaki, and I. Nishio. The behaviour of red cells in narrow tubes in vitro as a model of the microcirculation. *British Journal of Haematology*, 1996.

[90] A. Kyrylyuk, M. Hermant, T. Schilling, B. Klumperman, C. Koning, and P. Van Der Schoot. Controlling electrical percolation in multicomponent carbon nanotube dispersions. *Nature Nanotechnology*, 2011.

[91] Y. Lamarre, M. Romana, X. Waltz, M. Lalanne-Mistrih, B. Tressières, L. Divialle-Doumdo, M-D. Hardy-Dessources, J. Vent-Schmidt, M. Petras, C. Broquere, F. Maillard, V. Tarer, M. Etienne-Julan, and P. Connes. *Haematologica*, 2012.

[92] M. Lang and S. Block. Resource Letter: LBOT-1: Laser-based optical tweezers. *American Journal of Physics*, 2003.

[93] L. Lanotte, J. Mauer, S. Mendez, D. Fedosov, J. Fromental, V. Claveria, F. Nicoud, G. Gompper, and M. Abkarian. Red cells' dynamic morphologies govern blood shear thinning under microcirculatory flow conditions. *Proceedings of the National Academy of Sciences*, 2016.

[94] C. Lapoumeroulie, P. Connes, S. El Hoss, R. Hierso, K. Charlot, N. Lemonne, J. Elion, C. Le Van Kim, M. Romana, and M-D. Hardy-Dessources. New insights into red cell rheology and adhesion in patients with sickle cell anaemia during vaso-occlusive crises. *British Journal of Haematology*, 2018.

[95] C. Lawrence and M. Fabry. Erythrocyte sedimentation rate during steady state and painful crisis in sickle cell anemia. *The American Journal of Medicine*, 1986.

[96] K. Lee, M. Kinnunen, A. Danilina, V. Ustinov, S. Shin, A. Priezzhev, A. Danilina, I. Meglinski, M. Kinnunen, V. Ustinov, S. Shin, I. Meglinski, and A. Priezzhev. Characterization at the individual cell level and in whole blood samples of shear stress preventing red blood cells aggregation. *Journal of Biomechanics*, 2016.

[97] K. Lee, M. Kinnunen, M. Khokhlova, E. Lyubin, A. Priezzhev, I. Meglinski, and A. Fedyanin. Optical tweezers study of red blood cell aggregation and disaggregation in plasma and protein solutions. *Journal of Biomedical Optics*, 21(3).

[98] K. Lee, E. Shirshin, N. Rovnyagina, F. Yaya, Z. Boujja, A. Priezzhev, and C. Wagner. Dextran adsorption onto red blood cells revisited: single cell quantification by laser tweezers combined with microfluidics. *Biomedical Optics Express*, 2018.

[99] D. Leighton and A. Acrivos. The shear-induced migration of particles in concentrated suspensions. *Journal of Fluid Mechanics*, 1987.

[100] R. Lima, T. Ishikawa, Y. Imai, M. Takeda, Sh. Wada, and T. Yamaguchi. Radial dispersion of red blood cells in blood flowing through glass capillaries: The role of hematocrit and geometry. *Journal of Biomechanics*, 2008.

[101] E. Loiseau, G. Massiera, S. Mendez, P. Martinez, and M. Abkarian. Microfluidic study of enhanced deposition of sickle cells at acute corners. *Biophysical Journal*, 2015.

[102] N. Maeda, Y. Suzuki, J. Tanaka, and N. Tateishi. Erythrocyte flow and elasticity of microvessels evaluated by marginal cell-free layer and flow resistance. *American Journal of Physiology-Heart and Circulatory Physiology*, 1996.

[103] N. Maharshak, Y. Arbel, I. Shapira, S. Berliner, R. Ben-Ami, S. Yedgar, G. Barshtein, and I. Dotan. Increased strength of erythrocyte aggregates in blood of patients with inflammatory bowel disease. *Inflammatory Bowel Diseases*, 2009.

[104] A. Maklygin, A. Priezzhev, A. Karmenyan, S. Nikitin, I. Obolenskii, and A. Lugovtsov. Measurement of interaction forces between red blood cells in aggregates by optical tweezers. *Quantum Electronics*, 2012.

[105] J. L. McWhirter, H. Noguchi, and G. Gompper. Flow-induced clustering and alignment of vesicles and red blood cells in microcapillaries. *Proceedings of the National Academy of Sciences*, 2009.

[106] J. L. Mcwhirter, H. Noguchi, and G. Gompper. Deformation and clustering of red blood cells in microcapillary flows. *Soft Matter*, 2011.

[107] J. Liam McWhirter, Hiroshi Noguchi, and Gerhard Gompper. Flow-induced clustering and alignment of vesicles and red blood cells in microcapillaries. *Proceedings of the National Academy of Sciences*, 2009.

[108] E. Merrill, E. Gilliland, G. Cokelet, H. Shin, A. Britten, and R. Wells Jr. Rheology of human blood, near and at zero flow. *Biophysical Journal*, 1963.

[109] N Mohandas and E Evans. Mechanical properties of the red cell membrane in relation to molecular structure and genetic defects. *Annual Review of Biophysics and Biomolecular Structure*, 1994.

[110] P.A.G. Monro. The appearance of cell-free plasma and "grouping" of red blood cells in normal circulation in small blood vessels observed in vivo. *Biorheology*, 1963.

[111] E. Nader, P. Connes, Y. Lamarre, C. Renoux, P. Joly, M-D. Hardy-Dessources, G. Cannas, N. Lemonne, and S. Ballas. Plasmapheresis may improve clinical condition in sickle cell disease through its effects on red blood cell rheology. *American Journal of Hematology*, 2017.

[112] B. Namgung, L. H. Liang, and S. Kim. Physiological Significance of Cell-free Layer and Experimental Determination of its Width in Microcirculatory Vessels. In *Lecture Notes in Computational Vision and Biomechanics*. Springer, Dordrecht, 2014.

[113] G. Nash, T. Watts, C. Thornton, and M. Barigou. Red cell aggregation as a factor influencing margination and adhesion of leukocytes and platelets. *Clinical Hemorheology and Microcirculation*, 2008.

[114] B. Neu and H. Meiselman. Depletion-mediated red blood cell aggregation in polymer solutions. *Biophysical Journal*, 2002.

[115] G. Novacek, H. Vogelsang, D. Genser, G. Moser, A. Gangl, H. Ehringer, and R. Koppensteiner. Changes in blood rheology caused by Crohn's disease. *European Journal of Gastroenterology and Hepatology*, 1996.

[116] N. Peled, M. Kassirer, M. Kramer, O. Rogowski, D. Shlomi, B. Fox, A. Berliner, and D. Shitrit. Increased erythrocyte adhesiveness and aggregation in obstructive sleep apnea syndrome. *Thrombosis Research*, 2007.

[117] W. Pollack and R. Reckel. A reappraisal of the forces involved in hemagglutination. *Int. Archs Allergy appl. Immun.*, 1977.

[118] A. Popel and P. Johnson. Microcirculation and hemorheology. *Annual Review of Fluid Mechanics*, 2005.

[119] C. Pozrikidis. Numerical simulation of cell motion in tube flow. *Annals of Biomedical Engineering*, 2005.

[120] A. Priezzhev and K. Lee. Potentialities of laser trapping and manipulation of blood cells in hemorheologic research. *Clinical Hemorheology and Microcirculation*, 2016.

[121] A. Priezzhev, O. Ryaboshapka, N. Firsov, and I. Sirko. Aggregation and disaggregation of erythrocytes in whole blood: Study by backscattering technique. *Journal of Biomedical Optics*, 2002.

[122] Christophe Queguiner and Dominique Barthes-Biesel. Axisymmetric motion of capsules through cylindrical channels. *Journal of Fluid Mechanics*, 1997.

[123] S. Quint, A. Christ, A. Guckenberger, S. Himbert, L. Kaestner, S. Gekle, and C. Wagner. 3D tomography of cells in micro-channels. *Applied Physics Letters*, 2017.

[124] M. Radu and T. Schilling. Solvent hydrodynamics speed up crystal nucleation in suspensions of hard spheres. *EPL (Europhysics Letters)*, 2014.

[125] M. Rampling and J. Sirs. Rouleaux formation and the rate of packing of erythrocytes. *The Journal of Physiology*, 1971.

[126] M. Ravelojaona, L. Féasson, S. Oyono-Enguéllé, L. Vincent, B. Djoubairou, C. Ewa'sama Essoue, and L. Messonnier. Evidence for a profound remodeling of skeletal muscle and its microvasculature in sickle cell anemia. *American Journal of Pathology*, 2015.

[127] W. Reinke, P. Gaehtgens, and P. C. Johnson. Blood viscosity in small tubes: effect of shear rate, aggregation, and sedimentation. *American Journal of Physiology-Heart and Circulatory Physiology*, 1987.

[128] P. Saffman. The lift force on a small shpere in a slow shear flow. *Journal of Fluid Mechanics*, 1965.

[129] D. Samocha-Bonet, D. Lichtenberg, A. Tomer, S. Abu-Abeid, V. Deutsch, T. Mardi, Y. Goldin, A. Subchi, G. Shenkerman, H. Patshornik, I. Shapira, and S. Berliner. Enhanced erythrocyte adhesiveness/aggregation in obesity corresponds to low-grade inflammation. *Obesity Research*, 2003.

[130] G. Schmid-Schönbein, S. Usami, R. Skalak, and S. Chien. The interaction of leukocytes and erythrocytes in capillary and postcapillary vessels. *Microvascular Research*, 1980.

[131] T. W. Secomb, R. Skalak, N. Özkaya, and J. F. Gross. Flow of axisymmetric red blood cells in narrow capillaries. *Journal of Fluid Mechanics*, 1986.

[132] G. Segrè and A. Silberberg. Radial particle displacements in Poiseuille flow of suspensions. *Nature*, 1961.

[133] S. Shin, J. Nam, J. Hou, and J. Suh. A transient, microfluidic approach to the investigation of erythrocyte aggregation: The threshold shear-stress for erythrocyte disaggregation. *Clinical Hemorheology and Microcirculation*, 2009.

[134] M. Simmonds, J. Detterich, and P. Connes. Nitric oxide, vasodilation and the red blood cell. In *Biorheology*, 2014.

[135] S. Sinnapah, G. Cadelis, X. Waltz, Y. Lamarre, and P. Connes. Overweight explains the increased red blood cell aggregation in patients with obstructive sleep apnea. *Clinical Hemorheology and Microcirculation*, 2015.

[136] R. Skalak and P-I. Branemark. Deformation of red blood cells in capillaries. *Science*, 1969.

[137] R. Skalak, P. R. Zarda, K. M. Jan, and S. Chien. Mechanics of rouleau formation. *Biophysical Journal*, 1981.

[138] Richard Skalak and Cuijuan Zhu. Rheological aspects of red blood cell aggregation. *Biorheology*, 1990.

[139] A. Spann, J. Campbell, S. Fitzgibbon, A. Rodriguez, A. Cap, L. Blackbourne, and E. Shaqfeh. The effect of hematocrit on platelet adhesion: Experiments and simulations. *Biophysical Journal*, 2016.

[140] R. S. Sprague, M. L. Ellsworth, A. H. Stephenson, and A. J. Lonigro. Atp: the red blood cell link to no and local control of the pulmonary circulation. *American Journal of Physiology-Heart and Circulatory Physiology*, 1996.

[141] P. Steffen, C. Verdier, and C. Wagner. Quantification of depletion-induced adhesion of red blood cells. *Physical Review Letters*, 2013.

[142] S. Sutera and R. Skalak. The history of Poiseuille's law. *Applied Mechanics and Engineering*, 1930.

[143] Y. Suzuki, N. Tateishi, M. Soutani, and N. Maeda. Flow behavior of eruthrocytes in microvessels and glass capillaries: effects of erythrocyte deformation and erythrocyte aggregation. *International Journal of Microcirculation*, 1996.

[144] K. Svoboda and S. Block. Biological applications of optical forces. *Annual Review of Biophysics and Biomolecular Structure*, 2002.

[145] G. Tangelder, H. Teirlinck, D. Slaaf, and R. Reneman. Distribution of blood platelets flowing in arterioles. *The American Journal of Physiology*, 1985.

[146] G. Tomaiuolo. Biomechanical properties of red blood cells in health and disease towards microfluidics. *Biomicrofluidics*, 2014.

[147] G. Tomaiuolo, L. Lanotte, R. D'Apolito, A. Cassinese, and S. Guido. Microconfined flow behavior of red blood cells. *Medical Engineering and Physics*, 2016.

[148] G. Tomaiuolo, D. Rossi, S. Caserta, M. Cesarelli, and S. Guido. Comparison of two flow-based imaging methods to measure individual red blood cell area and volume. *Cytometry Part A*, 2012.

[149] J. Tripette, T. Alexy, M-D. Hardy-Dessources, D. Mougenel, E. Beltan, T. Chalabi, R. Chout, M. Etienne-Julan, O. Hue, H. Meiselman, and P. Connes. Red blood cell aggregation, aggregate strength and oxygen transport potential of blood are abnormal in both homozygous sickle cell anemia and sickle-hemoglobin C disease. *Haematologica*, 2009.

[150] K. S. Turitsyn and S. S. Vergeles. Wrinkling of vesicles during transient dynamics in elongational flow. *Phys. Rev. Lett.*, 2008.

[151] J. Vent-Schmidt, X. Waltz, M. Romana, M. Hardy-Dessources, N. Lemonne, M. Billaud, M. Etienne-Julan, and P. Connes. Blood thixotropy in patients with sickle cell anaemia: Role of haematocrit and red blood cell rheological properties. *PLOS ONE*, 2014.

[152] B. Vincent. The calculation of depletion layer thickness as a function of bulk polymer concentration. *Colloids and Surfaces*, 1990.

[153] B. Vincent, J. Edwards, S. Emmett, and A. Jones. Depletion flocculation in dispersions of sterically-stabilised particles ("soft spheres"). *Colloids and Surfaces*, 1986.

[154] Henry Wang and Richard Skalak. Viscous flow in a cylindrical tube containing a line of spherical particles. *Journal of Fluid Mechanics*, 1969.

[155] R Wells and H Schmid-Schönbein. Red cell deformation and fluidity of concentrated cell suspensions. *Journal of Applied Physiology*, 1969.

[156] R L Whitmore. A theory of blood flow in small vessels. *Journal of Applied Physiology*, 1967.

[157] F. Yu, É. Franceschini, B. Chayer, J. Armstrong, H. Meiselman, and G. Cloutier. Ultrasonic parametric imaging of erythrocyte aggregation using the structure factor size estimator. *Biorheology*, 2009.

[158] H. Zhao, E. Shaqfeh, and V. Narsimhan. Shear-induced particle migration and margination in a cellular suspension. *Physics of Fluids*, 2012.

[159] A. Zimran, A. Bashkin, D. Elstein, B. Rudensky, R. Rotstein, M. Rozenblat, T. Mardi, D. Zeltser, V. Deutsch, I. Shapira, and S. Berliner. Rheological determinants in patients with Gaucher disease and internal inflammation. *American Journal of Hematology*, 2004.

[160] M. Zurita-Gotor, J. Bławzdziewicz, and E. Wajnryb. Swapping trajectories: A new wall-induced cross-streamline particle migration mechanism in a dilute suspension of spheres. *Journal of Fluid Mechanics*, 2007.

CHAPTER 7

Platelet dynamics in blood flow

Jawaad Sheriff and Danny Bluestein

Department of Biomedical Engineering, Stony Brook University, Stony Brook, USA

CONTENTS

PLATELETS are the preeminent blood cell in thrombus formation in injured or diseased blood vessels and blood-recirculating cardiovascular devices. In flowing blood, they interact with red blood cells, marginate towards the cell-free layer, tumble in free-flowing plasma, interact with reactive surfaces or subendothelial constituents, upon which they tether, translocate, roll, and eventually become firmly adhered via receptor-ligand bonds. Platelets also interact with other platelets and blood cells, forming aggregates and contribute to a growing thrombus. In this chapter, we describe both the external physical environment and intraplatelet dynamics that guide platelet motion, deformation and binding, and explore several single-scale and multiscale continuum- and particle-based models developed to describe and predict various phenomena leading up to thrombus formation.

7.1 INTRODUCTION

Thrombosis is a physiological hemostatic response to prevent significant blood loss after vascular injury. However, thrombosis associated with cardiovascular diseases is the leading cause of mortality, accounting for 1 in 4 deaths worldwide in 2010 [137]. It is the common underlying mechanism of myocardial infarction (heart attack), ischemic stroke, and venous thromboembolism [228]. Thrombotic complications are also prevalent in patients implanted with prosthetic heart valves and ventricular assist devices [191] as therapy for valvular disease and advanced heart failure, respectively.

Central to hemostasis and thrombosis is the platelet. In humans, platelets are anucleate cells about 2 to 5 μm in diameter and 0.5 to 1 μm in thickness, with a lifespan of up to 10 days in the circulation [229]. Their discoid shape consists of lipid bilayer membrane with a rugose appearance enveloping a spectrin-based shell, a peripheral circumferential coil of microtubules, a cytoskeletal network of contractile actin microfilaments, and a viscous cytoplasm. Vesicles and organelles such as the protein-rich α granules, mineral-containing dense bodies, electron-dense clusters, lysosomes, glycosomes, and mitochondria are housed in the cytoplasm [229]. Due to their small size and shape, they are pushed by flowing blood to the periphery of the blood vessel wall, composed of endothelial cells, allowing them to rapidly detect and respond to vascular damage. In the so-called cell-depleted layer, platelets interact with surface proteins on the blood vessel wall and soluble factors released into blood, and quickly bind, spread, secrete, and interact with each other to form a fibrin plug to cover the damaged endothelium. As the platelets spread, their morphology changes from their resting discoid shape and start to project spiny filopods and sheet-like lamellipods, which recruit additional platelets and secure a firm adhesion to the injury site, respectively. The progression of recruitment leads to a growing thrombus. Initial events in flow-mediated thrombosis are summarized in Figure 7.1.

While the events leading up to flow-mediated thrombosis involve a complex interplay between biochemical events and the mechanical forces that influence them, it is only recently that the latter's influence on platelets is being realized. The following sections describe the physiological and pathological flow conditions in which platelets travel, their subsequent physical response (i.e. activation, adhesion, aggregation, and thrombus formation), and computational models developed to understand and describe such behavior.

7.1.1 Flow conditions in physiology, pathology, and cardiovascular devices

Virchow's triad, first postulated in 1886, states thrombus formation can be caused by alterations in hypercoagulability, endothelial dysfunction or vascular injury, and blood flow

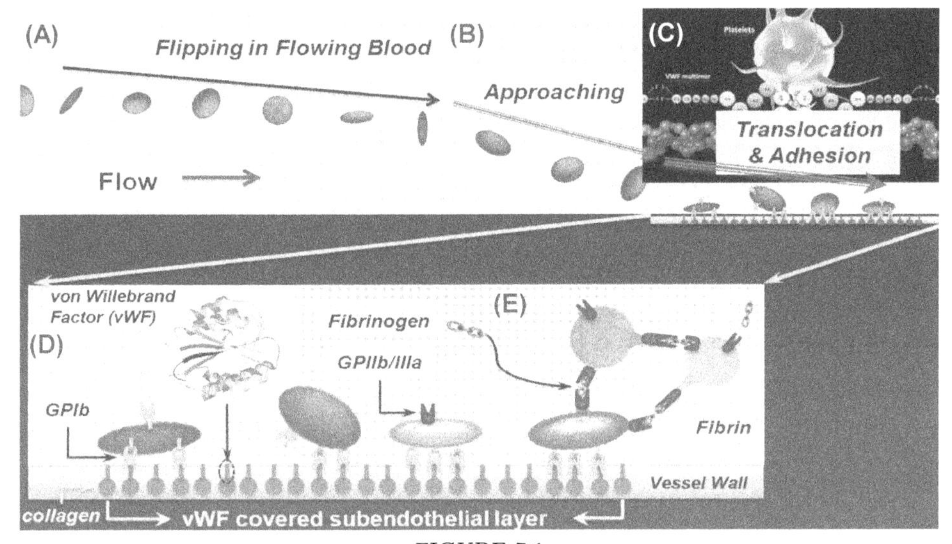

FIGURE 7.1
Initial physical events in injury- and flow-mediated thrombosis: (A) migration of marginated platelets to cell-free layer, (B) recruitment of platelets to reactive surface or injured endothelium, (C) translocation of platelets on immobilized von Willebrand factor, (D) activation and stable adhesion of platelets, and (E) recruitment of flowing platelets to adhered platelets to form growing aggregate.

(typically stagnation). The former two have been extensively studied, while the latter has been investigated more in recent years [218]. The advent of cardiovascular implantable devices has led to a redefinition of Virchow's triad, with an interaction between hypercoagulability, foreign surfaces, and pathologically higher flow conditions [203].

Flow is an important consideration for experimental and computational models of platelet behavior as it i) guides stabilization at vascular interfaces for efficient clot formation, ii) affects rheology, vascular architecture, and red blood cell (RBC) properties, iii) influences drug delivery efficacy, iv) triggers molecular unfolding of receptors and ligands, v) induces changes in cell morphology, vi) independently activates cellular reactants, vii) controls kinetics of tissue factor initiated coagulation and fibrin formation, viii) is affected by thrombus growth, and ix) plays a role in inflammatory cell response [254]. Blood flowing through the vasculature is characterized as Poiseuille flow, with a parabolic velocity profile that peaks at the center of the blood vessel. This approximation can be applied to most events in the circulation, despite the pulsatile (non-steady) arterial flow and non-Newtonian behavior at lower venous shear rates. Physiological flow is laminar with Reynolds (Re) numbers well below the threshold for turbulence [72]. Turbulent flow is only observed in severe stenosis, arteriovenous fistulas, coarctation [17], mechanical heart valves [206], and ventricular assist devices [139].

The shear rate $\dot{\gamma}$ is defined as the change in flow velocity as a function of the distance from the wall and is typically expressed in inverse seconds (s^{-1}). The shear stress τ is the tangential force per unit area of contact between the layers of fluid due to the pressure differences in the vasculature, and is represented as pascal (Pa) in SI units or dyne/cm^2 in the CGS system (1 Pa = 10 dyne/cm^2). For Newtonian fluids (see Chapter 1), $\tau = \mu\dot{\gamma}$ where μ is the fluid viscosity in Pa·s or dyne/cm^{2}s (or poise, P). The latter is termed wall shear stress and ranges between less than 1 dyne/cm^2 in veins to over 60 dyne/cm^2 in arterioles in normal physiology. Shear rates are minimal at the center of blood vessels, and more pronounced towards the wall as a result of wall friction [118]. In larger vessels, blood behaves as a Newtonian fluid, whereas in the microcirculation, interactions between RBCs are mediated by fibrinogen and increased viscosity, leading to a non-Newtonian behavior at shear rates below 200 s^{-1} [184]. Blood viscosity is generally accepted to be between 3 and 4 cP in the arterial circulation, with increasing values at lower shear rates [182] (see Chapter 1). The cell-depleted plasma layer near the wall can be considered an ideal Newtonian fluid, with a viscosity of 1.1-1.4 cP [182, 192]. In pathological conditions generated in diseased blood vessels and prosthetic cardiovascular devices, platelets are exposed to an extreme range of shear rates and stresses that promote thrombosis. Hemodynamic parameters in physiological vessels and some common pathologic conditions are summarized in Table 7.1.

TABLE 7.1
Hemodynamic parameters in blood vessels and devices (adapted from [17]).

Vessel/Condition	Shear Rate (s^{-1})	Shear Stress (dyn/cm^2)	References
Normal Physiology			
Ascending aorta	300	2-10	[72]
Large arteries	300-800	10-30	[72]
Coronary artery (LCA/RCA)	300-1500	10-60	[102, 245]
Carotid artery	250	10	[72]
Arterioles	500-1600	20-60 [72, 118]	
Capillary	200-2000	high[a]	[72]
Postcapillary venules	50-200	1-2	[123]
Veins	20-2000	0.8-8	[118]
Disease Conditions			
Coronary stenosis (LAD)	$5000 - 10^5$[b]	-	[95]
Carotid stenosis	-	40-360	[108]
Aortic coarctation[g]	-	140- > 1000[c]	[109, 121]
Atrioventricular fistula[g]	-	100-1000[d]	[201]
Deep vein thrombosis[g]	0-200[e]	-	[8]
Prosthetic Devices			
Mechanical heart valves[g]	-	up to 6000	[206]
Left ventricular assist devices[g]	5000 to > 10^5[f]	0- > 1000	[30, 67]

[a] Difficult to define in capillaries
[b] Higher shear rates are due to recirculation zones
[c] Flow can be turbulent
[d] Flow is turbulent
[e] Lower shear rates found in valve pocket
[f] Higher shear rates lead to acquired von Willebrand disease
[g] All shear rates and stresses are on the wall, except for these conditions

7.1.2 Pathological flow conditions in diseases and devices

Prothrombotic conditions exist in complex flow patterns that may develop in atherosclerotic lesions, prosthetic devices such as artificial heart valves or blood pumps, and anastomoses [82]. Increased fluid momentum within narrowed vessels or small gaps in devices may create jets into downstream expansions, leading to separation of fluid layers. This may generate low shear regions near the vessel wall or device surface, complex three-dimensional recirculation zones that markedly increase platelet residence time, or areas of stagnation [72]. Transitions in the cross-sectional area of the lumen, or gaps in devices, generate acceleration that pulls on the suspended blood cells and plasma proteins. This elongational flow can also induce a coagulation response at much lower thresholds than at constant shear rate [200]. Flow in the physiological vasculature is mostly laminar, but may transition to turbulent conditions in severe stenoses, prosthetic heart valves, or ventricular assist devices.

Atherosclerotic lesions develop at arterial branch points characterized by low shear and disturbed flow, particularly in the coronary circulation [93], that affect the adhesive properties of platelets, leukocytes, as well as morphology and function of endothelial cells. Disturbed flow at arterial branch points and growing atherosclerotic lesions affect the atherogenic process, which in turn induces a shear-dependent acceleration of atherosclerosis and produces stenoses [105, 160]. Alterations in blood flow include acceleration at the apex of a stenosis, and flow separation, eddy formation, flow reversal, and turbulence post-stenosis, with the decelerating subsequent low shear zone promoting progressive accumulation and aggregation of platelets [94]. Shear gradients characterized by shear acceleration followed by rapid deceleration promote deposition of platelets on reactive surfaces, which is not prevented by aspirin, clopidogrel, or thrombin inhibitors [161]. Stenotic regions of atherosclerotic vessels are typically on the order of 1 cm long, where narrowing of the vessel promotes a sharp increase in wall shear stress. Pathological shear rates range in the order of 10^4 s^{-1} [199, 226] to 10^5 s^{-1} [231] for such vessels, with peak rates found just upstream of the stenosis throat [231]. In the final stage of atherosclerosis, platelet enzymes may make plaques vulnerable to rupture by creating fissures in their fibrous caps, exposing a lipid-rich core and underlying extracellular matrix, leading to further platelet adhesion, activation, and aggregation. Arterial thrombosis progresses under shear rates of 1000 s^{-1} to 5000 s^{-1}, and up to 10^4-10^5 s^{-1} in stenoses [43]. The developed thrombus leads to acute coronary syndromes, ischemic strokes, and critical leg ischemia [114, 135]. Thrombi initiated at the apex of the stenosis, with higher flow velocity, are rich in platelets (white thrombi), while distal to the atherothrombotic stenosis, lower flow velocity and stagnation zones contain fewer platelets and more RBCs within a fibrin-rich clot (red thrombi) [135]. Deposition of platelets is also enhanced when they are trapped in recirculation zones and rapidly transported to flow reattachment points downstream of the stenosis [136, 190]. Low post-stenotic perfusion pressure promotes blood viscosity, platelet microemboli, and activated leukocytes to reduce flow in the microcirculation, triggering myocardial infarction (or heart attacks) [135].

On the other end of the shear spectrum, venous valve dysfunction creates zones of poor mixing with stasis, leading to valve thrombosis. Upstream of the failing venous valve, pressurization of the great veins triggers massive vessel distension and likely drives endothelial dysfunction [17]. In addition to flow stasis, deep vein thrombosis is associated with circulating or monocyte-produced tissue factor. Clots formed under such conditions are sometimes greater than 10 cm long and are rich in RBCS with platelet-rich regions. Shear rates associated with venous thrombosis are typically 100-200 s^{-1}, and drop to 0-50 s^{-1} for deep vein thrombosis [43]. At pathologically low shear rates (< 100 s^{-1}), RBCs can adhere to activated platelets and neutrophils [69].

In congestive heart failure (CHF), patients have an increased risk of venous thromboembolism, stroke, and sudden death [31]. Sudden cardiac death may also have a similar intracardiac or intracoronary thrombotic origin, and such complications in CHF patients is attributed to a prothrombotic state of a yet-unknown etiology [159]. Heart failure patients have increased whole blood aggregation, platelet-derived adhesion molecules, and higher mean platelet volume [31]. Dilated cardiac chambers, poor contractility, wall abnormalities, and atrial fibrillation promote intracardiac blood flow stasis and potentiate thromboembolic events [31]. Ventricular assist devices are indicated for use in patients with advanced heart failure. However, they generate non-physiological conditions in the circulatory system, resulting in complications such as thromboembolism, inflammation, and infection [1]. Shear rates approach 10^5 s^{-1} in left ventricular assist devices [19], with shear stresses exceeding 1000 dyne/cm^2 [67].

Prosthetic heart valves provide therapy for patients suffering from degenerative valvular heart disease, which include calcific aortic valve disease, aortic stenosis, and mitral regurgitation [41]. One category of prosthetic heart valves, mechanical heart valves, is especially prone to triggering platelet activation due to disturbed flow fields [244], particularly through cavitation (water hammer and squeeze flow effect), high Reynolds shear stress ($>$ 200 dyne/cm^2), stagnant flow [116], vortex shedding [14], large pressure drops and reduction of cardiac output [89], and recirculation zones [242]. In mechanical heart valves, shear stresses may approach 6000 dyne/cm^2 [206].

The role of mechanical forces generated by flowing blood on platelets under physiologic and pathologic conditions has been increasingly studied since 1960 [118]. In vitro methods to expose platelets to fluid flow and surface interactions include annular, tubular, and parallel plate flow chambers [53], as well as Couette, cone-and-plate [66], and capillary viscometers [35, 195]. Recent advances in microfluidic devices, which are miniaturized parallel plate chambers, have improved the micro- and nano-scale capabilities for studying platelets under shear flow [16, 53]. Techniques to measure platelet material properties and response to forces include micropipette aspiration [77, 146], micropost arrays [129, 158], microfluidics [215], atomic force microscopy [122, 177], optical tweezers [110], dual biomembrane force probe [101], and dielectrophoresis-mediated electrodeformation [128]. The latter provides a minimal contact approach that reduces the risk of surface-initiated platelet activation. However, these methods either do not directly measure dynamic platelet material properties under flow conditions or only focus on their local microenvironment (i.e. binding to extracellular matrix-like substrate) [32]. The following sections present a brief overview of platelet dynamics and physical response to physiologic and pathologic flow conditions as observed in vitro and in vivo, as well as some of the numerical models developed to describe and predict flow-mediated prothrombotic events.

7.2 PLATELET MOTION IN FREE FLOW

7.2.1 Platelet margination

Margination of platelets towards the injured vessel wall is critical for hemostasis and thrombosis, and is strongly affected by local hematocrit, flow and shear rates, fluid viscosity, vessel geometry, and RBC aggregation and deformability [48,52]. Margination is weak at low shear rates, increases with higher flow rates, and decreases again at very high shear rates [49]. Collisions with flowing RBCs enhance the motion of platelets 2 to 3 orders of magnitude greater than from Brownian motion alone [70, 219]. Under most arterial flow conditions, which include pulsatility, near-wall platelet concentrations are 2 to 3 times greater than at the center of the vessel [48, 82, 211], with peak platelet concentrations between 5 and 10 μm away from the wall and attributed to hydrodynamic lift or repulsion between platelets and the wall [48]. Margination requires a minimum shear rate between 430 s^{-1} and 780 s^{-1} [214]. Transport of platelets to the vessel wall or growing thrombus is driven by RBCs in a phenomenon of enhanced diffusivity, characterized by two proposed mechanisms: i) local mixing due to particle rotation and ii) shear-induced collision diffusion where particle interactions induce fluid displacement [255]. Effective diffusivity is strongly dependent on hematocrit and shear rate, and increases platelet deposition rate [100]. The effective diffusivity, D_{eff}, is described by:

$$D_{eff} = D_{sf} + 0.15a^2\dot{\gamma}\phi_p(1-\phi_p)^{0.8}, \tag{7.1}$$

where D_{sf} is the static diffusivity of the species of interest, a is the RBC radius, $\dot{\gamma}$ is the shear rate, ϕ_p is the hematocrit. For example, effective diffusivity of platelets increases by four orders of magnitude, 10^{-13} m^2/s to 10^{-8} m^2/s, when transitioning from quiescent plasma to shear rates of more than 10^4 s^{-1} [18].

7.2.2 Platelet motion in the cell-free layer

While several studies have observed marginated platelet motion once they interact with immobilized proteins on the damaged endothelium or extracellular matrix (see Section 7.4), few have studied their behavior away from reactive surfaces. Under low shear rates, translocating platelets are observed to flip near the wall [210,248], with their rotational trajectories resembling those of oblate spheroids flowing in viscous linear shear flow, a behavior termed as "Jeffery's orbit" [96] (see Chapter 1), describing the angle ϕ and its derivative $\dot{\phi}$ between the axis of symmetry and the flow direction:

$$\phi = arctan\left(\frac{b}{a}tan\left(\frac{ab\dot{\gamma}t}{a^2+b^2}\right)\right) \tag{7.2}$$

$$\dot{\phi} = \frac{\dot{\gamma}}{a^2+b^2}\left(a^2sin^2(\phi)+b^2cos^2(\phi)\right), \tag{7.3}$$

where t is time, and a and b are the major and minor axes of the ellipsoid. These measures can be non-dimensionalized with respect to the fluid shear rate [155], and Jeffery's orbit requires modification to account for wall effects that the original formulation does not consider [151].

Brownian motion is typically not considered for shear flows above 5 s^{-1}, but may play a role in recirculation zones and stagnation points found downstream of stenoses or at vessel branching sites, respectively [152]. Furthermore, platelets may collide with each other in regions of secondary flow, but these collisions are not thought to be necessary to initiate platelet activation [192].

7.3 INTRAPLATELET DYNAMICS AND SHAPE CHANGE DURING SHEAR-MEDIATED ACTIVATION

Shear-mediated activation of circulating platelets is typically associated with pathological conditions found in cardiovascular diseases and prosthetic blood-contacting devices. It is a multifaceted response characterized by morphological changes, secretion from platelet α and dense granules, scrambling of membrane phospholipids, procoagulant activity at the membrane, enhanced activation of surface integrins, initiation of internal signaling pathways, and interaction between surface receptors and ligands that lead to adhesion and aggregation [36,223]. In the traditional view of thrombosis, platelet activation occurs after adhesion to the injured vessel wall or thrombus, and requires the presence of soluble agonists [97], long exposure to high fluid shear rates, and binding of the GPIb receptor to von Willebrand factor (vWF) immobilized on the damaged endothelium [117] to initiate mechanotransduction. In recent years, this perspective has shifted in part due to the occurrence of platelet activation in the absence of the endothelium in implantable cardiovascular devices, as well as the ineffective interaction of cleaved large molecular weight vWF multimers with platelets under device-associated "hypershear" conditions [149,203]. Cell-cell collisions are not essential for shear-mediated platelet activation, which is controlled by fluid shear stress, or the magnitude of hydrodynamic force applied, rather than shear rate [192].

The level of platelet activation under flow conditions has been defined as the percentage of β-thromboglobulin released from the α granules [233], percentage of serotonin released from the dense granules [84], or the rate of thrombin generated on the platelet surface [98,99]. The latter is correlated with the shear "dose" - an accumulation of the shear stress and exposure time during platelet circulation through physiological and pathological conditions [13, 138, 163, 194, 205]. The Lagrangian stress accumulation history of platelets typically is presented in a power law form, with varying contributions of shear stress and exposure time, and is simplified into a linear form for computational fluid dynamics (CFD) simulations [138].

7.3.1 Resting and activated platelet morphology in the free flow

Resting platelets are typically described as oblate spheroids [63], and this approximation is often used in computational hydrodynamics models [27,131]. The submembrane platelet cortex includes spectrin, actin, myosin, and intermediate filaments [79], which provide tension to the platelet surface and result in a "wrinkling" effect on the lipid bilayer [229]. The lipid bilayer membrane unfolds to provide a greater surface area during activation and spreading, and buffers changes in surface membrane tension, helping to dampen shear-induced platelet activation due to rapid blood flow fluctuations [179]. The peripheral microtubule ring, or marginal band, stretches the membranes and provides the discoid shape of the platelet and consists of several stabilized and dynamic microtubules that wrap around the equator numerous times [157, 230]. Upon activation, the microtubule coil disassembles, reduces in diameter, and disappears in proximity to the edge [230]. It is not deemed essential for platelet activation and function [91].The cytoplasmic actin cytoskeletal network additionally secures the discoid appearance and actively aids in platelet spreading [81]. At the periphery of the cytoskeleton and just below the lipid membrane, compressed spectrin-rich networks are interdigitated with GPIb-IX-actin binding protein complexes that connect to curved filamentous actin (F-actin) emanating from a central oval core of radially-projected cross-linked F-actin. F-actin makes up approximately 40% of actin in the resting platelet, with the remainder in soluble form [81]. Upon activation, cytoplasmic F-actin levels increase markedly, with 0.22 mM of the 0.55 mM total concentration polymerized into an average of 2000 filaments, each approximately 1.1 μm long, and allows projections (filopods and pseudopods) from the platelet [81]. Upon activation, the protein gelsolin fragments actin filaments by a factor of 10, reducing them into 100 nm length within seconds of activation [80], and the platelet assumes a spherocytic form [79]. The spectrin network subsequently swells and allows for the protrusion of spindle-like filopods or sheet-like lamellipods with additional actin-assembling signals [80].

Several studies have described platelet morphology in response to external agonists. However, the overwhelming majority have either been performed under static conditions or necessitated platelet deposition on glass slides for microscopic observation [7,50,85,115,170]. These measurements, however, are not representative of morphological changes due to pathological flow conditions characteristic of cardiovascular diseases and devices. In vivo samples from patients with arterial stenosis yield a large number of platelets that have transformed into spherical shapes, extended pseudopods, and have occasional organelle centralization [4]. GPIIbIIIa receptors (integrin $\alpha_{IIb}\beta_3$) on the surface of these platelets are redistributed and relocated to the pseudopod extremities, but the platelets are maintained in a state of reversible activation [4]. In stroke patients, platelets undergo significant cytoskeletal rearrangement and activation [169].

A few in vitro studies have observed morphological changes of platelets under controlled flow conditions. Under physiological conditions, there may be considerable variability in platelet shape, where some small population of platelets may show hallmarks of activation. This subpopulation has cylindrical filopodia 50 to 200 nm in diameter extended radially or tangentially from the edge of the discoid body, and a slight taper with hemispherical ends [7]. Platelets exposed to physiological and slightly pathological shear stresses (1-70 dyne/cm^2) for up to 4 minutes in a cone-plate-Couette viscometer maintain their discoid shape (Figure 7.2 (A)), but extend filopodia that progressively grow in length with increasing shear stress (Figure 7.2 (B-D)). These filopodia range in length from 0.24 to 2.74 μm, with diameters of 0.06 to 0.73 μm at the discoid surface [171]. Shorter exposures, 60 ms, to 250 dyne/cm^2 yield significant increases in filopod lengths (Figure 7.2 (E-G)) [195]. At higher non-physiological shear stresses, platelets undergo significant morphological changes [84, 178]. Shear stresses of 570 dyne/cm^2 for 700 ms transform platelets from smooth discs to spherical or dendritic shapes, with about one-third yielding granule and organelle centralization. When the stress is increased beyond 1080 dyne/cm^2, with exposure times as low as 113 ms, platelets exhibit full shape change, where approximately 30% of the population have severe damage characterized by leaky membranes and total breakdown of the microtubule band and actin cytoskeleton [233, 234]. Elongational stresses similar to those found in the descending aorta during diastole may also induce significant platelet activation, where membrane stretching may allow a conformational change in the platelet receptors and increase their affinity for ligand binding, as well as support increased ion permeability [174].

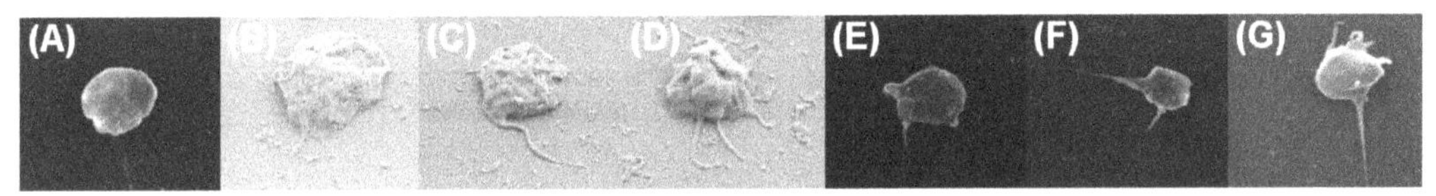

FIGURE 7.2
Scanning electron microscopy images of shear-activated free-flowing platelets. (A) Resting platelets have a discoid shape with a smooth surface. Filopods form and grow under physiological shear stresses of (B) 10 dyne/cm^2 (2 min exposure), (C) 50 dyne/cm^2 (3 min exposure), and slightly pathological shear stress of (D) 70 dyne/cm^2 (4 min exposure). Filopods also form under hypershear stresses for much shorter exposure times: (E) 50 dyne/cm^2 (300 ms exposure), (F) 250 dyne/cm^2 (60 ms exposure), and (G) 500 dyne/cm^2 (63 ms exposure). Figures 7.2(A,E-G) adapted with permission [195].

Non-physiological shear stresses and short exposure times found in cardiovascular devices, such as VADs, may also induce platelets to shed their $\alpha_{IIb}\beta_3$, GPIbα, and GPVI receptors, leading to platelet dysfunction and bleeding complications [22–24]. At sufficiently high enough shear rates (10500 s^{-1}, 315 dyne/cm^2), procoagulant microparticles are formed and released from the platelet surface by an exocytotic budding process [88].

Lipid microdomains, also known as rafts, play a crucial role in transforming discoid non-adhesive platelets into sticky dendritic bodies. These very small domains, 20-50 nm, are tightly-packed sphingolipid and cholesterol-based structures with lateral mobility [15], which influences membrane fluidity [203]. Stimulation with agonists such as fibrinogen and thrombin receptor activating peptide (TRAP) promotes the distribution of cholesterol-rich membrane domains towards the tips of filopodia, leading to downstream spreading on fibrinogen and TRAP-induced aggregation [83]. These lipid rafts also mediate the localization of the GPIb-IX-V complex to aid in platelet adhesion and subsequent outside-in signaling [197]. The effect of the lateral diffusion of lipid rafts on membrane fluidity under flow shear stress has yet to be investigated, but it has been proposed that they modulate shear stress mechanochemical signal transduction in a manner similar to other cell systems [134].

7.3.2 Material properties of resting and activated platelets

Resting platelets have viscoelastic material properties, with Young's modulus of $1.7 \pm 0.6 \times 10^2$ Pa and viscous modulus of $1.0 \pm 0.5 \times 10^3$ Pa [77]. Resting platelet membranes have an approximate strength of 2.09×10^{-4} N/m [146]. Material properties of shear-activated platelets are generally extrapolated from those obtained under static conditions, due to the former's dynamic nature. Platelets activated on glass slides have elastic moduli ranging between 1 and 10 kPa, and can withstand loading forces up to 3 nN on areas less than 100 nm in diameter [177]. Forces generated within activated platelets include those from actin polymerization, myosin contraction, and membrane tension [176]. The central region of surface-activated platelets contains primarily granula, small protein-filled vesicles, and cytosol, with moduli ranging between 1.5 and 4 kPa. The surrounding inner filamentous zone contains actin polymers overlapping with myosin, with stiffness ranging from 4 to 10 kPa, corresponding with dense myosin and filamentous actin network. The outer filamentous zone contains actin filament bundles and microtubules, with stiffness ranging from 10 to 40 kPa. The cortical, submembrane region has short actin filaments with similar stiffness, with areas up to 50 kPa [177]. In stroke patients, the Young's modulus is reduced approximately 40% [169].

7.4 FLOW-MEDIATED PLATELET ADHESION

Platelet adhesion is the first step in thrombus formation in the vascular wall injury model. It is characterized by three phases:

i translocation and tethering of platelets to recruit them to the site of vascular injury,

ii irreversible adhesion mediated by one or more surface integrins, and

iii cohesion or aggregation that is critically dependent on activation of integrin $\alpha_{IIb}\beta_3$ [47]. This latter stage is discussed in detail in Section 7.5.

Platelets and the reactive vessel wall cannot interact unless the distance between the two is lower than 100 nm [184]. Below this threshold, long-range forces, which include electrostatic interactions, increase and at 10 nm, intermolecular bonds form between the platelet receptor and immobilized ligand on the wall. An increasing shear rate reduces the interaction time and the possibility of platelet recruitment [184]. Shear stress on an adherent platelet depends on its size, the surrounding fluid viscosity, and its proximity to the wall, and affects initial platelet-wall contact, translocation, firm adhesion, and the lifetime of a formed adhesive bond [21].

The interaction of the GPIbα subunit of the GPIb-IX-V platelet surface receptor complex with the A1 domain of vWF is highly studied in vitro due to its status as the fastest bond found thus far in biology and prominent role in shear-mediated platelet adhesion [227]. The vWF is provided by both the plasma and subendothelium [221], with ultralarge vWF (molecular weight above 20000 kDa) primarily released from Weibel-Palade bodies in endothelial cells and platelet α granules into the circulation [78]. Other ligands such as fibrinogen, fibronectin, and collagen have been studied to a lesser extent, and play important roles in the later stages of adhesion and thrombus formation or under low shear rates [94]. The following sections focus on the GPIbα-vWF A1 interaction.

7.4.1 Physical parameters of flow-mediated platelet adhesion

Platelet interaction with the vessel wall involves a catch and release process, which regulates rolling, skipping, or firm adhesion, and provides the foundation of thrombus formation.

A high proportion of platelets translocate on the vessel wall and on the surface of thrombi before firm adhesion [119]. Platelet recruitment from bulk flow under high shear is characterized by a rapid association (on-rate) necessary for capturing platelets from the cell-free layer of flowing blood [187]. Complementary debonding between GPIbα and vWF occurs at a fast dissociation rate, resulting in platelet translocation over the vessel wall. These bonds, comprising a vWF multimer sandwiched between two GPIbα domains, are estimated to be 128 nm long [154]. The high tensile strength between platelets and the wall is attributed to large numbers of bonds, high complex density on platelet surface, and vWF multivalency, and permits sustained tethering and rolling under rapid blood flow conditions [47, 241]. Contact adhesion is increasingly dependent on the GPIbα-vWF interaction as shear rates increase from low physiological (50 s^{-1}) to high physiological (1500 s^{-1}) and pathological (4×10^4 s^{-1}) conditions [94], and may result in formation of tether-like structures parallel to flow direction [92, 94, 143, 181]. The thin membrane tether formation is shear-dependent, with mean lengths ranging from 3.23 μm to 16.55 μm and tether growth rate ranging from 0.04 μm/s to 8.39 μm/s over shear rates of 150 s^{-1} to 10^4 s^{-1}, and does not depend on platelet activation. These tethers regulate the stop-start phase of platelet translocation on vWF [46]. Tether lifetime peaks between 10 and 20 dyne/cm^2 [237], and rolling velocities reach a minimum at shear stresses within this range [34]. At 1500 s^{-1}, mean platelet translocation velocities range from 8.9 ± 1.0 to 12 ± 4 μm/s and are regulated by integrin $\alpha_{IIb}\beta_3$ activity [130]. At shear rates above 500-800 s^{-1}, only the GPIbα-vWF A1 interaction has a sufficient on-rate to initiate platelet adhesion [187], with a clear threshold at 800 s^{-1} delineating the switch from fibrinogen-mediated to vWF-mediated binding [218]. At very high shear stresses beyond 60 dyne/cm^2, vWF-GPIbα interaction is inefficient at inducing platelet activation and requires sudden accelerations in blood flow [73]. This platelet responsiveness is $\alpha_{IIb}\beta_3$-vWF interaction dependent and is accompanied by co-stimulation of the P2Y$_1$ ADP receptor and influx of extracellular calcium [73]. Beyond 5000 $\alpha_{IIb}\beta_3$, or at sharp shear rate gradients, GPIbα-vWF interaction can be sufficient for unstable thrombus formation and results in no more than weak intracellular signaling in platelets [161, 185].

7.4.2 Morphological changes under shear-mediated platelet adhesion

Under flow conditions, platelets undergo several morphological changes during the adhesion process (Figure 7.3). Translocating platelets change their shape in a shear-dependent manner [142]. Shear-mediated adhesion alters the platelet shape by remodeling the actin-myosin cytoskeleton (via the Rac1 and Rho-kinase pathways) and microtubule polymerization (via Ran-binding protein 10), which also promote formation of filopodia and lamellipodia after adhesion [127, 150]. Below shear rates of 600 s^{-1}, initial shape change involves membrane tether or filopodia extensions. Between 1800 and 5000 s^{-1}, platelets adopt a spherical morphology with multiple filopodia. At pathological shear rates of 10^4 to 2×10^4 s^{-1}, platelets retract these filopodia and develop a smooth ball-like appearance, which increases rolling velocities by 3- to 8-fold [142]. During prolonged exposure to vWF, signals generated during translocation promote a rolling spherical phenotype with numerous filopodia [120, 142, 241]. In addition to rolling, sliding rotational movement has been observed over shear rates from 50 s^{-1} to 40000 s^{-1} [181]. As contact adhesion increases with shear, tether-like structures parallel to flow direction may form [92, 94, 143, 181]. These small membrane protrusions facilitate the sliding mechanisms and further reduce tensile stress on adhesive bonds [92]. Thin membrane tether formation is shear-dependent and activation-independent, with mean lengths ranging from 3.23 μm to 16.55 μm, with some tethers approaching 30 μm in length, and tether growth rate ranging from 0.04 μm/s to 8.39 μm/s over shear rates of 150 s^{-1}

to 10^4 s^{-1} [143, 181]. These tethers regulate the stop-start phase of platelet translocation on immobilized vWF [46] and arrest platelets via discrete adhesion points (DAPs) on the membrane, with an area of 0.05 to 0.25 μm^2 [181]. These DAPs temporarily remain stationary during forward platelet body movement. For shear rates 2×10^3 s^{-1} to 4×10^4 s^{-1}, sometimes a single DAP anchors the platelet to the surface. Detachment of platelet bodies from tethers occurs at 6×10^3 s^{-1} and above, with an increase in severed tethers above 10^4 s^{-1} [181]. During firm adhesion, platelets also extend extremely long, negatively charged membrane strands up to 250 μm downstream of the firmly adhered and activated platelets, called flow-induced protrusions (FLIPRs), whose fragments activate rolling monocytes and neutrophils [212].

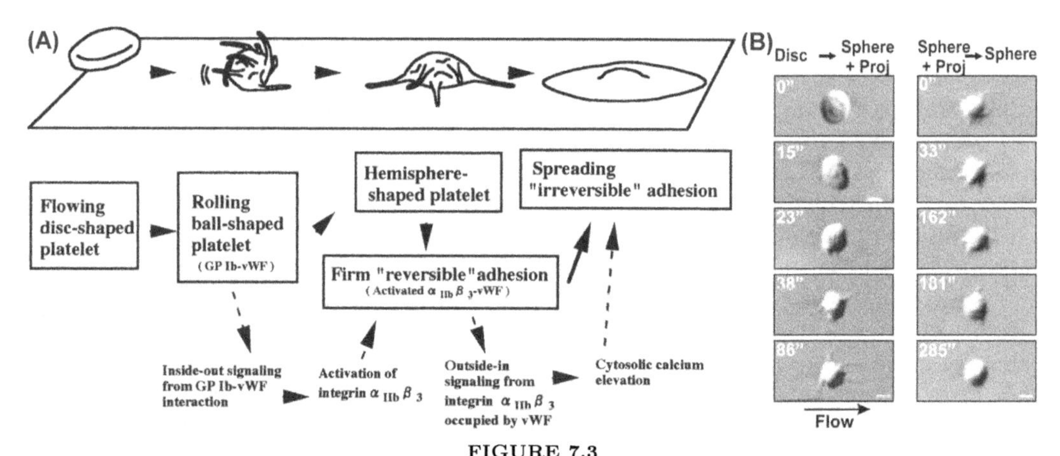

FIGURE 7.3
Platelet morphological changes during adhesion. (A) Discoid platelets form tethers and filopods during sustained rolling on immobilized vWF via the GPIbα receptor, may assume a spherical morphology. After activation of integrin $\alpha_{IIb}\beta_3$, platelets become stably adhered, and begin "irreversible" adhesion after cytosolic calcium levels become elevated [120]. (B) During translocation platelets undergo morphological changes in a shear-dependent manner, where discoid platelets transform into spheres with projections at lower shear rates, and spheres with projections to spheres only at higher shear rates [142]. Scale bars are 1 μm. Figures are adapted with permission.

7.4.3 Platelet spreading under flow conditions

The secondary phase of platelet adhesion is mediated by one or more surface integrins and guided by spatial cues in the platelet's immediate microenvironment. In this phase, the platelet actin cytoskeleton is spatially organized and distributed, and filopodia can span and spread over 5 μm [113]. The combination of collagen and fibrinogen increases growth of closely contacting areas in a fast-spreading regime, and is independent of type of ligand interaction. Platelet filopodia dynamically change their lateral and vertical distribution during spreading, with integrin α_2 and GPVI regulating $\alpha_{IIb}\beta_3$-mediated platelet spreading on fibrinogen [124]. At arterial rates of shear, the GTPase protein Rac1 provides platelet aggregate stability on collagen, plays a critical role in stable thrombus development at vascular injury sites, and is essential for sheet-like lamellipodia formation in platelets [144]. vWF plays an important role in spreading under shear, as indicated by the significantly-reduced numbers of spread and contacting platelets above shear rates of 1300 s^{-1} in blood obtained from patients with von Willebrand disease, with severely decreased thrombus volume and thrombus heights [220].

The tertiary phase of adhesion starts with formation of a platelet monolayer after spreading, with unactivated platelet recruitment to spread platelets mediated by binding of plasma fibrinogen to activated integrin $\alpha_{IIb}\beta_3$ receptors on adherent platelets under low shear and vWF-GPIb under high shear, with subsequent stabilization with vWF-$\alpha_{IIb}\beta_3$. This process is described in the next Section 7.5.

7.5 FLOW-MEDIATED PLATELET AGGREGATION

Under flow conditions, aggregation occurs under three distinct mechanisms (Figure 7.4):

i low-intermediate shear conditions ($< 10^3$ s^{-1}), which is exclusively mediated by fibrinogen and integrin $\alpha_{IIb}\beta_3$, with locally generated soluble agonists promoting platelet shape change and integrin affinity;

ii high shear conditions ($10^3 - 10^4$ s^{-1}), platelet-platelet interactions progressively more vWF-dependent with roles for both GPIbα and $\alpha_{IIb}\beta_3$ in discoid aggregates, with an identifiable two-stage aggregation process; and

iii very high shear rates ($> 10^4$ s^{-1}, shear stress at 400 dyne/cm^2) exclusively mediated by vWF-GPIb bonds independent of platelet activation and $\alpha_{IIb}\beta_3$ adhesive function with marked platelet membrane deformation [92, 94, 185]. Beyond 2×10^4 s^{-1} (800 dyne/cm^2), adherent platelets on surface-immobilized or membrane-bound vWF become elongated and serve as aggregate cores that can persist for several minutes [185].

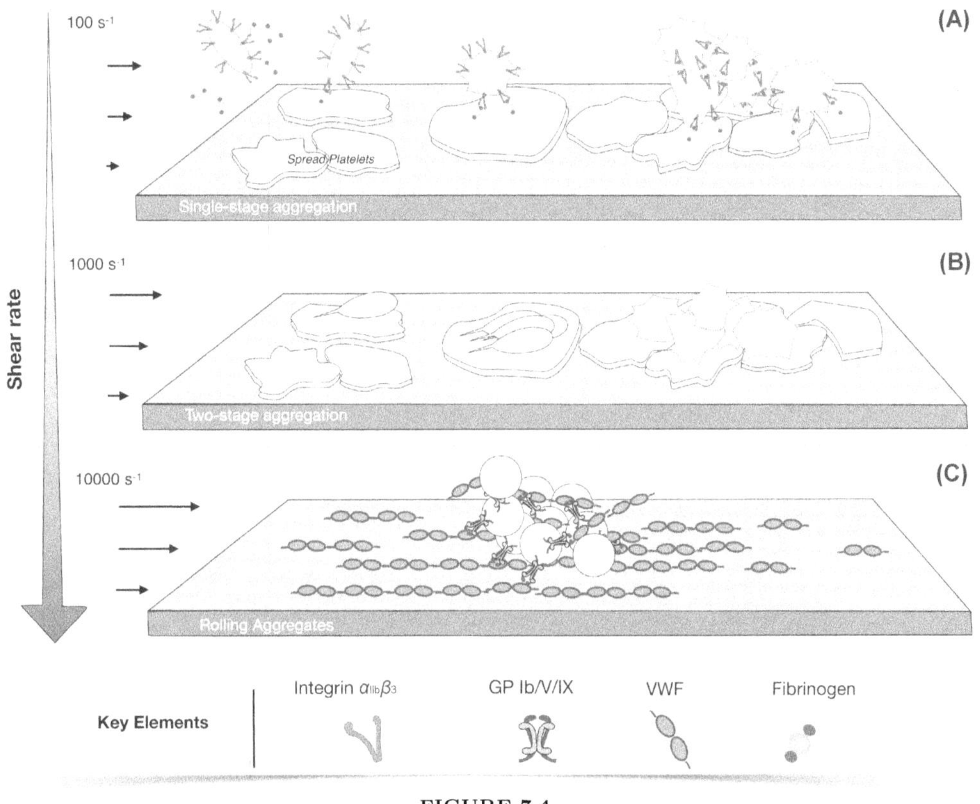

FIGURE 7.4
Shear rate dictates the platelet aggregation behavior. (A) Below 1000 s^{-1}, integrin $\alpha_{IIb}\beta_3$-fibrinogen interaction drives stable aggregation between shape changed platelets. (B) A two-stage aggregation process between discoid platelets is mediated by membrane tethers and is dependent on the adhesive function of GPIbα and $\alpha_{IIb}\beta_3$ for shear rates between 1000 and 10000 s^{-1}. (C) At shear rates greater than 10000 s^{-1}, aggregation is exclusively mediated by vWF-GPIbα interaction independent of $\alpha_{IIb}\beta_3$ or platelet activation. Image courtesy of Dr. Manouk Abkarian.

A large number of studies have focused on aggregation as the tertiary phase of thrombus formation at sites of vascular injury or in stenosed vessels, with a smaller number of efforts focused on aggregation in free-flowing blood.

7.5.1 Flow-mediated platelet aggregation and thrombus initiation

Platelet aggregation and thrombus growth may be primarily driven by rheology-dependent aggregation mechanisms, with soluble agonists playing a secondary role in aggregation stabilization [145]. Aggregation on adhered platelets occurs in two phases: (i) a reversible shear-mediated aggregation between discoid platelets facilitated by membrane tethers involving platelet activation and adhesion without the need for soluble agonists; and (ii) a subsequent stable phase associated with platelet shape change and granule release [143]. These membrane tethers are smooth cylinders of the lipid membrane pulled from the surface due to hemodynamic drag forces and are similar to those found on translocating platelets [46,143]. The tethers maintain close proximity to the platelets, which may facilitate autocrine and paracrine stimulation by locally-generated agonists [92], and are able to restructure via an activation-dependent mechanism with subsequent localized cytoskeletal remodeling [161]. These tethers extend upon exposure to shear acceleration (elongational force), and remodel and contract under shear deceleration, with the latter allowing strengthening contact between discoid platelets [161]. Upon recruitment into developing aggregates, a high proportion of platelets remain minimally activated by retaining their discoid morphology, do not elicit a sustained calcium response, and do not release α granule contents [145]. Under high shear alone (greater than 5400 s^{-1}), aggregation only occurs if initiation of activation by GPIbα interaction with vWF is accompanied by concurrent binding of the latter with $\alpha_{IIb}\beta_3$ [74]. In stenotic regions, vortex formation and recirculation zones become apparent below the growing aggregate, and serve as a secondary process to accelerate platelet aggregation [217], leading to subsequent thrombus formation. During thrombus development, hemodynamic perturbations that lead to flow separation, temporal shear gradients, and turbulent flow become more prevalent and subject platelets to increased aggregation events (Figure 7.5) [161].

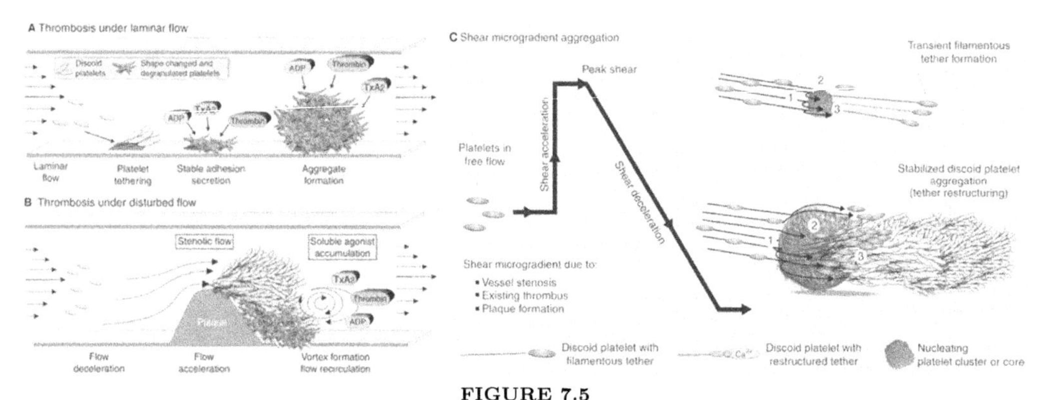

FIGURE 7.5
Local shear conditions drive aggregate formation. (A) Under laminar flow conditions above 1000 s^{-1}, discoid platelets are recruited and tethered on an immobilized vWF surface, whereupon soluble agonists promote calcium flux, shape change, degranulation, and formation of stable aggregates. (B) Disturbed flow due to vessel geometry (i.e. downstream flow vortices due to stenosis) creates shear gradients that promote membrane tether restructuring between discoid platelets, downstream agonist accumulation, and subsequent platelet activation (Adapted with permission [94]). (C) Shear microgradients are marked by acceleration of platelets towards an obstacle (i.e. stenosis, thrombus, or plaque), exposure to peak shear stress at its apex, and deceleration that promotes aggregate formation and growth (Reprinted with permission [161]).

The lag phase lasts 175 to 300 s for shear rates 500 to 5000 s^{-1} [10], allowing plasma proteins to adsorb to collagen surfaces to support platelet adhesion [18]. The lag time is further lengthened to capture hundreds of platelets before shear activation [10]. Increasing shear rate during this period decreases the lag time, reflecting enhanced protein transport, vWF adsorption on collagen, and mural platelet activation. Rapid thrombus growth during the second phase is characterized by uneven, fingerlike projections into the lumen [231], with rates varying widely at different locations on a large thrombus [18]. Local wall shear rate changes by more than 10-fold in the vicinity of the growing thrombus [10].Thrombus growth rates during this period increase with very high shear rates up to 25000 s^{-1}, beyond which they gradually decrease for rates up to 10^5 s^{-1} [18]. Exceedingly large drag forces are present due to continuing platelet capture and thrombus growth at shear rates above 10^5 s^{-1} and stress above 2000 dyne/cm^2. The thrombus has an inner core of highly activated platelets that are stabilized by thrombin generation and fibrin polymerization, and an outer shell of minimally activated, unstably attached discoid platelets that are sensitive to local flow conditions [207]. At physiological shear rates, such as 1500 s^{-1}, initial thrombus growth relies completely on vWF, and fibrinogen accumulates predominantly inside the growing thrombus over time and acts as core adhesive ligand, increasing thrombus strength and height. Growing thrombus surfaces are occupied constantly by vWF [141]. The third phase yields either full occlusion or cycling of a large thrombus between embolization and growth, with durations on the order of 10-20 minutes [10, 165]. Insoluble fibrin imparts strength to the thrombus to resist embolism up to wall shear stress of 2400 dyne/cm^2 [17].

Recent studies have examined the interaction and contraction forces of individual platelets in clots and aggregates via the $\alpha_{IIb}\beta_3$-fibrinogen or $\alpha_{IIb}\beta_3$-fibrin bonds. Single platelets contract instantaneously when activated by fibrinogen contact, completing contraction within 15 min. The average maximum contractile force for individual platelets is 25-29 nN [122, 158], with forces beyond 30 nN [158] and up to 70 nN for highly contractile platelets, with elasticity of 10 kPa [122]. Clots can be stiffened through direct reinforcement by platelets and strain stiffening of fibrin due to the tension of platelet contraction.

7.5.2 Platelet aggregation in free flow

Several studies have analyzed platelet aggregation independent of adhesion. Platelets in sheared platelet-rich plasma are typically assessed in a turbidimetric aggregometer after ADP-stimulation [35]. Pulsed exposure of plasma to physiological shear stresses, with 1 s in between pulses, results in greater aggregation than exposure to continuous shear stress at the same magnitudes and duration, suggesting stress loading plays an important role in shear-induced aggregation [209]. Post-stenotic low flow immediately after a short duration of pathologically high shear stress also enhances platelet aggregation without significant activation [247], while pathologically high shear stress (100 dyne/cm^2) requires durations greater than 20 seconds for irreversible aggregation [246]. At venous wall shear rates and stresses (100 s^{-1}, 1-2 dyne/cm^2), platelets aggregate due to $\alpha_{IIb}\beta_3$-fibrinogen interactions without the need for vWF [17].

7.6 FLOW-MEDIATED SURFACE RECEPTOR AND MEMBRANE BEHAVIOR

Under flow conditions, activation of several surface integrins and receptors allows platelets to participate in adhesion and aggregation. The primary receptor-ligand interactions under shear involve GPIb with vWF, $\alpha_{IIb}\beta_3$ with vWF, $\alpha_{IIb}\beta_3$ with fibrinogen, GPVI with collagen, and $\alpha_2\beta_1$ with collagen.

7.6.1 Physical conditions for receptor-ligand interactions leading to adhesion

Adhesion is the first step in thrombus formation (Section 7.4), and largely depends on interaction of GPIb with vWF. GPIb is part of the GPIb-IX-V complex, and the surface of resting platelets contains 25000 copies of GPIb-IX [62] and 12000 copies of GPV [11]. Platelets bind preferentially to vWF because of the latter's very large multimeric structure and binding sites, efficiency at capturing platelets from the bulk flow, shear-dependent changes from a globular to linear form, with subsequent exposure of receptor binding sites, and the ability of the GPIb-vWF bond to transmit $\alpha_{IIb}\beta_3$-activation signals and increase affinity of $\alpha_{IIb}\beta_3$-vWF bonds [106]. The availability of vWF for binding platelets progresses through a two-step process: (i) elongation from compact to linear form and (ii) tension-dependent local transition to a state with high affinity for GPIbα, where the tension exceeds 21 pN [64]. While shear flow is speculated as the primary precursor for elongation, elongational flow has been proposed as a regulator of vWF conformation [200, 251], with critical shear rates of 10^4 s^{-1} assumed to drive transition of the vWF globule to a stretched configuration, and a linear array of A1 receptors oriented in the direction of the shear stress field [198]. The unfolded vWF favors self-association [42] which promotes trapping of flowing platelets and initiating adhesion [208], and makes it available to bind to multiple GPIbα receptors [9]. Elongated insoluble vWF fibers (length $> 100\mu$m) increase their thickness to greater than 10 μm when shear rate increases from 5000 s^{-1} to 10^4 s^{-1}, with elastic modulus of approximately 50 MPa [86]. Shear also promotes reversible clustering of GPIbα receptors, with binding to vWF required at 1600 s^{-1} and not required at 10^4 s^{-1} [68]. This clustering requires translocation to lipid rafts. Soluble vWF and GPIb are subject to fluid forces on the order of 0.1 pN at 80 dyne/cm^2, and this increases to 1 pN when the two are bound together [192].

The approximate GPIbα-vWF A1 bond length is 78 nm (50 nm for GPIb stalk and 28 nm for vWF thickness) [154]. At a shear rate of 1000 s^{-1}, the force required to attach the platelet to a surface is estimated to be 97 pN, whereas it is 6 pN between flowing platelets [193]. The GPIbα-vWF A1 binding is biphasic, with prolonged (catch) and shortened (slip) lifetime bonds. Two theories serve to explain GPIbα-vWF A1 binding under hydrodynamic forces: i) Below 25 pN, a catch bond is formed with prolonged lifetime and decreased off-rate [237] and ii) forces cause the bond to switch between two slip bonds, forming a "flex-bond", with one having a much higher on-rate than the other [111]. For catch bonds, shear force may enhance binding interaction by inducing conformational changes in receptors or ligands, while in slip bonds, shear forces lower the energy barrier between bound and free states [140]. The GPIbα receptor has several regions, and the stalk region acts as a mechanosensor that unfolds at pulling forces between 5 and 20 pN, similar to that of catch and flex bonds, and further induces conformational changes in GPIbβ and GPIX [250]. Qiu and coworkers present an excellent review of how platelets mechanosense their environment [176]. Inside the platelet, the binding of the cytoplasmic tail of GPIbα to actin via filamin A supports the integrity of the platelet membrane at high shear rates of 5000 s^{-1} to 4×10^4 s^{-1} [37]. Furthermore, vWF A1 domain-mediated pulling on GPIb-IX induces unfolding of a mechanosensitive domain in GPIbα, which may contribute to platelet mechanosensing or vWF-platelet shear resistance [250]. Forces to rupture single bonds such as GPIbα-vWF [45], $\alpha_{IIb}\beta_3$-fibrinogen, and $\alpha_2\beta_1$-collagen [61] range between 10 to 100 pN, with loading rates between 10^2 to 10^4 pN/s [43]. The stiffness threshold for adhesion-mediated platelet activation is 5 kPa [107].

Immobilized vWF also induces GPIbα shedding under shear rates 250 s^{-1} to 2000 s^{-1}, with the extent of shedding reduced with increasing shear and increased linearly with time, with shear-induced adhesion a requisite for this behavior. This has potential implications for regulation of platelet function and thrombus growth under pathological conditions [25].

Platelet recruitment to injured sites is also enhanced by the cooperativity of vWF and collagen, which in concert recruit 2- to 18-fold more platelets under flow than collagen or vWF alone, respectively [164]. Integrin $\alpha_2\beta_1$ and GPVI work synergistically to recruit platelets to collagen and initiate platelet activation [78], and both are required to trigger inside-out activation of integrin $\alpha_{IIb}\beta_3$-mediated spreading on fibrinogen [124]. Irreversible adhesion requires involvement of integrins $\alpha_{IIb}\beta_3$ and $\alpha_2\beta_1$, which have characteristically slow bond dissociation rates. Activation of $\alpha_{IIb}\beta_3$, triggered by vWF-GPIb binding [183] and subsequent intracellular signaling events and other soluble platelet agonists (i.e. ADP) and subendothelial components (i.e collagen), allows for irreversible spreading [47]. Thrombus formation is typically slow on purified vWF, taking several minutes [187]. Integrin $\alpha_2\beta_1$ is in an active form that allows binding to subendothelial collagen, immobilizes the platelet and allows GPVI to interact with the Gly-Pro-Hyp sequence on collagen, leading to collagen-mediated platelet activation. On purified vWF, thrombus formation in this route takes seconds [186].

Gain-of-function mutations in vWF A1 or GPIbα are distinguished by enhanced bond lifetimes and spontaneous plasma clearance of large vWF [44, 45], while loss-of-function mutations in vWF A1 lead to shorter bond lifetimes and hemostatic defects [20]. Extreme shear stresses can lead to acquired vWD, where patients with aortic stenosis or mitral valve regurgitation have reduced levels of functional vWF [12, 87], and degradation of vWF is observed in continuous flow left ventricular assist devices [38].

7.6.2 Physical conditions for receptor-ligand interactions leading to aggregation and thrombus formation

The primary players in platelet aggregation are integrin $\alpha_{IIb}\beta_3$, and the ligands fibrinogen (at lower shear rates), fibronectin, and vWF (at higher shear rates). Resting platelets contain approximately 80000 copies of $\alpha_{IIb}\beta_3$ [166]. Integrins have alpha and beta subunits with a large ectodomain, single transmembrane domain, and short cytoplasmic tail, and interact with each other to form a conformation of a large "head" and two "legs". The ectodomain of inactive integrins has a "closed" conformation in which the head is less than 5 nm from the plasma membrane, while activated integrins have an "open" conformation with the head 19 nm away from the plasma membrane, allowing for higher affinity ligand binding [176]. Adhesion to subendothelial proteins vWF and collagen trigger inside-out signaling events leading to conformational changes within $\alpha_{IIb}\beta_3$ [188]. Further induction of conformational changes trigger outside-in signaling, leading to protein tyrosine kinase activation and cytoskeleton reorganization, with clustering of $\alpha_{IIb}\beta_3$ receptors and incorporation into large cytoskeletal complexes containing various structural proteins and signaling molecules [47]. Myosin acts as a direct bridge between the β_3 subunit of the integrin and the actin cytoskeleton and may serve as an alternative to $\alpha_{IIb}\beta_3$ mechanosensing [176]. External forces can extend integrins and trigger their activation on the order of sub-seconds, with external forces transmitted from the ligand to the integrin [172, 173]. An intermediate affinity state of $\alpha_{IIb}\beta_3$ is induced by ligand engagement of GPIbα via a mechanosignaling pathway and primes the receptor for outside-in signaling and further transition to an extended open configuration for high affinity ligand binding [22]. Limited integrin activation results in loose and often unstable adherence of platelets to a growing thrombus via GPIb-IX-V [161].

Fibrinogen, vWF, and fibronectin have similar binding affinities to activated integrin $\alpha_{IIb}\beta_3$. At shear rates below 1000 s^{-1}, the high molar ratio in plasma of fibrinogen allows it to be the dominant ligand [75, 92]. Beyond 1000 s^{-1}, aggregation initiation becomes progressively dependent on vWF and fibronectin [92]. The role of fibrinogen is to stabilize thrombi and is linked to its capacity to bridge platelets via the integrin $\alpha_{IIb}\beta_3$ receptors, and fibrin formation anchors the thrombus to the wall. The bond length of $\alpha_{IIb}\beta_3$-fibrinogen-$\alpha_{IIb}\beta_3$ is approximately 67.5 nm [71]. $\alpha_{IIb}\beta_3$-fibrinogen interactions behave like classical slip bonds when pulling forces are between 5 and 50 pN, with average bond lifetimes decreasing with increasing force. These bonds do not resemble the catch bonds of GPIbα-vWF [132]. Debonding strengths between adsorbed fibrinogen and $\alpha_{IIb}\beta_3$ is 50-80 pN for loading rates of 10-100 nN/s, but only once a critical binding epitope is available on adsorbed fibrinogen [3]. Once the fibrinogen is converted to insoluble fibrin, the binding strength and frequency of interaction with $\alpha_{IIb}\beta_3$ increases [133]. $\alpha_{IIb}\beta_3$ activation is a reversible process, requiring persistent platelet signaling to keep the integrin in an activated, pro-adhesive conformation [36]. Single platelet pairs have a mean rupture force of 1.5 nN between non-activated or weakly-activated platelets, and up to 2.6 nN on highly activated platelet pairs, as determined by single cell force spectroscopy [162].

7.7 NUMERICAL IMPLEMENTATIONS OF PLATELET DYNAMICS

A variety of computational models have sought to model platelet behavior and prothrombotic events in cardiovascular diseases and devices. However, discrete approaches such as individual cell tracking and platelet responses are generally limited to the study of fundamental platelet mechanisms at very small time and length scales due to finite computational power. Most of these models study surface- and injury-mediated responses rather than shear-induced thrombosis [139]. Cardiovascular device simulations typically use massless rigid spherical particles representing platelets to determine the Lagrangian stress history of the platelets as a proxy for risk of thrombosis. These particles range in number up to several hundred thousand, and have a diameter of 3 μm [6]. Those that do consider platelet motion at the device level typically mimic platelet motion along Lagrangian velocity trajectories using neutrally buoyant point particles, rigid spheres with fixed diameters [5], or meshed, rigid, linear elastic ellipsoids [243], and calculate their shear stress histories. The level of platelet activation is determined by use of empirically-derived "damage" models, which correlate platelet activation to the shear dose history of the platelet (i.e. function of stress and time) [138, 194]. These simulations typically neglect the presence of other blood components (RBCs, WBCs, plasma proteins) and assume that platelets are initially homogeneously distributed at the flow inlet. The number of platelets used depends on computational resources available. However, the past two decades have seen a marked increase in the number of models that consider platelet physiology and transport under physiologic and pathologic flow conditions. The majority of these models use continuum approaches to model platelet motion and interactions with other platelets or blood cells on the macroscopic scale of the vasculature or cardiovascular devices. Some continuum models consider the multiscale nature of events leading to thrombus formation, with a role for both shear and chemical species in platelet activation, adhesion, aggregation, and thrombus growth. These models typically take into account mass transport, diffusivities, and wall reactivity [90]. However, these models often cannot resolve the smaller spatiotemporal scales, such as the mesoscopic cells and their interaction with surrounding fluid, or microscopic sub-platelet events (i.e. receptor-ligand interactions and cytoskeletal changes). Recent advances in high performance computing (HPC) have allowed incorporation of particle-based approaches to describe molecular-scale intraplatelet changes and platelet-plasma interactions in response to external mechanical stimuli [249]. These particle-based approaches use a combination of mesoscopic dissipative particle dynamics (DPD) to model heterogeneous fluids, microscopic and atomistic molecular dynamics (MD) to describe receptor-ligand interactions, and coarse-grained MD (CGMD) to inform on the dynamic morphologic nature of platelets in respond to fluid flow [240]. The following sections provide a brief overview of platelet biomechanical models at various stages of thrombogenesis.

7.7.1 Platelet transport and margination

Platelet-fluid interaction problems tracking motion and interactions of discrete elements representing platelets are typically limited to small vessels or microfluidic devices [60], and use approaches such as immersed boundary method [59], force-coupling [167], cellular Potts model [235], hybrid lattice kinetic Monte Carlo-lattice Boltzmann [56], and a variety of particle methods [54, 65, 103, 104, 156, 168, 204, 216].

Two-dimensional lattice Boltzmann immersed boundary method studies of platelet margination and its mechanisms have described the process using the continuum drift-diffusion equation, where drift is hypothesized to arise from wall-platelet interactions (i.e. lift force) and shear-induced diffusivity is due to their interactions/collisions with red blood cells, with margination timescale independent of initial RBC distribution [39,40]. Expansion into three-dimensional models using a Stokes flow boundary-integral method confirms the appropriateness of the drift-diffusion approach with the observation that margination is diffusional [252, 253]. In a three-dimensional lattice Boltzmann spectrin-link (LB-SL) method, the hematocrit has a strong effect on margination, and stiff spherical platelets marginate faster than ellipsoidal and disk forms, where the former two interact with the vessel wall and the latter tend to position themselves adjacent to RBCs at the edge of the depletion zone [180]. RBC distribution examined using a three-dimensional immersed boundary method shows anisotropic behavior, where cavities allow platelets to marginate faster towards the wall, with anisotropic platelet clusters formed in the cell-free layer [222]. Mehrabadi et al. derived a margination length scaling law and using direct numerical simulation (DNS), showed that margination length in straight channels increases cubically with channel height and is independent of shear rate, and that RBC rigidity has a greater impact than its size on margination at 10^4 s^{-1} [148]. DNS can overcome limitations of the phenomenological continuum models to accurately model RBC-enhanced platelet diffusion and margination relevant to experimental observations under shear flow, but is impractical due to computational cost and difficulty implementing in complex geometries [147]. This is addressed with a diffusion with free-escape boundary condition (DFEB) model, which solves for the time evolution of platelet concentration at the boundary of the cell-laden region and RBC-free layer using a continuum mass transfer equation that considers platelet concentration, lateral diffusivity, and RBC-free layer thickness [147].

Platelet transport and margination have also been modeled using a DPD approach. Three-dimensional coarse-grained deformable RBCs and non-deformable platelets (i.e. their lipid bilayer and cytoskeleton) are constructed on two-dimensional triangulated networks, with particles connected via wormlike chain bonds and long-range repulsive potentials [51], and allowed to interact via dissipative and random forces with DPD fluid passing through a stenotic microchannel, where higher levels of constriction and wall shear rates lead to significantly enhanced platelet margination [238].

7.7.2 Flow-induced platelet deformation

Until recently, most models of platelet transport have considered platelets as rigid two- or three-dimensional spheres or disks. However, experimental observations have long shown that platelets deform due to both flow conditions and participation in activation, aggregation, and adhesion processes. The resting platelet shape, determined mainly by a marginal band stretching the cortex, was mathematically modeled by several groups [189, 213] and enhanced to obtain a realistic shape and maintain an equilibrium between the tensioned surface and strongly bent microtubules, which are modeled as semiflexible polymers that are inextensible but with finite bending rigidity [157]. The continuum three-dimensional Subcellular Element Langevin (SCEL) approach, which couples cells modeled by subcellular elements with fluid flow and substrates using the Langevin equation, simulates the three-dimensional motion and deformation of viscoelastic platelets in shear flow [210], and was experimentally validated using platelet-rich plasma flowing in a rectangular capillary. This model was expanded into a hybrid platelet model composed of subcellular elements (SCE) representing the cytoskeletal network and continuum description of the lipid bilayer, and subsequently coupled with a lattice Boltzmann model (LBM) of blood flow using the immersed boundary method (IBM) to generate platelet motion and deformation under fluid shear [232]. This approach demonstrated that stiffer platelets have smaller deformations and shorter pause times when adhering to a vessel wall [232].

Particle-based approaches have also been used to simulate platelet deformation under flow conditions. Deformation of platelets passing through a stenotic microchannel was simulated using a DPD approach, where platelets consist of particles on a triangular mesh composing an ellipsoid shape that interact with each other via harmonic bonds that confer elastic ability to recover from deformation and dihedral angles between adjacent triangles that grant the membrane with in-plane bending stiffness [204, 205]. The platelets are embedded in a DPD fluid that mimics viscous properties of plasma, and fictitious particles are used to enforce a no-slip boundary condition. Each particle is subject to Brownian, repulsive, and dissipative forces, among others, as well as attractive forces exerted by the wall. This approach allows platelets to self-orbit, deform, and collide with other platelets and the wall, although the compressibility of DPD systems causes a breakdown at higher Re [204, 205]. CGMD allows discretization of platelets into distinct zones that permit simulation of complex shape change and pseudopodia formation that platelets undergo during activation. In one such model, spring-connected particles were used to model a homogeneous elastic lipid bilayer and cytoskeletal structure, consisting of a filamentous core and filament bundles, and homogeneous cytoplasm particles described the organelle zone [171]. Non-bonded interactions between the various discrete particles are described using the Lennard-Jones potential, and the dynamics of filopod formation are enabled by exploring the parameter space of the CGMD model, where length and thickness parameters of the growing filopod are incremented while keeping other parameters constant to maintain structural integrity. A refined version of this model consists of an elastic bilayer membrane that deforms under strain with CGMD interactions; rigid CGMD filamentous carbon-70 core, from which a network of α-helical actin filaments mimics the spring-loaded molecular mechanism that mediates contractility of the cytoskeleton using an MD force field; and non-bonded particles that fill the cytoplasmic space and whose rheology is modeled using the Morse potential (Figure 7.6) [249]. A hybrid force field describing dynamic fluid-platelet interaction allows the CGMD-based platelet model to deform when immersed in a flowing viscous DPD plasma [248]. This approach indicates that neglecting the platelet deformability overestimates the stress on the platelet membrane by approximately 2.6 times, and in turn erroneously predicts platelet activation levels under viscous shear flow. Mapping of the hemodynamic stresses under Couette flow indicates that the deformable membrane and cytoplasm absorb more stress than the actin filaments (Figure 7.7), and it is proposed that the stress distribution in the filament network indicates the constituents which are most likely to spontaneously form filopodia bundles [249].

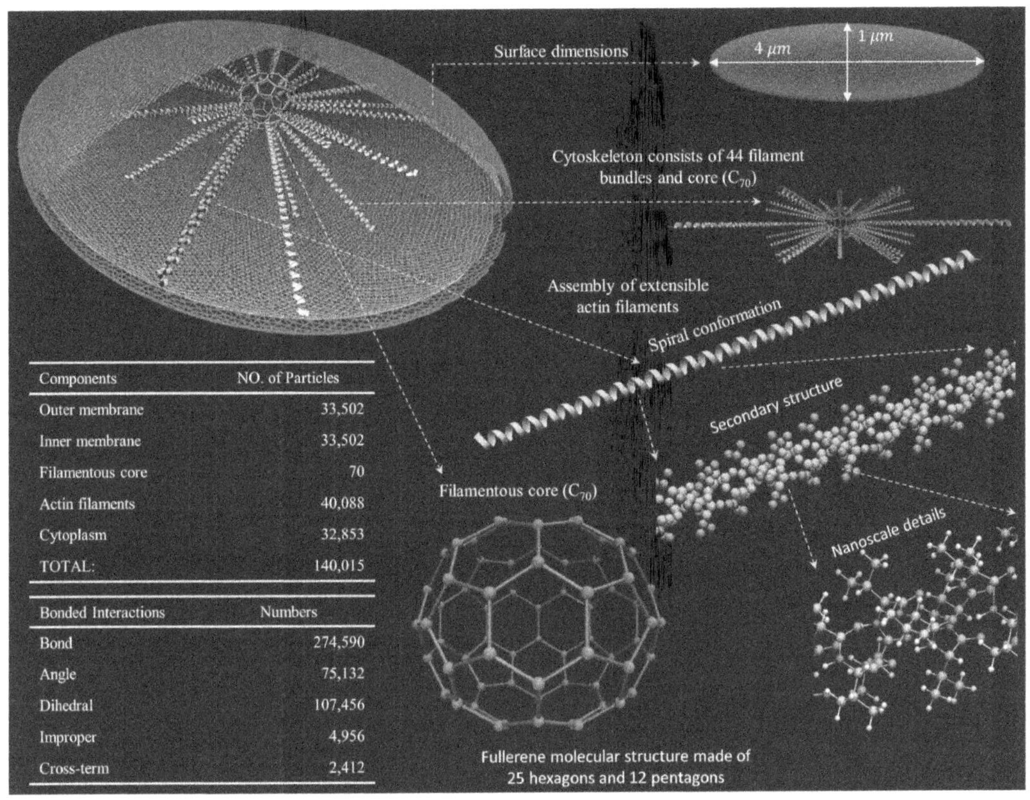

Components	NO. of Particles
Outer membrane	33,502
Inner membrane	33,502
Filamentous core	70
Actin filaments	40,088
Cytoplasm	32,853
TOTAL:	140,015

Bonded Interactions	Numbers
Bond	274,590
Angle	75,132
Dihedral	107,456
Improper	4,956
Cross-term	2,412

FIGURE 7.6
A CGMD particle-based model of a deformable platelet. The platelet consists of an elastic bilayer membrane supported by protrusible α-helical actin filaments anchored on a rigid filamentous core with a carbon-70 structure. Non-bonded particles fill the cytoplasmic space between the membrane and cytoskeleton. Reprinted with permission [249].

7.7.3 Flow-mediated platelet deposition and adhesion

An early model of flow-mediated platelet deposition used the Monte Carlo method to model convective diffusion of thromboactive molecules that activated platelets, modeled as spheres, under no or low shear rates to predict platelet deposition patterns [2]. This approach predicted deposition probability that was dependent on the platelet's activation state, distance from the wall, and wall thrombogenicity. In tortuous venules with low flow, two-dimensional rigid ellipsoidal platelets participated in transport, collision, receptor-ligand interaction, and shear-induced activation in a mesoscale discrete mesh method, and platelets were activated if they exceed a critical shear stress threshold or collided with other platelets [26, 27]. The discrete element method was also applied to coronary stent thrombosis initiation [28, 29].

Using parallel-plate flow chamber studies, Mody et al. developed and validated a two-dimensional analytical platelet flipping model to describe the force mechanics and motion

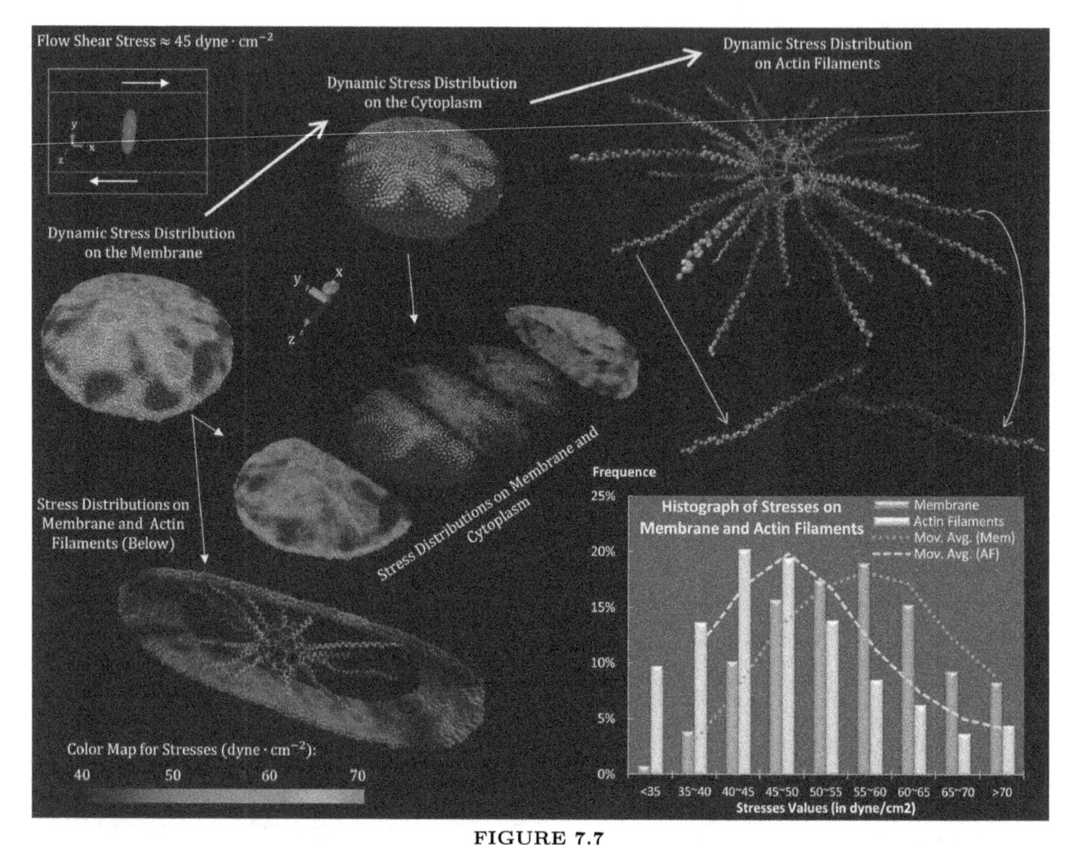

FIGURE 7.7
Couette flow results in mechanotransduction of shear stresses on the constituents of the deformable platelet, with heterogeneous shear distribution on the membrane and actin filaments. Reprinted with permission [249].

of tethered platelets on surface-bound vWF [151, 155]. In this model, the platelet is a thin plate, hinged at one end, and extends infinitely in a direction normal to flow and velocity gradient directions. The near-surface flow is approximated as linear shear flow, and the platelet flipping hydrodynamic problem is decomposed into: a (i) fence problem, with linear shear flow around a stationary fence inclined at angle α to an infinite rigid wall; and (ii) hinge problem, with flow due to rotation at angular velocity ω of a hinged plate oriented at angle α to a planar surface. For shear stress of 0.2 to 8 dyne/cm^2, platelet attachment only occurs in orientations predicted to result in compression along platelet length and bond being formed, and suggest that hydrodynamic compressive forces play an important role in tether bond formation. A suspension-level multiscale model of adhesion in channel flow incorporated rigid oblate spheroid platelets that marginate from core flow consisting of deformable RBCs to a wall coated with vWF, and platelet GPIb-vWF bonds are modeled as Hookean springs with spring constant of 10 pN/m [175]. These simulations determine that fluid dynamics play a significant role in bond-level kinetics, free-flowing platelet reactivity with the wall, RBC migration, and platelet margination [175]. MD simulations of GPIba-vWF A1 bond mechanics involved under such flow conditions predict maximum bond strength of 62 pN [196]. Adhesive dynamics simulations for single platelet capture [151, 155] and platelet translocation on vWF [151, 155, 224, 225] have single molecule resolution on molecular length scales and involve time-consuming computations, which are not feasible for clotting involving hundreds to millions of platelets [43]. Furthermore, these models do not account for the deformability of platelets and the loading mechanics due to tether pulling at pathologically high shear rates. For clotting, simpler adhesion models based on apparent rates of attachment and detachment are typically used [55].

7.7.4 Flow-mediated aggregation and thrombus formation

Cell-cell binding in several models used for platelet transport (Section 7.7.1) implement attractive potentials between particles [167] and rules for cell binding and unbinding [56]. Interaction of platelet pairs is modeled by boundary-integral methods [153, 154, 225]. Cell-cell binding is represented by attractive potentials between particles [167], prescribed rules for binding and unbinding [56], coarse-grained elastic links between platelets [202], or Monte Carlo treated formation and debonding of individual vWF- GPIbα interactions [154, 225]. These models are extensively described elsewhere [55, 60, 224, 236]. In addition, hybrid molecular-level force fields simulating non-bonded and bonded interactions between $\alpha_{IIb}\beta_3$ and fibrinogen have been used in DPD-CGMD models [76].

Several approaches have attempted to model flow-mediated platelet aggregation. Population balance models suggest only a small fraction of platelet collisions results in platelet binding, with an efficiency of approximately 0 for shear rates below 3000 s^{-1} and 0.01 for a shear rate of 8000 s^{-1}, with an increase of one order of magnitude if chemical agonists are introduced [84]. Fogelson developed a microscopic scale continuum model of aggregation in small arterioles (50 μm diameter), where neutrally buoyant platelets are attributed a mass and volume in the fluid space they occupy, and move in the fluid via convection and diffusion [57]. The local fluid is a viscous incompressible fluid modeled by the Navier-Stokes equations, and platelets are chemically activated by ADP. In this discrete-platelet model, an elastic link between two contacting cohesive activated platelets is created to resist motions that would separate the platelets and subsequently generate stresses that influence fluid motion [57]. This model was updated to account for adhesion to the injured vessel wall, and bridges the platelet microscale and vessel macroscale by means of a closure approximation [58].

A multiparticle adhesive dynamics model [112] was expanded to simulate multiplatelet adhesive dynamics (PAD) and particle motion under shear flow. This model comprises two distinct and interconnected algorithms: (i) adhesive dynamics calculations and (ii) hydrodynamic mobility calculations [153]. The approach involves calculation of all forces and torques (i.e. gravitational force, bond spring forces, and repulsion) acting on each particle in the system of oblate spheroidal platelets, followed by a numerical solution of relevant fluid mechanics equations that solve the mobility problem for the particles and external far-field flow, as well as rigid body motions. The hydrodynamics calculations involve creeping flow motion of two rigid oblate spheroid particles in a semi-infinite three-dimensional domain, with discretized rigid platelet surfaces. For flow at 10^4 s^{-1}, the particle-particle hydrodynamic interactions were predicted using the completed double layer-boundary integral equation method (CDL-BIEM) to solve the integral representation of the Stokes equation. Inclusion of short-range repulsive forces accounts for nonspecific short-range interactions such as electrostatic repulsions that dominate when cell membranes, glycocalices, or glycoproteins contact each other, and become dominant at distances $<$20 nm between interacting surfaces. Application of this model to two unactivated platelets bridged by GPIbα-vWF-GPIbα, treated as individual linear springs, shows that bond formation rate has a piecewise dependence on the fluid shear rate, with sharp transition at 7200 s^{-1}. The kinetics and binding efficiency are driven by the size of vWF, with large multimers dominating [154].

Multiple-platelet continuum models of aggregation consider platelets in terms of their number densities. One approach uses variable flow energies to describe the activation states of platelets, which are allowed to interact with each other, fibrinogen, and the vessel wall through a discrete stochastic cellular Potts model at the microscale, with macroscale blood flow dynamics described by Navier-Stokes equations [235]. Each position of discretized space is occupied by fluid, platelet mass, or other cell types. In another approach, interplatelet bond stress development is tracked in macroscale platelet thrombosis in atherosclerotic arteries using an Oldroyd-B-like evolution equation [57,58,60]. Interactions among platelets and coagulation chemistry also play a role in growing thrombi, whose porosity is dependent on the number density of bound platelets [125, 126]. This comprehensive approach utilizes the finite difference approach to discretize equations relating to coagulation biochemistry, chemical activation and deposition of platelets, and two way interaction between plasma dynamics and the platelet plug, whose resistance to flow is determined by a term added to the Navier-Stokes momentum equation [33]. Larger-scale examination of individual platelet behavior in a growing thrombus (i.e. platelet-platelet or platelet-wall interactions) may also be performed using a force coupling method (FCM), where thousands of platelets are treated as rigid spherical Lagrangian particles and two-way coupled to the background flow. The FCM is affected by local hydrodynamics and can be incorporated in the Navier-Stokes equation via a particle body-force term [239]. Attractive or repulsive forces between platelets can also be described by a Morse potential [240].

7.7.5 Considerations for numerical models

Continuum and particle-based models of platelet dynamics under flow conditions are powerful tools for the study of phenomena that are difficult to observe in vivo and in vitro, and may be able to accurately describe and predict events leading to thrombosis. However, considering appropriate model types, algorithms, and parameters is critical to successful outcomes. Some key points to consider in selection of an appropriate model:

i *Prothrombotic event of interest* – Platelets interact with other cell types, surrounding fluid, ligands, biochemical agonists, and surfaces during their journey from the RBC-rich core flow to thrombus formation and propagation.

ii *Flow conditions* – Flow shear stresses, shear rates, and patterns play important roles in platelet motion and their activation, adhesion, aggregation, and receptor-ligand interactions, as described in the preceding sections.

iii *Scale* – Intraplatelet events and extraplatelet dynamics occur across several spatiotemporal scales, from microscopic events, such as intraplatelet dynamics and platelet membrane deformations, occurring on the order of nm and ps, to mesocopic events, such as thrombus propagation, occurring on the order of mm and s (or even min) length and time scales [240]

iv *Computational resources and time* – Even with the advent of large scale HPC clusters, certain large-scale MD and continuum simulations become computationally prohibitive, necessitating improvement of computational efficiency by either spatial coarse-graining of the platelet and/or fluid models, or incorporating multiple time-stepping (MTS) algorithms, with different time step sizes for different phenomena [245].

Thus, basic knowledge of the platelet phenomena being modeled and a careful examination of available continuum- and particle-based models will enhance their computational efficiency, and in turn, their accuracy and predictive capabilities.

ACKNOWLEDGEMENTS

This work was supported by a grant from the National Institutes of Health (NIH) NHLBI U01HL131052 (D. Bluestein).

Bibliography

[1] A. Adzic, S.R. Patel, and S. Maybaum. Impact of adverse events on ventricular assist device outcomes. *Curr. Heart Fail. Rep.*, 10:89–100, 2013.

[2] K. Affeld, L. Goubergrits, U. Kertzscher, L.Gadischke, and A. Reininger. Mathematical model of platelet deposition under flow conditions. *Int. J. Artif. Organs*, 27:699–708, 2004.

[3] A. Agnihotri, P. Soman, and C.A. Siedlecki. Afm measurements of interactions between the platelet integrin receptor gpiibiiia and fibrinogen. *Colloids Surf. B Biointerfaces*, 71:138–147, 2009.

[4] N. Ajzenberg, A.T. Talab, J. M. Masse, A. Drouin, K. Jondeau, H. Kobeiter, D. Baruch, and E.M. Cramer. Platelet shape change and subsequent glycoprotein redistribution in human stenosed arteries. *Platelets*, 16:13–18, 2005.

[5] Y. Alemu and D. Bluestein. Flow-induced platelet activation and damage accumulation in a mechanical heart valve: numerical studies. *Artif. Organs*, 31:677–688, 2007.

[6] A. Aliseda, V.K. Chivukula, P. McGah, A.R. Prisco, J.A. Beckman, G.J. Garcia, N.A. Mokadam, and C. Mahr. Lvad outflow graft angle and thrombosis risk. *ASAIO J.*, 63:14–23, 2017.

[7] R.D. Allen, L.R. Zacharski, S.T. Widirstky, R. Rosenstein, L.M. Zaitlin, and D.R. Burgess. Transformation and motility of human platelets: details of the shape change and release reaction observed by optical and electron microscopy. *J. Cell Biol.*, 83:126–142, 1979.

[8] F. Bajd, J. Vidmar, A. Fabjan, A. Blinc, E. Kralj, N. Bizjak, and I. Sersa. Impact of altered venous hemodynamic conditions on the formation of platelet layers in thromboemboli. *Thromb. Res.*, 129:158–163, 2012.

[9] A. Barg, R. Ossig, T. Goerge, M.F. Schneider, H. Schillers, H. Oberleithner, and S.W. Schneider. Soluble plasma-derived von willebrand factor assembles to a haemostatically active filamentous network. *Thromb. Haemost.*, 97:514–526, 2007.

[10] D.L.Jr Bark and D.N. Ku. Wall shear over high degree stenoses pertinent to atherothrombosis. *J. Biomech.*, 43:2970–2977, 2010.

[11] W. Bergmeier, C.L. Piffath, T. Goerge, S.M. Cifuni, Z.M. Ruggeri, J. Ware, and D.D. Wagner. The role of platelet adhesion receptor gpibalpha far exceeds that of its main ligand, von willebrand factor, in arterial thrombosis. *Proc. Natl. Acad. Sci USA*, 45:16900–5, 2006.

[12] J.L. Blackshear, E.M. Wysokinska, R.E. Safford, C.S. Thomas, B.P. Shapiro, S. Ung, M.E. Stark, P. Parikh, G.S. Johns, and D. Chen. Shear stress-associated acquired von willebrand syndrome in patients with mitral regurgitation. *J. Thromb. Haemost.*, 12:1966–1974, 2014.

[13] D. Bluestein, L. Niu, R.T. Schoephoerster, and M.K. Dewanjee. Fluid mechanics of arterial stenosis: relationship to the development of mural thrombus. *Ann. Biomed. Eng.*, 25:344–356, 1997.

[14] D. Bluestein, E. Rambod, and M. Gharib. Vortex shedding as a mechanism for free emboli formation in mechanical heart valves. *J. Biomech. Eng.*, 122:125–134, 2000.

[15] S. Bodin, H. Tronchere, and B. Payrastre. Lipid rafts are critical membrane domains in blood platelet activation processes. *Biochim. Biophys. Acta*, 1610:247–257, 2003.

[16] B.R. Branchford, C. J. Ng, K.B. Neeves, and J. Di Paola. Microfluidic technology as an emerging clinical tool to evaluate thrombosis and hemostasis. *Thromb. Res.*, 136:13–19, 2015.

[17] L.F. Brass and S.L. Diamond. Transport physics and biorheology in the setting of hemostasis and thrombosis. *J. Thromb. Haemost.*, 14:906–917, 2016.

[18] L.D.C. Casa and D.N. Ku. Thrombus formation at high shear rates. *Annu. Rev. Biomed. Eng.*, 19:415–433, 2017.

[19] C.H. Chan, I.L. Pieper, S. Fleming, Y. Friedmann, G. Foster, K. Hawkins, C.A. Thornton, and V. Kanamarlapudi. The effect of shear stress on the size, structure, and function of human von willebrand factor. *Artif. Organs*, 38:741–750, 2014.

[20] J. Chen, H. Zhou, A. Diacovo, X.L. Zheng, J. Emsley, and T.G. Diacovo. Exploiting the kinetic interplay between gpibalpha-vwf binding interfaces to regulate hemostasis and thrombosis. *Blood*, 124:3799–807, 2014.

[21] S. Chen and T. A. Springer. Selectin receptor-ligand bonds: Formation limited by shear rate and dissociation governed by the bell model. *Proc. Natl. Acad. Sci. USA*, 98:950–955, 2001.

[22] Y. Chen, L.A. Ju, F. Zhou, J. Liao, L. Xue, Q.P. Su, D. Jin, Y. Yuan, H. Lu, S.P. Jackson, and C. Zhu. An integrin alphaiibbeta3 intermediate affinity state mediates biomechanical platelet aggregation. *Nat. Mater.*, 18:760–769, 2019.

[23] Z. Chen, N.K. Mondal, J. Ding, J. Gao, B.P. Griffith, and Z.J. Wu. Shear-induced platelet receptor shedding by non-physiological high shear stress with short exposure time: glycoprotein ibalpha and glycoprotein vi. *Thromb. Res.*, 135:692–698, 2015.

[24] Z. Chen, N.K. Mondal, J. Ding, S.C. Koenig, M.S. Slaughter, B.P. Griffith, and Z.J. Wu. Activation and shedding of platelet glycoprotein iib/iiia under non-physiological shear stress. *Mol. Cell Biochem.*, 409:93–101, 2015.

[25] H. Cheng, R. Yan, S. Li, Y. Yuan, J. Liu, C. Ruan, and K. Dai. Shear-induced interaction of platelets with von willebrand factor results in glycoprotein ibalpha shedding. *Am. J. Physiol. Heart Circ. Physiol.*, 297:H2128–35, 2009.

[26] J.K. Chesnutt and H.C. Han. Tortuosity triggers platelet activation and thrombus formation in microvessels. *J. Biomech. Eng.*, 133:121004, 2011.

[27] J.K. Chesnutt and H.C. Han. Platelet size and density affect shear-induced thrombus formation in tortuous arterioles. *Phys. Biol.*, 10:056003, 2013.

[28] J.K. Chesnutt and H.C. Han. Simulation of the microscopic process during initiation of stent thrombosis. *Comput. Biol. Med.*, 56:182–191, 2015.

[29] J.K. Chesnutt and H.C. Han. Computational simulation of platelet interactions in the initiation of stent thrombosis due to stent malapposition. *Phys. Biol.*, 13:016001, 2016.

[30] W.C. Chiu, G. Girdhar, M. Xenos, Y. Alemu, J.S. Soares, S. Einav, M. Slepian, and D. Bluestein. Thromboresistance comparison of the heartmate ii ventricular assist device with the device thrombogenicity emulation-optimized heartassist 5 vad. *J. Biomech. Eng.*, 136:021014, 2014.

[31] I. Chung and G.Y. Lip. Platelets and heart failure. *Eur. Heart J.*, 27:2623–31, 2006.

[32] J.C. Ciciliano, R. Tran, Y. Sakurai, and W.A. Lam. The platelet and the biophysical microenvironment: lessons from cellular mechanics. *Thromb. Res.*, 133:532–537, 2014.

[33] S. Cito, M.D. Mazzeo, and L. Badimon. A review of macroscopic thrombus modeling methods. *Thromb. Res.*, 131:116–124, 2013.

[34] L.A. Coburn, V. S. Damaraju, S. Dozic, S.G. Eskin, M.A. Cruz, and L.V. McIntire. Gpibalpha-vwf rolling under shear stress shows differences between type 2b and 2m von willebrand disease. *Biophys. J.*, 100:304–312, 2011.

[35] G. Colantuoni, J.D. Hellums, J.L. Moake, and C.P. Alfrey Jr. The response of human platelets to shear stress at short exposure times. *Trans. Am. Soc. Artif. Intern. Organs*, 23:626–631, 1977.

[36] J.M. Cosemans, A. Angelillo-Scherrer, N.J. Mattheij, and J.W. Heemskerk. The effects of arterial flow on platelet activation, thrombus growth, and stabilization. *Cardiovasc. Res.*, 99:342–352, 2013.

[37] S.L. Cranmer, K.J. Ashworth, Y. Yao, M.C. Berndt, Z.M. Ruggeri, R.K. Andrews, and S.P. Jackson. High shear-dependent loss of membrane integrity and defective platelet adhesion following disruption of the gpibalpha-filamin interaction. *Blood*, 117:2718–2727, 2011.

[38] S. Crow, D. Chen, C. Milano, W. Thomas, L. Joyce, V. Piacentino, R. Sharma, J. Wu, G. Arepally, D. Bowles, J. Rogers, and N. Villamizar-Ortiz. Acquired von willebrand syndrome in continuous-flow ventricular assist device recipients. *Ann. Thorac. Surg.*, 90:1263–1269, 2010.

[39] L. Crowl and A.L. Fogelson. Computational model of whole blood exhibiting lateral platelet motion induced by red blood cells. *Int. J. Numer. Method Biomed. Eng.*, 26:471–487, 2010.

[40] L. Crowl and A.L. Fogelson. Analysis of mechanisms for platelet near-wall excess under arterial blood flow conditions. *J. Fluid Mech.*, 676:348–375, 2011.

[41] G.D. Dangas, J.I. Weitz, G. Giustino, R. Makkar, and R. Mehran. Prosthetic heart valve thrombosis. *J. Am. Coll. Cardiol.*, 68:2670–2689, 2016.

[42] K.M. Dayananda, I. Singh, N. Mondal, and S. Neelamegham. Von willebrand factor self-association on platelet gpibalpha under hydrodynamic shear: effect on shear-induced platelet activation. *Blood*, 116:3990–3998, 2010.

[43] S.L. Diamond. Systems analysis of thrombus formation. *Circ. Res.*, 118:1348–1362, 2016.

[44] T.A. Doggett, G. Girdhar, A. Lawshe, J.L. Miller, I.J. Laurenzi, S.L. Diamond, and T.G. Diacovo. Alterations in the intrinsic properties of the gpibalpha-vwf tether bond define the kinetics of the platelet-type von willebrand disease mutation, gly233val. *Blood*, 102:152–160, 2003.

[45] T.A. Doggett, G. Girdhar, A. Lawshe, D.W. Schmidtke, I.J. Laurenzi, S.L. Diamond, and T.G. Diacovo. Selectin-like kinetics and biomechanics promote rapid platelet adhesion in flow: the gpib(alpha)-vwf tether bond. *Biophys. J.*, 83:194–205, 2002.

[46] S.M. Dopheide, M.J. Maxwell, and S.P. Jackson. Shear-dependent tether formation during platelet translocation on von willebrand factor. *Blood*, 99:159–167, 2002.

[47] S.M. Dopheide, C.L. Yap, and S.P. Jackson. Dynamic aspects of platelet adhesion under flow. *Clin. Exp. Pharmacol. Physiol.*, 28:355–363, 2001.

[48] E.C. Eckstein, D.L. Bilsker, C.M. Waters, J.S. Kippenhan, and A.W. Tilles. Transport of platelets in flowing blood. *Ann. N.Y. Acad. Sci.*, 516:442–452, 1987.

[49] E.C. Eckstein, A.W. Tilles, and F.J. Millero. Conditions for the occurrence of large near-wall excesses of small particles during blood flow. *Microvasc. Res.*, 36:31–39, 1988.

[50] G. Escolar, M. Krumwiede, and J.G. White. Organization of the actin cytoskeleton of resting and activated platelets in suspension. *Am. J. Pathol.*, 123:86–94, 1986.

[51] D.A. Fedosov, B. Caswell, and G.E. Karniadakis. Systematic coarse-graining of spectrin-level red blood cell models. *Comput. Methods Appl. Mech. Eng.*, 199:29–32, 2010.

[52] D.A. Fedosov, M. Dao, G.E. Karniadakis, and S. Suresh. Computational biorheology of human blood flow in health and disease. *Ann. Biomed. Eng.*, 42:368–387, 2014.

[53] S. Feghhi and N.J. Sniadecki. Mechanobiology of platelets: techniques to study the role of fluid flow and platelet retraction forces at the micro- and nano-scale. *Int. J. Mol. Sci.*, 12:9009–9030, 2011.

[54] N. Filipovic, M. Kojic, and A. Tsuda. Modelling thrombosis using dissipative particle dynamics method. *Philos. Trans. A Math Phys. Eng. Sci.*, 366:3265–3279, 2008.

[55] M.H. Flamm and S.L. Diamond. Multiscale systems biology and physics of thrombosis under flow. *Ann. Biomed. Eng.*, 40:2355–2364, 2012.

[56] M.H. Flamm, T. Sinno, and S.L. Diamond. Simulation of aggregating particles in complex flows by the lattice kinetic monte carlo method. *J. Chem. Phys.*, 134:034905, 2011.

[57] A.L. Fogelson. Continuum models of platelet aggregation: Formulation and mechanical properties. *SIAM J. Appl. Math.*, 52:1089–1110, 1992.

[58] A.L. Fogelson and R.D. Guy. Platelet-wall interactions in continuum models of platelet thrombosis: formulation and numerical solution. *Math Med. Biol.*, 21:293–334, 2004.

[59] A.L. Fogelson and R.D. Guy. Immersed-boundary-type models of intravascular platelet aggregation. *Computer Methods in App. Mech. and Eng.*, 197:2087–2104, 2008.

[60] A.L. Fogelson and K.B. Neeves. Fluid mechanics of blood clot formation. *Annu. Rev. Fluid Mech.*, 47:377–403, 2015.

[61] C.M. Franz, A. Taubenberger, P.H. Puech, and D.J. Muller. Studying integrin-mediated cell adhesion at the single-molecule level using afm force spectroscopy. *Sci STKE*, 406:pl5, 2007.

[62] M.M. Frojmovic. Platelet aggregation in flow: differential roles for adhesive receptors and ligands. *Am. Heart J.*, 135:S119–S131, 1998.

[63] M.M. Frojmovic and R. Panjwani. Geometry of normal mammalian platelets by quantitative microscopic studies. *Biophys. J.*, 16:1071–1089, 1976.

[64] H. Fu, Y. Jiang, D. Yang, F. Scheiflinger, W.P. Wong, and T.A. Springer. Flow-induced elongation of von willebrand factor precedes tension-dependent activation. *Nat. Commun.*, 8:324, 2017.

[65] C. Gao, P. Zhang, G. Marom, Y. Deng, and D. Bluestein. Reducing the effects of compressibility in dpd-based blood flow simulations through severe stenotic microchannels. *J. Comp. Phys.*, 335:812–827, 2017.

[66] G. Girdhar and D. Bluestein. Biological effects of dynamic shear stress in cardiovascular pathologies and devices. *Expert Rev. Med. Devices*, 5:167–181, 2008.

[67] G. Girdhar, M. Xenos, Y. Alemu, W.C. Chiu, B.E. Lynch, J. Jesty, S. Einav, M.J. Slepian, and D. Bluestein. Device thrombogenicity emulation: a novel method for optimizing mechanical circulatory support device thromboresistance. *PLoS One*, 7:e32463, 2012.

[68] E. Gitz, C.D. Koopman, A. Giannas, C.A. Koekman, D.J. van den Heuvel, H. Deckmyn, J.W. Akkerman, H.C. Gerritsen, and R.T. Urbanus. Platelet interaction with von willebrand factor is enhanced by shear-induced clustering of glycoprotein ibalpha. *Haematologica*, 98:1810–1818, 2013.

[69] M.S. Goel and S.L. Diamond. Adhesion of normal erythrocytes at depressed venous shear rates to activated neutrophils, activated platelets, and fibrin polymerized from plasma. *Blood*, 100:3797–3803, 2002.

[70] H.L. Goldsmith and J.C. Marlow. Flow behavior of erythrocytes .2. Particle motions in concentrated suspensions of ghost cells. *J. Colloid and Int. Sci.*, 71:383–407, 1979.

[71] H.L. Goldsmith, F. A. McIntosh, J. Shahin, and M.M. Frojmovic. Time and force dependence of the rupture of glycoprotein iib-iiia-fibrinogen bonds between latex spheres. *Biophys. J.*, 78:1195–1206, 2000.

[72] H.L. Goldsmith and V.T. Turitto. Rheological aspects of thrombosis and haemostasis: basic principles and applications. *Thromb. Haemost.*, 55:415–435, 1986.

[73] I. Goncalves, W.S. Nesbitt, Y. Yuan, and S.P. Jackson. Importance of temporal flow gradients and integrin alphaiibbeta3 mechanotransduction for shear activation of platelets. *J. Biol. Chem.*, 280:15430–7, 2005.

[74] S. Goto, Y. Ikeda, E. Saldivar, and Z.M. Ruggeri. Distinct mechanisms of platelet aggregation as a consequence of different shearing flow conditions. *J. Clin. Invest.*, 101:479–486, 1998.

[75] H.R. Gralnick, S.B. Williams, and B.S. Coller. Fibrinogen competes with von willebrand factor for binding to the glycoprotein iib/iiia complex when platelets are stimulated with thrombin. *Blood*, 64:797–800, 1984.

[76] P. Gupta, P. Zhang, J. Sheriff, D. Bluestein, and Y. Deng. A multiscale model for recruitment aggregation of platelets by correlating with in vitro results. *Cell Mol Bioeng*, 12(4):327–343, 2019.

[77] J.H. Haga, A J. Beaudoin, J.G. White, and J. Strony. Quantification of the passive mechanical properties of the resting platelet. *Ann. Biomed. Eng.*, 26:268–77, 1998.

[78] C.E. Hansen, Y. Qiu, O.J.T. McCarty, and W.A. Lam. Platelet mechanotransduction. *Annu. Rev. Biomed. Eng.*, 26:268–77, 1998.

[79] J.H. Hartwig. The platelet: form and function. *Semin. Hematol.*, 43:S95–S100, 2006.

[80] J.H. Hartwig, K. Barkalow, A. Azim, and J. Italiano. The elegant platelet: signals controlling actin assembly. *Thromb. Haemost.*, 82:392–398, 1999.

[81] J.H. Hartwig and M. DeSisto. The cytoskeleton of the resting human blood platelet: structure of the membrane skeleton and its attachment to actin filaments. *J. Cell Biol.*, 112:407–425, 1991.

[82] J.J. Hathcock. Flow effects on coagulation and thrombosis. *Arterioscler. Thromb. Vasc. Biol.*, 26:1729–37, 2006.

[83] H.F. Heijnen, M. Van Lier, S. Waaijenborg, Y. Ohno-Iwashita, A.A. Waheed, M. Inomata, G. Gorter, W. Mobius, J.W. Akkerman, and J.W. Slot. Concentration of rafts in platelet filopodia correlates with recruitment of c-src and cd63 to these domains. *J. Thromb. Haemost.*, 1:1161–73, 2003.

[84] J.D. Hellums. 1993 Whitaker Lecture: Biorheology in thrombosis research. *Ann. Biomed. Eng.*, 22:445–455, 1994.

[85] M.E. Hensler, M. Frojmovic, R.G. Taylor, R.R. Hantgan, and J.C. Lewis. Platelet morphologic changes and fibrinogen receptor localization. Initial responses in adp-activated human platelets. *Am. J. Pathol.*, 141:707–719, 1992.

[86] B.A. Herbig and S.L. Diamond. Pathological von willebrand factor fibers resist tissue plasminogen activator and adamts13 while promoting the contact pathway and shear-induced platelet activation. *J. Thromb. Haemost.*, 13:1699–1708, 2015.

[87] M.J. Hollestelle, C.M. Loots, A. Squizzato, T. Renne, B.J. Bouma, P.G. de Groot, P.J. Lenting, J.C. Meijers, and V.E. Gerdes. Decreased active von willebrand factor level owing to shear stress in aortic stenosis patients. *J. Thromb. Haemost.*, 9:953–958, 2011.

[88] P.A. Holme, U. Orvim, M.J. Hamers, N.O. Solum, F.R. Brosstad, R.M. Barstad, and K.S. Sakariassen. Shear-induced platelet activation and platelet microparticle formation at blood flow conditions as in arteries with a severe stenosis. *Arterioscler. Thromb. Vasc. Biol.*, 17:646–653, 1997.

[89] T. Hong and C.N. Kim. A numerical analysis of the blood flow around the bileaflet mechanical heart valves with different rotational implantation angles. *Journal of Hydrodynamics*, 23:607–614, 2011.

[90] H. Hosseinzadegan and D.K. Tafti. Prediction of thrombus growth: Effect of stenosis and reynolds number. *Cardiovasc. Eng. Technol.*, 8:164–181, 2017.

[91] J.E.Jr. Italiano, W. Bergmeier, S. Tiwari, H. Falet, J.H. Hartwig, K.M. Hoffmeister, P. Andre, D.D. Wagner, and R.A. Shivdasani. Mechanisms and implications of platelet discoid shape. *Blood*, 101:4789–96, 2003.

[92] S.P. Jackson. The growing complexity of platelet aggregation. *Blood*, 109:5087–95, 2007.

[93] S.P. Jackson. Arterial thrombosis–insidious, unpredictable and deadly. *Nat. Med.*, 17:1423–36, 2011.

[94] S.P. Jackson, W.S. Nesbitt, and E. Westein. Dynamics of platelet thrombus formation. *J. Thromb. Haemost.*, 7:17–20, 2011.

[95] A. Javadzadegan, A.S. Yong, M. Chang, A.C. Ng, J. Yiannikas, M.K. Ng, M. Behnia, and L. Kritharides. Flow recirculation zone length and shear rate are differentially affected by stenosis severity in human coronary arteries. *Am. J. Physiol. Heart Circ. Physiol.*, 304:H559–66, 2013.

[96] G.B. Jeffery. The motion of ellipsoidal particles in a viscous fluid. *Proceedings of the Royal Society of London Series a-Containing Papers of a Mathematical and Physical Character*, 102:161–179, 1922.

[97] L.K. Jennings. Mechanisms of platelet activation: need for new strategies to protect against platelet-mediated atherothrombosis. *Thromb. Haemost.*, 102:248–257, 2009.

[98] J. Jesty and D. Bluestein. Acetylated prothrombin as a substrate in the measurement of the procoagulant activity of platelets: elimination of the feedback activation of platelets by thrombin. *Anal. Biochem.*, 272:64–70, 1999.

[99] J. Jesty, W. Yin, P. Perrotta, and D. Bluestein. Platelet activation in a circulating flow loop: combined effects of shear stress and exposure time. *Platelets*, 14:143–149, 2003.

[100] A. Jordan, T. David, S. Homer-Vanniasinkam, A. Graham, and P. Walker. The effects of margination and red cell augmented platelet diffusivity on platelet adhesion in complex flow. *Biorheology*, 41:641–653, 2004.

[101] L. Ju, Y. Chen, K. Li, Z. Yuan, B. Liu, S.P. Jackson, and C. Zhu. Dual biomembrane force probe enables single-cell mechanical analysis of signal crosstalk between multiple molecular species. *Sci. Rep.*, 7:14185, 2017.

[102] J. Jung, A. Hassanein, and R.W. Lyczkowski. Hemodynamic computation using multiphase flow dynamics in a right coronary artery. *Ann. Biomed. Eng.*, 34:393–407, 2006.

[103] H. Kamada, K. i Tsubota, M. Nakamura, S. Wada, T. Ishikawa, and T. Yamaguchi. A three-dimensional particle simulation of the formation and collapse of a primary thrombus. *Int. J. for Num. Methods in Biomed. Eng.*, 26:488–500, 2010.

[104] H. Kamada, Y. Imai, M. Nakamura, T. Ishikawa, and T. Yamaguchi. Computational study on thrombus formation regulated by platelet glycoprotein and blood flow shear. *Microvasc. Res.*, 89:95–106, 2013.

[105] Z.S. Kaplan and S.P. Jackson. The role of platelets in atherothrombosis. *Hematology Am. Soc. Hematol. Educ. Program*, 2011:51–61, 2011.

[106] A. Kasirer-Friede, M.R. Cozzi, M. Mazzucato, L. De Marco, Z.M. Ruggeri, and S.J. Shattil. Signaling through gp ib-ix-v activates alpha iib beta 3 independently of other receptors. *Blood*, 103:3403–11, 2004.

[107] M.F. Kee, D.R. Myers, Y. Sakurai, W.A. Lam, and Y. Qiu. Platelet mechanosensing of collagen matrices. *PLoS One*, 10:e0126624, 2015.

[108] S. Kefayati, J.S. Milner, D.W. Holdsworth, and T.L. Poepping. In vitro shear stress measurements using particle image velocimetry in a family of carotid artery models: effect of stenosis severity, plaque eccentricity, and ulceration. *PLoS One*, 9:e98209, 2014.

[109] Z. Keshavarz-Motamed, J. Garcia, and L. Kadem. Fluid dynamics of coarctation of the aorta and effect of bicuspid aortic valve. *PLoS One*, 8:e72394, 2013.

[110] S. Khan, A. Jesacher, W. Nussbaumer, S. Bernet, and M. Ritsch-Marte. Quantitative analysis of shape and volume changes in activated thrombocytes in real time by single-shot spatial light modulator-based differential interference contrast imaging. *J. Biophotonics*, 4:600–609, 2011.

[111] J. Kim, C.Z. Zhang, X. Zhang, and T.A. Springer. A mechanically stabilized receptor-ligand flex-bond important in the vasculature. *Nature*, 466:992–995, 2010.

[112] M.R. King and D.A. Hammer. Multiparticle adhesive dynamics. Interactions between stably rolling cells. *Biophys. J.*, 81:799–813, 2001.

[113] A. Kita, Y. Sakurai, D.R. Myers, R. Rounsevell, J.N. Huang, T.J. Seok, K. Yu, M.C. Wu, D.A. Fletcher, and W. A. Lam. Microenvironmental geometry guides platelet adhesion and spreading: a quantitative analysis at the single cell level. *PLoS One*, 6:e26437, 2011.

[114] K. Kottke-Marchant. Importance of platelets and platelet response in acute coronary syndromes. *Cleve. Clin. J. Med.*, 76:S2–S7, 2009.

[115] M.J. Kraus, E.F. Strasser, and R. Eckstein. A new method for measuring the dynamic shape change of platelets. *Transfus. Med. Hemother.*, 37:306–310, 2010.

[116] S. Krishnan, H.S. Udaykumar, J.S. Marshall, and K.B. Chandran. Two-dimensional dynamic simulation of platelet activation during mechanical heart valve closure. *Ann. Biomed. Eng.*, 34:1519–34, 2006.

[117] M.H. Kroll, T.S. Harris, J.L. Moake, R.I. Handin, and A.I. Schafer. Von willebrand factor binding to platelet gpib initiates signals for platelet activation. *J. Clin. Invest.*, 88:1568–73, 1991.

[118] M.H. Kroll, J.D. Hellums, L.V. McIntire, A.I. Schafer, and J.L. Moake. Platelets and shear stress. *Blood*, 88:1525–1541, 1996.

[119] S. Kulkarni, S.M. Dopheide, C.L. Yap, C. Ravanat, M. Freund, P. Mangin, K.A. Heel, A. Street, I.S. Harper, F. Lanza, and S.P. Jackson. A revised model of platelet aggregation. *J. Clin. Invest.*, 105:783–791, 2000.

[120] M. Kuwahara, M. Sugimoto, S. Tsuji, H. Matsui, T. Mizuno, S. Miyata, and A. Yoshioka. Platelet shape changes and adhesion under high shear flow. *Arterioscler. Thromb. Vasc. Biol.*, 22:329–34, 2002.

[121] J.F.Jr. LaDisa, C. Alberto Figueroa, I.E. Vignon-Clementel, H.J. Kim, N. Xiao, L.M. Ellwein, F.P. Chan, J.A. Feinstein, and C.A. Taylor. Computational simulations for aortic coarctation: representative results from a sampling of patients. *J. Biomech. Eng.*, 133:091008, 2011.

[122] W.A. Lam, O. Chaudhuri, A. Crow, K.D. Webster, T.D. Li, A. Kita, J. Huang, and D.A. Fletcher. Mechanics and contraction dynamics of single platelets and implications for clot stiffening. *Nat. Mater.*, 10:61–66, 2011.

[123] M.B. Lawrence, L.V. McIntire, and S.G. Eskin. Effect of flow on polymorphonuclear leukocyte/endothelial cell adhesion. *Blood*, 70:1284–90, 1987.

[124] D. Lee, K.P. Fong, M.R. King, L.F. Brass, and D.A. Hammer. Differential dynamics of platelet contact and spreading. *Biophys. J.*, 102:472–82, 2012.

[125] K. Leiderman and A.L. Fogelson. Grow with the flow: a spatial-temporal model of platelet deposition and blood coagulation under flow. *Math. Med. Biol.*, 28:47–84, 2011.

[126] K. Leiderman and A.L. Fogelson. The influence of hindered transport on the development of platelet thrombi under flow. *Bull. Math. Biol.*, 75:1255–1283, 2013.

[127] C. Leon, A. Eckly, B. Hechler, B. Aleil, M. Freund, C. Ravanat, M. Jourdain, C. Nonne, J. Weber, R. Tiedt, M.P. Gratacap, S. Severin, J.P. Cazenave, F. Lanza, R. Skoda, and C. Gachet. Megakaryocyte-restricted myh9 inactivation dramatically affects hemostasis while preserving platelet aggregation and secretion. *Blood*, 110:3183–91, 2007.

[128] S.L. Leung, Y. Lu, D. Bluestein, and M.J. Slepian. Dielectrophoresis-mediated electrodeformation as a means of determining individual platelet stiffness. *Ann. Biomed. Eng.*, 44:903–13, 2016.

[129] X.M. Liang, S.J. Han, J.A. Reems, D. Gao, and N.J. Sniadecki. Platelet retraction force measurements using flexible post force sensors. *Lab. Chip*, 10:991–998, 2010.

[130] B. Lincoln, A.J. Ricco, N.J. Kent, L. Basabe-Desmonts, L.P. Lee, B.D. MacCraith, D. Kenny, and G. Meade. Integrated system investigating shear-mediated platelet interactions with von willebrand factor using microliters of whole blood. *Anal. Biochem.*, 405:174–83, 2010.

[131] A.L. Litvinenko, A.E. Moskalensky, N.A. Karmadonova, V.M. Nekrasov, D.I. Strokotov, A.I. Konokhova, M.A. Yurkin, E.A. Pokushalov, A.V. Chernyshev, and V.P. Maltsev. Fluorescence-free flow cytometry for measurement of shape index distribution of resting, partially activated, and fully activated platelets. *Cytometry A*, 89:1010–1016, 2016.

[132] R.I. Litvinov, V. Barsegov, A.J. Schissler, A.R. Fisher, J.S. Bennett, J.W. Weisel, and H. Shuman. Dissociation of bimolecular alphaiibbeta3-fibrinogen complex under a constant tensile force. *Biophys. J.*, 100:165–73, 2011.

[133] R.I. Litvinov, D.H. Farrell, J.W. Weisel, and J.S. Bennett. The platelet integrin alphaiibbeta3 differentially interacts with fibrin versus fibrinogen. *J. Biol. Chem.*, 291:7858–67, 2016.

[134] D.A. Los and N. Murata. Membrane fluidity and its roles in the perception of environmental signals. *Biochim. Biophys. Acta*, 1666:142–57, 2004.

[135] G.D. Lowe. Virchow's triad revisited: abnormal flow. *Pathophysiol. Haemost. Thromb.*, 33:455–457, 2003.

[136] G.D. Lowe, A. Saniabadi, A. Turner, P. Lieberman, J. Pollock, and J. Drury. Studies on haematocrit in peripheral arterial disease. *Klin. Wochenschr.*, 64:969–74, 1986.

[137] R. Lozano, M. Naghavi, K. Foreman, S. Lim, K. Shibuya, V. Aboyans, J. Abraham, T. Adair, R. Aggarwal, S.Y. Ahn, M. Alvarado, H.R. Anderson, L.M. Anderson, K. G. Andrews, C. Atkinson, L.M. Baddour, S. Barker-Collo, D.H. Bartels, M.L. Bell, E.J. Benjamin, D. Bennett, K. Bhalla, B. Bikbov, A. Bin Abdulhak, G. Birbeck, F. Blyth, I. Bolliger, S. Boufous, C. Bucello, M. Burch, P. Burney, J. Carapetis, H. Chen, D. Chou, S.S. Chugh, L.E. Coffeng, S.D. Colan, S. Colquhoun, K.E. Colson, J. Condon, M.D. Connor, L.T. Cooper, M. Corriere, M. Cortinovis, K.C. de Vaccaro, W. Couser, B.C. Cowie, M.H. Criqui, M. Cross, K.C. Dabhadkar, N. Dahodwala, D. De Leo, L. Degenhardt, A. Delossantos, J. Denenberg, D.C. Des Jarlais, S.D. Dharmaratne, E.R. Dorsey, T. Driscoll, H. Duber, B. Ebel, P.J. Erwin, P. Espindola, M. Ezzati, V. Feigin, A.D. Flaxman, M.H. Forouzanfar, F.G. Fowkes, R. Franklin, M. Fransen, M.K. Freeman, S.E. Gabriel, E. Gakidou, F. Gaspari, R.F. Gillum, D. Gonzalez-Medina, Y.A. Halasa, D. Haring, J.E. Harrison, R. Havmoeller, R.J. Hay, B. Hoen, P.J. Hotez, D. Hoy, K.H. Jacobsen, S.L. James, R. Jasrasaria, S. Jayaraman, N. Johns, G. Karthikeyan, N. Kassebaum, A. Keren, J.P. Khoo, L.M. Knowlton, O. Kobusingye, A. Koranteng, R. Krishnamurthi, M. Lipnick, S.E. Lipshultz, S.L. Ohno, J. Mabweijano, M.F. MacIntyre, L. Mallinger, L. March, G.B. Marks, R. Marks, A. Matsumori, R. Matzopoulos, B.M. Mayosi, J.H. McAnulty, M.M. McDermott, J. McGrath, G.A. Mensah, T.R. Merriman, C. Michaud, M. Miller, T.R. Miller, C. Mock, A.O. Mocumbi, A.A. Mokdad, A. Moran, K. Mulholland, M.N. Nair, L. Naldi, K.M. Narayan, K. Nasseri, P. Norman, M. O'Donnell, S.B. Omer, K. Ortblad, R. Osborne, D. Ozgediz, B. Pahari, J.D. Pandian, A.P. Rivero, R.P. Padilla, F. Perez-Ruiz, N. Perico, D. Phillips, K. Pierce, C.A. Pope, E. Porrini, F. Pourmalek, M. Raju, D. Ranganathan, J.T. Rehm, D.B. Rein, G. Remuzzi, F.P. Rivara, T. Roberts, F.R. De Leon, L.C. Rosenfeld, L. Rushton, R.L. Sacco, J.A. Salomon, U. Sampson, E. Sanman, D.C. Schwebel, M. Segui-Gomez, D.S. Shepard, D. Singh, J. Singleton, K. Sliwa, E. Smith, A. Steer, J.A. Taylor, B. Thomas, I.M. Tleyjeh, J.A. Towbin, T. Truelsen, E.A. Undurraga, N. Venketasubramanian, L. Vijayakumar, T. Vos, G.R. Wagner, M. Wang, W. Wang, K. Watt, M.A. Weinstock, R. Weintraub, J.D. Wilkinson, A.D. Woolf, S. Wulf, P.H. Yeh, P. Yip, A. Zabetian, Z.J. Zheng, A.D. Lopez, C.J. Murray, M.A. AlMazroa, and Z.A. Memish. Global and regional mortality from 235 causes of death for 20 age groups in 1990 and 2010: a systematic analysis for the global burden of disease study 2010. *Lancet*, 380:2095–128, 2012.

[138] G. Marom and D. Bluestein. Lagrangian methods for blood damage estimation in cardiovascular devices–how numerical implementation affects the results. *Expert. Rev. Med. Devices*, 13:113–122, 2016.

[139] A.L. Marsden, Y. Bazilevs, C.C. Long, and M. Behr. Recent advances in computational methodology for simulation of mechanical circulatory assist devices. *Wiley Interdiscip Rev Syst Biol Med*, 6:169–188, 2014.

[140] B.T. Marshall, M. Long, J.W. Piper, T. Yago, R.P. McEver, and C. Zhu. Direct observation of catch bonds involving cell-adhesion molecules. *Nature*, 423:190–193, 2003.

[141] H. Matsui, M. Sugimoto, T. Mizuno, S. Tsuji, S. Miyata, M. Matsuda, and A. Yoshioka. Distinct and concerted functions of von willebrand factor and fibrinogen in mural thrombus growth under high shear flow. *Blood*, 100:3604–10, 2002.

[142] M.J. Maxwell, S.M. Dopheide, S.J. Turner, and S.P. Jackson. Shear induces a unique series of morphological changes in translocating platelets: effects of morphology on translocation dynamics. *Arterioscler. Thromb. Vasc. Biol.*, 26:663–9, 2006.

[143] M.J. Maxwell, E. Westein, W.S. Nesbitt, S. Giuliano, S.M. Dopheide, and S.P. Jackson. Identification of a 2-stage platelet aggregation process mediating shear-dependent thrombus formation. *Blood*, 109:566–76, 2007.

[144] O.J. McCarty, M.K. Larson, J.M. Auger, N. Kalia, B.T. Atkinson, A.C. Pearce, S. Ruf, R.B. Henderson, V.L. Tybulewicz, L.M. Machesky, and S.P. Watson. Rac1 is essential for platelet lamellipodia formation and aggregate stability under flow. *J. Biol. Chem.*, 280:39474–84, 2005.

[145] J.D. McFadyen and S.P. Jackson. Differentiating haemostasis from thrombosis for therapeutic benefit. *Thromb. Haemost.*, 110:859–67, 2013.

[146] B. McGrath, G. Mealing, and M.R. Labrosse. A mechanobiological investigation of platelets. *Biomech. Model. Mechanobiol.*, 10:473–84, 2011.

[147] M. Mehrabadi, D.N. Ku, and C.K. Aidun. A continuum model for platelet transport in flowing blood based on direct numerical simulations of cellular blood flow. *Ann. Biomed. Eng.*, 43:1410–21, 2015.

[148] M. Mehrabadi, D.N. Ku, and C.K. Aidun. Effects of shear rate, confinement, and particle parameters on margination in blood flow. *Phys. Rev. E*, 93:023109, 2016.

[149] A.L. Meyer, D. Malehsa, C. Bara, U. Budde, M.S. Slaughter, A. Haverich, and M. Strueber. Acquired von willebrand syndrome in patients with an axial flow left ventricular assist device. *Circ. Heart Fail.*, 3:675–81, 2010.

[150] I. Meyer, S. Kunert, S. Schwiebert, I. Hagedorn, J.E. Italiano, S. Dutting, B. Nieswandt, S. Bachmann, and H. Schulze. Altered microtubule equilibrium and impaired thrombus stability in mice lacking ranbp10. *Blood*, 120:3594–602, 2012.

[151] N.A. Mody and M.R. King. Three-dimensional simulations of a platelet-shaped spheroid near a wall in shear flow. *Phys. Fluids*, 17:113302, 2005.

[152] N.A. Mody and M.R. King. Influence of brownian motion on blood platelet flow behavior and adhesive dynamics near a planar wall. *Langmuir*, 23:6321–8, 2007.

[153] N.A. Mody and M.R. King. Platelet adhesive dynamics. Part i: Characterization of platelet hydrodynamic collisions and wall effects. *Biophys. J.*, 95:2539–55, 2008.

[154] N.A. Mody and M.R. King. Platelet adhesive dynamics. Part ii: High shear-induced transient aggregation via gpibalpha-vwf-gpibalpha bridging. *Biophys. J.*, 95:2556–74, 2008.

[155] N.A. Mody, O. Lomakin, T.A. Doggett, T.G. Diacovo, and M.R. King. Mechanics of transient platelet adhesion to von willebrand factor under flow. *Biophys. J.*, 88:1432–43, 2005.

[156] D. Mori, K. Yano, K. Tsubota, T. Ishikawa, S. Wada, and T. Yamaguchi. Simulation of platelet adhesion and aggregation regulated by fibrinogen and von willebrand factor. *Thromb. Haemost.*, 99:108–115, 2008.

[157] A.E. Moskalensky, M.A. Yurkin, A.R. Muliukov, A.L. Litvinenko, V.M. Nekrasov, A.V. Chernyshev, and V.P. Maltsev. Method for the simulation of blood platelet shape and its evolution during activation. *PLoS Comput. Biol.*, 14:e1005899, 2018.

[158] D.R. Myers, Y. Qiu, M.E. Fay, M. Tennenbaum, D. Chester, J. Cuadrado, Y. Sakurai, J. Baek, R. Tran, J.C. Ciciliano, B. Ahn, R.G. Mannino, S.T. Bunting, C. Bennett, M. Briones, A. Fernandez-Nieves, M.L. Smith, A.C. Brown, T. Sulchek, and W.A. Lam. Single-platelet nanomechanics measured by high-throughput cytometry. *Nat. Mater.*, 16:230–235, 2017.

[159] R. Narang, J.G. Cleland, L. Erhardt, S.G. Ball, A.J. Coats, A.J. Cowley, H.J. Dargie, A.S. Hall, J.R. Hampton, and P.A. Poole-Wilson. Mode of death in chronic heart failure. A request and proposition for more accurate classification. *Eur. Heart J.*, 17:1390–403, 1996.

[160] W.S. Nesbitt, P. Mangin, H.H. Salem, and S.P. Jackson. The impact of blood rheology on the molecular and cellular events underlying arterial thrombosis. *J Mol Med (Berl)*, 84:989–95, 2006.

[161] W.S. Nesbitt, E. Westein, F.J. Tovar-Lopez, E. Tolouei, A. Mitchell, J. Fu, J. Carberry, A. Fouras, and S.P. Jackson. A shear gradient-dependent platelet aggregation mechanism drives thrombus formation. *Nat. Med.*, 15:665–73, 2009.

[162] T.H. Nguyen, R. Palankar, V.C. Bui, N. Medvedev, A. Greinacher, and M. Delcea. Rupture forces among human blood platelets at different degrees of activation. *Sci. Rep.*, 6:25402, 2016.

[163] M. Nobili, J. Sheriff, U. Morbiducci, A. Redaelli, and D. Bluestein. Platelet activation due to hemodynamic shear stresses: damage accumulation model and comparison to in vitro measurements. *ASAIO J.*, 54:64–72, 2008.

[164] U.M. Okorie and S.L. Diamond. Matrix protein microarrays for spatially and compositionally controlled microspot thrombosis under laminar flow. *Biophys. J.*, 91:3474–81, 2006.

[165] A. Para, D. Bark, A. Lin, and D. Ku. Rapid platelet accumulation leading to thrombotic occlusion. *Ann. Biomed. Eng.*, 39:1961–71, 2011.

[166] P. Perutelli, P. Marchese, and P.G. Mori. The glycoprotein iib/iiia complex of the platelets. An activation-dependent integrin. *Recenti. Prog. Med.*, 83:100–104, 1992.

[167] I. Pivkin, P. Richardson, and G. Karniadakis. Blood flow velocity effects and role of activation delay time on growth and form of platelet thrombi. *Proc. Natl. Acad. Sci USA*, 103:17164–9, 2006.

[168] I. Pivkin, P. Richardson, and G. Karniadakis. Effect of red blood cells on platelet aggregation. *IEEE Eng. Med. Biol. Mag.*, 28:32–37, 2009.

[169] J.N. Du Plooy, A. Buys, W. Duim, and E. Pretorius. Comparison of platelet ultrastructure and elastic properties in thrombo-embolic ischemic stroke and smoking using atomic force and scanning electron microscopy. *PLoS One*, 8:e69774, 2013.

[170] S. Posch, I. Neundlinger, M. Leitner, P. Siostrzonek, S. Panzer, P. Hinterdorfer, and A. Ebner. Activation induced morphological changes and integrin alphaiibbeta3 activity of living platelets. *Methods*, 60:179–85, 2013.

[171] S. Pothapragada, P. Zhang, J. Sheriff, M. Livelli, M.J. Slepian, Y. Deng, and D. Bluestein. A phenomenological particle-based platelet model for simulating filopodia formation during early activation. *Int. J. Numer. Method Biomed. Eng.*, 31:e02702, 2015.

[172] E. Puklin-Faucher, M. Gao, K. Schulten, and V. Vogel. How the headpiece hinge angle is opened: New insights into the dynamics of integrin activation. *J. Cell Biol.*, 175:349–60, 2006.

[173] E. Puklin-Faucher and M.P. Sheetz. The mechanical integrin cycle. *J. Cell Sci.*, 122:179–186, 2009.

[174] N.B. Purvis and T.D. Giorgio. The effects of elongational stress exposure on the activation and aggregation of blood platelets. *Biorheology*, 28:355–67, 1991.

[175] Q.M. Qi, E. Dunne, I. Oglesby, I. Schoen, A.J. Ricco, D. Kenny, and E.S.G. Shaqfeh. In vitro measurement and modeling of platelet adhesion on vwf-coated surfaces in channel flow. *Biophys. J.*, 116:1136–1151, 2019.

[176] Y. Qiu, J. Ciciliano, D.R. Myers, R. Tran, and W.A. Lam. Platelets and physics: How platelets "feel" and respond to their mechanical microenvironment. *Blood Rev.*, 29:377–86, 2015.

[177] M. Radmacher, M. Fritz, C.M. Kacher, J.P. Cleveland, and P.K. Hansma. Measuring the viscoelastic properties of human platelets with the atomic force microscope. *Biophys. J.*, 70:556–567, 1996.

[178] J.M. Ramstack, L. Zuckerman, and L.F. Mockros. Shear-induced activation of platelets. *J. Biomech.*, 12:113–25, 1979.

[179] D. Raucher and M.P. Sheetz. Characteristics of a membrane reservoir buffering membrane tension. *Biophys. J.*, 77:1992–2002, 1999.

[180] D.A. Reasor, M. Mehrabadi, D.N. Ku, and C.K. Aidun. Determination of critical parameters in platelet margination. *Ann. Biomed. Eng.*, 41:238–249, 2013.

[181] A.J. Reininger, H.F. Heijnen, H. Schumann, H.M. Specht, W. Schramm, and Z.M. Ruggeri. Mechanism of platelet adhesion to von willebrand factor and microparticle formation under high shear stress. *Blood*, 107:3537–45, 2006.

[182] R.S. Rosenson, A. McCormick, and E.F. Uretz. Distribution of blood viscosity values and biochemical correlates in healthy adults. *Clin. Chem.*, 42:1189–95, 1996.

[183] Z.M. Ruggeri. Von willebrand factor and fibrinogen. *Curr. Opin. Cell Biol.*, 5:898–906, 1993.

[184] Z.M. Ruggeri. Platelet adhesion under flow. *Curr. Opin. Cell Biol.*, 16:58–83, 2009.

[185] Z.M. Ruggeri, J.N. Orje, R. Habermann, A.B. Federici, and A.J. Reininger. Activation-independent platelet adhesion and aggregation under elevated shear stress. *Blood*, 108:1903–1910, 2006.

[186] B. Savage, F. Almus-Jacobs, and Z.M. Ruggeri. Specific synergy of multiple substrate-receptor interactions in platelet thrombus formation under flow. *Cell*, 94:657–66, 1998.

[187] B. Savage, E. Saldivar, and Z.M. Ruggeri. Initiation of platelet adhesion by arrest onto fibrinogen or translocation on von willebrand factor. *Cell*, 84:289–97, 1996.

[188] B. Savage, S.J. Shattil, and Z.M. Ruggeri. Modulation of platelet function through adhesion receptors. a dual role for glycoprotein iib-iiia (integrin alpha iib beta 3) mediated by fibrinogen and glycoprotein ib-von willebrand factor. *J. Biol. Chem.*, 267:11300–6, 1992.

[189] S.Dmitrieff, A. Alsina, A. Mathur, and F.J. Nedelec. Balance of microtubule stiffness and cortical tension determines the size of blood cells with marginal band across species. *Proc. Natl. Acad. Sci USA*, 114:4418–4423, 2017.

[190] S.Einav and D. Bluestein. Dynamics of blood flow and platelet transport in pathological vessels. *Ann. N.Y. Acad. Sci.*, 1015:351–366, 2004.

[191] P. Shah, U.S. Tantry, K.P. Bliden, and P.A. Gurbel. Bleeding and thrombosis associated with ventricular assist device therapy. *J. Heart Lung Transplant.*, 36:1164–1173, 2017.

[192] H. Shankaran, P. Alexandridis, and S. Neelamegham. Aspects of hydrodynamic shear regulating shear-induced platelet activation and self-association of von willebrand factor in suspension. *Blood*, 101:2637–45, 2003.

[193] H. Shankaran and S. Neelamegham. Hydrodynamic forces applied on intercellular bonds, soluble molecules, and cell-surface receptors. *Biophys. J.*, 86:576–588, 2004.

[194] J. Sheriff, J.S. Soares, M. Xenos, J. Jesty, and D. Bluestein. Evaluation of shear-induced platelet activation models under constant and dynamic shear stress loading conditions relevant to devices. *Ann. Biomed. Eng.*, 41:1279–96, 2013.

[195] J. Sheriff, P.L. Tran, M. Hutchinson, T. DeCook, M.J. Slepian, D. Bluestein, and J. Jesty. Repetitive hypershear activates and sensitizes platelets in a dose-dependent manner. *Artif. Organs*, 40:586–95, 2016.

[196] S. Shiozaki, S. Takagi, and S. Goto. Prediction of molecular interaction between platelet glycoprotein ibalpha and von willebrand factor using molecular dynamics simulations. *J. Atheroscler. Thromb.*, 23:455–64, 2016.

[197] C.N. Shrimpton, G. Borthakur, S. Larrucea, M.A. Cruz, J.F. Dong, and J.A. Lopez. Localization of the adhesion receptor glycoprotein ib-ix-v complex to lipid rafts is required for platelet adhesion and activation. *J. Exp. Med.*, 196:1057–66, 2002.

[198] C.A. Siedlecki, B.J. Lestini, K.K. Kottke-Marchant, S.J. Eppell, D.L. Wilson, and R.E. Marchant. Shear-dependent changes in the three-dimensional structure of human von willebrand factor. *Blood*, 88:2939–50, 1996.

[199] J.M. Siegel, C.P. Markou, D.N. Ku, and S.R. Hanson. A scaling law for wall shear rate through an arterial stenosis. *J. Biomech. Eng.*, 116:446–51, 1994.

[200] C.E. Sing and A. Alexander-Katz. Elongational flow induces the unfolding of von willebrand factor at physiological flow rates. *Biophys. J.*, 98:L35–7, 2010.

[201] S. Sivanesan, T.V. How, R.A. Black, and A. Bakran. Flow patterns in the radiocephalic arteriovenous fistula: an in vitro study. *J. Biomech.*, 32:915–25, 1999.

[202] T. Skorczewski, L.C. Erickson, and A.L. Fogelson. Platelet motion near a vessel wall or thrombus surface in two-dimensional whole blood simulations. *Biophys. J.*, 104:1764–72, 2013.

[203] M.J. Slepian, J. Sheriff, M. Hutchinson, P. Tran, N. Bajaj, J.G.N. Garcia, S. Scott Saavedra, and D. Bluestein. Shear-mediated platelet activation in the free flow: Perspectives on the emerging spectrum of cell mechanobiological mechanisms mediating cardiovascular implant thrombosis. *J. Biomech.*, 50:20–25, 2017.

[204] J.S. Soares, C. Gao, Y. Alemu, M. Slepian, and D. Bluestein. Simulation of platelets suspension flowing through a stenosis model using a dissipative particle dynamics approach. *Ann. Biomed. Eng.*, 41:2318–33, 2013.

[205] J.S. Soares, J. Sheriff, and D. Bluestein. A novel mathematical model of activation and sensitization of platelets subjected to dynamic stress histories. *Biomech. Model Mechanobiol.*, 12:1127–41, 2013.

[206] F. Sotiropoulos, T. Bao Le, and Anvar Gilmanov. Fluid mechanics of heart valves and their replacements. *Annu. Rev. Fluid Mech.*, 48:259–283, 2016.

[207] T.J. Stalker, E.A. Traxler, J. Wu, K.M. Wannemacher, S.L. Cermignano, R. Voronov, S.L. Diamond, and L.F. Brass. Hierarchical organization in the hemostatic response and its relationship to the platelet-signaling network. *Blood*, 121:1875–85, 2013.

[208] E. Di Stasio and R. De Cristofaro. The effect of shear stress on protein conformation: Physical forces operating on biochemical systems: The case of von willebrand factor. *Biophys. Chem.*, 153:1–8, 2010.

[209] S.P. Sutera, M.D. Nowak, J.H. Joist, D.J. Zeffren, and J.E. Bauman. A programmable, computer-controlled cone plate viscometer for the application of pulsatile shear-stress to platelet suspensions. *Biorheology*, 25:449–459, 1988.

[210] C.R. Sweet, S. Chatterjee, Z. Xu, K. Bisordi, E.D. Rosen, and M. Alber. Modelling platelet-blood flow interaction using the subcellular element langevin method. *J. R. Soc. Interface*, 8:1760–71, 2011.

[211] G.J. Tangelder, H.C. Teirlinck, D.W. Slaaf, and R.S. Reneman. Distribution of blood platelets flowing in arterioles. *Am. J. Physiol.*, 248:H318–23, 1985.

[212] C. Tersteeg, H.F. Heijnen, A. Eckly, G. Pasterkamp, R.T. Urbanus, C. Maas, I.E. Hoefer, R. Nieuwland, R.W. Farndale, C. Gachet, P.G. de Groot, and M. Roest. Flow-induced protrusions (fliprs): a platelet-derived platform for the retrieval of microparticles by monocytes and neutrophils. *Circ. Res.*, 114:780–91, 2014.

[213] J.N. Thon, H. Macleod, A.J. Begonja, J. Zhu, K.C. Lee, A. Mogilner, J.H. Hartwig, and J.E. Italiano. Microtubule and cortical forces determine platelet size during vascular platelet production. *Nat. Commun.*, 3:852, 2012.

[214] A.W. Tilles and E.C. Eckstein. The near-wall excess of platelet-sized particles in blood flow: its dependence on hematocrit and wall shear rate. *Microvasc. Res.*, 33:211–23, 1987.

[215] L.H. Ting, S. Feghhi, N. Taparia, A.O. Smith, A. Karchin, E. Lim, A.S. John, X. Wang, T. Rue, N.J. White, and N.J. Sniadecki. Contractile forces in platelet aggregates under microfluidic shear gradients reflect platelet inhibition and bleeding risk. *Nat. Commun.*, 10:1204, 2019.

[216] A. Tosenberger, F. Ataullakhanov, N. Bessonov, M. Panteleev, A. Tokarev, and V. Volpert. Modelling of thrombus growth in flow with a dpd-pde method. *J. Theor. Biol.*, 337:30–41, 2013.

[217] F.J. Tovar-Lopez, V.M. Dominguez-Hernandez, P. Diez-Garcia Mdel, and V.M. Araujo-Monsalvo. Finite-element analysis of the effect of basic hip movements on the mechanical stimulus within a proximal femur. *Rev. Invest. Clin.*, 66:S32–8, 2014.

[218] L. Tran, K. Mottaghy, S. Arlt-Korfer, C. Waluga, and M. Behbahani. An experimental study of shear-dependent human platelet adhesion and underlying protein-binding mechanisms in a cylindrical couette system. *Biomed. Tech. (Berl)*, 62:383–392, 2017.

[219] V.T. Turitto, H.J. Weiss, and H.R. Baumgartner. The effect of shear rate on platelet interaction with subendothelium exposed to citrated human blood. *Microvasc. Res.*, 19:352–65, 1980.

[220] V.T. Turitto, H.J. Weiss, and H.R. Baumgartner. Platelet interaction with rabbit subendothelium in von willebrand's disease: altered thrombus formation distinct from defective platelet adhesion. *J. Clin. Invest.*, 74:1730–41, 1984.

[221] V.T. Turitto, H.J. Weiss, T.S. Zimmerman, and II. Sussman. Factor viii/von willebrand factor in subendothelium mediates platelet adhesion. *Blood*, 65:823–31, 1985.

[222] K. Vahidkhah, S.L. Diamond, and P. Bagchi. Platelet dynamics in three-dimensional simulation of whole blood. *Biophys. J.*, 106:2529–40, 2014.

[223] H.H. Versteeg, J.W. Heemskerk, M. Levi, and P.H. Reitsma. New fundamentals in hemostasis. *Physiol. Rev.*, 93:327–58, 2013.

[224] W. Wang and M.R. King. Multiscale modeling of platelet adhesion and thrombus growth. *Ann. Biomed. Eng.*, 40:2345–54, 2012.

[225] W. Wang, N.A. Mody, and M.R. King. Multiscale model of platelet translocation and collision. *J. Comput. Phys.*, 244:223–235, 2013.

[226] H.J. Weiss. Flow-related platelet deposition on subendothelium. *Thromb. Haemost.*, 74:117–22, 1995.

[227] P.J. Wellings and D.N. Ku. Mechanisms of platelet capture under very high shear. *Cardiovasc. Eng. Tech.*, 3:161–170, 2012.

[228] A.M. Wendelboe and G.E. Raskob. Global burden of thrombosis: Epidemiologic aspects. *Circ. Res.*, 118:1340–7, 2016.

[229] J.G. White. Platelet structure. In A.D. Michelson, editor, *Platelets*, pages 45–73. Academic Press, San Diego, CA, USA, 2007.

[230] J.G. White and G.H. Rao. Microtubule coils versus the surface membrane cytoskeleton in maintenance and restoration of platelet discoid shape. *Am. J. Pathol.*, 152:597–609, 1998.

[231] D.M. Wootton and D.N. Ku. Fluid mechanics of vascular systems, diseases, and thrombosis. *Annu. Rev. Biomed. Eng.*, 1:299–329, 1999.

[232] Z. Wu, Z. Xu, O. Kim, and M. Alber. Three-dimensional multi-scale model of deformable platelets adhesion to vessel wall in blood flow. *Philos. Trans A Math Phys. Eng. Sci.*, 372:20130380, 2014.

[233] L.J. Wurzinger, P. Blasberg, and H. Schmid-Schonbein. Towards a concept of thrombosis in accelerated flow: rheology, fluid dynamics, and biochemistry. *Biorheology*, 22:437–50, 1985.

[234] L.J. Wurzinger, R. Opitz, M. Wolf, and H. Schmid-Schonbein. Ultrastructural investigations on the question of mechanical activation of blood platelets. *Blut*, 54:97–107, 1987.

[235] Z. Xu, N. Chen, M.M. Kamocka, E.D. Rosen, and M. Alber. A multiscale model of thrombus development. *J. R. Soc. Interface*, 5:705–22, 2008.

[236] Z. Xu, M. Kamocka, M. Alber, and E.D. Rosen. Computational approaches to studying thrombus development. *Arterioscler. Thromb. Vasc. Biol.*, 31:500–505, 2011.

[237] T. Yago, J. Lou, T. Wu, J. Yang, J.J. Miner, L. Coburn, J.A. Lopez, M.A. Cruz, J.F. Dong, L.V. McIntire, R.P. McEver, and C. Zhu. Platelet glycoprotein ibalpha forms catch bonds with human wt vwf but not with type 2b von willebrand disease vwf. *J. Clin. Invest.*, 118:3195–207, 2008.

[238] A. Yazdani and G.E. Karniadakis. Sub-cellular modeling of platelet transport in blood flow through microchannels with constriction. *Soft Matter*, 12:4339–51, 2016.

[239] A. Yazdani, H. Li, J.D. Humphrey, and G.E. Karniadakis. A general shear-dependent model for thrombus formation. *PLoS Comput. Biol.*, 13:e1005291, 2017.

[240] A. Yazdani, P. Zhang, J. Sheriff, M.J. Slepian, Y. Deng, and D. Bluestein. Multiscale modeling of blood flow-mediated platelet thrombosis. In W. Andreoni and S. Yip, editors, *Handbook of Materials Modeling: Applications: Current and Emerging Materials*, pages 1–32. Springer International Publishing, 2018.

[241] Y. Yuan, S. Kulkarni, P. Ulsemer, S.L. Cranmer, C.L. Yap, W.S. Nesbitt, N. Mistry I. Harper, S.M. Dopheide, S.C. Hughan, D. Williamson, C. de la Salle, H.H. Salem, F. Lanza, and S.P. Jackson. The von willebrand factor-glycoprotein ib/v/ix interaction induces actin polymerization and cytoskeletal reorganization in rolling platelets and glycoprotein ib/v/ix-transfected cells. *J. Biol. Chem.*, 274:36241–51, 1999.

[242] B.M. Yun, D.B. McElhinney, S. Arjunon, L. Mirabella, C.K. Aidun, and A.P. Yoganathan. Computational simulations of flow dynamics and blood damage through a bileaflet mechanical heart valve scaled to pediatric size and flow. *J. Biomech.*, 47:3169–77, 2014.

[243] B.M. Yun, J. Wu, H.A. Simon, S. Arjunon, F. Sotiropoulos, C.K. Aidun, and A.P. Yoganathan. A numerical investigation of blood damage in the hinge area of aortic bileaflet mechanical heart valves during the leakage phase. *Ann. Biomed. Eng.*, 40:1468–85, 2012.

[244] M.S. Zakaria, F. Ismail, M. Tamagawa, A.F.A. Aziz, S. Wiriadidjaja, A.A. Basri, and K.A. Ahmad. Review of numerical methods for simulation of mechanical heart valves and the potential for blood clotting. *Med. Biol. Eng. Comput.*, 55:1519–1548, 2017.

[245] J.M. Zhang, T. Luo, S.Y. Tan, A.M. Lomarda, A.S. Wong, F.Y. Keng, J.C. Allen, Y. Huo, B. Su, X. Zhao, M. Wan, G.S. Kassab, R.S. Tan, and L. Zhong. Hemodynamic analysis of patient-specific coronary artery tree. *Int. J. Numer. Method Biomed. Eng.*, 31:e02708, 2015.

[246] J.N. Zhang, A.L. Bergeron, Q. Yu, C. Sun, L. McBride, P.F. Bray, and J.F. Dong. Duration of exposure to high fluid shear stress is critical in shear-induced platelet activation-aggregation. *Thromb. Haemost.*, 90:672–8, 2003.

[247] J.N. Zhang, A.L. Bergeron, Q. Yu, C. Sun, L.V. McIntire, J.A. Lopez, and J.F. Dong. Platelet aggregation and activation under complex patterns of shear stress. *Thromb. Haemost.*, 88:817–21, 2002.

[248] P. Zhang, C. Gao, N. Zhang, M.J. Slepian, Y. Deng, and D. Bluestein. Multiscale particle-based modeling of flowing platelets in blood plasma using dissipative particle dynamics and coarse grained molecular dynamics. *Cell Mol. Bioeng.*, 7:552–574, 2014.

[249] P. Zhang, L. Zhang, M.J. Slepian, Y. Deng, and D. Bluestein. A multiscale biomechanical model of platelets: Correlating with in-vitro results. *J. Biomech.*, 50:26–33, 2017.

[250] W. Zhang, W. Deng, L. Zhou, Y. Xu, W. Yang, X. Liang, Y. Wang, J.D. Kulman, X.F. Zhang, and R. Li. Identification of a juxtamembrane mechanosensitive domain in the platelet mechanosensor glycoprotein ib-ix complex. *Blood*, 125:562–9, 2015.

[251] X. Zhang, K. Halvorsen, C.Z. Zhang, W.P. Wong, and T.A. Springer. Mechanoenzymatic cleavage of the ultralarge vascular protein von willebrand factor. *Science*, 324:1330–4, 2009.

[252] H. Zhao and E.S.G. Shaqfeh. Shear-induced platelet margination in a microchannel. *Phys. Rev. E*, 83:061924, 2011.

[253] H. Zhao, E.S.G. Shaqfeh, and V. Narsimhan. Shear-induced particle migration and margination in a cellular suspension. *Phys. Fluids*, 24:011902, 2012.

[254] J.J. Zwaginga, G. Nash, M.R. King, J.W. Heemskerk, M. Frojmovic, M.F. Hoylaerts, K.S. Sakariassen, and S.S.C. of the Isth Biorheology Subcommittee of 2006. Flow-based assays for global assessment of hemostasis. Part 1: Biorheologic considerations. *J. Thromb. Haemost.*, 4:2486–7, 2006.

[255] A.L. Zydney and C.K. Colton. Augmented solute transport in the shear-flow of a concentrated suspension. *Physicochemical Hydrodynamics*, 10:77–96, 1988.

CHAPTER 8

Blood suspension in a network

Sylvie Lorthois

IMFT, CNRS, Univ Toulouse, Toulouse, France

CONTENTS

THE MAIN FUNCTION of the blood vascular system in higher vertebrates is to ensure and regulate the delivery of oxygen and nutrients to every cell in the peripheral tissues, as well as the removal of their metabolic waste. In this system, the arteries carry blood away from the heart through a divergent arborescence; the capillaries are the main sites for exchanges between blood and tissue and have therefore to supply the entire volume of the organism; and the veins carry blood back to the heart through a convergent arborescence. Functionally, these compartments are organized in series: blood flow proceeds from the upstream arteries to the downstream veins through the capillaries.

In the microcirculation, however, their spatial organization is much more complex, with sizes, lengths, shapes and connectivity of vessels that are very specific to the perfused tissue. Since its first descriptions by Jean-Marie Poiseuille almost two centuries ago [44], this complexity was widely characterized, as summarized and illustrated in Part 1 of the present chapter.

The physics of blood flow in these small vessels is dominated by the interactions between blood cells, plasma and blood vessel walls ([18]). This leads to tremendous spatial and temporal variations and to a large variability of the associated suspension dynamics from vessel to vessel. The comprehension of the basic mechanisms underlying the emergence of this tremendous heterogeneity at the scale of microvascular networks is still largely incomplete. They involve strong couplings between network architecture and blood flow dynamics, some of which will be described in the present chapter. Moreover, its physiological role remains largely unknown.

Many excellent reviews on blood flow in microvascular networks are available, mainly based on the contribution of pioneers who investigated superficial two-dimensional networks, such as the cremaster muscle or the mesentery, through a combination of intravital microscopy, model experiments and mathematical modeling (*e.g.* [19, 34, 45, 51, 63]. In the last decade, a renewed interest in the field has been driven by the facilitated access to microfluidics and High Performance Computing, which enabled novel studies focusing on the collective behavior of RBCs in smaller, often model, networks and their associated emergent properties at network scale (*e.g.* [4, 5, 10, 32, 58, 67]. The main advantages of this strategy are that 1) the focus on relatively small networks enables us to spatially resolve the individual RBCs and 2) the network architecture can be controlled, thus eliminating any change induced by biological compensatory mechanisms (*e.g.* short term diameter variations). As a result, it provides a wealth of information on the physical mechanisms by which network architecture impacts blood flow dynamics. Finally, technological breakthroughs (*e.g.* X-ray microtomography, confocal or multi-photon microscopy) enabled the anatomical or functional imaging of highly complex three-dimensional microvascular networks, including the microvascular network of the brain (*e.g.* [8, 24, 43, 74]). The latter is essential to a large variety of physiological processes in the central nervous system and plays a prominent role in the associated mechanisms leading to disease (stroke, neurodegenerative diseases...). This attracted a large and dynamic community of investigators devoted to understanding the relationships between processes occurring at the cellular and molecular levels in the brain (*e.g.* neurovascular coupling) and large scale hemodynamic, transport and regulation properties (*e.g.* [6, 40, 59, 69]).

The goal of the present chapter is not to provide an exhaustive and synthetic review of the state-of-the art. Rather, the current understanding of the physical ingredients involved in the couplings between network architecture and blood flow dynamics, and associated limitations, will be underlined. In line with this objective, the present chapter will be organized as follows. First, key elements about the architectural organization of microvascular networks will be presented in Section 8.1. Second, the suspension behavior of blood in simple components of a network, from single vessels to bifurcations, will be considered in Section 8.2. Third, how these mechanisms contribute to the heterogeneity of blood flow at the scale of microvascular networks will be discussed in Section 8.3.

8.1 BACKGROUND ELEMENTS ABOUT THE ARCHITECTURAL ORGANIZATION OF MICROVASCULAR NETWORKS AND IMPACT ON BLOOD FLOW

Microvascular networks are able to ensure adequate blood supply, feeding every cell in peripheral tissues under a wide range of physiologic conditions and varying metabolic needs. As stated in the Introduction, the capillary vessels, the smallest vessels of 4-10 μm diameters, are the main sites for exchanges between blood and tissue and have to supply the entire volume of the organism. Healthy capillary networks, despite being highly variable among organs [3], are therefore ubiquitously dense, mesh-like, interconnected structures, which are space-filling above a characteristic length-scale [20, 22, 36], as illustrated in Figures 8.1 and 8.2. This characteristic length-scale is of order $25\mu m$-$75\mu m$ in the case of the brain, as demonstrated by Lorthois and Cassot [36]. This roughly corresponds to the mean length of capillary vessels and ensures that no point in the tissue is on average further than half this characteristic length from the nearest vessel, consistent with the diffusion-limited distance

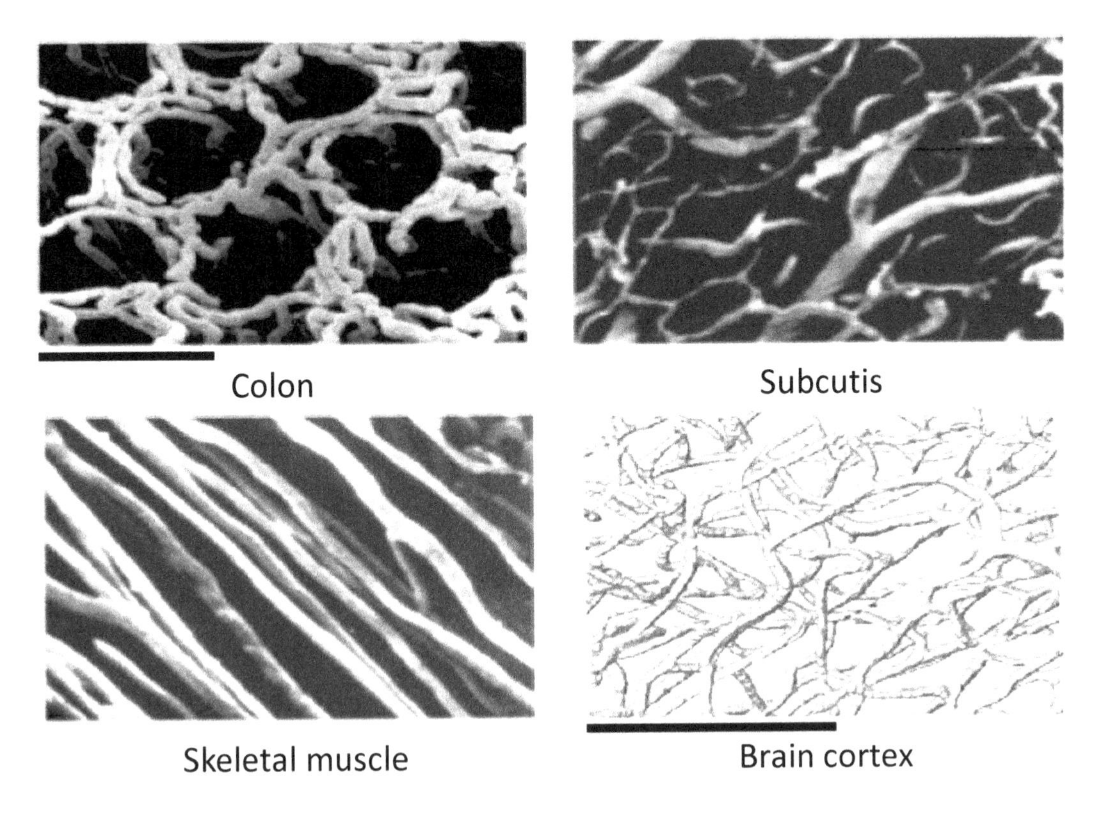

FIGURE 8.1
Close-up of the capillary network of various organs: colon, skin or muscle acquired by scanning electron microscopy, from [76] or brain acquired by confocal microscopy, from [36], with permissions. Scale bars represent 200 μm.

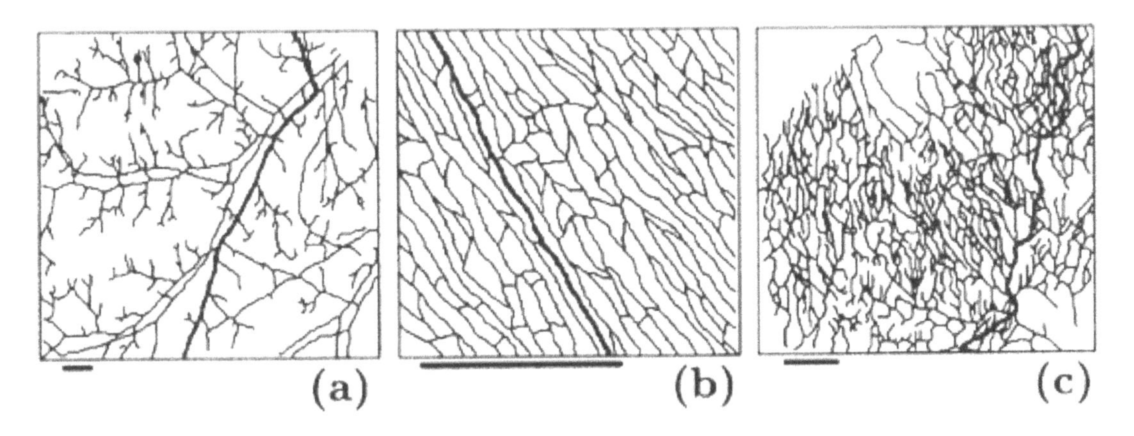

FIGURE 8.2
Typical skeletonized images of two vascular networks at various spatial resolutions: (a) Normal subcutaneaous arterioles and venules; (b) normal subcutaneaous capillary network; (c) tumor network. In (a) and (c), the spatial resolution is not sufficient to visualize the capillary vessels. Scale bars represent 500 μm. In (a) and (c), the spatial resolution is not sufficient to visualize the capillary vessels. From [22] with permission.

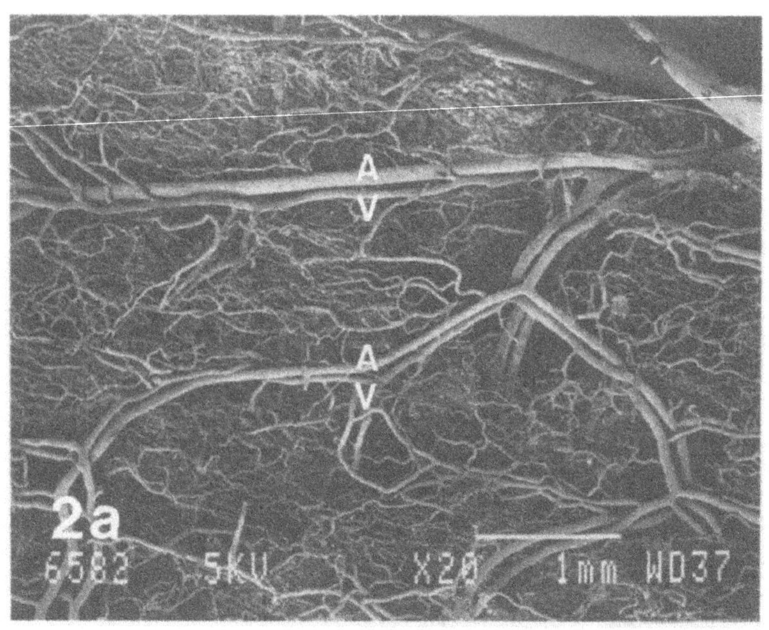

FIGURE 8.3
The blood supply of the human temporalis muscle: many venules (V) run alongside arterioles (A). Scanning electron microscopy, from [9], Journal of Anatomy, Anatomical Society. Reprinted with permission from Wiley Publishing, Inc. Scale bar represents 1000 μm.

of oxygen transport in the oxygen consuming tissue. This homogeneous mesh-like structure gives the capillaries an extremely large surface area facilitating their vital role in nutrient exchange [45].

While arterioles, capillaries and venules are organized in series from a functional point of view (blood flow proceeds from the upstream arteries to the downstream veins through the capillaries), their spatial organization is much more complex. First, arterial and venous trees are hierarchical quasi-fractal branching structures (Figure 8.2a) covering a wide range of diameter scales, from centimeters at the level of the heart to tens of micrometers at their micro-vascular (arteriolar or venular) extremities. Second, arterial and venous trees interdigitate, with numerous veins observed paralleling the arteries at various scales [2], as illustrated in Figure 8.3. Third, despite their close proximity in space, arterial and venous trees are not directly connected to each other, except in pathological situations. Instead, they are connected by their distal extremities through the capillary network, the smallest vessels of 4-10 μm diameters (Figure 8.4, right). Their microvascular parts, *i.e.*, the arteriolar and venular trees, as well as the capillary network, are thus embedded within the organ to be supplied [50] (Figure 8.4, left). Fourth, there is a marked variability of capillary patterns among organs, as illustrated in Figure 8.1. Finally, many pathological conditions, such as atherosclerosis, cancers, arterio-venous malformations, infections, stroke, hypertension, diabetes, obesity and Alzheimer's disease, as well as normal aging, induce changes to vessels' morphology or spatial organization (*e.g.* Figure 8.2c).

Not surprisingly, blood flow in such complex networks is highly heterogeneous. First, due to spatial constraints, the main feeding and draining vessels are necessarily closer to some parts of the region to be supplied and more distant from others. The resulting dispersion in the length of microvascular pathways between these feeding and draining vessels leads to dispersion in pressure drops per length along these pathways [34], so that higher

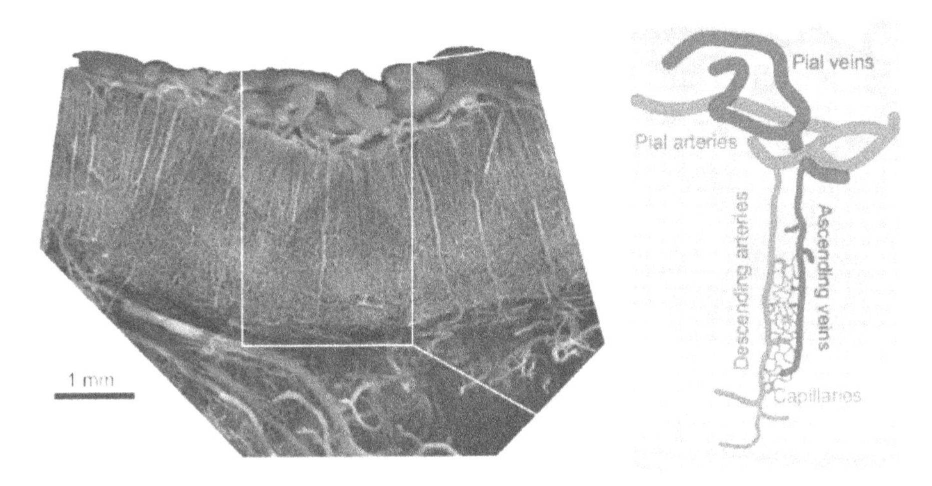

FIGURE 8.4
The blood supply of the cerebral cortex: Left: Scanning electron micrograph of a vascular corrosion cast from the monkey visual cortex (primary visual cortex). Arteries are shaded in red and veins are blue. Capillaries, in light grey, fill the whole volume of the cortex. Right: Schematic representation of the cortical vasculature. Arterioles and venules are connected by their distal extremities through the capillary vessels, in green. From Schmid et al. (2017) [61], with permission.

flow velocities are expected on shorter pathways [50]. Second, most of these flow pathways involve at least an arteriole, a capillary and a venule, *i.e.*, vessel diameters typically ranging from about 4 μm to 100 μm. This represents half to ten times the characteristic size of single red blood cells (RBCs), *i.e.*, a very large range in RBC confinement. Third, microvascular networks exhibit huge spatial and temporal variations in RBC concentration, with some vessels containing no blood cells and some others containing almost no suspending plasma, as illustrated in Figure 8.5 [13,44,52,77]. Thus, various RBC deformations and flow regimes, from single-file or zipper RBC flows in narrow capillaries to highly packed RBCs in larger vessels, can coexist in a given network [21,33]. This induces a large variability of the associated suspension dynamics from vessel to vessel. Finally, microvascular networks exhibit striking temporal dynamics, including extreme unsteady events, where a vessel or a series of vessels are temporarily filled or depleted with RBCs.

In summary, strong couplings between network architecture and RBC flow dynamics underlie the emergence of a tremendous hemodynamic heterogeneity at the scale of microvascular networks. While the comprehension of the mechanisms involved is still largely incomplete, several basic mechanisms associated with smaller scales have been shown to play a role, as described in the next section.

8.2 BASIC MECHANISMS OF FLOW STRUCTURATION IN MICROVASCULAR NETWORKS

Here, we consider RBC flows in the elementary bricks of a network, from single vessels to individual bifurcations. Following Pries et al. (1996) [51], we focus on regimes that can be encountered in physiology, that is, dilute to concentrated RBC suspensions flowing in vessels with diameter typically ranging from about 4 μm to 100 μm, in the limit of large shear rates. In these regimes, the Reynolds and Womerseley numbers are both much lower than one, so that blood flow in each vessel is governed by the pressure gradient and viscous shear stresses, and not by inertial forces.

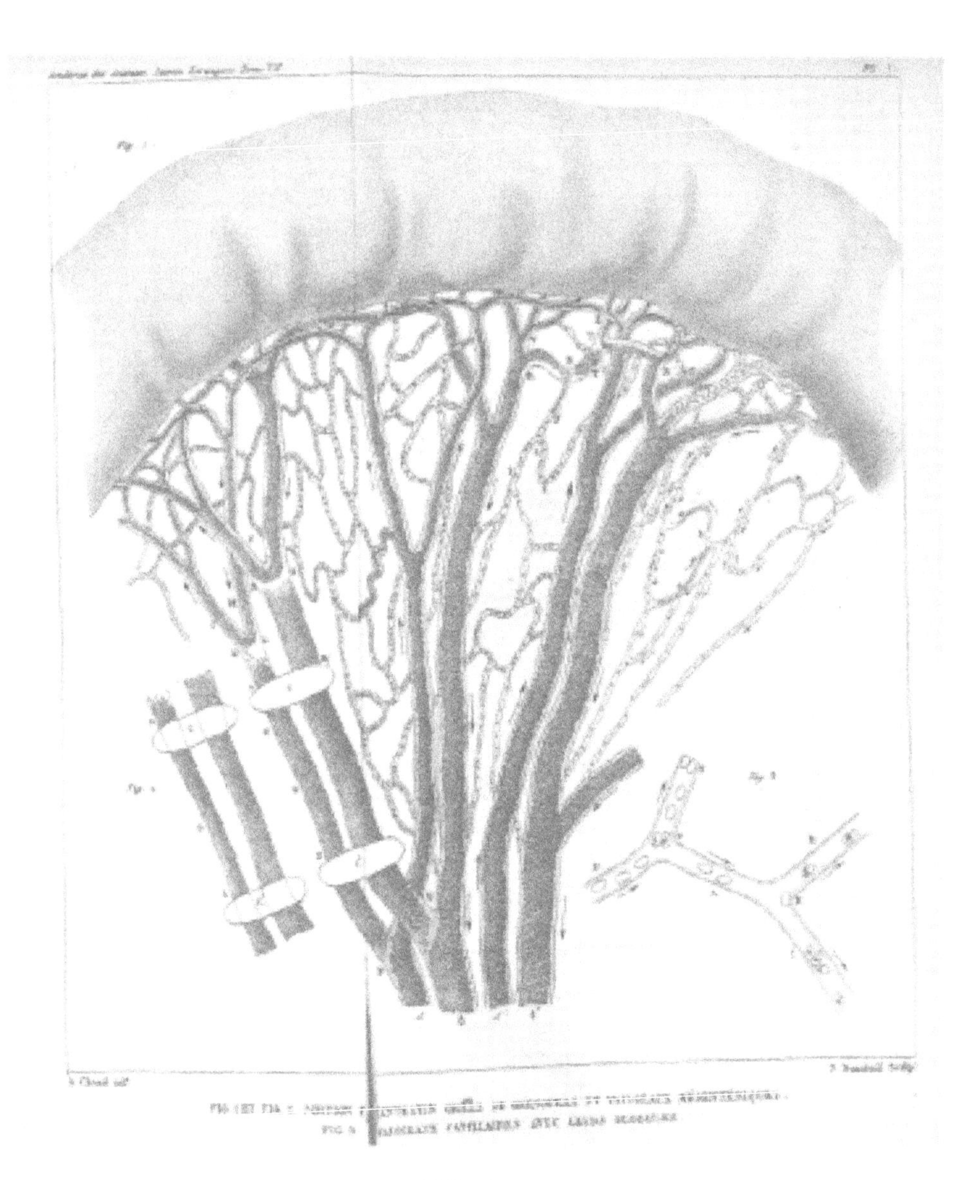

FIGURE 8.5
Heterogeneity of RBC concentration at vessel and network scale, as illustrated by Poiseuille. Original caption (translated from French): *"Fig. 1 and Fig. 2: Portion of frog's small intestine and mesenteric vessels. Fig.3. Capillary vessels and their RBCs"*. From [44].

8.2.1 Structuration and rheology at vessel scale

Details about the collective dynamics of RBCs as deformable cells interacting with the suspending fluid and the vessel walls in a confined environment have been given in Chapters 3, 4, 5 and 6. Here, we focus on the main aspects of RBC flow structuration needed to interpret the rheology of blood flow at vessel scale. The central concepts for that purpose are those of apparent and relative viscosities[(i)]. These concepts have been defined precisely

(i)The relative viscosity (sometimes also referred to as the relative apparent viscosity) is the apparent viscosity normalized by the viscosity of plasma, the suspending fluid.

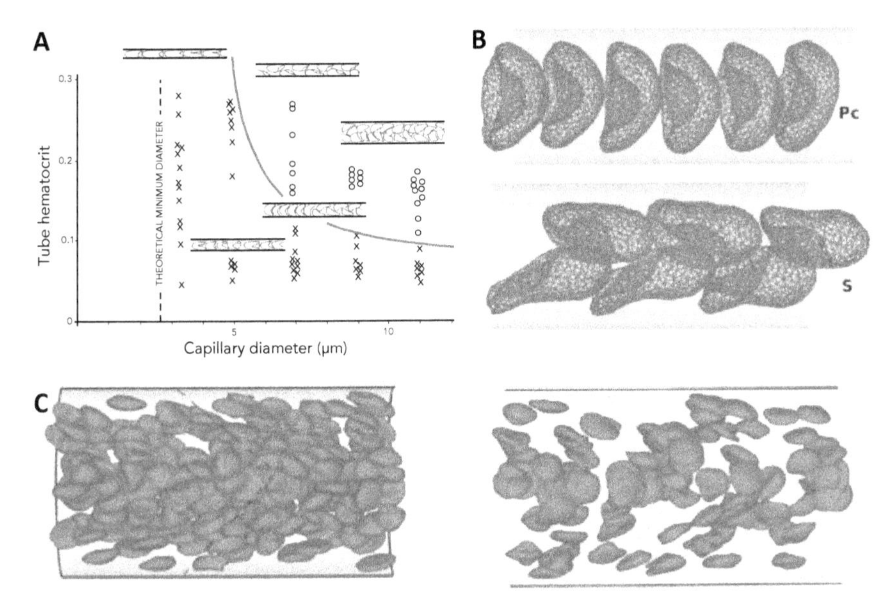

FIGURE 8.6
Typical flow structures at vessel scale. A: Transition from single-file (x) to multi-file (o) as a function of hematocrit for various tube diameters, adapted from [21]. The zipper configuration appears for $7\mu m$ tubes at high hematocrit. B: It is retrieved by numerical simulations of the collective motion of RBCs, which also capture the cell-free layer, from [33]. C: Central cut-plane snapshots along the tube axis for $D = 40\mu m$ at $H_T = 0.3$. Left: half-tube images; Right: thin slices across the cut. CFL thickness is shown by dashed lines parallel to the walls. From [31], with permission.

in the past, but they are often used without further pointing out that they represent spatially and temporally averaged effective parameters characterizing the global viscous dissipation at vessel scale. Adopting this point of view provides an efficient alternative to a vision where the corpuscular nature of blood flow prevents the introduction of any viscosity concept[(ii)]. In particular, as emphasized by Fung (1981) [17], *"apparent and relative viscosities are not intrinsic properties of the blood; they are properties of the blood and blood vessel interaction".*

These mechanical interactions generally result in the formation of a plasma layer or a region of reduced hematocrit adjacent to the vessel walls, the so-called cell-free or cell-depleted layer (see *e.g.* [51, 63]. This layer has been observed *in vitro* or *in vivo* for a large range of diameters and hematocrits and results in an increased concentration of red cells in the central region of the vessel. For the smallest vessels, with diameters typically below 10 μm or slightly larger, RBCs flow in single-file at small hematocrits and progressively adopt a zipper configuration for increasing hematocrits (Figure 8.6A), as confirmed by recent computational fluid dynamics simulations that include a realistic representation of their mechanical properties (Figure 8.6B). For larger vessels, RBCs do not usually flow in single file, adopting a much more disordered configuration that still conserves a cell-free layer, as apparent in Figure 8.6C. Noteworthy, in *in vitro* situations for which the largest number of experimental data points is available, the thickness of the cell-free layer normalized by the radius of the tube[(iii)] exhibits a striking change of behavior for tube diameters around 10 μm (Figure 8.7): below, this relative thickness increases with the tube diameter, while it

[(ii)]In the regime of microcirculation, the size of RBCs is not negligible compared to the size of blood vessels, which breaks off the continuum assumption.

[(iii)]For clarity, and when possible, *"vessel'* will be subsequently used to denote *in vivo* situations while *"tube"* or *"channel"* will be used to denote *in vitro* situations as appropriate. The latter term will mainly be

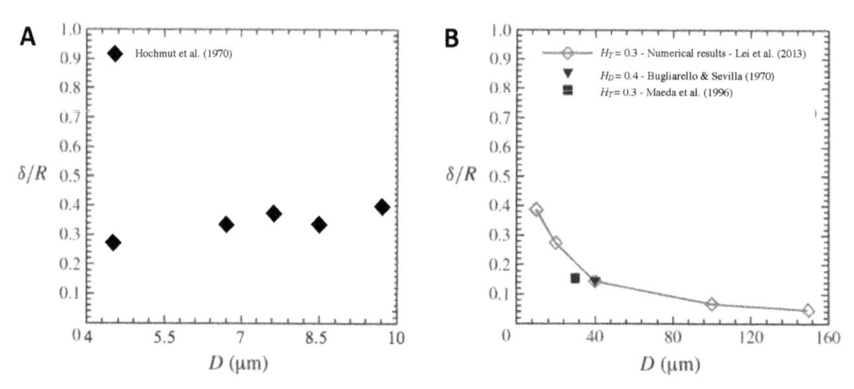

FIGURE 8.7
Thickness of the cell-free layer normalized by the tube radius as a function of tube diameter. Left. *In vitro* data obtained with human blood at physiological reservoir hematocrit by Hochmuth et al. (1970) [25] in the high shear limit in tubes with diameter below $10\mu m$. The local tube hematocrit is unknown. Right. Compilation of experimental data and *in silico* results obtained for a tube hematocrit of 0.3, figure adapted from [31], with permission.

decreases above, as supported by numerical results. Many authors thus consider the cell-free layer thickness to be roughly constant in the latter case (see *e.g.* [63]).

The presence of this cell-free layer strongly influences the overall flow behavior of blood in small vessels. Its first consequence is the bluntness of the velocity profile, which flattens in the central region and is steeper near the walls compared to a parabolic Poiseuille profile (Figure 8.8). This can be qualitatively understood by considering the following idealized situation: the steady, fully-developed axisymmetric pressure-driven flow of a continuous fluid with spatially variable viscosity $\mu(r)$ in a rectilinear tube[(iv)]. In such a flow, the pressure is uniform in the cross-section of the tube and the pressure gradient $\frac{\partial p}{\partial x}$ is invariant along the tube axis [23]. The mechanical equilibrium of a cylindrical fluid element of radius r and length l thus writes:

$$\pi r^2 l \frac{\partial p}{\partial x} = 2\pi r l \mu(r) \frac{\partial u}{\partial r}(r), \tag{8.1}$$

where the left-hand term corresponds to the sum of pressure forces applied on the upstream and downstream faces of the element ($\pi r^2(p_{upstream} - p_{downstream})$) and the right-hand term corresponds to the friction force on its outer surface ($2\pi r l \tau(r)$, $\tau(r)$ denoting the shear stress).

As a result, the product of the viscosity and the shear rate is inversely proportional to the radius, the constant of proportionality being controlled by the pressure difference applied to the tube ($\mu(r)\frac{\partial u}{\partial r}(r) = \frac{1}{2}\frac{\partial p}{\partial x}\frac{1}{r}$). For a spatially uniform viscosity, this results in a linear variation of the shear rate as a function of the radius, so that the velocity profile is parabolic (Poiseuille flow). In the case of blood, because of the cell-free layer, the viscosity locally decreases near the wall and locally increases in the central region. Thus, the velocity gradients must increase near the wall and decrease in the center, which explains the blunting of the velocity profile. For a given pressure gradient, this enables it to accommodate a larger flow rate than Poiseuille flow, which demonstrates a reduced global

used in relationship to microfluidic devices where the channels are usually rectangular or square, and not cylindrical.

(iv)This constitutes a very crude approximation of the situation as the local viscosity results from an interplay between the flow field and the shear-thinning properties of blood, so that the local viscosity is dependent on the local stresses. For the structure of the flow solution for power-law or Bingham fluids, see Section 4.6.3 in [23].

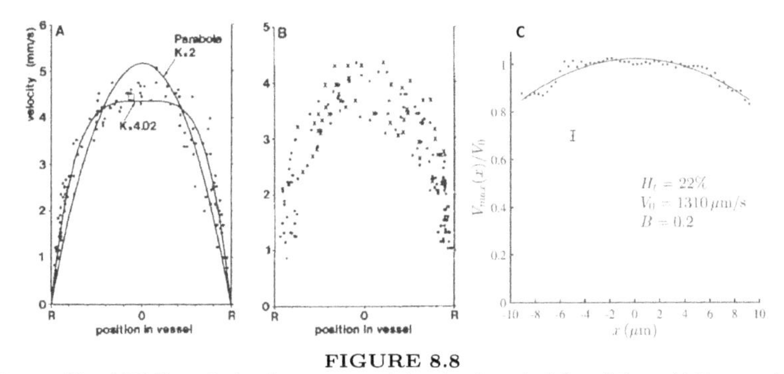

FIGURE 8.8
A and B: Velocity profile of RBCs and platelets measured *in vivo* in arterioles of the rabbit mesentery. A. Velocity profile of blood platelets flowing in a $32\mu m$ arteriole during the systolic phase. In addition, the best fits to $V(r) = V_{max}\left(1 - (r/R)^K\right)$, where R is the radius of the arteriole and V_{max} is the maximal velocity, are obtained for $K = 0.42$. Poiseuille velocity profile is plotted for comparison ($K = 2$). B. Velocity profiles of blood platelets (dots) and RBCs (crosses) flowing in a $24\mu m$ arteriole during the systolic phase. From [72], with permission. C. RBC velocity profile measured *in vitro* in the centerplane of a RBC maximal velocity profiles in a $20 \times 20\mu m^2$ square channel, normalized by the maximal velocity. Dots: experimental data. Solid line: best fit to $V = V_{max}\left(1 - B\left(x - H/2\right)^2\right)$), where H is the channel height. This profile demonstrates RBC slip at wall for decreasing channel size and increasing hematocrit. From [58], with permission.

viscous dissipation induced by the cell-free layer. This should materialize into a reduction of the apparent viscosity.

Before going further, it is worth going back to the concept of apparent viscosity. Let's define the apparent viscosity of blood in a given tube as the viscosity of the Newtonian fluid flowing in the same tube with the same flow rate when the same pressure difference is applied to both ends of the tube. Such a definition makes it clear why Fung (1981) [17] writes "*There are as many definitions for apparent viscosities as there are good formulas for well-defined problems. Examples are: [...] channel flow [...] and flow in a cylindrical tube. But if a vessel system has a geometry such that the theoretical problem for homogeneous fluid flow has not been solved, then we cannot derive an apparent viscosity for flow in such a system*".

For a cylindrical tube of length L and diameter D, the theoretical solution for a homogeneous Newtonian fluid of viscosity μ can be obtained from the well-known Poiseuille equation:

$$\mu = \frac{\pi D^4 \Delta P}{128 L Q} \tag{8.2}$$

where ΔP is the time-averaged pressure difference applied to both ends of the tube and Q is the time-averaged flow rate [35]. Thus, the apparent viscosity μ_{app} can be deduced from experiments where the blood flow is measured as a function of the pressure drop by identification with the previous equation. This does not imply a parabolic velocity profile (nor any other assumption on the shape of the velocity profile) for the flowing blood.

In addition, if this complex flow problem can be approximated by a simpler problem for which an analytical solution relating the blood flow to the pressure drop can be found, then, an analytical approximation of the apparent viscosity as a function of its parameters can be obtained. To illustrate this, let's now consider an even simpler two-fluid annular model for blood flow, the outer fluid having the viscosity of plasma μ_p and the inner core fluid having a larger viscosity μ_c [54, 63]. From the above reasoning, the velocity profile is parabolic in each region. Using the no-slip condition at wall and the continuity of shear stress at the

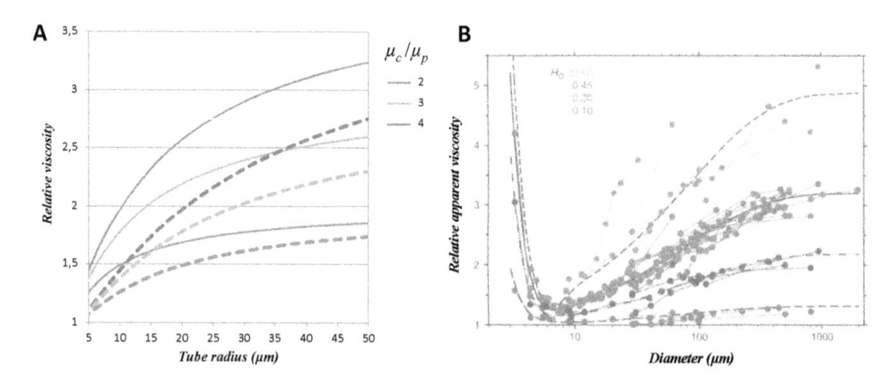

FIGURE 8.9
A. Relative viscosity as a function of tube radius for a two-fluid model with various values of the inner core to plasma layer viscosity ratio. Continuous lines: $\delta = 1\mu m$; dashed lines: $\delta = 2\mu m$. B. Experimental results compiled by [48], reprinted with permission from [50].

interface between the two fluids, this yields

$$v(r) = \begin{cases} \frac{\Delta P a^2}{4L}\left(\frac{\lambda^2 - r^2/a^2}{\mu_c} + \frac{1-\lambda^2}{\mu_p}\right) & \text{when} \quad r < \lambda a, \\ \frac{\Delta P a^2}{4L}\frac{1-r^2/a^2}{\mu_p} & \text{when} \quad r \geq \lambda a, \end{cases} \tag{8.3}$$

where a is the tube radius, δ is the thickness of the cell-free-layer and $\lambda = 1 - \delta/a$ is the ratio of the core radius to the tube radius. The total flow rate is obtained by integration of this velocity profile:

$$Q = Q_p + Q_c = \frac{\pi D^4 \Delta P}{128 \mu_p L}\left[1 - \lambda^4\left(1 - \frac{\mu_p}{\mu_c}\right)\right], \tag{8.4}$$

which gives:

$$\mu_{app} = \frac{\mu_p}{1 - \lambda^4\left(1 - \frac{\mu_p}{\mu_c}\right)}. \tag{8.5}$$

This apparent viscosity normalized by the viscosity of plasma is plotted in Figure 8.9A for constant values of δ, which roughly corresponds to the situation encountered in tubes with diameters above 10 μm. It shows a strong decrease of the relative viscosity with decreasing tube diameters, the order of magnitude of which is consistent with the *in vitro* experimental results displayed in Figure 8.9B. Initially reported by Fåhraeus and Lindqvist (1931), this well-known counterintuitive effect characteristic of blood flow in the microcirculation[(v)] is thus captured by the very crude analysis presented above. Thus, the second major consequence of the presence of the cell-free layer is the Fåhraeus-Lindqvist effect.

The inversion of the Fåhraeus-Lindqvist effect for tubes with diameters below 10 μm can be understood in the same way. In this case, the single-file motion of RBCs can be crudely approximated as a rigid-body motion, for which μ_c becomes infinite. As a result, from Eq. 8.5, when the cell-free layer vanishes because the tube diameter is so small that RBCs are maximally deformed, infinite apparent viscosity is obtained [28]. If we recall that, in this diameter range, the thickness of the cell-free layer also decreases (Figure 8.7A), the apparent viscosity then decreases with increasing tube diameter.

(v) Now called the Fåhraeus-Lindqvist effect.

More refined analytical models have been proposed (*e.g.* models where the central core is a Casson fluid [54], or models where the single-file flow of rigid or deformable particles are considered, *e.g.* [65, 70], but each of them provides predictions for limited diameter or hematocrit ranges, so that it is difficult to use them for studying blood flow in networks where larger ranges must be considered. By contrast, numerous experiments performed by several investigators in glass tubes[(vi)] allowed the apparent viscosity to be determined in such a large range (Symbols in Figure 8.9B). From these experimental results, Pries et al. (1992) [48] subsequently developed the following empirical relationship for the dependence of the relative apparent viscosity on the tube diameter and hematocrit:

$$\mu_{rel} = 1 + (\mu_{45} - 1))\,\frac{(1 - H_D)^C - 1}{(1 - 0.45)^C - 1}, \tag{8.6}$$

where

$$\mu_{45} = 220\exp(-1.3D) + 3.2 - 2.44\exp(-0.06D^{0.645}) \tag{8.7}$$

is the relative apparent *in vitro* blood viscosity for a discharge hematocrit $H_D = 0.45$, and D is the lumen diameter in micrometers[(vii)].

This latter equation, represented by Lines in 8.9B, was designed to asymptotically approach, with increasing diameter, a preset viscosity value at infinite tube diameter, to show a minimum in an intermediate diameter range and a steep increase in the smallest tubes. In the former, the parameter C describes the curvature of the relationship between relative apparent blood viscosity and hematocrit for a given tube diameter:

$$C = (0.8 + \exp(-0.075D))\left(-1 + \frac{1}{1 + 10^{-11}D^{12}}\right) + \frac{1}{1 + 10^{-11}D^{12}}. \tag{8.8}$$

With this expression, the relationship between viscosity and hematocrit is linear ($C = 1$) for tubes with diameter below $6\mu m$, consistent with theoretical predictions for single-file RBC flows for sufficiently small hematocrits (Fung 1981, [17]). For larger tubes ($C < 1$), the viscosity increases faster than the hematocrit because of the increased viscous dissipation induced by cell-cell interactions.

Noteworthy, *in vivo* experiments with direct measurement of the pressure difference in single vessels are extremely difficult and have been performed by only few investigators [34]. The results collected did not enable obtaining a comprehensive parametrization of the *in vivo* apparent viscosity as a function of the vessel diameter and hematocrit. Such a parametrization has nevertheless been indirectly obtained by considering blood flow at the scale of a whole microvascular network, as we shall see in Section 8.3.2.

The third consequence of the cell-free layer is the relative reduction of the tube hematocrit H_T, *i.e.*, the volume fraction of RBCs in blood within a given vessel, compared with the discharge hematocrit H_D, *i.e.*, the volume fraction of RBCs in blood flowing through the same vessel. This dynamic reduction of the microvascular hematocrit, or Fåhraeus effect, is quantified as the ratio between H_T and H_D. In single-file corpuscular approximations (*e.g.* [70]) or two-fluid models, this ratio is equal to the mean blood velocity divided by the mean RBC velocity, which is always greater than the mean blood velocity due to the

(vi) Mostly using human RBCs resuspended in plasma or whole human blood.

(vii) From dimensional analysis, one of the relevant non-dimensional parameters for the relative viscosity should be the ratio of the RBC characteristic size to the tube diameter. Thus, the previous equation can be adapted to other species by scaling the diameter-dependent terms with the cube root of the mean red blood cell volume [52]. Similarly, when *in vivo* dimensional data are available from rodent experiments, the resulting empirical expressions can be adapted to human blood [37, 58].

presence of the cell-free layer. In the same spirit as above, Pries et al. (1990) [52] introduced an empirical relationship for the Fåhraeus effect, adapted for human RBCs by Roman et al. (2016) [58], as follows:

$$\frac{H_T}{H_D} = H_D + (1 - H_D)(1 + 1.7\exp(-0.356D) - 0.6\exp(-0.009D)), \quad (8.9)$$

where the diameter D is once again expressed in micrometers[(viii)]. Whatever the hematocrit, the Fåhraeus effect is maximal (H_T/H_D is minimal) for a tube diameter $\sim 10\mu m$ in the same way as the cell-free layer.

The above empirical relationships do not include any dependence on the driving pressure gradient or, alternately, on the shear rate. The magnitude of the latter is often quantified by the ratio of mean blood velocity to vessel diameter, the so-called pseudo shear rate. From the above discussion, this implies that the shape of the velocity profile must be roughly independent on the driving pressure gradient. This is not straightforward[(ix)]. However, experimental results in glass tubes have shown that the apparent viscosity of RBCs resuspended in plasma is essentially constant whatever the tube diameter for pseudo-shear rates above $50s^{-1}$ by contrast to lower shear regimes, where RBC aggregates form [57]. Analysis of *in vivo* experimental data obtained at network scale have led to a similar result ([53], see Section 8.3.2). Thus, in physiological regimes, for which the pseudo-shear rate is typically above $50s^{-1}$, the flow rate in a given vessel is proportional to the pressure drop, the coefficient of proportionality $\frac{\pi D^4}{128\mu_{app}(D,H_D)L}$ being only dependent on the channel geometry and on the discharge hematocrit.

As a result, microvascular networks can be viewed as networks of interconnected tubes in which blood flow is described by a succession of linear relationships between the flow rate and the pressure drop[(x)] This, however, does not yield any heterogeneity of the discharge hematocrit. To understand this heterogeneity, the behavior of blood flow at diverging microvascular bifurcations must be considered. This is the object of the next section.

8.2.2 Phase separation at diverging microvascular bifurcations

Microvascular bifurcations are the smallest architectural components of a microvascular network where hematocrit heterogeneities can be observed. While in converging microvascular bifurcations, the discharge hematocrit in the outlet vessel o can be simply deduced from the discharge hematocrits in the two upstream vessels ($u1$ and $u2$) by mass conservation ($H_D^o Q^o = H_D^{u1} Q^{u1} + H_D^{u2} Q^{u2}$), this is not the case of diverging microvascular bifurcations. In such bifurcations, the distribution of RBCs and plasma is indeed non-proportional, one daughter vessel receiving a higher hematocrit than the feeding vessel, and the other one receiving a lower hematocrit [62]. This effect, known as the phase separation effect, has two distinct asymptotic behaviors depending on the diameters of vessels composing the bifurcation. In large blood vessels with uniform distribution of very small cells, the phase separation effect vanishes. In very narrow symmetric capillary branches, the cells always enter the vessel with the higher flow rate, so that the vessel with the lower flow rate is devoid of RBCs.

(viii) Note that the numerical values in the above equation have been updated here to correct for a misprint in [58]

(ix) For instance, in the case of a Bingham fluid, the radius of the central plug-flow region decreases with the driving pressure gradient, as a larger portion of the tube reaches the yield-stress of the fluid, see Section 4.6.3 in [23].

(x) In such a description, the singular pressure drops at the inlet regions of each vessel are neglected.

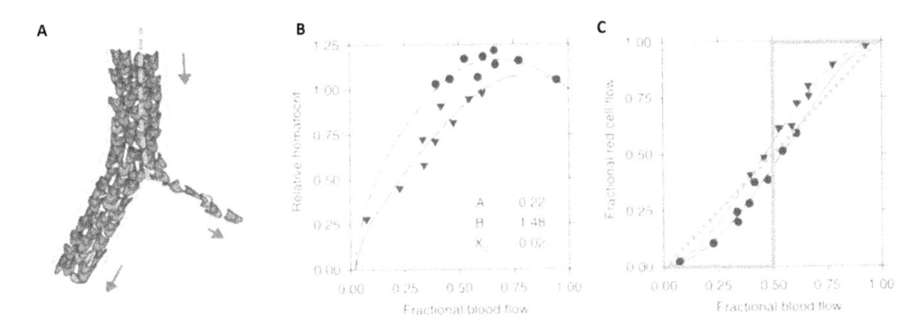

FIGURE 8.10
Phase separation. A: Schematic drawing of a microvascular bifurcation. The blue dotted line schematizes the fluid separating streamline. B. Hematocrit relative to that in the feeding vessel. C. Fractional erythrocyte flow in both daughter branches versus the fractional blood flow in that same branch. The symbols depict results of experimental measurements while the continuous lines and the parameters given correspond to the logit fit of the data obtained with Eq. 8.10, as reported in [51]. Blue dotted line: no phase separation; blue continuous line: asymptotic behavior for very narrow vessels. Adapted from [50], with permission.

Once again, if the upstream vessel is not too small, the main origin of the phase separation effect lies in the presence of the cell-free layer and its relative position with the fluid separating streamline, as illustrated on Figure 8.10A. The position of this fluid separating streamline is indeed controlled by the total flow rate partitioning between the daughter branches. For a symmetrical bifurcation with equal flow partitioning, it lies on the symmetry plane, so that the proportion of RBC flow entering both branches is equal to 50%. When the flow gets smaller in one of the side branches, the fluid separating streamline gets progressively closer to the side wall corresponding to this branch, in which the contribution of fluid originating from the cell-free layer thus increases (an effect which is often referred as plasma skimming). This non-linearly decreases the fraction of RBC flow entering this daughter branch. When the separating streamline lies within the cell-free layer, the low-flow branch is devoid of RBCs[(xi)].

The phase separation effect can be characterized in two different ways. The first one focuses on the deviation from unity of the relative discharge hematocrits (H_D^i/H_D^e) in both outlet branches i of the bifurcation compared to the one in the entry branch e (Figure 8.10B). However, mass conservation does not lead to a trivial relationship between these parameters, so that the results are difficult to interpret. Because phase separation vanishes when the proportion of RBCs flowing from the inlet branch to a given outlet branch is equal to the proportion of the blood flowing from the inlet branch to this same outlet branch, the second one focuses on the deviation of the fractional RBC flows in each daughter branch of the bifurcation ($FQ_{RBC}^i = Q_{RBC}^i/Q_{RBC}^e$), from the fractional blood flow entering the same branch ($FQ_{blood}^i = Q_{blood}^i/Q_{blood}^e$), see Figure 8.10C[(xii)]. Mass conservation obviously leads to $FQ_{RBC}^1 + FQ_{RBC}^2 = 1$ and $FQ_{blood}^1 + FQ_{blood}^2 = 1$ so that the relationship between FQ_{RBC}^2 and FQ_{blood}^2 (sometimes called the red cell distribution function for branch 2) can be deduced from that of branch 1 from a central symmetry around point (50%, 50%). In the first asymptotic regime described above (large vessels), $FQ_{RBC}^i = FQ_{blood}^i$ (blue dotted

---

[(xi)] In addition, RBCs may not follow the streamline associated with their mass (*e.g.* [5, 63]), so that the contribution of *plasma skimming* to phase separation can be modulated.

[(xii)] This latter representation yields smaller variations around the identity line than the former one, which exhibits larger variations of relative hematocrit around the constant value of one (Figure 8.10C vs. Figure 8.10B). In practice, following Roman et al. 2016 [58], it is therefore recommended to use both representations to compare a given parametric description of the phase separation effect to experimental data.

line in Figure 8.10C) while in the second one (small vessels), a step function is obtained (blue continuous line in Figure 8.10C).

Between these two regimes, time-averaged experimental data have been obtained *in vitro* and *in vivo* by many investigators (see [46, 58, 62] and references therein). Based on the previous experimental findings, Pries et al. (1989) [46] first demonstrated that, besides the fractional blood flow, the parameters primarily controlling the phase separation effect were: the feeding vessel to RBC size ratio, daughter vessels to feeding vessel size ratios and inlet discharge hematocrit, with minor influence of the bifurcation angle. They introduced an empirical relationship based on a logit function to reproduce 1) the observed sigmoid shape of the experimental measurements and 2) the threshold value X_0 for the fractional blood flow entering a given daughter branch below which no RBC enters this same branch [51]. This empirical relationship was later refined from the original *in vivo* experimental data in order to render predictions more robust for extreme combinations of input hematocrit and diameter distribution [49]. Adapted for human RBCs as before, it reads:

$$FQ^i_{RBC} = \begin{cases} 0 & \text{if} \quad FQ^i_{blood} < X_0, \\ 1 & \text{if} \quad FQ^i_{blood} > X_0, \\ \frac{1}{1+\exp\left[-\left(A+B\,\text{logit}\left(\frac{FQ^i_{blood}-X_0}{1-2X_0}\right)\right)\right]} & \text{else,} \end{cases} \tag{8.10}$$

where logit $= \ln\left(\frac{x}{1-x}\right)$ and X_0, A and B are non-dimensional parameters given by $X_0 = \underline{X_0} \times (1 - H_D)/D^e$ with $\underline{X_0} = 1.12\mu m$; $A = -\underline{A} \times \frac{D^i - D^j}{D^i + D^j} \times (1 - H_D)/D^e$ with $\underline{A} = 15.47\mu m$; and $B = 1 + \underline{B} \times (1 - H_D)/D^e$ with $\underline{B} = 8.13\mu m$. These non-dimensional parameters describe the threshold, asymmetry and sigmoidal shape (non-linearity) of the cell distribution function, respectively. All these parameters tend to zero in the limit of large vessel diameters or large hematocrits, so that phase separation vanishes. For given diameters, they vary linearly with the discharge hematocrit, with maximal separation in the limit of small hematocrits. In this limit, the threshold flow partitioning reaches 50% for $D_e = 2.24\mu m$, retrieving the behavior of very narrow symmetric bifurcations[xiii].

Noteworthy, the asymmetry parameter A equals zero for symmetric bifurcations, so that the cell distribution function is the same for both branches, and symmetrical around point (50%, 50%). For asymmetric bifurcations, A is negative for the larger branch and opposite for the smaller branch: for a given flow fraction, the smaller-diameter branch receives a higher hematocrit. Qualitatively, this behavior occurs because a small-diameter branch draws its flow from a small part of the periphery of the parent vessel, and more of its flow comes from the high-hematocrit core region of the parent vessel, compared to a large-diameter branch [63].

Following the same line of reasoning, asymmetry of phase separation can also be induced by asymmetry of the hematocrit profile in the feeding vessel. Such asymmetry can be generated by an upstream side branch. For example, a narrow side branch in the parent vessel may divert a small amount of plasma, creating a non-symmetric cell-free layer [46, 62, 66]. These authors however demonstrated a negligible impact on phase separation for upstream vessels lying further away from the bifurcation than 10 times the diameter of the entry branch, as the cells return to a symmetric position within such a distance. Noteworthy, this relaxation length might be larger for *in vitro* flows [38, 46] or under reduced global perfusion [39]. In this case, the resulting asymmetry of the hematocrit profile in the upstream

[xiii] Note, however, that such a diameter falls below the range of physiological diameters: while the form of the above parametrization (logit function with threshold) is very robust to describe phase separation at microvascular bifurcations, the quantitative dependence of parameters X_0, A and B as a function of the entry branch hematocrit and the bifurcation diameters needs further investigation [38, 55, 58].

branch of the bifurcation can lead to an inversion of the phase separation effect, as recently emphasized by direct numerical simulations of RBC collective behavior in networks [5].

8.3 BLOOD FLOW IN MICROVASCULAR NETWORKS

In the present section, we consider how the non-linear phenomena described in the previous section, which arise at vessel scale, contribute to the heterogeneity of blood flow at the scale of microvascular networks. Practically all hemodynamic parameters accessible to quantitative measurement in *in vivo* experiments indeed exhibit an extreme scatter among vessels for any type considered (arterioles, capillaries or venules), with coefficient of variations often of the same order (in the case of velocities) or even larger (in the case of hematocrit) than their mean values [20], as illustrated in Figure 8.11. *In vitro* experiments performed in microfluidic networks also demonstrate large spatial and temporal heterogeneities [14, 67, 68]. Such heterogeneities thus emerge independently of any biological regulation mechanism possibly involving variations of vessel diameters.

Section 8.3.1 first introduces a classical network approach for modeling time-averaged blood flow in networks. How this approach enables us to identify and parametrize the *in vivo* Fåhraeus-Lindquist effect as a function of diameter and discharge hematocrit is then presented in Section 8.3.2. Finally, the temporal dynamics of blood flow at network scale will be considered in Section 8.3.3.

8.3.1 A time-averaged network model for blood flow at network scale

If the complete architectural description of a given microvascular network is available, including the diameter and length of each micro-vessel and their adjacency list[(xiv)], and if appropriate boundary conditions are available at its entries and outlets, as detailed below, then a simple non-linear network model can be constructed based on the above time-averaged empirical descriptions of the Fåhraeus-Lindqvist effect (Eqs. 8.6 to 8.8), Fåhraeus effect (Eq. 8.9) and phase separation effect (Eq. 8.10).

In such a model, the microvascular network is viewed as a network of interconnected tubes ij, in which blood flow is described by a succession of linear relationships between the flow rate and the pressure drop:

$$Q_{ij} = G_{ij}(P_i - P_j), \tag{8.11}$$

where i and j denote the two extremities of vessel ij and G_{ij} its conductance:

$$G_{ij} = \frac{\pi D_{ij}^4}{128\mu_{app}(D_{ij}, H_{ij})L_{ij}}, \tag{8.12}$$

D_{ij} and L_{ij} its diameter and length, and H_{ij} is the discharge hematocrit in this vessel. Further, to simplify the writing of equations, the following convention may be adopted: inner vertices are defined as vertices where the pressure is unknown, and may include vertices where a flow rate boundary condition is imposed. Mass balance at each inner vertex i reads:

$$\sum_{j \in \mathcal{N}_{i,in}} G_{ij}(P_i - P_j) = q_i^s, \tag{8.13}$$

(xiv)The adjacency list of a graph with n vertices consists of n lists, one for each vertex v_i, $1 \le i \le n$, which contain the vertices to which v_i is adjacent.

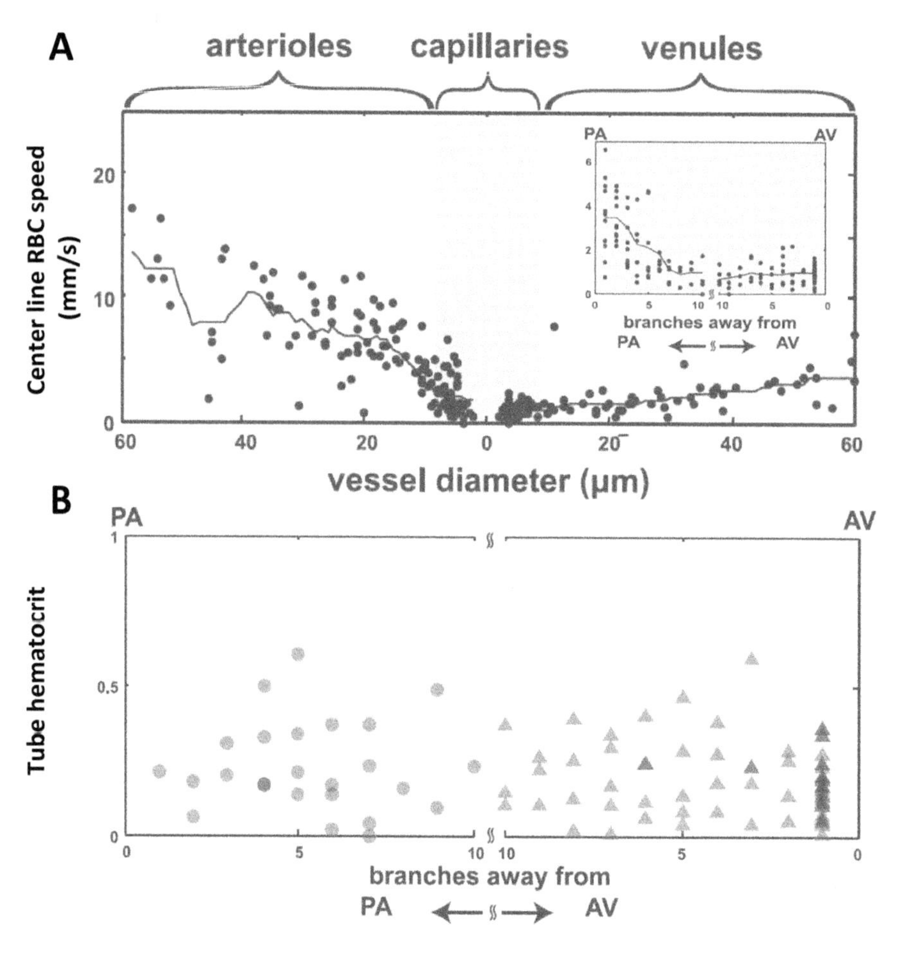

FIGURE 8.11
In vivo measurements of RBC speed and hematocrit by Two-Photon Laser Scanning Microscopy in mouse brains. A. RBC velocity at vessel centerline as a function of vessel diameter in arterioles, capillaries and venules. Inset: velocity in capillaries as a function of topological distance from upstream arteries (PA) and downstream veins (AV). B. Tube hematocrit in capillaries as a function of topological distance from upstream arteries (PA) and downstream veins (AV). From [60], with permission.

where $\mathcal{N}_{i,in}$ represents the set of inner neighboring vertices that are connected to i and q_i^s is a flow rate source term that describes any flow rate boundary conditions imposed at arteriolar inlets ($q_i^s > 0$) or venular outlets ($q_i^s < 0$). Alternately, when a pressure boundary condition P_s is imposed on a boundary, the corresponding vertex s is called an outer vertex and Eq. 8.13 at any vertex i connected to s may be rewritten:

$$\sum_{j \in \mathcal{N}_{i,in}} G_{ij}(P_i - P_j) + G_{is}P_i = G_{is}P_s. \tag{8.14}$$

If pressures or blood flows are known at all arteriolar entries and venular outlets of the network, this yields a sparse linear system, the unknown of which are the pressures at internal vertices, and the coefficients of which not only depend on the network architecture but also on the hematocrit distribution according to Eq. 8.12. This description holds for microfluidic networks or organs with continuous capillaries. The specific architecture of their endothelial cell lining indeed prevents any convective loss of plasma or circulating cells through the vessel walls [3][xv].

If the hematocrit distribution is known within the network, the corresponding sparse linear system can be solved to yield the pressures at all internal nodes, so that the flow rate in each vessel can be deduced from Eq. 8.11. Due to the phase separation effect, this distribution, however, results from an interplay with the flow distribution. To obtain a solution for this coupled non-linear problem, an iterative procedure is used, which includes the following steps [12, 52]:

1. The inlet discharge hematocrit is specified,

2. Guesses are made for the discharge hematocrit in each segment of the network. The initial guess can typically be a uniform value equal to the inlet hematocrit (which is equivalent to neglecting phase separation),

3. The effective blood viscosity in each tube is calculated using the corresponding empirical relationships (Eqs. 8.6 to 8.8). These can also be modified to represent the differences between *in vivo* and *in vitro* flows (see Section 8.3.2),

4. The sparse linear system (Eqs. 8.13 and 8.14) is solved to obtain the pressures at each internal vertex and deduce the flow rates in each vessel,

5. The discharge hematocrits in every tube are updated using the phase separation empirical relationship (Eq. 8.10). This relation gives the hematocrits in the two daughter segments as a function of the hematocrit in the parent segment, the flow rates, and the segment diameters. RBC mass conservation is self-contained in its formulation. Since the hematocrit in each segment depends on the phase separation at all upstream bifurcations, the hematocrit in a given segment is often computed after all upstream nodes have been considered. This involves a sorting of the nodes, which are then considered from high to low pressures along every flow pathway [37, 52]. The tube hematocrit is deduced *a posteriori* from the discharge hematocrit using the empirical relationship (Eq. 8.9).

6. The process is repeated until convergence.

Noteworthy, as the linear part of the problem (Step 4) yields a unique solution for the nodal pressures, convergence of the hematocrit distribution implies convergence of the pressure distribution, but not *vice versa*. Besides, this procedure can lead to numerical oscillations[xvi], so that a predictor-corrector scheme can be needed to obtain a converged solution [37], especially for large networks and high hematocrits. The question of whether such a converged solution is unique has not been fully investigated. However, from our experience, it seems that changing the initialization of the hematocrit distribution leads to stationary solutions for which the distributions of RBC flow rates ($H_{ij}Q_{ij}$) are very close. This suggests that convergence problems specifically arise from vessels with very low pressure drops

[xv]Such continuous barrier-forming capillaries are found ubiquitously in mammals, including in the central nervous system. Fenestrated capillaries are specifically located in the choroid plexus of the brain; several endocrine organs such as the pineal, pituitary, and thyroid glands; the hypothalamus; filtration sites in the kidneys; and absorptive areas of the intestinal tract, while sinusoidal capillaries can be found in the liver, spleen, bone marrow, and several endocrine organs, including the pituitary gland and the adrenal medulla [3].

[xvi]As all empirical laws used here are based on time-averaged experimental measurements, care must be taken if such oscillations are to be interpreted with regard to the temporal dynamics of RBC flows in networks.

(or equivalently very low flow), in which the exact value of the discharge hematocrit has almost no impact on time-averaged blood or RBC flows. As a result, the integrated parameters characterizing blood flow at the scale of the network, such as the total flow rate (or, equivalently, network resistance) or the distribution of transit times, are unaffected.

8.3.2 Identification of in vivo versus in vitro rheology

While *in vitro* experiments are invaluable to identify the basic mechanisms and understand their contributions across different scales, they often overlook many important aspects encountered in normal *in vivo* situations. For example, micro-vessels often are tortuous and their cross-section is neither circular nor constant along the vessel length, with corrugations due to the presence of endothelial cells. They may also be shorter than glass tubes used for *in vitro* experiments, so that singular pressure drops in their entrance region might play a role. Moreover, it is difficult to experimentally fabricate microchannels with wall properties mimicking the endothelial surface layer and its glycocalyx (glycoprotein coverage). Recent investigators have used silica-tubes grafted with polymer brushes [30] or even endothelialized PDMS(xvii) channels [75] to study the impact of the glycocalyx on RBC flow at channel scale but most of the experimental data used to quantify apparent viscosities as a function of tube diameter and hematocrit have been acquired in glass capillaries. Pries et al. (1990) [52] however, realized the value of data acquired *in vivo* throughout microvascular networks, such as the one presented in Figure 8.12A, for which a complete architectural description can be obtained and where flow rates and hematocrits can be measured in each vessel. Such datasets indeed combine vessels with a large range of hematocrit and diameter values. The wealth of information included can be used to quantify apparent viscosities in *in vivo* conditions(xviii). For that purpose, these authors first intuited that the parametrization of the Fåhraeus-Lindqvist effect *in vivo* could be deduced from the *in vitro* empirical description (Eq. 8.6) by correcting the latter to account for the effective reduction in vessel diameter associated with the glycocalyx, so that:

$$\mu_{rel}^{in\ vivo} = \mu_{rel}^{in\ vitro} \frac{D}{(D-W)^4}. \tag{8.15}$$

The value of W that yielded the smallest difference with the experimental datasets was between 3.5 and 4 μm, a quite large value compared to the typical diameter of microvascular vessels. They further refined this correction (Pries et al. 1994) [53] to account for increased dissipation induced by cell/wall interactions. Finally, considering that significant information on effective viscosity is only obtained in a diameter range covered by a sufficient number of vessels in the observed networks (4 to 40 μm), they adjusted the numerical values in Eq. 8.7 so that the final *in vivo* empirical relationship matches the *in vitro* viscosity for diameters above 50 μm, for which interactions between the flowing blood and the vessel walls have much less influence.

The resulting *in vivo* empirical relationship for the relative viscosity led to the smallest deviation between the computed and measured hematocrit and velocity distributions. Scaled to humans [53], it reads:

$$\mu_{rel}^{in\ vivo} = \left[1 + (\mu_{45}^{in\ vivo} - 1)\left(\frac{(1-H_D)^C - 1}{(1-0.45)^C - 1}\right)\left(\frac{D}{D-1.1}\right)^2\right]\left(\frac{D}{D-1.1}\right)^2, \tag{8.16}$$

(xvii) Poly(dimethylsiloxane).

(xviii) More generally, as later noted by Secomb et al. (2008) [64], *"Detailed sets of data, regarding a given system with many variables measured in a single instance, are generally much more valuable for model development than measurements of few parameters in multiple instances."*

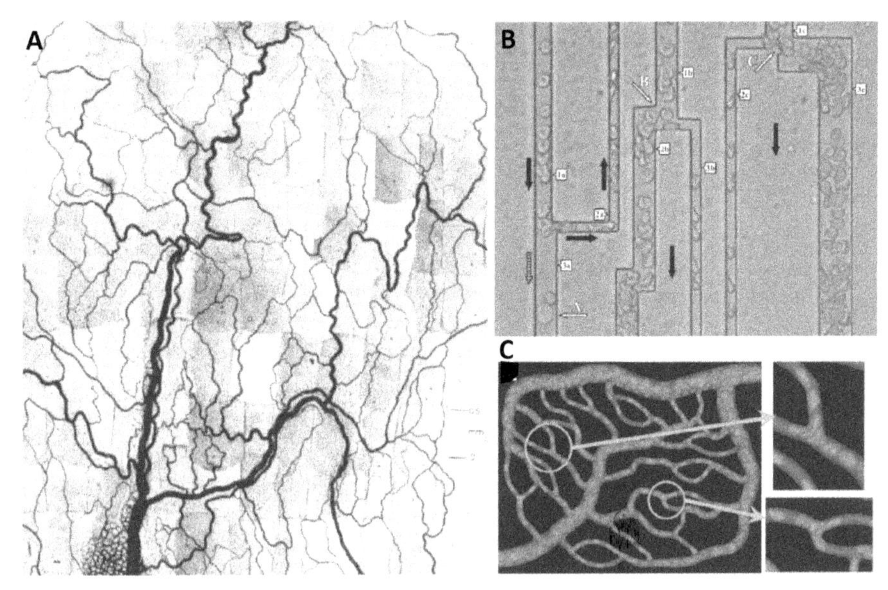

FIGURE 8.12
A. Photomontage of a mesenteric microvessel network with an area of $\sim 31mm^2$ recorded in $\sim$ 100 single fields of view [47], reprinted with permission from [50]. B. RBC distribution in a microfluidic network, from [67], with permission. Black arrows show flow direction in the corresponding channels. Note RBC jamming at bifurcation labeled by C. Also note an adherent white blood cell obstructing microchannel $3a$, causing flow cessation. C. Snapshots from direct simulations of collective RBC motion blood through a microvascular network. Right panels: RBCs are observed to linger (or, jam) at capillary bifurcations. From [5], with permission.

where C is given by Eq. 8.8 and

$$\mu_{45}^{in\ vivo} = 6\exp(-0.085D) + 3.2 - 2.44\exp(-0.06D^{0.645}). \tag{8.17}$$

Note that W is now reduced to 1.1 μm, consistent with recent *in vitro* measurements of the glycocalyx thicknesses ($0.7 \pm 0.2\mu m$, [75]). As displayed in Figure 8.13A, this viscosity decreases with decreasing vessel diameter only down to diameters of $\sim$ 20 to 30 μm. Its minimal values are much higher than those obtained from the *in vitro* viscosity parametrization (compare with Figure 8.9B). In addition to the specific geometric patterns of microvessels and the increased interactions between RBCs and glycocalyx, the presence of leucocytes in whole blood also contributes to this global increase of viscosity [52, 71]. These are removed in most *in vitro* experiments, but increase flow resistance *in vivo*, due to interactions with vessel walls, train formation[(xix)], or even transient occlusions of single capillaries [11].

This *in vivo* empirical relationship has been validated by comparing the results of the model predictions with extensive network experimental data, including velocity and hematocrit measurements acquired in three large networks (with up to $\sim$ 900 vessels) of the rat mesentery (Figure 8.12A). The global deviation only slightly increased by comparison with deviations resulting from measurement errors or uncertainty on parameters in Eq. 8.10, which describes phase separation[(xx)].

[(xix)]Microfluidic experiments performed with mixed RBC/leucocyte suspensions or whole blood have shown that trains of densely packed RBCs built up behind leucocytes, which move at reduced velocity in channels with capillary diameters [14]. This leads to enhanced RBC interactions, which may further increase global dissipation.

[(xx)]More recently, the same group used Bayesian analysis to revisit these results, showing the robustness of Eqs. 8.16 and 8.17 to the uncertain boundary conditions at small vessels, which, in addition to the major feeding arterioles and draining venules, inevitably crossed the limits of the observation areas [56]. With a

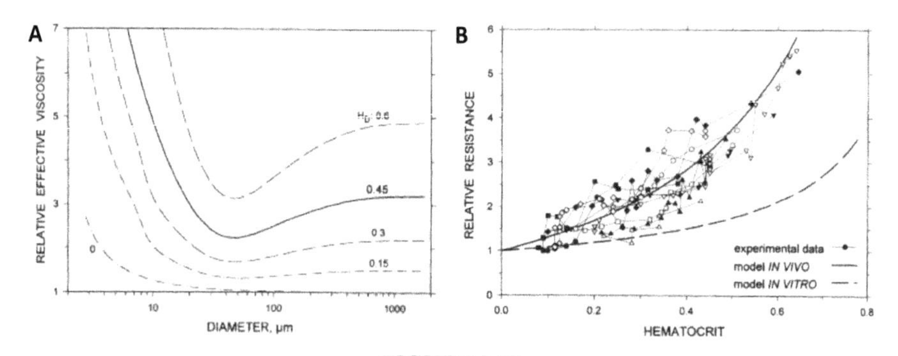

FIGURE 8.13
A. Relative effective blood viscosity as described by the *in vivo* empirical description (Eqs. 8.16, 8.17 and 8.8), from [53], with permission. B. Flow resistance across microvascular networks as a function of systemic hematocrit. Experimental data from 10 microvascular networks (different symbol for each network) together with predictions of the model described in Section 8.3.1 on the basis of *in vivo* and *in vitro* viscosity laws. From [53], with permission.

Besides, the addition of a shear-rate-dependent component led to substantial increases of this deviation, confirming the minor influence of shear-rate in physiological regimes.

To further test the validity of the results, an additional series of experiments was performed in which the pressure drop across the arteriolar inlet and venular outlet of these microvascular networks was directly measured as a function of the systemic hematocrit(xxi). This enabled them to deduce the overall network resistance as a function of the latter. The agreement with model predictions based on the *in vivo* empirical relationship was excellent (see Figure 8.13B), which was not the case for the *in vitro* relationship.

The *in vivo* acquisition of extensive hemodynamic datasets, including RBC velocities in a large number of vessels in three-dimensional organs, has been made possible by recent technological breakthroughs (*e.g.* [13,60]. In parallel, large-scale anatomical reconstructions have been obtained from post-mortem preparations (*e.g.* [8,24,74]. This recently enabled the confirmation of the statistical validity of this *in vivo* empirical description by comparing the velocity distributions predicted in arterioles, venules or capillaries of the cerebral cortex with experimental distributions [11]. New experimental data where the hemodynamic parameters and the network topology are acquired in the same network for all categories of vessels are nevertheless needed to perform comparisons on a vessel by vessel basis. Besides, new experiments in pathological situations are needed to understand how impaired mechanical RBC properties (*e.g.* in sickle cell disease) or reduced blood flow (reduced shear) conditions (*e.g.* in Alzheimer's disease) influence the blood rheology in microvascular networks.

8.3.3 Oscillatory behavior in microvascular networks

While the rheological description presented in the previous section is based on time-averaged experimental data and modeling studies, a very rich temporal dynamics of blood flow in microvascular networks has been described. Irregular and unpredictable velocity variations, including flow reversal, have already been documented by August Krogh(xxii), the 1920 Nobel Prize winner in Physiology and Medicine.

---

similar approach, they also proposed novel corrections for the parametrization of the phase separation effect [55].

(xxi) For these experiments, the systemic hematocrit was either increased to up to 0.65 by slow infusion of concentrated red blood cells obtained from another rat or lowered by successive hemodilution [53].

(xxii) *"In single capillaries, the flow may become retarded or accelerated for no visible cause; in capillary anastomoses, the direction of flow may change from time to time"* (A. Krogh, 1922, cited by [29]).

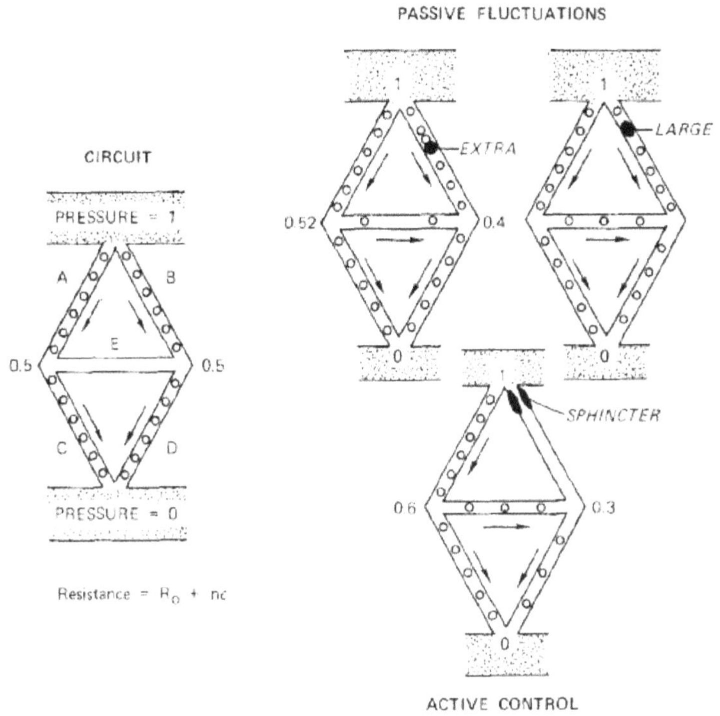

FIGURE 8.14
An idealized capillary flow circuit. Left: balanced. Right: balance upset by (a), an extra cell in branch B, (b), an extremely large cell in branch B, and (c), a sphincter contraction. From [16], with permission.

The question of whether these fluctuations might be spontaneous, originating entirely from within the network due to the non-linear flow properties of blood at the microscale, has been a long standing question. In fact, microvascular networks *in vivo* are also submitted to other physiological mechanisms, such as temporal variations of the perfusion pressure associated with heart beat [60], vasomotion (spontaneous variations in vessel diameter independent of heart beat or respiration, [1]) or compensatory mechanisms, which may also lead to fluctuations.

Separating these contributions in *in vivo* experiments is extremely difficult. Nevertheless, the theoretical possibility of spontaneous oscillations in narrow capillary vessels has been predicted more than 40 years ago [16]. Considering a model balanced circuit connecting a vessel at pressure 1 to another at pressure 0 (see Figure 8.14) and assuming that all branches A, B, C, D, E, are of equal length and diameter, he first argued that, for identical RBCs with an initially uniform distribution in all vessels, then the flow would be uniform in branches A, B, C, D. For symmetry reasons, there would be no flow in E. Supposing that branch B receives one extra RBC more than A would increase the pressure drop in B, disturb the balance and induce flow in branch E. The same would happen if branch B, instead of getting an extra cell, gets a cell that is larger than those in A. The flow in E, thus created, would continue unless the resistance is balanced again. Thus, because of the statistical spread of the red-cell sizes, continued fluctuation in branch E is expected.

Since, computational studies of the blood flow dynamics in microvascular networks with steady boundary conditions [7,12,26,27,29,73] have shown that the coupling between phase separation and the induced effective viscosity variations is sufficient to generate blood flow

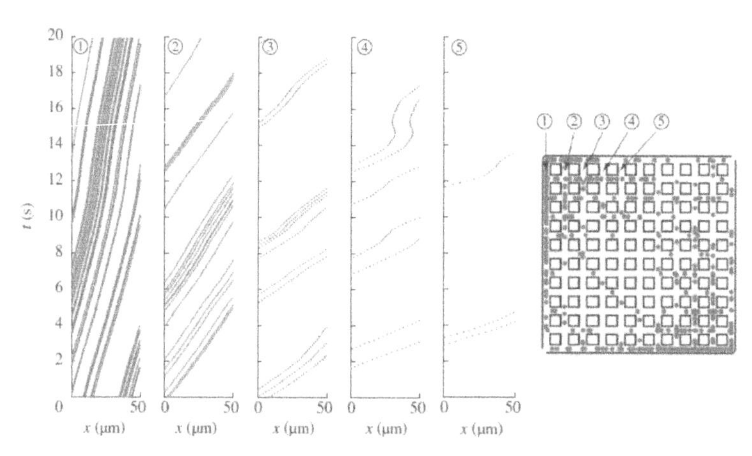

FIGURE 8.15
Unsteady transport of RBCs in a simple square network, as computed by a Lagrangian model where the position of each RBC is tracked individually. Left: Spatio-temporal diagrams of RBC locations (captured in intervals of $0.1s$) in five different capillaries, captured in intervals of $0.1s$. Right: The location of the five capillaries are indicated in the network (snapshot at $t = 20s$). From [41], with permission.

and hematocrit oscillations. These studies are all based on unsteady versions of the transport equations within each vessel, but rely on time-averaged descriptions of the phase separation and Fåhraeus-Lindqvist effect, similar to those presented in Section 8.2 or 8.3.2. Thus, such descriptions may not capture the whole complexity of the temporal dynamics of RBC flow. For example, considering the initial filling of a single bifurcation [(xxiii)] with identical RBCs, the following dynamics is expected: a first cell would randomly enter branch A or B. The increased resistance would decrease the blood flow, so that the second cell would enter the other branch, and so on and so forth. In this situation, phase separation can be described at each time by its narrow channel asymptotic limit (continuous line in Figure 8.10C). By contrast, on average, the same number of cells enters both branches A and B, so that no phase separation occurs in a time-averaged description[(xxiv)].

Alternately, simplified Lagrangian models where the position of each RBC is tracked individually have been introduced (*e.g.* [41,62]). Such models rely on analytical expressions relating the pressure drop at vessel scale to the number of RBCs within each vessel. Up to now, they have thus been restricted to narrow capillaries and single-file RBC flows. In this regime, they also require an instantaneous description of phase separation. For narrow capillaries with equal diameters, this corresponds to the above step function, or, equivalently, to assume that RBCs at bifurcations enter the branch with the higher local pressure gradient. Under these assumptions, a rich spatio-temporal dynamics, including RBC flow reversals, is obtained in idealized capillary networks (See Figure 8.15).

More sophisticated models requiring heavy computational resources enable us to study the collective dynamics of RBCs at network scale in networks including larger vessels [4,5]. The results also predict the onset of oscillations. In particular, depending on the upstream bifurcation, phase separation may oscillate between classical phase separation and phase separation reversals, which cannot be captured by a time-averaged description of phase separation [5]. They also suggest a strong impact of specific RBC structures induced by bifurcations in these vessels, such as jamming (see Figure 8.12C).

(xxiii) *i.e.* further simplifying the network of Figure 8.14 by only keeping branches A and B.

(xxiv) This might explain why the diameter for which X_0 reaches 50% in Eq. 8.10 lies below the physiological range of diameters (see Section 8.2.2).

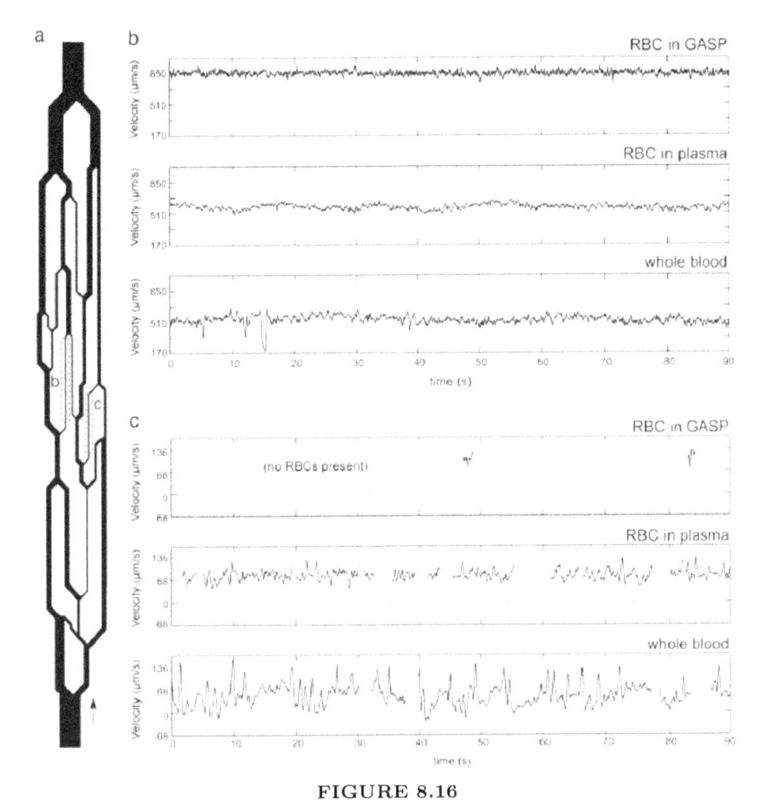

FIGURE 8.16
Spation-temporal dynamics of RBC or blood flow in the microfluidic network displayed on Panel (a). Panels (b) and (c) illustrate the oscillatory behavior in two channels exhibiting distinct dynamics (position of the channels is marked in panel (a)). For each channel, representative RBC velocity recordings are shown for three fluid samples: (i) purified RBCs suspended in buffer, (ii) purified RBCs suspended in plasma, and (iii) whole blood. From [15].

Recent *in vitro* experiments performed in microfluidic networks with steady boundary conditions also demonstrated large temporal heterogeneities of RBC flows, thus confirming their spontaneous character [15]. The oscillatory dynamics was self-sustained (lasting for the whole duration of an experiment) and highly reproducible (capillary blood flow oscillations occurred every time for different blood samples under a wide range of boundary conditions). It was strongly dependent on the channel position in the network and on the fluid sample used, as illustrated in Figure 8.16. In particular, RBC velocity recordings from two representative channels of capillary size (channels a and b in Figure 8.16a) show very different dynamics: the channel with the higher mean flow (channel b) always contained RBCs (Figure 8.16b). The other one had a significantly lower average RBC velocity, and often did not contain any RBCs due to plasma skimming (Figure 8.16c). For both channels, the increasing complexity of the fluid sample composition (from RBC suspended in buffer medium, to RBC suspended in plasma, to whole blood) resulted in significant changes of the oscillatory dynamics. In the fast-flowing channel, the flow of RBCs suspended in buffer was characterized by high-frequency, low amplitude oscillations (Figure 8.16b, top panel). Using plasma instead of buffer introduced low frequency oscillations (Figure 13b, middle panel). Additionally, for whole blood, high-amplitude fluctuations in RBC velocity (Figure 8.16b, lower panel) were associated with leucocytes flowing through the channel. The second channel with low flow was almost devoid of RBCs when buffer was used (Figure 8.16c, top panel). Suspending RBCs in autologous plasma changed the blood flow dynamics significantly. The channel remained filled with RBCs, emptying only occasionally (Figure 8.16c, middle panel) with oscillations of increased amplitude (Figure 8.16c, middle panel). For the whole blood sample, the channel contained RBCs almost continuously, and the oscillation amplitude increased significantly, with rapid spikes and flow reversals.

The authors interpret the impact of the fluid sample on the oscillatory dynamics as follows: in buffer medium, where RBC aggregation is inhibited, the transient imbalances of local hematocrits among channels are very small (ultimately, as small as a single cell) and are quickly corrected, generating low-amplitude, high-frequency oscillations. This interpretation is corroborated by experiments performed in smaller idealized networks containing a single loop with channel sizes slightly larger than capillary vessels [10]. With plasma, the formation of loose RBC aggregates in the sample makes the suspension less uniform. This intensifies the hematocrit heterogeneity by increasing the effective size of the particles being distributed at network bifurcations, leading to larger transient imbalances in local hematocrits that may take the system longer to equilibrate. This induces larger amplitude oscillations (Figures 8.16b and c, middle panels) and results in a significant shift of the dominant frequencies to lower frequencies. With whole blood, a major impact of leukocytes is demonstrated. Abrupt, transient fluctuations in local hematocrit (and therefore apparent viscosity) originate from the passage of trains of densely packed RBCs that built up behind leucocytes. The frequency of these large amplitude oscillations likely depends on the concentration of leukocytes and on the architecture of the microvascular network (particularly the presence of narrow capillaries of a sufficient length to enable the formation of RBC trains).

8.4 CONCLUSION

This chapter focused on the basic physical mechanisms by which the complex flow properties of blood at local scale lead to large spatio-temporal variations of all hemodynamic quantities throughout microvascular networks *in vitro* and *in vivo* . Based on a combination of theoretical work, in vitro experiments in single tubes and steady-state network flow simulations, these mechanisms had already been qualitatively understood by pioneers in the field, as illustrated by the many references to their work. They have mainly been confirmed, and also enriched, by more recent studies based on microfluidics and High Performance Computing. Such studies are however still very scarce by comparison with the wealth of studies conducted in single vessels or tubes. For example, to the best of my knowledge, no *in vitro* experiments performed in periodic model networks with channels of capillary size, similar to the one presented in Figure 8.15, right, have been published. For these type of channels or slightly larger, phase separation results are also very scarce [58], and only obtained in high-shear conditions using RBCs resuspended in simple buffer. Adding *e.g.* glutaraldehyde or dextran (or resuspending in plasma), and increasing the volume fraction of RBCs lead to considerable technical difficulties that limit our ability to obtain results about the time-dependent partitioning of RBCs. Such results can be obtained by analyzing the results of direct numerical simulations of a large number of RBCs flowing at bifurcations in networks (*e.g.* [5,32]. By contrast to experiments, in such simulations, 1) the RBCs can indeed be tracked and 2) RBC and suspending velocity fields can be obtained with unprecedented spatial and temporal resolution. However, they still need to be validated by careful comparisons with *in vitro* data. Combined with the potential of novel *in vivo* imaging techniques to acquire network architectural data as well as hemodynamic parameters in a large number of vessels, this will certainly yield a detailed fundamental comprehension of the impact of architecture on hemodynamics across scales in large microvascular networks.

Noteworthy, the reverse coupling, i.e. how blood flow profoundly shapes vascular network architecture during and after development, is also a fascinating problem for physicists. While beyond the scope of this chapter, this reverse coupling may in particular explain the coexistence of a space-filling capillary network and fractal-like arterioles and venules in microvascular networks (see *e.g.* [36] and references therein). The fundamental comprehension of these couplings will help understanding how microvascular networks adjust to various pathological conditions, including sudden obstructions of vessels or progressive reduction of microvascular density. This often postpones irreversible damage, making these pathological conditions intrinsically difficult to detect and to study. This is especially true in the clinics, where the spatial resolution of the available imaging techniques is strongly insufficient, so that hemodynamic data are averaged over regions which contain a very large number of vessels. Bridging the gap will require building upon the network-scale approaches described in the present chapter in order to derive homogenized models enabling descriptions of microvascular hemodynamics at a larger scale (see *e.g.*[42] and references therein).

ACKNOWLEDGEMENTS

I would like to gratefully acknowledge all participants in ERC BrainMicroFlow, and especially my former and current PhD students (Myriam Peyrounette, Adlan Merlo, Maxime Berg and Florian Goirand), postdocs (Amy Smith, Vincent Doyeux, Arthur Ghigo and Alexandra Vallet) and collaborators from Toulouse (Paul Duru, Yohan Davit, Michel Quintard and Frédéric Risso) and Cornell (Nozomi Nishimura and Chris Schaffer). I would like to thank Marianne Fenech for advising on some parts of this book chapter, and Manouk Abkarian and Annie Viallat for their invitation to contribute.

This work has been supported by the European Research Council under the European Union's Seventh Framework Programme (FP7/2007-2013)/ ERC grant agreement 615102, by the National Cancer Institute of the National Institutes of Health under Award Number R21CA214299 and by the National Institute of Neurological Disorders and Stroke of the National Institutes of Health under Grant Number 1RF1NS110054.

Bibliography

[1] C. Aalkjaer, D. Boedtkjer, and V. Matchkov. Vasomotion - what is currently thought?: Vasomotion. *Acta Physiologica*, 202(3):253–269, July 2011.

[2] A. Al-Kilani, S. Lorthois, Ti-H. Nguyen, F. Le Noble, A. Cornelissen, M. Unbekandt, O. Boryskina, L. Leroy, and V. Fleury. During vertebrate development, arteries exert a morphological control over the venous pattern through physical factors. *Physical Review E*, 77(5), 2008.

[3] H.G. Augustin and G.Y. Koh. Organotypic vasculature: From descriptive heterogeneity to functional pathophysiology. *Science*, 357(6353):eaal2379, August 2017.

[4] P. Balogh and P. Bagchi. Direct numerical simulation of cellular-scale blood flow in 3d microvascular networks. *Biophysical Journal*, 113(12), December 2017.

[5] P. Balogh and P. Bagchi. Analysis of red blood cell partitioning at bifurcations in simulated microvascular networks. *Physics of Fluids*, 30(5):051902, May 2018.

[6] P. Blinder, P.S. Tsai, J.P. Kaufhold, P.M. Knutsen, H. Suhl, and D. Kleinfeld. The cortical angiome: an interconnected vascular network with noncolumnar patterns of blood flow. *Nature Neuroscience*, 16(7):889–897, June 2013.

[7] R.T. Carr and M. Lacoin. Nonlinear dynamics of microvascular blood flow. *Annals of Biomedical Engineering*, 28(6):641–652, 2000.

[8] F. Cassot, F. Lauwers, C. Fouard, S. Prohaska, and V. Lauwers-Cances. A novel three-dimensional computer-assisted method for a quantitative study of microvascular networks of the human cerebral cortex. *Microcirculation*, 13(1):1–18, January 2006.

[9] L.K. Cheung. The blood supply of the human temporalis muscle: a vascular corrosion case study. *Journal of Anatomy*, 189(Pt 2):431, 1996.

[10] F. Clavica, A. Homsy, L. Jeandupeux, and D. Obrist. Red blood cell phase separation in symmetric and asymmetric microchannel networks: effect of capillary dilation and inflow velocity. *Scientific Reports*, 6:36763, November 2016.

[11] J.C. Cruz Hernández, O. Bracko, C.J. Kersbergen, V. Muse, M. Haft-Javaherian, M. Berg, L. Park, L.K. Vinarcsik, I. Ivasyk, D.A. Rivera, Y. Kang, M. Cortes-Canteli, M. Peyrounette, V. Doyeux, A. Smith, J. Zhou, G. Otte, J.D. Beverly, E. Davenport, Y. Davit, C.P. Lin, S. Strickland, C. Iadecola, S. Lorthois, N. Nishimura, and C.B. Schaffer. Neutrophil adhesion in brain capillaries reduces cortical blood flow and impairs memory function in alzheimer's disease mouse models. *Nature Neuroscience*, 22(3):413–420, March 2019.

[12] J.M. Davis and C. Pozrikidis. Self-sustained oscillations in blood flow through a honeycomb capillary network. *Bulletin of Mathematical Biology*, 76(9):2217–2237, September 2014.

[13] M. Desjardins, R. Berti, J. Lefebvre, S. Dubeau, and F. Lesage. Aging-related differences in cerebral capillary blood flow in anesthetized rats. *Neurobiology of Aging*, 35(8):1947–1955, August 2014.

[14] O. Forouzan, J.M. Burns, J.L. Robichaux, W.L. Murfee, and S.S. Shevkoplyas. Passive recruitment of circulating leukocytes into capillary sprouts from existing capillaries in a microfluidic system. *Lab on a Chip*, 11(11):1924, 2011.

[15] O. Forouzan, X. Yang, J.M. Sosa, J.M. Burns, and S.S. Shevkoplyas. Spontaneous oscillations of capillary blood flow in artificial microvascular networks. *Microvascular Research*, 84(2):123–132, September 2012.

[16] Y.-C. Fung. Stochastic flow in capillary blood vessels. *Microvascular Research*, 5(1):34–48, January 1973.

[17] Y. C. Fung. *Biomechanics: Mechanical Properties of Living Tissues.* Springer New York, New York, NY, 1981.

[18] Y. C Fung. *Biodynamics: Circulation.* Springer New York, New York, NY, 1984.

[19] Y.C. Fung and B.W. Zweifach. Microcirculation: mechanics of blood flow in capillaries. *Annual Review of Fluid Mechanics*, 3(1):189–210, 1971.

[20] P. Gaehtgens. Microcirculation—historical background and conceptual update. In *International Congress Series*, volume 1235, pages 3–13. Elsevier, 2002.

[21] P. Gaehtgens, C. Duhrssen, and K.H. Albrecht. Motion, deformation, and interaction of blood cells and plasma during flow through narrow capillary tubes. *Blood Cells*, 6:799–817, 1980.

[22] Y. Gazit, D.A. Berk, M. Leunig, L.T. Baxter, and R.K. Jain. Scale-invariant behavior and vascular network formation in normal and tumor tissue. *Physical Review Letters*, 75(12):2428–2431, September 1995.

[23] E. Guyon, J.-P. Hulin, L. Petit, and C.D. Mitescu. *Physical Hydrodynamics, 2nd Edn.* Oxford University Press, January 2015.

[24] S. Heinzer, T. Krucker, M. Stampanoni, R. Abela, E.P. Meyer, A. Schuler, P. Schneider, and R. Ḿuller. Hierarchical microimaging for multiscale analysis of large vascular networks. *NeuroImage*, 32(2):626–636, August 2006.

[25] R.M. Hochmuth, R.N. Marple, and S.P. Sutera. Capillary blood flow. I. Erythrocyte deformation in glass capillaries. *Microvascular Research*, 2:409–419, 1970.

[26] N.J. Karst, J.B. Geddes, and R.T. Carr. Model microvascular networks can have many equilibria. *Bulletin of Mathematical Biology*, 79(3):662–681, March 2017.

[27] N.J. Karst, B.D. Storey, and J.B. Geddes. Oscillations and multiple equilibria in microvascular blood flow. *Bulletin of Mathematical Biology*, 77(7):1377–1400, July 2015.

[28] M.F. Kiani and A.G. Hudetz. A semi-empirical model of apparent blood viscosity as a function of vessel diameter and discharge hematocrit. *Biorheology*, 28(1-2):65–73, 1991.

[29] M.F. Kiani, A.R. Pries, L.L. Hsu, I.H. Sarelius, and G.R. Cokelet. Fluctuations in microvascular blood flow parameters caused by hemodynamic mechanisms. *American Journal of Physiology - Heart and Circulatory Physiology*, 266(5):H1822–H1828, May 1994.

[30] L. Lanotte, G. Tomaiuolo, C. Misbah, L. Bureau, and S. Guido. Red blood cell dynamics in polymer brush-coated microcapillaries: A model of endothelial glycocalyx in vitro. *Biomicrofluidics*, 8:014104, 2014.

[31] H. Lei, D.A. Fedosov, B. Caswell, and G.E. Karniadakis. Blood flow in small tubes: quantifying the transition to the non-continuum regime. *Journal of Fluid Mechanics*, 722:214–239, May 2013.

[32] X. Li, A.S. Popel, and G.E. Karniadakis. Blood-plasma separation in Y-shaped bifurcating microfluidic channels: a dissipative particle dynamics simulation study. *Physical Biology*, 9(2):026010, April 2012.

[33] J. Liam McWhirter, H. Noguchi, and G. Gompper. Ordering and arrangement of deformed red blood cells in flow through microcapillaries. *New Journal of Physics*, 14(8):085026, August 2012.

[34] H. Lipowsky. Microvascular rheology and hemodynamics. *Microcirculation*, 12(1):5–15, February 2005.

[35] H.H. Lipowsky and B.W. Zweifach. Methods for the simultaneous measurement of pressure differentials and flow in single unbranched vessels of the microcirculation for rheological studies. *Microvascular Research*, 14(3):345–361, November 1977.

[36] S. Lorthois and F. Cassot. Fractal analysis of vascular networks: Insights from morphogenesis. *Journal of Theoretical Biology*, 262(4):614–633, February 2010.

[37] S. Lorthois, F. Cassot, and F. Lauwers. Simulation study of brain blood flow regulation by intra-cortical arterioles in an anatomically accurate large human vascular network: Part I: Methodology and baseline flow. *NeuroImage*, 54(2):1031–1042, January 2011.

[38] A. Merlo. *Ecoulement de suspensions de globules rouges dans des réseaux de micro-canaux: hétérogénéités et effets de réseau.* PhD thesis, Université de Toulouse, 2018.

[39] Y.C. Ng, B. Namgung, S.L. Tien, H.L. Leo, and S. Kim. Symmetry recovery of cell-free layer after bifurcations of small arterioles in reduced flow conditions: effect of RBC aggregation. *American Journal of Physiology - Heart and Circulatory Physiology*, 311(2):H487–H497, August 2016.

[40] N. Nishimura, C.B. Schaffer, B. Friedman, P.D. Lyden, and D. Kleinfeld. Penetrating arterioles are a bottleneck in the perfusion of neocortex. *Proceedings of the National Academy of Sciences*, 104(1):365–370, 2007.

[41] D. Obrist, B. Weber, A. Buck, and P. Jenny. Red blood cell distribution in simplified capillary networks. *Philosophical Transactions of the Royal Society A: Mathematical, Physical and Engineering Sciences*, 368(1921):2897–2918, June 2010.

[42] M. Peyrounette, Y. Davit, M. Quintard, and S. Lorthois. Multiscale modelling of blood flow in cerebral microcirculation: Details at capillary scale control accuracy at the level of the cortex. *PLOS ONE*, 13(1):e0189474, January 2018.

[43] F. Plouraboue, P. Cloetens, C. Fonta, A. Steyer, F. Lauwers, and J.-P. Marc-Vergnes. X-ray high-resolution vascular network imaging. *Journal of Microscopy*, 215(2):139–148, August 2004.

[44] J.M. Poiseuille. Recherches sur les causes du mouvement du sang dans les vaisseaux capillaires,. Séance publique du 28 décembre 1835, Académie des Sciences, 1835.

[45] A.S. Popel and P.C. Johnson. Microcirculation and hemorheology. *Annual Review of Fluid Mechanics*, 37(1):43–69, January 2005.

[46] A.R. Pries, K. Ley, M. Claassen, and P. Gaehtgens. Red cell distribution at microvascular bifurcations. *Microvascular Research*, 38(1):81–101, 1989.

[47] A.R. Pries, K. Ley, and P. Gaehtgens. Generalization of the Fåhraeus principle for microvessel networks. *American Journal of Physiology-Heart and Circulatory Physiology*, 251(6):H1324–H1332, 1986.

[48] A.R. Pries, D. Neuhaus, and P. Gaehtgens. Blood viscosity in tube flow: dependence on diameter and hematocrit. *American Journal of Physiology - Heart and Circulatory Physiology*, 263(6):H1770–H1778, December 1992.

[49] A.R. Pries, B. Reglin, and T.W. Secomb. Structural response of microcirculatory networks to changes in demand: information transfer by shear stress. *American Journal of Physiology - Heart and Circulatory Physiology*, 284(6):H2204–H2212, June 2003.

[50] A.R. Pries and T.W. Secomb. Blood flow in microvascular networks. *Comprehensive Physiology*, 2011.

[51] A.R. Pries, T.W. Secomb, and P. Gaehtgens. Biophysical aspects of blood flow in the microvasculature. *Cardiovascular research*, 32(4):654–667, 1996.

[52] A.R. Pries, T.W. Secomb, P. Gaehtgens, and J. F. Gross. Blood flow in microvascular networks. Experiments and simulation. *Circulation research*, 67(4):826–834, 1990.

[53] A.R. Pries, T.W. Secomb, T. Gessner, M.B. Sperandio, J.F. Gross, and P. Gaehtgens. Resistance to blood flow in microvessels in vivo. *Circulation Research*, 75:904–914, 1994.

[54] D. Quemada. Hydrodynamique sanguine : Hemorheologie et ecoulement du sang dans les petits vaisseaux. *Le Journal de Physique Colloques*, 37(C1):C1:9–22, January 1976.

[55] P.M. Rasmussen, T.W. Secomb, and A.R. Pries. Modeling the hematocrit distribution in microcirculatory networks: A quantitative evaluation of a phase separation model. *Microcirculation*, 25(3):e12445, 2018.

[56] P.M. Rasmussen, A.F. Smith, S. Sakadzic, D.A. Boas, A.R. Pries, T.W. Secomb, and L. Ostergaard. Model based inference from microvascular measurements: Combining experimental measurements and model predictions using a bayesian probabilistic approach. *Microcirculation*, 2016.

[57] W. Reinke, P. Gaehtgens, and P. C. Johnson. Blood viscosity in small tubes: effect of shear rate, aggregation, and sedimentation. *American Journal of Physiology-Heart and Circulatory Physiology*, 253(3):H540–H547, September 1987.

[58] S. Roman, A. Merlo, P. Duru, F. Risso, and S. Lorthois. Going beyond 20 μm-sized channels for studying red blood cell phase separation in microfluidic bifurcations. *Biomicrofluidics*, 10(3):034103, May 2016.

[59] S. Sakadzic, E.T. Mandeville, L. Gagnon, J.J. Musacchia, M.A. Yaseen, M.A. Yucel, J. Lefebvre, F. Lesage, A.M. Dale, K. Eikermann-Haerter, C. Ayata, V.J. Srinivasan, E.H. Lo, A. Devor, and D.A. Boas. Large arteriolar component of oxygen delivery implies a safe margin of oxygen supply to cerebral tissue. *Nature Communications*, 5:5734, December 2014.

[60] T.P. Santisakultarm, N.R. Cornelius, N. Nishimura, A.I. Schafer, R.T. Silver, P.C. Doerschuk, W.L. Olbricht, and C.B. Schaffer. In vivo two-photon excited fluorescence microscopy reveals cardiac- and respiration-dependent pulsatile blood flow in cortical blood vessels in mice. *AJP: Heart and Circulatory Physiology*, 302(7):H1367–H1377, April 2012.

[61] F. Schmid, M.J.P. Barrett, P. Jenny, and B. Weber. Vascular density and distribution in neocortex. *NeuroImage*, 2017.

[62] G.W. Schmid-Schönbein, R. Skalak, S. Usami, and S. Chien. Cell distribution in capillary networks. *Microvascular Research*, 19(1):18–44, January 1980.

[63] T.W. Secomb. Blood flow in the microcirculation. *Annual Review of Fluid Mechanics*, 49(1):443–461, January 2017.

[64] T.W. Secomb, D.A. Beard, J.C. Frisbee, N.P. Smith, and A.R. Pries. The role of theoretical modeling in microcirculation Research. *Microcirculation*, 15(8):693–698, January 2008.

[65] T.W. Secomb and J.F. Gross. Flow of red blood cells in narrow capillaries: role of membrane tension. *Int. J. Microcirc. Clin. Exp.*, 2(3):229–40, May 1983.

[66] J.M. Sherwood, D. Holmes, E. Kaliviotis, and S. Balabani. Spatial distributions of red blood cells significantly Alter Local Haemodynamics. *PLoS ONE*, 9(6):e100473, June 2014.

[67] S.S Shevkoplyas, S.C Gifford, T. Yoshida, and M.W. Bitensky. Prototype of an in vitro model of the microcirculation. *Microvascular Research*, 65(2):132–136, 2003.

[68] S.S. Shevkoplyas, T. Yoshida, S.C. Gifford, and M.W. Bitensky. Direct measurement of the impact of impaired erythrocyte deformability on microvascular network perfusion in a microfluidic device. *Lab on a Chip*, 6(7):914, 2006.

[69] A.Y Shih, . Blinder, P.S Tsai, B. Friedman, G. Stanley, P.D. Lyden, and D. Kleinfeld. The smallest stroke: occlusion of one penetrating vessel leads to infarction and a cognitive deficit. *Nature Neuroscience*, 16(1):55–63, December 2012.

[70] S.P. Sutera, V. Seshadri, P.A. Croce, and R.M. Hochmuth. Capillary blood flow II. Deformable model cells in tube Flow. *Microvascular Research*, 2:420–433, 1970.

[71] D.W. Sutton and G.W. Schmid-Schönbein. Elevation of organ resistance due to leukocyte perfusion. *American Journal of Physiology-Heart and Circulatory Physiology*, 262(6):H1646–H1650, June 1992.

[72] G.J. Tangelder, D.W. Slaaf, A.M. Muijtjens, T. Arts, M.G. oude Egbrink, and R.S. Reneman. Velocity profiles of blood platelets and red blood cells flowing in arterioles of the rabbit mesentery. *Circulation Research*, 59(5):505–514, November 1986.

[73] Y. Tawfik and R.G. Owens. A mathematical and numerical investigation of the hemodynamical origins of oscillations in microvascular networks. *Bulletin of Mathematical Biology*, 75(4):676–707, April 2013.

[74] P.S. Tsai, J.P. Kaufhold, P. Blinder, B. Friedman, P.J. Drew, H.J. Karten, P.D. Lyden, and D. Kleinfeld. Correlations of neuronal and microvascular densities in murine cortex revealed by direct counting and colocalization of nuclei and vessels. *Journal of Neuroscience*, 29(46):14553–14570, November 2009.

[75] D. Tsvirkun, A. Grichine, A. Duperray, C. Misbah, and L. Bureau. Microvasculature on a chip: study of the endothelial surface layer and the flow structure of red blood cells. *Scientific Reports*, 7(1), 2017.

[76] P. Vaupel. Tumor microenvironmental physiology and its implications for radiation oncology. *Seminars in Radiation Oncology*, 14(3):198–206, July 2004.

[77] A. Villringer, A. Them, U. Lindauer, K. Einhaupl, and U. Dirnagl. Capillary perfusion of the rat-brain cortex. An in vivo confocal microscopy study. *Circulation Research*, 75(4):55–62, 1990.

CHAPTER 9

White blood cell dynamics in micro-flows

Annie Viallat, Emmanuèle Helfer
Aix Marseille Univ, CNRS, CINAM, Marseille, France

Jules Dupire
L'Oreal Recherche & Innovation, Aulnay sous Bois, France

CONTENTS

THIS CHAPTER is devoted to the dynamics of white blood cells circulating in the vascular system. After a short introduction on the different white blood cell types, the paramount question of the capture of white blood cells on the endothelial wall, their adhesion and transmigration towards inflamed tissues will be very briefly reviewed. The chapter then focuses on the microcirculation of white blood cells in the pulmonary capillaries, a very important issue because, in some acute diseases, white blood cells are sequestered in the capillaries and release toxic products that seriously damage lung tissue.

9.1 INTRODUCTION - CIRCULATING WHITE BLOOD CELLS

White blood cells, also called leukocytes, are the cells of the immune system that help protect the body from infectious diseases and foreign invaders. WBCs are produced in the bone marrow and are found throughout the body, particularly in the blood and lymphatic system.

There are five main types of WBCs, neutrophils, eosinophils, basophils, monocytes and lymphocytes. Three types are rare in the bloodstream. Eosinophils, which participate in allergic reactions and fight multicellular parasites, reside in the tissues. Basophils, which are mainly responsible for the allergic and antigenic response, are the rarest WBCs (less than 0.5% of the total). Lymphocytes are mainly found in the lymphatic system. These cells do not participate significantly in microcirculation and will not be considered here.

The other two WBC types, neutrophils and monocytes, circulate in the bloodstream and are the primary foot soldier in the immune system. When a trauma occurs in tissue, signals of inflammation are given nearby in the body, which are recognised by neutrophils and monocytes. The cells then respond to the inflammatory stimulus by leaving the bloodstream and migrating to the site of inflammation. Neutrophils are the first to be recruited and have a very high microbicidal activity, while monocytes/ macrophages are recruited later.

Neutrophils are the most abundant WBC, constituting 60-70% of the WBCs of most mammals. They are the essential part of the innate immune system. In suspension, their average diameter is 8.85 μm. They have a multilobed nucleus, which consists of three to five lobes tethered together by flexible necks, such as pearls on a string. The nucleus occupies 20% of the cell volume. The multilobed characteristic of the nucleus probably results from an adaptation to the requirements of cell passage through narrow capillary beds and extravasation in tissues. The nucleus had to evolve to be mechanically as unobtrusive as possible. Neutrophils exist in two basic states. In the passive state, they circulate and deform in the vascular system without disturbing their environment. In the activated state, when they respond to an inflammatory stimulus, they actively develop forces that cause adhesion to the vascular wall and cell deformations, including the formation of pseudopods. At the site of infection, neutrophils, which possess powerful microbicidal activity, phagocyte bacteria. Phagocytosis is the process of uptake of microbes and particles followed by digestion and destruction of this material. After phagocytosing a few pathogens, neutrophils die and do not return to the microcirculation. Neutrophils are the most common cell type in the early stages of acute inflammation, being recruited within minutes of the trauma. The lifetime of a circulating human neutrophil is about three days. Due to their high reactivity to pathogens, neutrophils are generally not present in body cavities. Instead, they are produced and stored in large reserves in the bone marrow, ready for deployment in the circulation.

Monocytes constitute about 7% of all leukocytes in the human body. They originate in the bone marrow from myeloid progenitor cells and are then released into the peripheral blood. They circulate in the bloodstream for about one to three days, then migrate into the tissues by mechanisms similar to those of neutrophils and they differentiate into macrophages and dendritic cells. They are the largest type of WBCs, and their diameter can reach 15 to 30 μm. They have a large eccentrically placed nucleus, which is U-shaped - or kidney shaped and have a soft, spongy, three-dimensional appearance. The nucleus : cytoplasm ratio is approximately 3 : 1 or 2 : 1. Like neutrophils, they are phagocytic cells that respond to inflammation signals in the body and quickly reach the sites of infection or tissue damage. However, monocytes come after neutrophils and are on site after about 8 to 12 hours. There, they differentiate into macrophages, which strengthens the immune system's response.

9.2 MIGRATION TO SITES OF INFLAMMATION, THE LEUKOCYTE CASCADE ADHESION

The circulatory and migratory properties of neutrophils and monocytes allow efficient tissue surveillance for infectious pathogens and rapid accumulation at injury and infection sites. To be recruited into inflammatory foci, neutrophils and monocytes emigrate from the circulation in postcapillary venules to the surrounding tissues.

To succeed in this migration, the first challenge of the leukocyte is to reach the vascular wall. As seen in Chapters 3 and 4, red blood cells mainly circulate in the centre of the venules and their collision with leukocytes results in the margination of these latter towards the wall of the blood vessels. The following steps involve the active response of leukocytes and endothelial cells located in the vessel wall near the inflamation sites. Specific ligands and receptors are expressed on the cell surface, which are activated by the molecules of inflammation released by the inflamed tissues. These steps have been the subject of numerous studies over the past three decades. These studies have led to the creation of a paradigm called the leukocyte adhesion cascade. The reader will find excellent reviews in the literature on this topic (for example [28, 29, 32, 33, 52, 53]). This chapter will therefore not be devoted to this exciting and crucial issue, which involves strong cellular biological activity and many adhesion molecules, each with a particular function. We will limit ourselves here to a brief description of the main steps.

Leukocyte recruitment is mediated by a cascade of complex molecular and cellular events in response to molecular changes on the surface of blood vessels that signal injury or infection. It occurs in at least three stages: leukocyte rolling, adhesion and transmigration. These three steps were gradually completed and refined [33] by adding steps of slow rolling, adhesion strengthening, intraluminal crawling and paracellular and transcellular migration (see Figure 9.1). The three steps of the original leukocyte adhesion cascade are as follows.

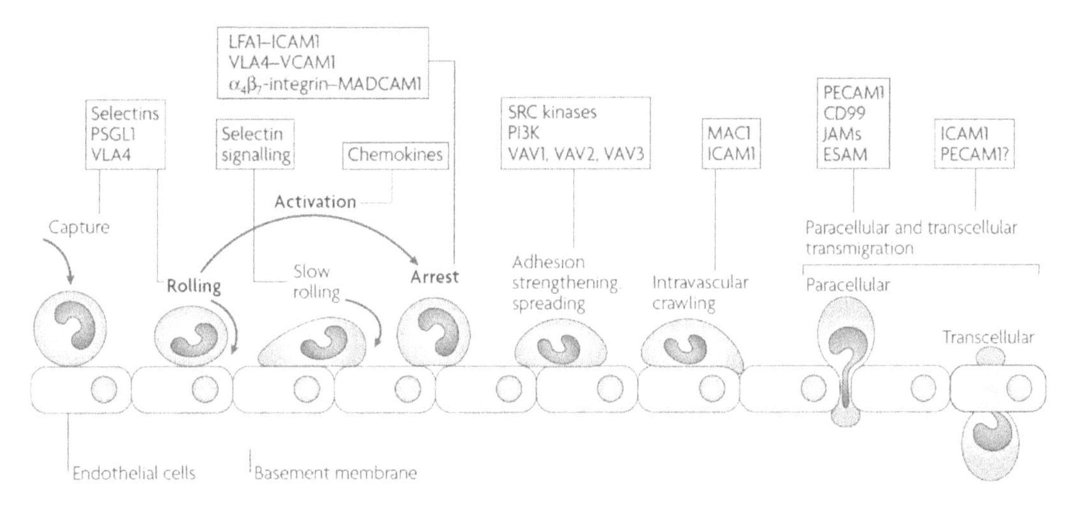

FIGURE 9.1
The leukocyte adhesion cascade: The original three steps are shown in bold: rolling mediated by selectins, activation mediated by chemokines, and arrest mediated by integrins [52]. Additional steps: capture, slow rolling, adhesion strengthening and spreading, intravascular crawling, and paracellular and transcellular transmigration. Key molecules involved in each step are indicated in boxes. ESAM, endothelial cell-selective adhesion molecule; ICAM1, intercellular adhesion molecule 1; JAM, junctional adhesion molecule; LFA1, lymphocyte function-associated antigen 1; MAC1, macrophage antigen 1; MADCAM1, mucosal vascular addressin cell-adhesion molecule 1; PSGL1, P-selectin glycoprotein ligand 1; PECAM1, platelet/endothelial-cell adhesion molecule 1; PI3K, phosphoinositide 3-kinase; VCAM1, vascular cell-adhesion molecule 1; VLA4, very late antigen 4. From [33]

First, leukocytes marginated near the endothelial surface touch the vessel endothelium, then tether and roll along through reversible and rapid interactions between adhesion molecules of the selectin family (cell surface proteins) expressed on endothelial cells or on leukocytes and their carbohydrate ligands expressed on cells. This step of slowing down and capturing triggers the leukocyte adhesion cascade. Molecules of the selectin family bind with exceptionally high on- and off-rates (which determine the speed with which bonds are formed and broken, respectively). L-selectin and P-selectin require shear stress to maintain adhesion since the shear stress does not inhibit, but actually promotes adhesion. This phenomenon is related to the 'catch bond' character of selectins: under a low force the bonds are weak and relatively short-lived; with force increase, the bond's strength (i.e. lifetime) increases, i.e. each bond strengthens as shear stress is applied. The result is that selectins are very efficient to slowdown and finally catch flowing leukocytes. Also the transport of selectin ligands relative to the selectins provided by the rolling motion of the cell allows the formation of new bonds before the old ones break.

Second, the rolling leukocyte is stimulated by chemokines or other chemotactic compounds, such as N-formyl-methionyl-leucyl-phenylalanine (fMLP), platelet activation factor (PAF) or leukotriene B4 (LTB4) that trigger cell activation. Activated leukocytes present surface integrins that bind with counter-receptors expressed by endothelial cells, namely members of the immunoglobulin superfamily, such as ICAM1 and VCAM1, expressed by endothelial cells. Indeed, integrins are 'activatable' receptors. It means that intracellular signalling through cell-surface molecules is necessary to increase their ligand-binding capability. The avidity of the cell adhesion mediated by integrins is regulated by two molecular parameters: integrin affinity and ligand binding valence. Integrin engagement leads to leukocyte arrest and firm adhesion to the vascular endothelium. The specificity of leukocyte arrest is attributed to the combined and differential expression of integrins and their ligands, as well as chemokines and their receptors.

Finally, the leukocyte spreads out over the epithelial cells thanks to an active process of reorganisation of its cytoskeleton. Then it extends pseudopods to get into the interendothelial slits between adjacent endothelial cells. This transmigration, called diapedesis, is driven by 'Platelet endothelial cell adhesion molecule' proteins that pull the cell through the slit. The last barrier to be crossed is the basement membrane, which requires a proteolytic digestion step of the membrane by the leukocyte. These last steps involve biological signalling with a high activity of the leukocyte.

9.3 MICROCIRCULATION IN THE CAPILLARY PULMONARY BED

In addition to the critical role played by activated leukocytes in the fight against infection, they are also of interest from a rheological point of view [29]. As their diameter is larger than the average diameter of a capillary, they must undergo large deformations to pass through the systemic or pulmonary microcirculation. Pressure drops in systemic capillaries are generally sufficient to deform and pass leukocytes without being sequestrated in the capillaries, but the passage through the pulmonary capillary bed is more challenging. Capillaries of the alveolar walls are narrow and strongly interconnected, so that the pressure drop driving a leukocyte through a single narrowed capillary segment is low, of the order of 30 Pa [26]. As a result, leukocytes are delayed in their passage through the lungs and their concentration in the lungs is generally more than 50 times higher than in the general circulation [7, 34].

Accumulation of leukocytes in the lungs is an advantage because pathogens, allergens and pollutants inhaled during breathing are numerous, thus increasing the risks in lung aggression. However, it is also a potential problem because the pulmonary capillary bed displays a complex architecture with many bifurcations and very narrow capillaries in which leukocytes must strongly deform and may be sequestered. This problem has been relatively little studied.

To pass from venules to arterioles blood cells must go through several pulmonary alveoli with many possible paths in each alveolus and a significant variability of lengths and sections of each capillary [7] (Figure 9.2). Pulmonary bed capillaries display diameters ranging between 2 and 15 μm ([10, 18, 25, 59]), whereas neutrophils measure 6 to 8 μm in diameter [10,18] and monocytes can reach 15 to 30 μm in diameter [57]. WBCs must therefore change their shape during their transit from pulmonary venules to pulmonary arterioles in the lung, when crossing from 8 to 17 alveolar walls and flowing through 40-100 capillary segments [7].

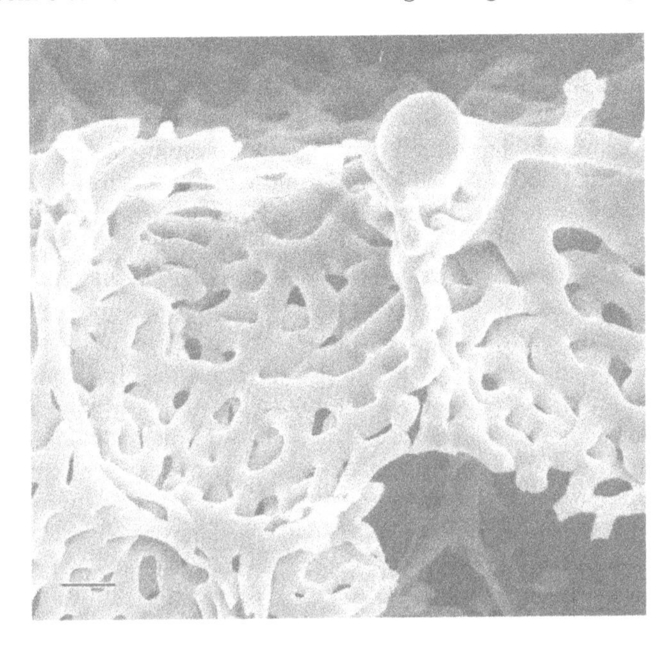

FIGURE 9.2
Scanning Electron Microscopy of vascular cast obtained on a rat(From [18]).

Most of the studies on WBC circulation focused on neutrophils, as the most common WBCs. It has been shown that their transit time in human, dog and rabbit lungs [8,9,11,25, 35–38] is prolonged compared to that of plasma [43] or of red blood cells (RBCs), the latter, being of the order of a few sec [24, 25, 34, 43]. This difference is thought to be due to the slow deformation of neutrophils compared to RBCs which do not contain a nucleus and are highly deformable. A more detailed observation of neutrophils in the capillaries showed that they move by hops [10, 16, 34]. The temporary arrests observed *in-vivo* occur in capillary segments rather than in junctions and were related to the fact that neutrophils must change their shape to enter into a narrower segment [16]. Despite this motion, there is usually no clogging of the microvasculature. It is obviously critical that WBC circulation must be properly maintained so that they can migrate out of the lung in response to inflammation or infection. Thus, deregulation of WBC activation and/or circulation can have dramatic effects as observed for some pathologies like sepsis, cancer, pulmonary disease or injury.

In acute inflammatory syndromes such as acute respiratory distress syndrome (ARDS) for instance, the key problem is the massive trapping of neutrophils in lung blood capillaries which causes fatal ischemia and lung injury [23]. The increased amount in neutrophils correlates with impaired lung function [1]. In severe sepsis, local infection is accompanied by systemic neutrophil activation, and the inappropriate activation and subsequent mis-positioning of neutrophils lead to multiple organ failure [4]. In these diseases, both adhesion and stiffness properties of the cells are modified, but their specific roles remain unclear.

Inflammatory molecules, like cytokines, are known to stiffen neutrophils, which thus may become unable to deform enough to flow in the narrow pulmonary capillaries [60, 63]. This behavior has been observed in WBCs from patients affected by ARDS [40, 41] as well as by sepsis or trauma [39]. Adhesion to vessel walls has been also reported in ARDS [20] as well as in the context of other inflammations [27]. However, several studies in which stiffening and adhesion pathways were varied indicate that WBC stiffening may be sufficient and necessary for the sequestration to occur, and that adhesion may develop at a later stage during cell arrest [40, 60]. Thus, to date, the mechanisms of cell trapping in acute inflammatory syndromes and, more generally, the dynamics of WBCs in pulmonary capillaries are still not well understood. The difficulty with *in-vivo* experiments is, first, that they are quite invasive, and, second, that it is hard to control and vary external parameters. Measurements of WBCs kinetics performed in capillary networks of various mammals (dogs, rabbits, rats) have clearly shown cell slowdowns and stops, and subsequent accumulation in the densely capillarised alveolar networks. However, they have not led to the establishment of clear correlations between these quantities [24, 30, 31]. So, well-controlled *in-vivo* studies that focus on changing one parameter at a time and providing comprehensive sets of data are lacking. Researchers have therefore turned to *in-vitro* approaches to obtain better controlled experimental conditions. First experiments consisted in micropipette aspiration of cells which allowed deducing cell rheological properties like elastic moduli and viscosity [14, 22]. However, this kind of experiment does not give high statistics due to its one-shot aspect (single-cell measurement). The rise of the micro-fluidic technology over the past 10-15 years, which introduced rapid flow control, combined with micro-fabrication techniques of flow chambers compatible with optical microscopy, opened the way to studies with more statistics, as multiple cells can be handled simultaneously or sequentially with the same setup. Another advantage is the diversity of possible experiments: cells can be subjected to a shear flow, forced to encounter obstacles or to pass through confined passageways mimicking, for example, blood vessels and capillaries. So far,*in-vitro* studies of the dynamics of WBCs in capillaries are still rare. Most of the experiments focused on the entry and movement of individual cells, either neutrophils or monocytes from cultured cell lines, in a single short constriction [15] or in a single long channel [15, 40, 42, 62]. Gabriele et al. also investigated the passage of monocytes through 50 subsequent constrictions and observed that cells can either pass rapidly through or be stopped for a long time [15]. In [42] the cell loss modulus was extracted from its entry in the constriction. Nishino et al, forced whole blood of patients with sepsis or trauma through parallel micro-channels to extract rheological properties of blood, but not the mechanical properties of individual cells [39]. Thus, *in-vitro* studies mainly consisted of quantifying entry and transit times in and through a narrow channel, but the data were little used to deduce mechanical properties of the cells with regard to their mechanisms of passage in narrow capillaries.

The main reason is that WBC mechanical properties are not easy to characterise and model. Although much work has been done over the years on designing a correct rheological model for the neutrophil [13, 19], there is not a unique simple model capable to reproduce fast and long-time cell responses both at low and high mechanical stresses. Various linear viscoelastic models have been developed, depending on which experiment has to be described (see [48] for a review). For example, the standard viscoelastic model of the passive neutrophil [46] was used to interpret the small deformation in the first seconds of micropipette aspiration, but was unable to describe correctly the recovery after ejection [54]; the Newtonian liquid-drop model of a viscous material with a shell cortical tension and no elasticity [14] was not compatible with rapid deformations observed experimentally; addition of an elastic element in the Maxwell fluid model [12, 51] led to a better capture of the short and long-time behaviors [60]; it was later upgraded to a non-linear viscoelastic model [64]; the compound drop model considered tensions of cell and nuclear membranes and viscosities of nucleus and cytoplasm [21, 55] but it was difficult to apply to neutrophils with multi-lobed nuclei. In contrast with the so numerous rheological models on cell deformation, few have been proposed for the flow of WBCs through the capillary network [2, 49, 50]. Though they are still under development to elucidate factors like friction, cell-cell interactions, etc., they are now ready for comparison with experiments on WBCs flowing through a biomimetic capillary structure.

The rest of this chapter is devoted to study the dynamics of THP1 monocytes in a micro-fluidic narrow channel network that mimics the pulmonary capillary bed. THP1 cells are human monocyte lineage cells derived from an acute monocytic leukemia patient [56], and are commonly used as an *in-vitro* cell model. We use drugs acting on the actin network to produce cells with different mechanical properties, in order to understand their role in cell transport in the pulmonary bed. A simple mechanical model involving membrane tension, short elastic relaxation time and a highly viscous contribution is used to describe the cell dynamics.

9.3.1 The biomimetic channel network

The micro-fluidic bio-mimicking device consists of a network of micro-channels made up of a series of 50 columns of 7 pillars arranged in staggered rows, all pillars having the same shape and being regularly spaced. The positioning of the pillars was chosen to obtain horizontal channels of 15 μm in length l. Several widths w were prepared, namely 14.6 μm, 12.7 μm, 11.9 μm, 11.1 μm, 10.6 μm, 9 μm and 8.6 μm. The height h of the channel was set equal to 9.3 μm. The diagonal distance between pillars was set equal to w (see Figure 9.3). This device mimics the average length and width of human pulmonary capillaries, the average number of capillaries crossed by a blood cell, and the multiplicity of possible pathways between pulmonary venules and arterioles. The total pressure drop applied between the inlet and outlet of the micro-channel network ΔP_t is in the physiological range (10, 20, 30 and 40 mbar). Flow simulations using Comsol software (Figure 9.3B) show that the pressure variation within a network mesh is $\Delta P_h = 1.7\Delta P_{mesh}$ and $\Delta P_d = 0.7\Delta P_{mesh}$, where ΔP_h and ΔP_d are the pressure drops in a horizontal channel and in a diagonal channel, respectively, and ΔP_{mesh} is the pressure drop within a mesh. The network consisting of 50 meshes, ΔP_{mesh} is equal to $1/50^{th}$ of the total pressure drop in the network: $\Delta P_{mesh} = \Delta P_t/50$. THP1 monocytes have an average diameter of $14.8\mu m \pm 1.45\mu$m and are therefore slightly pre-stressed in height in the 9.3-μm high inlet channel before being further deformed as they pass through the micro-channels.

9.3.2 Monocytes reach a steady-state periodic dynamic in the network

Monocytes of size smaller than that of the channels display periodic trajectory, the periodic mesh being twice the unit mesh (Figure 9.4 A). Their trajectory alternates right and left direction at each bifurcation following the horizontal channel. The velocity is also periodic. These small objects follow the flow streamlines. Strikingly, large monocytes exhibit a different behavior (Figure 9.4 B). In the first part of the network, they have a non-periodic trajectory and a motion by hops. Their velocity and shape change over the meshes and, further down the network, they gradually reach a steady-state regime (same dynamics in all meshes). There, cell trajectories are periodic with alternating right and left turns at the exit of the horizontal channels without stopping. Both velocity and shape deformation also become periodic. In the following, the focus is given on these large cells.

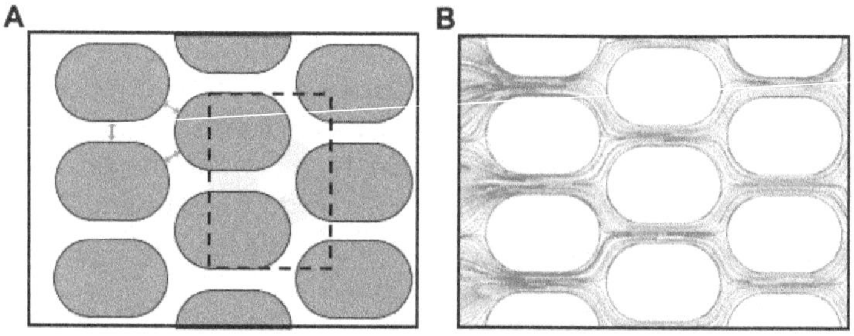

FIGURE 9.3
Description of the micro-channel network. A) Pillars are positioned so that the minimal distance between their extremities (tilted channel in a bifurcation or in a junction) equals the width of the straight horizontal channel, as illustrated by the blue arrows. The dashed rectangle highlights the mesh unit of the network. The pressure drops in the orange (horizontal) and green (tilted) channels are 1.7 ΔP_{mesh} and $0.7\Delta P_{mesh}$, respectively, where ΔP_{mesh} is the pressure drop within the mesh. B) Comsol simulation of the streamlines in the device (the first three pillar series are displayed).

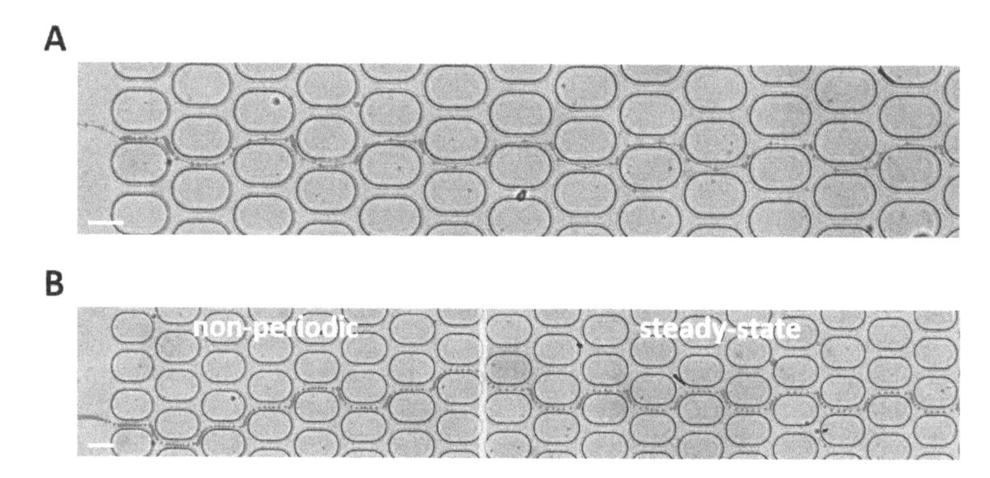

FIGURE 9.4
Typical trajectories of monocytes in the micro-channel network. A small monocyte (A) has a periodic trajectory. B) A large monocyte has first a non-periodic trajectory before reaching a steady-state regime. The cells were automatically tracked using Matlab software and the red dots correspond to the position of their centre of mass. Scale bars: 20 μm.

In the transient regime the time to pass through a mesh is very variable and may be as long as a few seconds; it sometimes alternates between large and small values. Along these first meshes of the network the passage time globally decreases and the average cell velocity increases until it reaches the steady-state regime. Looking closely on the cell velocity during passage through a mesh, the minimum cell velocity is reached during the stretching phase in the bifurcation (Figure 9.5). Figure 9.5 also shows the evolution of the cell shape, characterised by its reduced volume ν, a dimensionless number defined as the ratio of the cell volume V to the volume of the sphere with the same surface area A

$$\nu = \frac{3V}{4\pi(\frac{A}{4\pi})^{3/2}} \tag{9.1}$$

($\nu \leq 1$, $\nu = 1$ for a sphere).

The maximum of cell elongation is reached in the horizontal channel and the minimum at its exit. The mesh-to-mesh evolution of the cell deformation during the transient regime

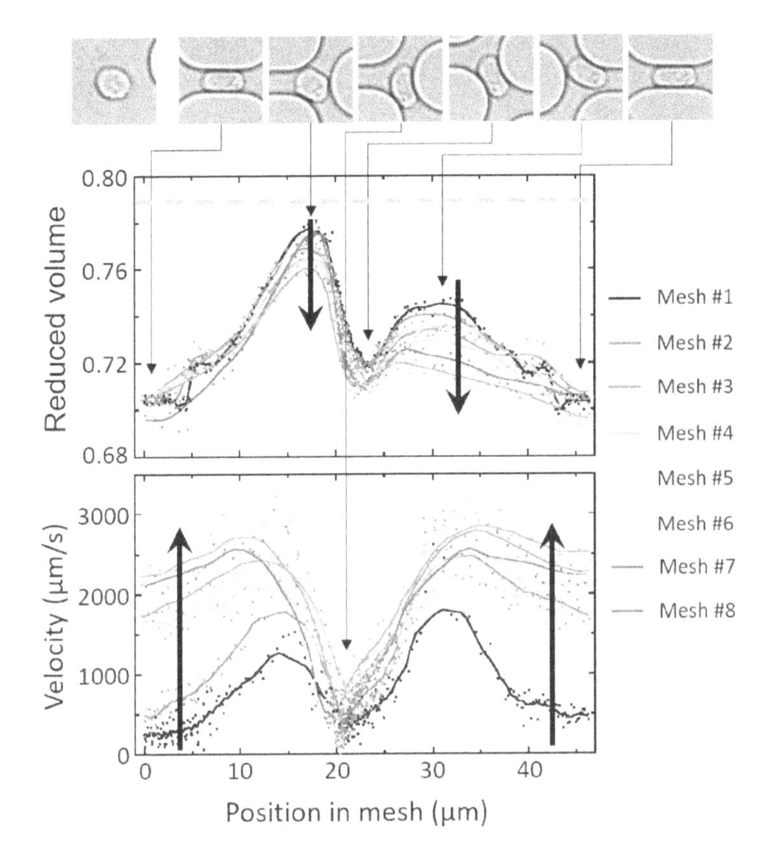

FIGURE 9.5
Evolution of cell shape and velocity during the transient regime. Reduced volume ν (middle) and instantaneous velocity v (bottom) as function of the cell position in the mesh. The lines are derived from sliding averages over 11 points. The wide arrows show the evolution of ν and v as the cell passes through 8 successive meshes before reaching the steady-state regime. Top) Images of the cell at min and max values of ν and min value of v. The far left image shows the cell before it enters the micro-channel network, with $\nu = 0.79$ (indicated by the dashed grey line).

shows that the cells are increasingly stretched, especially in the diagonal channel. In the steady-state regime, the cells barely slow down at the entrance of the horizontal channels. In some fast cases, the crossing of a mesh can take only a few tens of milliseconds.

9.3.3 The mechanical properties of monocytes affect their dynamics in the network

The remarkable progressive adaptation of cell shapes along the successive meshes of the micro-channel network leads to rapid and efficient cell movement at the final steady-state regime. The question therefore arises of whether this adaptation behavior is due to an active response of the cells or whether it is purely due to their complex rheological properties. This led to modulate the cell rheology, and more specifically, the elasticity of the actin cortex by treating cells with Latrunculin A (lat-cells) or jasplakinolide (jasp-cells) that respectively inhibits and activates actin polymerisation. Consequently, Latrunculin A reduces cell membrane tension whereas jasplakinolide increases it.

Strikingly, both treatments significantly affect the length of the transient regime, the cell deformation within the network, and the mesh passage time (Figure 9.6 and Figure 9.7).

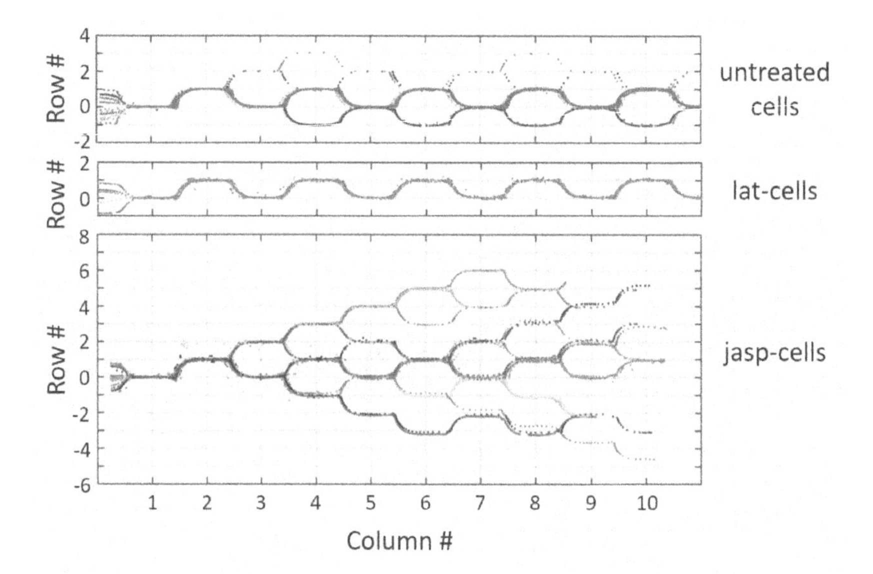

FIGURE 9.6
Trajectories of cells treated or not with drugs in the first 10 meshes of the network. Untreated cells first display a random regime and eventually adopt a regular trajectory (top); lat-cells immediately adopt steady-state motion (middle); jasp-cells display a very long transient regime, some never switch to steady-state motion. Scale bars are 20 μm

TABLE 9.1
Percentage of cells reaching the steady-state regime

Lat	**Untreated**	**Jasp**
95% (n=38)	85% (n=73)	46% (n=100)

As shown in Figure 9.6 and Table 9.1, lat-cells reach a steady-state periodic regime immediately after passing through the second horizontal channel, whereas most jasp-cells do not reach the steady-state. The cell kinematics is also affected. The entrance time into the first channel and the cell velocity in the steady-state regime strongly increases and decreases, respectively, from lat- to untreated to jasp-cells. These points will be discussed later. Figure 9.7 displays time-lapses of the three types of cells where each specific behavior, including cell deformation, is clearly illustrated.

Lat-cells smoothly adapt their shape when they pass through the mesh. At the exit of the horizontal channel, when their front end reaches the pillar, the part of the cell closest to a diagonal channel enters it, thus becoming the new cell front. The rear of the cells is reshaped by the shear of the fluid flowing in the second diagonal channel and takes on an asymmetric shape.

In contrast, jasp-cells retain their sausage shape at the exit of the horizontal channel and tumble into the bifurcation. There, in the non-steady regime, the cells can completely clog the two diagonal channels, and hang in the bifurcation for a long time, in a 'stagnation' position, their front and rear becoming undifferentiated. Finally, they return to circulation after one of the two diagonal channels unclogs, thanks to a slight random displacement of the cellular position. This long sequestration may be associated with an increase in cellular elongation.

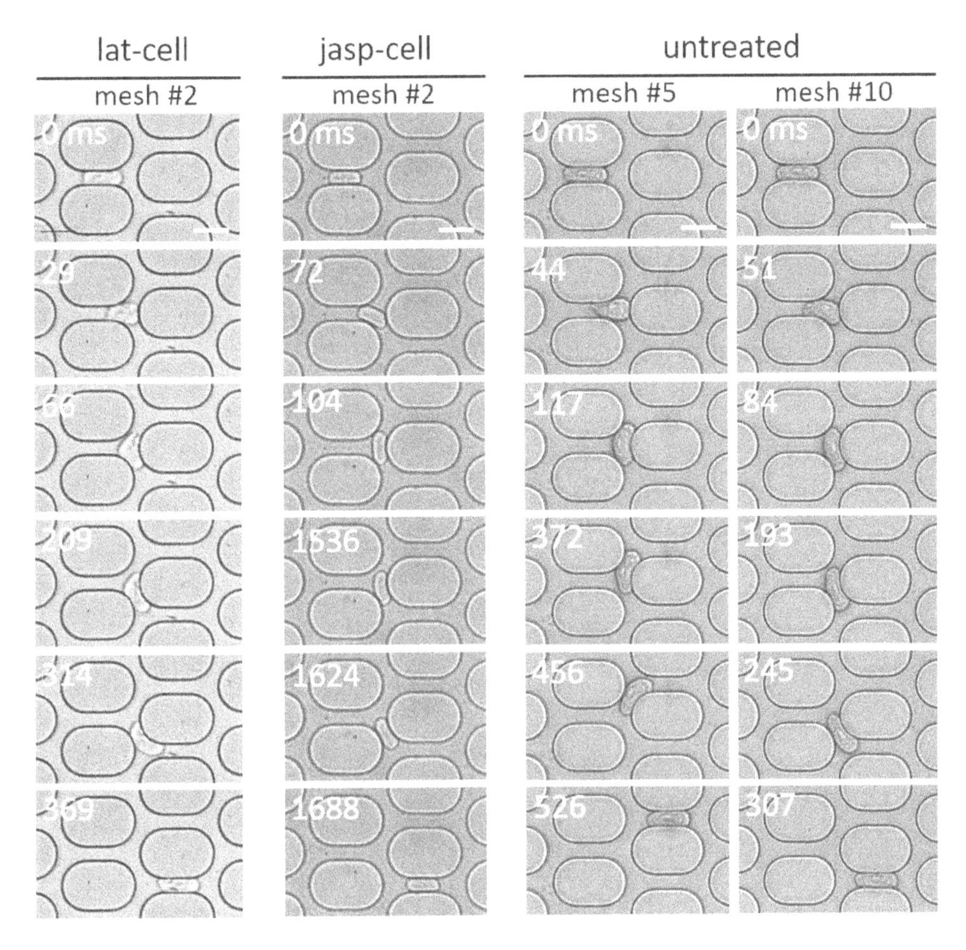

FIGURE 9.7
Timelapses of cells treated or not with drugs passing through meshes. Elapsed time in msec, time 0 is set when the cell is in the middle of the horizontal channel of the mesh. Scale bars are 20 μm

Most untreated cells reach the steady-state regime. However, they are deformable at a slower rate than lat-cells. In the non-steady regime, the cells exiting the horizontal channel with a symmetrical shape crash onto the next pillar and then deform into a quasi-symmetrical shape, which may clog the two diagonal channels as reported for jasp-cells. But farther in the network, they slightly turn at the exit of the horizontal channel, thus helping the cell front to enter in a diagonal channel. The rear of the cell then tumbles in the bifurcation but does not fill the entire space and does not clog the other diagonal channel.

9.3.4 Relevant mechanical models for monocyte dynamics

In order to understand how the mechanical parameters of monocytes govern their circulation in the network, the cell entrance in the network, i.e. its entrance in the first horizontal channel, was analysed and the relevant involved mechanical parameters were determined. First, it was noted that a pressure drop higher than a critical one, ΔP_c must be applied to drive a cell into the channel, thus revealing a non-zero cell cortical tension. Above ΔP_c the cell elongates like a sausage (Figure 9.8A) and slowly enters the channel. The cell elongation ratio, $\epsilon(t) = \frac{L(t)-L(0)}{L(0)}$, where $L(0)$ is the cell diameter before entry in the network and $L(t)$ is the cell length along the channel axis at time t, is displayed in Figure 9.8B, for a typical untreated cell. The curve initially shows a fast non-linear regime at small deformation (typically $\epsilon \leq 0.1$), typical of a viscoelastic material, followed by a slow viscous flow (up to $\epsilon \geq 0.3$) at constant velocity. Thus, these cells are considered as viscoelastic drops with a surface tension τ_0 and an internal composite material described by the modified Maxwell model which consists in a viscoelastic Kelvin-Voigt body (dashpot of viscosity η_1 and spring of elasticity κ_1) and an additional dashpot (viscosity η_2) for the long-time viscous dissipation (Figure 9.8C)[17,61].

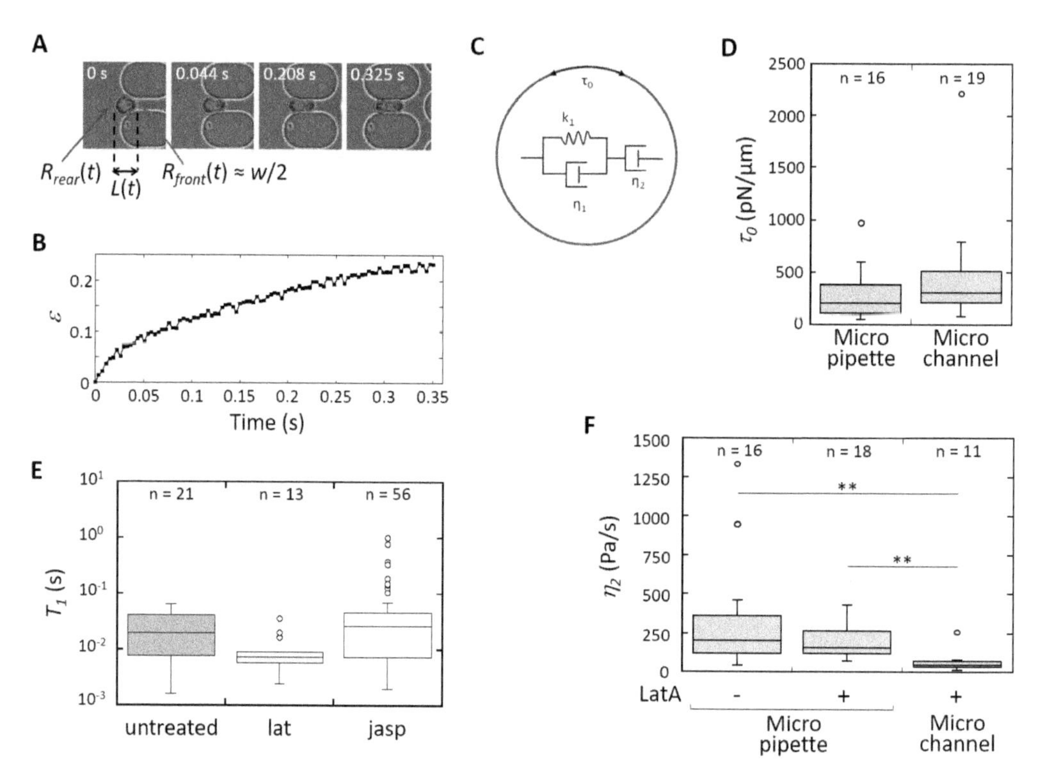

FIGURE 9.8
Mechanical properties of the cells. A) Timelapse of a cell entering the first horizontal micro-channel of a network. The cell contour is detected (in red) and the total length L of the deformed cell is measured. The cell front and rear are approximated with circles (in green and blue, respectively). The front radius R_{front} is half the channel width. B) Cell elongation ratio ϵ while it enters the channel. C) Rheological model of the cell as a viscoelastic liquid combining a Kelvin-Voigt solid (viscosity η_1 and elastic modulus k_1) in series with a dashpot (viscosity η_2), and with a cortical tension τ_0. D) Cell cortical tension τ_0 derived from cell entry in micropipettes at 23°C (left) and in micro-channels at 37°C (right). E) Characteristic time T_1 of the elastic deformation, derived from cell entry in micro-channels for untreated, lat- and jasp-cells. F) Viscosity η_2 measured by micropipette aspiration (at 23°C) and by cell entry in micro-channels (at 37°C). Cells were treated or not with LatrunculinA. **: $p < 10^{-3}$.

The cell cortical tension generated by the actin cortex can be determined from the critical pressure required to enter a constriction. Here, ΔP_c was measured as the minimal pressure allowing a cell to enter in the first horizontal channel of the network or to enter in a micropipette. In the case of a cylindrical pipette of radius R_p, ΔP_c relates to the cell surface tension τ_0 by the Laplace law [14,22]:

$$\Delta P_c = 2\tau_0(\frac{1}{R_p} - \frac{1}{R_{cell}}) \tag{9.2}$$

where R_{cell} is the cell radius. In the case of the rectangular micro-channel of the network, the channel height h, which is smaller than the cell diameter fixes the vertical curvature at the front and at the rear of the cell (Figure 9.8A). ΔP_c thus writes as:

$$\Delta P_c = \tau_0(\frac{2}{d} - \frac{1}{R_{rear}}) \tag{9.3}$$

where R_{rear} is the horizontal radius at the rear of the cell, d is the half of the channel width w. Using Eqs 9.2 and 9.3, the surface tension of untreated cells was measured in both types of experiments, at 37°C in the channel network and at 23°C in micropipettes (Figure 9.8D). The median values of τ_0 are 300 pN/μm and 196 pN/μm for micro-channel and pipette methods, respectively, with no significant effect of temperature. This result is in the range of those previously reported for THP1 cells in reference [45]. In contrast, lat-cells entered in the micro-channels for applied pressure drops as low as 0.3 mbar (i.e. $\tau_0 < 8.5pN\mu$m), confirming that Latrunculin A destroyed the actin cortex. From the modified Maxwell model [17], the mechanical constitutive law of the cell is

$$(1 + \frac{\eta_1}{\eta_2})\frac{d\sigma}{dt} + \frac{k_1}{\eta_2}\sigma = k_1\frac{d\epsilon}{dt} + \eta_1\frac{d^2\epsilon}{dt^2} \tag{9.4}$$

where the effective applied stress σ for a flowing cell is determined by the pressure balance between the critical pressure drop ΔP_c and the pressure drop in the horizontal channel ΔP_h

$$\sigma = \Delta P_h - \Delta P_c = \Delta P_h - \tau_0(\frac{2}{d} - \frac{1}{R_{rear}}) \tag{9.5}$$

Note that σ is constant only when the second term of the right member of Eq 9.5 is constant. It is the case during the initial deformation stage of large cells as long as $R_{rear} >> d/2$ so that $1/R_{rear} << 2/d$, or of cells small enough to little deform during their entrance in the channel because $R_{cell} \approx R_{rear} \approx d/2$, or either, of cells treated with Latrunculin A for which τ_0 is very small, so that ΔP_h dominates the right term of Eq 9.5. Integration of Eq 9.4 was possible in these cases and led to the relation :

$$\epsilon(t) = \frac{\Delta P_h}{k_1}(1 - e^{\frac{k_1 t}{\eta_1}}) + \frac{\Delta P_h}{\eta_2}t \tag{9.6}$$

$\epsilon(t)$ curves in the first horizontal micro-channel were fitted by Eq. 9.6, which yielded the parameters η_1, k_1, and $T_1 = \eta_1/k_1$. No significant difference was found between untreated, lat- and jasp-cells. The Voigt elasticity k_1 and the viscosity η_1 are of the order of 10^3 Pa and 10 Pa.s, respectively. As shown in Figure 9.8E the raising time for the elastic deformation T_1 is very short, of the order of 10 to 20 ms, indicating that the cells can undergo fast small deformations. The viscosity η_2 was determined during the entrance of lat-cells in the network at 37°C for which the effective applied stress σ remains constant (Eq 9.5), and from complementary micropipette experiments (at 23°C) both for untreated cells and for lat-cells (Figure 9.8F). The median viscosity is 203 Pa.s for untreated cells and 156 Pa.s for lat-cells at 23°C. These values are in excellent agreement with that of 185 Pa.s reported in [45] for monocytes and close to that reported for granulocytes (135 Pa.s) [14]. Also in agreement with this latter work, we observed a strong temperature dependence of the viscosity. The viscosity of lat-cells decreased from 156 Pa.s at 23°C to 45 Pa.s measured in micro-channels at 37°C. This mechanical analysis shows that the main parameter affected by actin-drug treatment is the cortical tension, which is therefore responsible for the very different dynamics presented by lat-, untreated and jasp-cells. In particular, the differences in the cell velocity within the first horizontal channel come from the effective stress σ felt by the cells, which decreases when the cell membrane tension increases (according to Eq 9.5).

The cell transit into the bifurcation following the horizontal channel is governed by the capillary number $Ca = \eta_{out} v_{fluid}/\tau_0$, where η_{out} and v_{fluid} are the suspending fluid viscosity and velocity, respectively. Ca increases from jasp-cells to untreated cells to lat-cells. As known for long [44], the shape of droplets in linear flows is determined by the nonlinear coupling of their restoring internal tension forces and the deforming hydrodynamic forces, and their dynamics depend strongly on the strength of the surrounding flow rates, on surface tension, and on the droplet to external fluid viscosity ratio. At the lowest Ca, for jasp- and untreated cells, the surface tension resists cell deformation and reduces the ability of viscous force to deform the cells, which keep a large curvature at their rear. A part of the work done by the viscous force is converted into surface energy, thus slowing the cell in the bifurcation, as computed for droplets in Refs. [6] and [58]. In line with our observations on cells, these works report that a droplet hangs in the bifurcation during a long period, the lower Ca the longer the hanging time.

9.3.5 Towards the periodic steady-state

In the periodic steady-state regime, the cell exits the horizontal channel asymmetrically and quickly engages in a diagonal channel. This fast engagement prevents the cell from hanging for a random time at the centre of the bifurcation and directs its periodic trajectory. As shown in Figure 9.7 for a lat-cell, this directional asymmetry is associated with a periodic asymmetric shaping of the cell rear by viscous shear forces in the bifurcation. Even when cell shapes are similar in two successive horizontal channels, the periodic change observed from one mesh to the next one in the inclination of the segments connecting two points at the cell rear shows that the cell is strongly sheared along alternatively the right and the left wall of the horizontal channel (see Figure 9.9 superimposed images of a cell flowing at the end of the channel network in the steady state regime and see inserts). Therefore, an asymmetric shear force alternately directs the cell into the right and left diagonal channels in successive meshes. This is due to a stronger friction along the side of the channel belonging to the pillar of the previous bifurcation on which the cell crashed. The spontaneous left turn of the cell at the network exit illustrates this directional asymmetry.

The key elements for reaching the steady-state regime are therefore to present a shape asymmetry and/or to experience a shear in the horizontal channel, for example because of a difference in friction between the different walls of the channel. Indeed, in such cases the cell turns at the exit of the horizontal channel and directly engages in a diagonal channel without randomly hanging in the bifurcation.

It is in the bifurcations and the adjacent diagonal channels, where the stress is not symmetric, that the cells can deform asymmetrically. Moreover, there, the cells are pushed towards the downstream pillar, so that the thin lubrication fluid film between the cell and the downstream pillar can be partially drained to create an asymmetric friction between the cell and the channel walls. Bifurcations and adjacent diagonal channels are also the place where the cells stay the longest, thus allowing both consequent cell deformation and fluid drainage. The asymmetry must then be maintained until the next mesh after having crossed the horizontal channel, thus requiring this last step to be short enough to prevent the relaxation of the asymmetry.

To achieve the degree of asymmetry required for the steady-state regime, the cell must remain for an optimal time T^*_{asym} in the bifurcation and diagonal channel. If the residence time in the bifurcation and diagonal channel of the mesh n, $T_{bif}(n)$, is less than T^*_{asym},

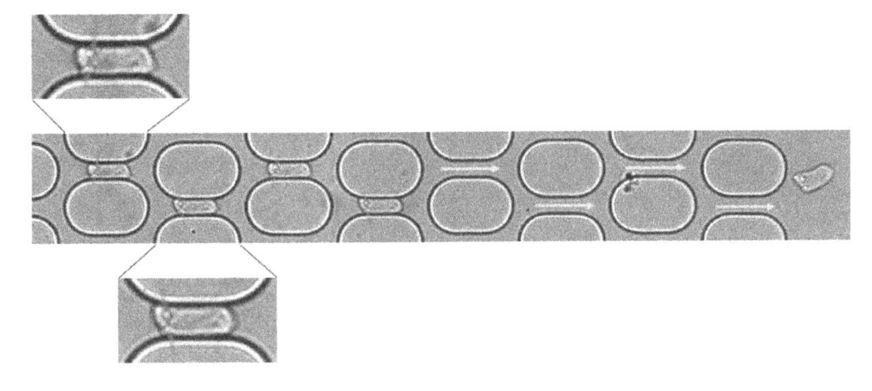

FIGURE 9.9
Asymmetric motion in the channels. Superimposed images of a cell passing through the network. The cell rear displays an asymmetric shape that alternates left and right along the successive horizontal channels. Red lines drawn between two bright dots in the cell (zoomed in the insets) highlight the rear asymmetry.

then the required deformation is not achieved. The cell will progress in the network after leaving the stagnation point of the bifurcation but will be likely to be blocked in the next mesh n+1 and therefore present a random trajectory. If $T_{bif}(n) \approx T^*_{asym}$, the cell has the optimal asymmetry to engage in the diagonal channel of mesh n+1, where it remains long enough to develop a new asymmetry mirroring that of mesh n, sufficiently marked not to block the cell at the next stagnation point. If $T_{bif}(n) > T^*_{asym}$, for example when the cell hangs for a long time at the stagnation point of the bifurcation n, it develops a higher degree of asymmetry than the one associated with T^*_{asym}, which causes it to pass very quickly through the bifurcation n+1 and engage in the diagonal channel n+1. In this case, $T_{bif}(n+1) < T_{bif}(n)$, which reduces the degree of asymmetry developed in mesh n+1 and slows down the cell as it passes the bifurcation n+2. This alternating strong and weak asymmetry is associated with fast and slow transits in successive meshes. Gradually, this behavior converges towards a stable dynamics with an optimal asymmetry developed during the steady-state residence time in the bifurcation, T_{bif-SS} which is equal to T^*_{asym}.

This reasoning is supported by examining the orders of magnitude of the cell deformation that can be reached during the residence time in the bifurcation and the diagonal channel in the steady-state regime, T_{bif-SS}. Experimentally, this time is of the order of 0.1s. As shown in the previous section, the cell deformation results from two mechanisms:

- The Kelvin-Voigt viscoelasticity allows a rapid and reversible deformation of the order of 10%. The characteristic time $T_1 \approx 0.02s$ is short, less than T_{bif-SS}. This indicates that the cells have the ability to stretch and compress rapidly to adapt to the geometrical changes within the mesh. However, as T_1 is of the order of the passage time through the horizontal channel, the asymmetric deformation cannot be fully maintained during the cell passage in the horizontal channel. Thus, this means that another mechanism is at play to keep a cell asymmetric deformation over a time longer than T_1.

- The viscous creep associated with the very high viscosity η_2 is a slower process. The time required to reach an elongation ratio ϵ in the bifurcation and diagonal channel is:

$$T \approx \frac{\eta_2 \epsilon}{\sigma_{bif}}, \tag{9.7}$$

where σ_{bif} is the effective stress experienced by the cell in the bifurcation/diagonal channel. According to Eq. 9.7, at the end of the time T_{bif-SS}, the deformation ratio $\approx T_{bif-SS}\frac{\sigma_{bif}}{\eta_2}$ is of the order of 10% if one takes $T_{bif-SS} \approx 0.1s$, $\eta_2 \approx 40Pa.s$, and $\sigma_{bif} \approx$

$40Pa$. This deformation hardly relaxes in the next horizontal channel because the passage time in it is much shorter than T_{bif-SS} (typically by a factor 3 or 4).

It is therefore the cell elements that have the high η_2 viscosity that allow the cell to asymmetrically deform by roughly 10% in the bifurcation and diagonal channel. As η_2 is high this deformation does not have the time to relax during the short cell passage in the next horizontal channel and prevents the cell to hang a long time in the following bifurcation.

Therefore, reaching a regular steady-state regime without being blocked in the channel network requires a remarkable adjustment of the mechanical properties of the cell. A rapid viscoelastic deformation time is required to adapt the cell shape in the network but it is also necessary that a slow viscous deformation can be established in order to create the asymmetry necessary to negotiate turns. Let's note that although a low cortical tension allows the cell to quickly modulate its shape to adapt to the geometry of the network, the risk is that the cell cannot resist the hydrodynamic stress in the channels and break or loose small vesicles in the flow. However, a too high tension does not allow sufficient levels of cell deformation, as seen for jasp-cells, which then progress by hops with long hanging periods during which they can finally adhere to the capillary wall. Therefore, untreated cells present the adequate mechanical combination combining cell integrity and regular progression.

9.3.6 Steady-state. Dynamics of cell transport

In the steady-state, a cell can be considered as progressing from mesh to mesh with an average mesh velocity v_{cell}. This velocity results from the balance between the driving hydrodynamic force due to the pressure difference applied in a mesh, $F_d \approx hw\Delta P_{mesh}$, and the friction force of the cell membrane on the channel walls. This latter force per unit of surface of lateral membrane (membrane close to a wall) is the shear stress exerted by the lubrication layer between the wall and the cell membrane $F_f \approx -S_f k_{vis} v_{cell}$, where S_f is the surface area of the cell close to the walls (estimated by considering that the cell has an almost parallelepipedic shape of length L, height h and width w: $S_f = 2L(h+w)$) and k_{vis} is the viscous coefficient due to the shearing force in the lubrication layer between the cell membrane and the channel wall ($k_{vis} = \eta_{out}/\delta$, where δ is the thickness of the lubrication layer). At constant v_{cell} the sum of forces is zero so that:

$$\frac{hw}{S_f} = k_{vis}\frac{v_{cell}}{\Delta P_{mesh}} \tag{9.8}$$

hw/S_f is plotted versus $v_{cell}/\Delta P_{mesh}$ in Figure 9.9 for cells transiting in channels of width $w < 11\mu m$. A linear dependence is observed with different slopes for lat-, untreated and jasp-cells, thus disclosing increasing friction from lat- to untreated to jasp-cells. Experimental data of k_{vis} and δ found by taking $\eta_{out} = 10^{-3}Pa.s$ are given in Table 9.2.

Surprisingly, the curves converge to a non-zero value when $v_{cell}/\Delta P_{mesh}$ tends to 0, in a behavior equivalent to a solid friction:

$$\frac{hw}{S_f} = k_{vis}\frac{v_{cell}}{\Delta P_{mesh}} + k_{sol} \tag{9.9}$$

where $k_{sol} \approx 0.09$. This term may be due to our simplified approach, which considers the average values of speed and friction within a mesh while these quantities vary greatly within a mesh.

The lubrication theory states that the thickness of the lubrication layer δ is inversely proportional to the tension gradient in the membrane [3]. It is written as:

$$\tau_F \approx \eta_{out} v_{cell} (\frac{\delta}{R_{cell} c_0})^{\frac{3}{2}} \tag{9.10}$$

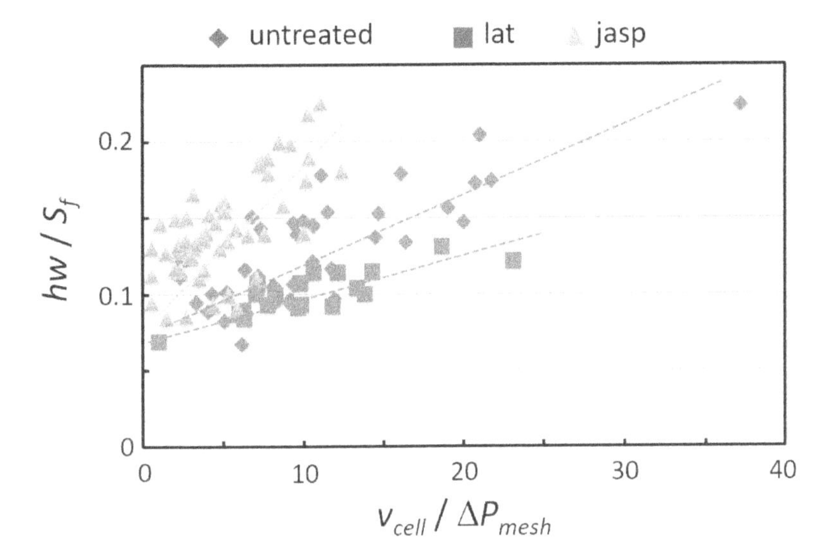

FIGURE 9.10
Experimental determination of k_{vis} andk_{sol} based on Eq.9.9.k_{vis} is the curve slope; k_{sol} is the curve value at $\frac{v_{cell}}{\Delta P_{mesh}} = 0$

where τ_F is the tension at the cell front and c_0 is a numerical pre-factor that differs very slightly in the literature ($c_0 = 2.123$ in Ref. [47] and $c_0 = 2.05$ in Ref. [5]). Values of the front cell tension τ_F, calculated from Eq. 9.10 with $v_{cell} \approx 1mm.s^{-1}$, are shown in Table 9.2. These values are in excellent agreement with the values measured from cell entrance in the micro-channel network and from micropipettes. The tension of lat-cells is very low while the tension of untreated cells is of the order of 10^{-4}N/m as reported in Figure 9.8D. Let's note that the value of the tension estimated from the lubrication layer is that of the front of the cell, higher than that in the rear part. Let's also note that the tension of jasp-cells is twice that of untreated cells, thus explaining their difficulty to reach the steady-state regime.

TABLE 9.2
Lubrication parameters and surface tension of cells flowing in the steady-state regime.

Cell	k_{vis} $(mPa.s.m^{-1})$	$\delta(nm)$	$\tau_F(N/m)$
Lat	2.41 ± 0.09	415 ± 162	$1.96\ 10^{-6}$
Untreated	4.00 ± 1.18	250 ± 74	$4.19\ 10^{-4}$
Jasp	7.59 ± 2.29	132 ± 40	$1.09\ 10^{-3}$

9.4 CONCLUSION

The WBC circulation in the vascular system of the lung is a crucial phenomenon for healthy immune watch. The monocyte mechanical properties are remarkably well adapted to its circulation. It is known that WBC nuclei are highly deformable and often multi-lobed the (case of neutrophils) to squeeze into the smallest capillaries, but it was not suspected how finely the cell membrane tension is adjusted. Tension is high enough to prevent the cell from breaking under the hydrodynamic elongational stresses, but low enough to allow a suitable deformation of the cell that prevents it from remaining blocked in a branching connection, and limits the risk of cell adhesion to the endothelium and capillary obstruction. The remarkable combination of mechanical properties enable WBCs to adapt rapidly and optimise their shape for passing the channels and take smooth turns at bifurcations.

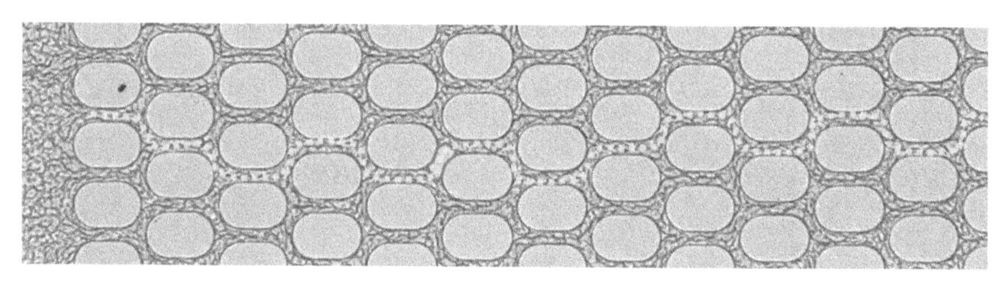

FIGURE 9.11
Motion of a THP1 cell in a suspension of concentrated red blood cells (RBCs) in the micro-channel network. The cell displays a trajectory (in yellow) and a shape (in red) similar to those observed in absence of RBCs.

It must be noted that the study presented above focused on the dynamics of individual WBCs. However, physiologically, they flow in a highly concentrated suspension of RBCs which are susceptible to affecting the WBC behavior. It is reasonable to think that the role of the crowded RBCs may be equivalent to an increase of the effective viscosity of the suspending fluid. Figure 9.11 indeed shows a qualitatively similar trajectory and deformation of a THP1 cell passing through the network while surrounded by RBCs. However, the behavior of deformable suspensions is tricky and this question could lead to interesting surprises.

ACKNOWLEDGMENTS

The author's research on the dynamics of monocytes in biomimetic lung capillaries benefited from the support of the project Projet ANR-09-BLAN-ChipCellTrap of the French National Research Agency (ANR). The authors warmly thank the students who have worked in this area: Frauke Beyer and Adeline Peignier, and our collaborators: Dr. Patrick Tabeling Dr. Mathilde Reyssat, Dr. Clemence Vergne, dr. Fabrice Monti, Dr. Jacques Magnaudet, Dr. Melanie Leroux, and Dr. Pierre-Henri Puech.

Bibliography

[1] John M Adams, Carl J Hauser, David H Livingston, Robert F Lavery, Zoltan Fekete, and Edwin A Deitch. Early trauma polymorphonuclear neutrophil responses to chemokines are associated with development of sepsis, pneumonia, and organ failure. *Journal of Trauma and Acute Care Surgery*, 51(3):452–457, 2001.

[2] Mark Bathe, Atsushi Shirai, Claire M Doerschuk, and Roger D Kamm. Neutrophil transit times through pulmonary capillaries: the effects of capillary geometry and fmlp-stimulation. *Biophysical Journal*, 83(4):1917–1933, 2002.

[3] Francis Patton Bretherton. The motion of long bubbles in tubes. *Journal of Fluid Mechanics*, 10(2):166–188, 1961.

[4] KA Brown, SD Brain, JD Pearson, JD Edgeworth, SM Lewis, and DF Treacher. Neutrophils in development of multiple organ failure in sepsis. *The Lancet*, 368(9530):157–169, 2006.

[5] Robijn Bruinsma. Rheology and shape transitions of vesicles under capillary flow. *Physica A: Statistical Mechanics and its Applications*, 234(1-2):249–270, 1996.

[6] Andreas Carlson, Minh Do-Quang, and Gustav Amberg. Droplet dynamics in a bifurcating channel. *International Journal of Multiphase Flow*, 36(5):397–405, 2010.

[7] Claire M Doerschuk. Mechanisms of leukocyte sequestration in inflamed lungs. *Microcirculation*, 8(2):71–88, 2001.

[8] Claire M Doerschuk, Michael F Allard, and James C Hogg. Neutrophil kinetics in rabbits during infusion of zymosan-activated plasma. *Journal of Applied Physiology*, 67(1):88–95, 1989.

[9] Claire M Doerschuk, Michael F Allard, Bridget A Martin, A MacKenzie, AP Autor, and JC Hogg. Marginated pool of neutrophils in rabbit lungs. *Journal of Applied Physiology*, 63(5):1806–1815, 1987.

[10] CM Doerschuk, N Beyers, HO Coxson, B Wiggs, and JC Hogg. Comparison of neutrophil and capillary diameters and their relation to neutrophil sequestration in the lung. *Journal of Applied Physiology*, 74(6):3040–3045, 1993.

[11] CM Doerschuk, GP Downey, DE Doherty, D English, RP Gie, M Ohgami, GS Worthen, PM Henson, and JC Hogg. Leukocyte and platelet margination within microvasculature of rabbit lungs. *Journal of Applied Physiology*, 68(5):1956–1961, 1990.

[12] Cheng Dong, R Skalak, Kuo-Li Paul Sung, GW Schmid-Schonbein, and Shu Chien. Passive deformation analysis of human leukocytes. *Journal of Biomechanical Engineering*, 110(1):27–36, 1988.

[13] Jeanie L Drury and Micah Dembo. Hydrodynamics of micropipette aspiration. *Biophysical Journal*, 76(1):110–128, 1999.

[14] E Evans and A Yeung. Apparent viscosity and cortical tension of blood granulocytes determined by micropipet aspiration. *Biophysical Journal*, 56(1):151–160, 1989.

[15] Sylvain Gabriele, Marie Versaevel, Pascal Preira, and Olivier Théodoly. A simple microfluidic method to select, isolate, and manipulate single-cells in mechanical and biochemical assays. *Lab on a Chip*, 10(11):1459–1467, 2010.

[16] Sarah A Gebb, Jacquelyn A Graham, Christopher C Hanger, Patricia S Godbey, Ronald L Capen, Claire M Doerschuk, and WW Wagner Jr. Sites of leukocyte sequestration in the pulmonary microcirculation. *Journal of Applied Physiology*, 79(2):493–497, 1995.

[17] Karine Guevorkian, Marie-Josée Colbert, Mélanie Durth, Sylvie Dufour, and Françoise Brochard-Wyart. Aspiration of biological viscoelastic drops. *Physical Review Letters*, 104(21):218101, 2010.

[18] WG Guntheroth, Daniel L Luchtel, and Isamu Kawabori. Pulmonary microcirculation: tubules rather than sheet and post. *Journal of Applied Physiology*, 53(2):510–515, 1982.

[19] Marc Herant, William A Marganski, and Micah Dembo. The mechanics of neutrophils: synthetic modeling of three experiments. *Biophysical Journal*, 84(5):3389–3413, 2003.

[20] Mark Hirsh, Eugenia Mahamid, Yulia Bashenko, Irina Hirsh, and Michael M Krausz. Overexpression of the high-affinity fcgamma receptor (cd64) is associated with leukocyte dysfunction in sepsis. *Shock (Augusta, Ga.)*, 16(2):102–108, 2001.

[21] RM Hochmuth, HP Ting-Beall, BB Beaty, D Needham, and R Tran-Son-Tay. Viscosity of passive human neutrophils undergoing small deformations. *Biophysical Journal*, 64(5):1596–1601, 1993.

[22] Robert M Hochmuth. Micropipette aspiration of living cells. *Journal of Biomechanics*, 33(1):15–22, 2000.

[23] James C Hogg. Neutrophil kinetics and lung injury. *Physiological Reviews*, 67(4):1249–1295, 1987.

[24] James C Hogg, T McLean, BA Martin, and B Wiggs. Erythrocyte transit and neutrophil concentration in the dog lung. *Journal of Applied Physiology*, 65(3):1217–1225, 1988.

[25] JC Hogg, HO Coxson, ML Brumwell, N Beyers, CM Doerschuk, W MacNee, and BR Wiggs. Erythrocyte and polymorphonuclear cell transit time and concentration in human pulmonary capillaries. *Journal of Applied Physiology*, 77(4):1795–1800, 1994.

[26] Yaqi Huang, Claire M Doerschuk, and Roger D Kamm. Computational modeling of rbc and neutrophil transit through the pulmonary capillaries. *Journal of Applied Physiology*, 90(2):545–564, 2001.

[27] Geoffrey C Ibbotson, Christopher Doig, Jaswinder Kaur, Varinder Gill, Lena Ostrovsky, Todd Fairhead, and Paul Kubes. Functional α 4-integrin: a newly identified pathway of neutrophil recruitment in critically ill septic patients. *Nature Medicine*, 7(4):465, 2001.

[28] Beat A Imhof and Michel Aurrand-Lions. Adhesion mechanisms regulating the migration of monocytes. *Nature Reviews Immunology*, 4(6):432, 2004.

[29] Roger D Kamm. Cellular fluid mechanics. *Annual Review of Fluid Mechanics*, 34(1):211–232, 2002.

[30] Wolfgang M Kuebler, GE Kuhnle, Joachim Groh, and Alwin E Goetz. Leukocyte kinetics in pulmonary microcirculation: intravital fluorescence microscopic study. *Journal of Applied Physiology*, 76(1):65–71, 1994.

[31] Wolfgang M Kuebler, Gerhard EH Kuhnle, and Alwin Eduard Goetz. Leukocyte margination in alveolar capillaries: interrelationship with functional capillary geometry and microhemodynamics. *Journal of Vascular Research*, 36(4):282–288, 1999.

[32] Kathryn Prame Kumar, Alyce J Nicholls, and Connie HY Wong. Partners in crime: neutrophils and monocytes/macrophages in inflammation and disease. *Cell and Tissue Research*, 371(3):551–565, 2018.

[33] Klaus Ley, Carlo Laudanna, Myron I Cybulsky, and Sussan Nourshargh. Getting to the site of inflammation: the leukocyte adhesion cascade updated. *Nature Reviews Immunology*, 7(9):678, 2007.

[34] DC Lien, WW Wagner, RL Capen, C Haslett, WL Hanson, SE Hofmeister, PM Henson, and GS Worthen. Physiologic neutrophil sequestration in the canine pulmonary circulation. *J. Appl. Physiol*, 62:1236–1243, 1987.

[35] W MacNee, BA Martin, BR Wiggs, AS Belzberg, and JC Hogg. Regional pulmonary transit times in humans. *Journal of Applied Physiology*, 66(2):844–850, 1989.

[36] BA Martin, BR Wiggs, S Lee, and JC Hogg. Regional differences in neutrophil margination in dog lungs. *Journal of Applied Physiology*, 63(3):1253–1261, 1987.

[37] BA Martin, Joanne L Wright, Harvey Thommasen, and JC Hogg. Effect of pulmonary blood flow on the exchange between the circulating and marginating pool of polymorphonuclear leukocytes in dog lungs. *The Journal of Clinical Investigation*, 69(6):1277–1285, 1982.

[38] AL Muir, M Cruz, BA Martin, H Thommasen, A Belzberg, and JC Hogg. Leukocyte kinetics in the human lung: role of exercise and catecholamines. *Journal of Applied Physiology*, 57(3):711–719, 1984.

[39] Masato Nishino, Hiroshi Tanaka, Hiroshi Ogura, Yoshiaki Inoue, Taichin Koh, Kieko Fujita, and Hisashi Sugimoto. Serial changes in leukocyte deformability and whole blood rheology in patients with sepsis or trauma. *Journal of Trauma and Acute Care Surgery*, 59(6):1425–1431, 2005.

[40] Pascal Preira, Jean-Marie Forel, Philippe Robert, Paulin Nègre, Martine Biarnes-Pelicot, Francois Xeridat, Pierre Bongrand, Laurent Papazian, and Olivier Theodoly. The leukocyte-stiffening property of plasma in early acute respiratory distress syndrome (ards) revealed by a microfluidic single-cell study: the role of cytokines and protection with antibodies. *Critical Care*, 20(1):8, 2015.

[41] Pascal Preira, Thomas Leoni, Marie-Pierre Valignat, Annemarie Lellouch, Philippe Robert, Jean-Marie Forel, Laurent Papazian, Guillaume Dumenil, Pierre Bongrand, and Olivier Thé. Microfluidic tools to investigate pathologies in the blood microcirculation. *International Journal of Nanotechnology*, 9(3):529, 2012.

[42] Pascal Preira, Marie-Pierre Valignat, José Bico, and Olivier Théodoly. Single cell rheometry with a microfluidic constriction: Quantitative control of friction and fluid leaks between cell and channel walls. *Biomicrofluidics*, 7(2):024111, 2013.

[43] RG Presson Jr, JA Graham, CC Hanger, PS Godbey, SA Gebb, RA Sidner, RW Glenny, and WW Wagner Jr. Distribution of pulmonary capillary red blood cell transit times. *Journal of Applied Physiology*, 79(2):382–388, 1995.

[44] JM Rallison. The deformation of small viscous drops and bubbles in shear flows. *Annual Review of Fluid Mechanics*, 16(1):45–66, 1984.

[45] Fabienne Richelme, Anne Marie Benoliel, and Pierre Bongrand. Aspiration of thp1 into a micropipette. Mechanical deformation of monocytic thp-1 cells: occurrence of two sequential phases with differential sensitivity to metabolic inhibitors. *Experimental Biology Online*, 2(5):1–14, 1997.

[46] GW Schmid-Schönbein, KL Sung, H Tözeren, R Skalak, and S Chien. Passive mechanical properties of human leukocytes. *Biophysical Journal*, 36(1):243–256, 1981.

[47] Timothy W Secomb, R Skalak, N Özkaya, and JF Gross. Flow of axisymmetric red blood cells in narrow capillaries. *Journal of Fluid Mechanics*, 163:405–423, 1986.

[48] Atsushi Shirai. Modeling neutrophil transport in pulmonary capillaries. *Respiratory Physiology & Neurobiology*, 163(1-3):158–165, 2008.

[49] Atsushi Shirai, Ryo Fujita, and Toshiyuki Hayase. Simulation model for flow of neutrophils in pulmonary capillary network. *Technology and Health Care*, 13(4):301–311, 2005.

[50] Atsushi Shirai, Sunao Masuda, and Toshiyuki Hayase. 3-d numerical simulation of flow of a neutrophil for the retention time in a moderate constriction of a rectangular microchannels. In *Proceedings of ASME 2006 Summer Bioengineering Conference*, 2006.

[51] Richard Skalak, Cheng Dong, and Cheng Zhu. Passive deformations and active motions of leukocytes. *Journal of Biomechanical Engineering*, 112(3):295–302, 1990.

[52] Timothy A Springer. Traffic signals for lymphocyte recirculation and leukocyte emigration: the multistep paradigm. *Cell*, 76(2):301–314, 1994.

[53] Prithu Sundd, Maria K Pospieszalska, Luthur Siu-Lun Cheung, Konstantinos Konstantopoulos, and Klaus Ley. Biomechanics of leukocyte rolling. *Biorheology*, 48(1):1–35, 2011.

[54] KL Sung, Cheng Dong, Geert W Schmid-SchOenbein, S Chien, and Richard Skalak. Leukocyte relaxation properties. *Biophysical Journal*, 54(2):331–336, 1988.

[55] R Tran-Son-Tay, H-C Kan, HS Udaykumar, E Damay, and W Shyy. Rheological modelling of leukocytes. *Medical and Biological Engineering and Computing*, 36(2):246–250, 1998.

[56] Shigeru Tsuchiya, Michiko Yamabe, Yoshiko Yamaguchi, Yasuko Kobayashi, Tasuke Konno, and Keiya Tada. Establishment and characterization of a human acute monocytic leukemia cell line (thp-1). *International Journal of Cancer*, 26(2):171–176, 1980.

[57] SY Wang, KL Mak, LY Chen, MP Chou, and CK Ho. Heterogeneity of human blood monocyte: two subpopulations with different sizes, phenotypes and functions. *Immunology*, 77(2):298, 1992.

[58] Yuli Wang, Minh Do-Quang, and Gustav Amberg. Viscoelastic droplet dynamics in a y-shaped capillary channel. *Physics of Fluids*, 28(3):033103, 2016.

[59] Ewald R Weibel, André Frédérick Cournand, and Dickinson W Richards. *Morphometry of the Human Lung*, volume 1. Springer, 1963.

[60] G Scott Worthen, BIII Schwab, Elliot L Elson, and Gregory P Downey. Mechanics of stimulated neutrophils: cell stiffening induces retention in capillaries. *Science*, 245(4914):183–186, 1989.

[61] Pei-Hsun Wu, Dikla Raz-Ben Aroush, Atef Asnacios, Wei-Chiang Chen, Maxim E Dokukin, Bryant L Doss, Pauline Durand-Smet, Andrew Ekpenyong, Jochen Guck, Nataliia V Guz, et al. A comparison of methods to assess cell mechanical properties. *Nature Methods*, 15:491–498, 2018.

[62] Belinda Yap and Roger D Kamm. Mechanical deformation of neutrophils into narrow channels induces pseudopod projection and changes in biomechanical properties. *Journal of Applied Physiology*, 98(5):1930–1939, 2005.

[63] Kazuo Yoshida, Ryoichi Kondo, Qin Wang, and Claire M Doerschuk. Neutrophil cytoskeletal rearrangements during capillary sequestration in bacterial pneumonia in rats. *American Journal of Respiratory and Critical Care Medicine*, 174(6):689–698, 2006.

[64] Chunfeng Zhou, Pengtao Yue, and James J Feng. Simulation of neutrophil deformation and transport in capillaries using newtonian and viscoelastic drop models. *Annals of Biomedical Engineering*, 35(5):766–780, 2007.

CHAPTER 10

Inertial microfluidics and its applications in hematology

Wonhee Lee

KAIST, Daejeon, Republic of Korea

CONTENTS

10.1 INTRODUCTION

THE INERTIAL MICROFLUIDICS FIELD has emerged from the finding of the importance of fluid inertia in microfluidic flows [4, 27, 66, 79, 134]. Microfluidic flows had been considered as low-Reynolds number flows or Stokes flow due to the small length scale of microchannels. The Reynolds number ($Re = \rho_f U H/\mu_f$, where ρ_f is fluid density, U is the average velocity, H is the hydraulic diameter of the channel and μ_f is the fluid viscosity) of microfluidic flows are typically smaller than 1 [108]. As Re is the dimensionless number describing the ratio between inertial forces and viscous forces, inertial forces were often ignored to solve microscale fluid dynamics. However, Re in microfluidic flows can be as large as a few 100 with high flow velocity ($U > 0.1$ m/s). Under such conditions, many interesting inertial effects have been discovered from microfluidic systems and were used in many applications [4, 27, 66, 79, 134]. Inertial effects that are most widely studied in microfluidic systems would be inertial migration and secondary flows such as Dean flows. Inertial migration refers to the cross-stream migration of microparticles due to inertial lift forces, which eventually leads to particle focusing into equilibrium positions. The equilibrium positions or the focusing positions can differ by particle properties including size and deformability, which allows hydrodynamic microparticle separations. Unlike in the case of low-Reynolds number microfluidic systems, channel geometry of inertial microfluidics is an important control parameter of microfluidic functions because geometries like curved structures can generate secondary flows orthogonal to the main flow direction, which had been used for fluid and particle manipulations. In the first part of this chapter, we have summarized the theoretical background of inertial microfluidics that is important for applications in the field of hematology. For more details of inertial microfluidic physics, there are several well-written review papers available [4,27,66,79,134].

In the second part, we reviewed two main applications of inertial microfluidics in hematology: separations and single cell analysis. Blood consists of many components of varying size from tens of microns to sub-microns; red blood cells (RBCs), white blood cells (WBCs), platelets and other rare cells such as bacteria and circulating tumor cells (CTCs) are microscale size, and plasma contains various small molecules of nanoscale size. These components contain much important information indicating physiological states, and separation of the components is often the most important first step of appropriate diagnosis and treatment [66,100,130]. As conventional label-free methods for separation, centrifugation and mechanical filtration have been widely used due to a simple process and large-scale operation. However, these methods have disadvantages regarding purity and yield. Other widely used conventional sorting techniques are fluorescent-activated cell sorting (FACS) and magnetic-activated cell sorting (MACS), which are antibody-based techniques. FACS allows high purity and yield for cell separation, however, disadvantages such as large instrumentation, cell damage, limited throughput, and high costs limit the applications [66]. MACS have lower purity and yield than FACS but provide higher throughput. The difficulty of magnetic particle removal after separation is another downside. Both techniques have common issues associated with labeling: the complexity of the process, cost, and possible interference with cell functions. Microfluidic systems for cell sorting and separation are often suggested as solutions to the limitations of the conventional methods due to advantages of small sample volume, low cost, portability, disposability, high efficiency, and diverse separation principles. Depending on the separation principles, microfluidic separation techniques can be categorized into two types: active separation that requires external force fields for particle manipulation and passive separation that relies on channel structure and hydrodynamic forces. The active separation methods such as dielectrophoresis [34,122], acoustophoresis [61,106], magnetophoresis [33,91] and optophoresis [89] generally offer high separation efficiency and sensitivity, however, have drawbacks such as low throughput and high device complexity. The passive separation methods such as pinch flow fractionation [113,129], hydrophoresis [13,14], inertial and Dean flow fractionation [29,60,120,125] and deterministic lateral displacement can mitigate the drawbacks of active methods. Among the passive techniques, inertial microfluidics has attracted much attention recently due to its extremely high-throughput with fast flow velocity (up to ~ 1 m/s). Inertial microfluidics have also been widely used as means to align or position cells for single-cell analysis. Flow cytometry has been a popular tool for single cell analysis such as expression level of the cell surface and intracellular molecules, characterization and detection of different cell types. Flow cytometry requires tightly focused cell stream for accurate measurements, where inertial microfluidic cell focusing can be a promising method to replace existing cell focusing apparatus. In addition to flow cytometry, there has been a rapidly growing interest in microfluidics technology as an approach to building new types of single cell analysis tools with higher sensitivity and resolution. Physical biomarkers associated with cytoskeletal and membrane composition and organization have been reported as powerful label-free cell analyzer [19, 21, 22]. The mechanical factors such as deformability can be directly used for cell state identifiers using inertial microfluidics approach, for example, single-stream 3D particle focuser and microfluidic cell stretcher [16, 25]. Since focused cells and its deformability information was visualized and analyzed similarly to traditional flow cytometry, the inertial microfluidic device could be demonstrated as a high-speed single cell analysis platform.

10.2 PHYSICS OF INERTIAL MICROFLUIDICS

10.2.1 Inertial focusing of particles at finite-*Re* flows

10.2.1.1 Inertial focusing in straight channels

Inertial lift forces

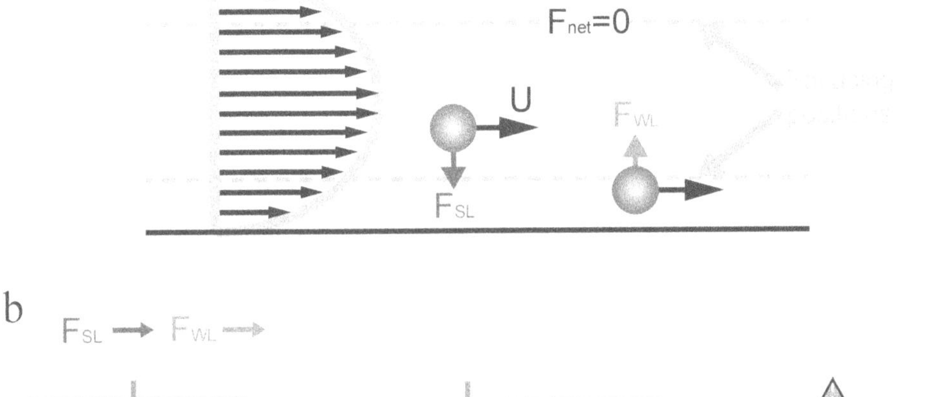

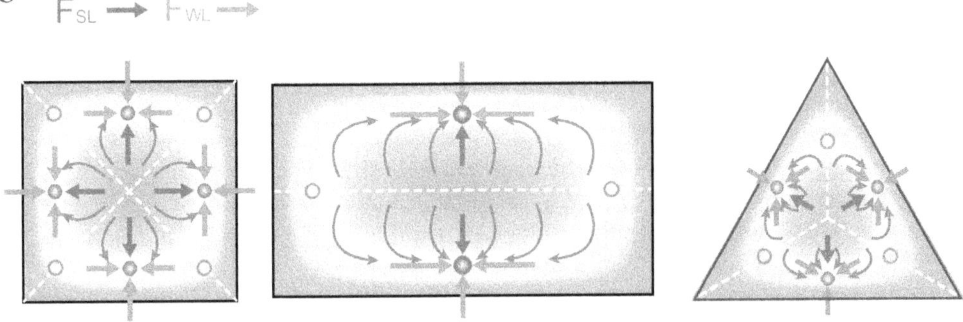

FIGURE 10.1

Mechanism of inertial focusing. (a) Two major lift forces opposing each other: the shear gradient lift force pushes the particle towards the wall (purple arrow) and the wall effect lift force pushes the particle away from the wall (green arrow). The balance of two lift forces determines focusing positions ($F_{net} = 0$). (b) The locations of focusing positions can be altered by the shape of the channel cross-section. Channel cross-sections can be divided into several basins of attraction (white lines) which include each attractor, that is, focusing positions (solid blue circles). Randomly distributed particles at the inlet follow the trajectories of force fields (blue arrows) and focus to the attractors. The stable focusing positions are determined by the lift forces and force map.

Within finite-Reynolds-number flows where both inertial and viscous forces are important, spherical particles can experience inertial lift forces perpendicular to the main flow direction, in addition to viscous drag forces [4, 27, 79, 134]. Two different lift forces are recognized main players determining inertial focusing of microparticles: the shear gradient lift force and the wall effect lift force (Figure 10.1a) [5,44]. In microfluidic channels, flow velocity profiles in the cross-section of the channel typically have parabolic shapes that have a gradient in shear rate. The shear gradient lift force pushes particles towards the walls under these simple parabolic velocity profiles. Near the wall region, wall effect lift force pushes particle away from the wall and the opposing two lift forces form equilibrium positions. It took almost a decade to develop the fluid mechanic theory to explain the inertial focusing since it was first observed by Segre and Silberberg [104]. It is still difficult to provide a simple intuitive explanation for the lift forces. According to Ho and Leal [44], the wall effect lift is an interaction of a disturbance stresslet and its wall reflection with bulk shear and the shear gradient lift is an interaction of a disturbance stresslet with the curvature of velocity. This theory also suggests that the shear gradient lift force can be expressed by the product of shear rate and the gradient of shear rate [44]. While a simple parabolic velocity profile has a unidirectional gradient of shear towards the wall, complex cross-section shape may allow a complex velocity profile with varying shear gradient direction, which can lead to alteration of the direction of shear gradient lift force towards the channel center [4].

The mismatch between the motion of a particle and the surrounding fluid leads to weak lift forces. Near the wall of microfluidic channels, particles generally lag the surrounding fluids due to the influence of the walls. The slip-shear lift force (or Saffman lift force) occurs when a particle falls behind or leads ahead of the fluid and its direction is towards the maximum relative velocity. When the particles lead the fluid flow, the slip-shear lift force is towards the channel walls, whereas when the particles lag the fluid flow, it is pointed towards the channel center [103]. In addition, the rotational difference between the particle and underlying flow induces the slip-spin lift forces towards the channel center [102]. Since the magnitude of slip shear and slip-spin is scaling with one and three orders higher of a/H (a is the particle size) compared to the wall-effect and shear gradient lift forces, these minor lift forces are not applicable in most cases with small a/H. The minor lift forces may become important in special circumstances such as a microfluidic system with external forces acting on the particle to lag or lead fluid and non-neutrally buoyant particles in a vertical flow [57].

Channel cross-section dependency

Inertial focusing positions are sensitive to the cross-sectional shape of the microchannel because both shear gradient lift force and wall-effect lift forces are strongly influenced by the cross-sectional shape of microfluidic channels. In general, focusing positions are determined by the balance of two major lift forces. In circular tubes with rotational symmetry, two lift forces are always in opposite directions and the focusing positions appear in the form of an annulus at $\sim 0.6r_c$ (r_c is the radius of the channel) from the center of the channel, which was first observed in macroscale tubing [104]. Inertial focusing in the microchannel has been extensively studied with a rectangular cross-section simply because the widely adapted microchannel fabrication techniques based on soft lithography lead to such cross-sectional shapes. Rectangular microchannels exhibit interesting inertial migration patterns depending on the aspect ratio (Figure 10.1b). Four stable focusing positions are typical in square channels following the channel symmetry [28]. With the changes of aspect ratio, the number of stable focusing positions changes with the changes of velocity profile shape. The blunt velocity profile along the wide channel walls leads to the dominant shear gradient lift force towards the long channel walls. Here, particles first migrate towards the long channel walls by shear gradient lift and then focus towards the middle of the long channel walls by wall effect lift. Consequently, the rectangular channel with high or low aspect ratio shows a reduction in the number of focusing positions from four to two [40]. Recent studies revealed even more intriguing features of inertial focusing in triangular channels [56]. Three focusing positions near each channel face were found for triangular channels and these focusing positions changed with vertex angles and Re. Especially, the top two focusing positions were found to be shifting away from the apex with increasing Re. Currently, the mechanism of focusing position shift is not fully understood.

The connection of channels with different cross-sections allows focused particle streams to change their positions in the channel, which can lead to interesting applications. For example, kinetic separation of particles can be achieved by connecting rectangular channels with two different aspect ratios [135]. First, particles are focused at two side focusing positions in the high aspect ratio rectangular channel. Then, particles focus to the center of the channel in the low aspect ratio channel. Larger particles migrate faster than smaller particles to the new focusing positions because of the strong dependency of the magnitude of lift forces on particle size, which allows kinetic separation of particles. Similarly, particles can be transferred from one solution to another at the time scale as fast as 1 ms by transferring particles from one focusing position to another [41].

The channel cross-section can be divided into basins of attraction including each attractor, or focusing position. The basins of attraction are indicated as the areas divided by white dashed lines in Figure 10.1b. Particles within each basin of attraction at the inlet will migrate into the corresponding focusing position after following the force map (blue lines in Figure 10.1b). Mapping of focusing positions in one channel to basins of attraction in the connecting channel allows not only a change of positions of focused particles but also enables control of accessible focusing positions [56]. Single stream particle focusing was achieved by connecting channels with various cross-section shapes (high-aspect ratio rectangle to triangle to half-circle). Both focusing positions in the rectangular channel map to the basins of attraction of the top focusing positions in the triangular channel and access to the bottom focusing position is restricted in this specific example. Then, the particles from top focusing positions in the triangular channel migrate into the top focusing position in the half-circular channel. This method suggests the channel cross-sectional shape as an important control parameter for inertial microfluidic particle manipulations.

Particle size and Reynolds number dependency

The focusing positions change with particle size and Re in a straight rectangular channel. The inertial lift force is strongly dependent on particle size ($F_L \propto \rho U^2 a^4/H^2$, for $a/H \ll 1$ [62], $F_L \propto \rho U^2 a^3/H$ near the channel center and $F_L \propto \rho U^2 a^6/H^4$ near the channel wall for a/H between 0.05 and 0.2 [28]. Apart from strong size dependency of lift force, focusing positions barely change with a particle size in case of $a/H \ll 1$. However, when particle size becomes comparable to the channel dimension ($a/H \sim 0.1 - 1$), steric effects dominate over the lift forces and the focusing positions shift towards the channel center as the particle size increases [28].

In general, the focusing positions change only slightly depending on Re; the focusing positions get closer to channel walls with increasing Re because the increase in shear gradient lift force is larger than the increase in wall effect lift force [5, 40]. More noticeably, two unstable focusing positions located near the short channel faces can be stabilized in case of rectangular channels as the Re is increased [18, 72]. Liu et al. found additional splitting of focusing positions near the long channel faces from a rectangular channel with small particle size or high Re [72]. Recently, we found the focusing position shift depending on particle size and Re in triangular channels [55]. The inertial focusing position shift was investigated using isosceles right triangular channels and the number and the locations of focusing positions were found to be changing depending on particle size and Re. The top focusing positions shift towards the apex with decreasing Re number and they can merge into a single focusing position near the apex in the case of large a/H. At the same Re condition, the focusing positions shift towards the apex with increasing particle size. The mechanism of the focusing position shift is currently not fully understood. Nevertheless, the strong size-dependent focusing position shift was utilized for separation applications and highly efficient microparticle and cell separations were demonstrated.

10.2.1.2 Inertial focusing in non-straight channels

Spiral and serpentine channels

Fluid inertia can induce secondary flows perpendicular to the main flow direction in non-straight channels. The most widely studied secondary flow is Dean flow in curved channels, which is formed by the uneven fluid inertia across the channel because of the difference in fluid velocity between the center of the channel and near the walls. Since the fluid inertia is larger at the center of the channel than the fluid near the channel walls, a centrifugal pressure gradient is created in the radial direction of the curving channel. By the conservation of mass, faster-moving fluid near the center of the channel experiences a force outward and fluid near the wall will re-circulate inward [4,27,79,134]. Therefore, two symmetrical counter-rotating vortices (Dean vortices) are created on top and bottom of the channel (Figure 10.2a). The magnitude and shape of the Dean vortices can be estimated with the Dean number given by $De = (\rho_f UH/\mu_f)(H/2R)^{1/2} = Re(H/2R)^{1/2}$ where R is the radius of curvature [23,24]. Dean number increases with the smaller curvature of radius as well as larger channel dimension and higher flow rate. Increase in Dean number is related to speed and shape of Dean flow velocity which will end up shifting the symmetric vortices towards the outer wall of the channel [6]. Generally, the inertia-induced secondary flows are noticeable when the Dean number is greater than 1 [7].

Inertial lift forces

Dean flows were first recognized for their effectiveness in mixing fluids, producing hydrodynamic focusing and modifying the curvature of fluid interfaces [76,110]. Dean flows are also useful for manipulating particles because Dean flows apply viscous drag (Dean drag force: F_D) on particles. The Dean drag force aids inertial lift force and allows efficient methods for concentrating cells and particles in certain positions in microfluidic channels, which was successfully implemented for continuous particle or cell separation in a high-throughput manner using non-straight channel geometries such as spiral [7,42,60,78,105,120,126,127] and serpentine channels [29,30,131–133].

Dean drag force can significantly change inertial focusing positions of straight channels. Finding exact focusing positions within curved channel geometry is not a simple task and mostly done empirically because inertial lift forces and Dean drag forces have not only different magnitudes and directions depending on channel positions but also different scaling with the size of the particles and flow velocity. Generally, low aspect-ratio rectangular channels are used to maximize the effect of Dean drag force, the focusing positions in these curving channels being located near the inner wall (Figure 10.2a), which can be interpreted as shifting from original focusing positions straight channel due to Dean drag force. More specifically, in the rectangular cross-section spiral channel, two symmetrical Dean vortices are placed in the horizontal direction which creates a force balance between inertial lift force and Dean drag force. As the Dean number increases(e.g., channel curvature increases), the focusing position of particles starts to shift closer to the inner wall. For different size particles, the focusing position of relatively larger particles shifts further closer

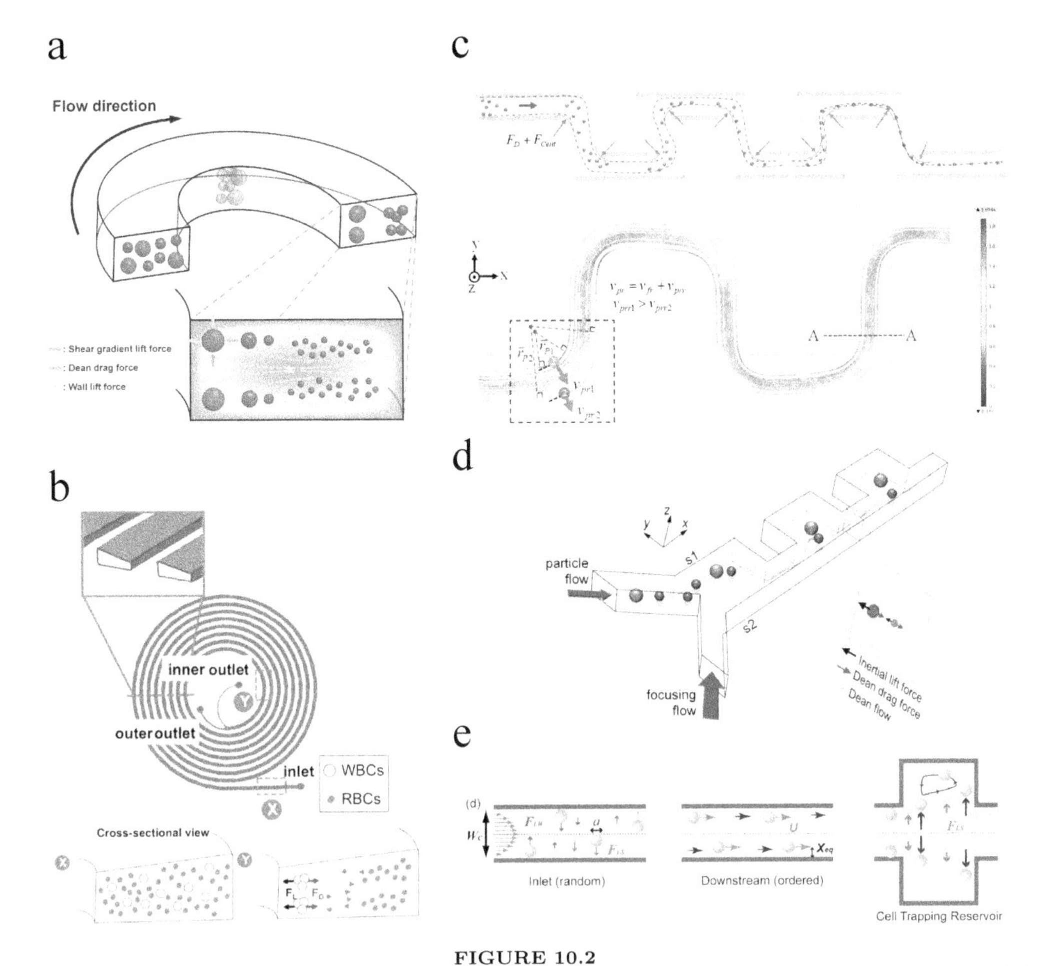

FIGURE 10.2
(a) Schematic illustration of inertial focusing in a curved microchannel. Particles' focusing positions are shifted towards the inner wall corners by the balance of wall effect lift force and Dean drag force. Larger particles are pushed more toward the inner wall by the dominant Dean drag force. Particles smaller than a certain size cannot focus and circulate within the core of Dean vortices. (b) Spiral channel with trapezoid cross-section allows highly efficient size-based separation [123]. Different focusing positions of RBCs and WBCs are illustrated with altered Dean vortex shape. (c) Inertial focusing in a symmetric serpentine channel [131]. Difference in density of fluidic and particle leads to particle migration towards the channel center. (d) Expansion-contraction structure creates secondary flow similar to Dean flow [64]. Additional drag force due to the secondary flow allows size dependent focusing positions. (e) Particle trapping in vortices at expansion-contraction channel [50]. Particles migrate and enter the cell trapping reservoirs by dominant shear gradient lift force due to the sudden absence of the side wall. Larger particles are more susceptible to the lift forces, which allows selective trapping.

to the inner wall due to dominant Dean drag influence near the top and bottom channel walls. While large particles have stable focusing positions, small particles' motion may be entirely dominated by Dean drag and eventually get trapped at the core of Dean vortices located at the center of channel width. This broad distribution of small particles can lead to low separation efficiency when using the rectangular cross-sectional spiral channel in particle/cell separation [60]. To increase the spacing between the distribution of two different size particles, spiral channels with trapezoidal cross-section were devised (Figure 10.2b) [123]. The spiral channel with trapezoidal shape cross-section deforms the shape of the velocity field and forms strong Dean vortex cores skewed toward the outer wall with larger channel depth which results in a dramatic shift of distributions of small particles toward the outer wall without affecting the focusing position of large particles. Therefore, a greater difference in equilibrium positions between two different particles results in higher separation efficiency [123].

Although the spiral channels have a great potential for cell/particle separation applications, the difficulties in parallelization limit its throughput. Preferably, a linear channel structure is suitable for parallelization design. Therefore, regarding concise design and easy parallelization, a serpentine channel with linear structure was studied as an alternative to spiral channels (Figure 10.2c) [131]. Repeated alteration of channel curvature complicates the analysis of secondary flows and particle motion in the serpentine channels and the inertial focusing mechanism in serpentine channels is not completely understood at this point. Asymmetric serpentine channels can be mostly understood with a similar theory of spiral channel. Differences are the smaller radius of curvature and disturbance from the shorter curving channel which has a minor effect on overall focusing. Asymmetric serpentine channels are known to achieve particle focusing and separation within a shorter length than a straight channel due to the aid of secondary flows [30,39]. Symmetric curved channels have opposing channel segments that lead to the Dean flows counteracting each other, which may reduce the overall effect of Dean drag and separation efficiency. In the case of channels with a sharp turn, a centrifugal force (F_{Cent}) on non-neutrally buoyant particles was suggested as an additional force changing particle trajectory (Figure 10.2c) [131]. The F_{Cent} can be given as $F_{Cent} = (\rho_p - \rho_f)\pi a^3 3\nu_{pt}^2/3a$ where ρ_p and ν_{pt} are the density and tangential velocity of particles, respectively. Symmetric serpentine channels were also applied to separate two different size particles/cells [131,132].

Expansion-contraction channels

Not only the curvatures in the channels like spiral and serpentine channels but also series of expansion-contraction in straight channel design can efficiently generate secondary flows. Especially, sudden expansion-contraction structure leads to vortex formation, which allows particle manipulation including separation and trapping [94]. Flow along series of expansion-contraction geometries results in two important features: Dean-like secondary flows in the center of the channel cross-section and a horizontal vortex in the expansion-contraction region (Figure 10.2d and e) [50,64]. The properties of the secondary flows and vortex are dependent on various factors including flow velocity, expansion-contraction and main channel dimensions, surface roughness, and the angle and roundness of the orifice corner [94].

Similar to the case of curved channels, drag force due to the secondary flow combined with inertial lift forces changes inertial focusing patterns in a straight channel. In the expansion channel regions, the wall-induced lift forces toward the channel center suddenly become weaker because the side walls have disappeared. The particles are affected dominantly by the shear gradient lift forces and the secondary flows toward the channel walls. In the contraction regions, the Dean-like secondary flows change their rotational directions and push the particle towards the channel center. Passing through series of the expansion and contraction channels allows fast particles focusing. This focusing mechanism leads to different focusing positions for different size particles because the shear gradient lift and drag forces are strongly dependent on particle size [64]. Large particles are focused at the side of the expansion-contraction region since the large particles are affected dominantly by the inertial lift forces. On the other hand, small particles are influenced dominantly by the Dean-like secondary flows, which lead to focusing position near the opposite side walls and enabling the size-dependent separation of particles or cells.

In addition to secondary flows, the vortex in the expansion-contraction regions can be used as particle or cell trapping reservoirs (Figure 10.2e) [26,50,107]. Multiple microscale laminar vortices, Moffatt's corner eddy flow [84], are created by flowing a stream of particle-laden fluid through parallel channels composed of a series of expansion-contraction reservoirs at a sufficiently high Re [50]. When the pre-focused particles through the high aspect ratio rectangular channel enter the expansion channel region, they migrate away from the channel center due to the sudden absence of the side walls [93]. In this situation, the larger particles are influenced by the shear gradient lift forces (proportional to a^3) more and migrate toward the walls further than the smaller particles. Under certain conditions, particles can get trapped into the vortex [50]. The lateral migration along the particle size causes the particles to move across the streamline to the vortex core through the detached boundary (separatrix) and remain isolated from the orbit of the vortex. In this way, the particles larger than a critical diameter (a_c) are selectively trapped and the particles smaller than a_c flow through without trapping, which allows a size-dependent separation. The high flow rate is required for vortex generation, which assists in the shift of the particle equilibrium position toward the side walls [5], making the focused streams of flowing particles/cells close to the separatrix at the reservoir and improving the capture efficiency.

10.2.2 Particle effects on inertial focusing

10.2.2.1 Particle's shape and deformability

Particle's shape Particles or cells having other than spherical shape rotate in a shear flow with a certain rotation axis (Figure 10.3a) [80]. When the flow rate and Re_p are low ($Re_p < 0.3$), the particles rotate randomly or do not rotate. As the flow rate and Re_p increase, the particles preferentially rotate around the axis of the highest vorticity perpendicular to the long face of the channel. This uniform rotational mode is a tumbling motion or periodic flipping. Jeffery has previously observed the rotational motion of ellipsoidal particles in the flow but did not predict a single axis of rotation [51]. However, particle orbits are preceded about a stable rotational axis in a shear or parabolic flow when a little inertia is added for fluid or particle [58, 109]. Masaeli et al. have presented the convergence of rotational modes to a single-plane mode with increasing Re_p [80]. The period of rotational motion dependent on the particle shape is expressed by $T = 2(A_R + 1/A_R)\pi/\dot{\gamma}$ where A_R is the aspect ratio of the particle and $\dot{\gamma}$ is the local shear rate.

The dominant tumbling rotational motion of the non-spherical particle provides a mechanism for how the particles of different shapes are focused at unique focusing positions [80]. When the major axis aligns in the direction perpendicular to the wall plane, wall-effect lift increases substantially due to the closer distance, which makes the particle push away from the wall. In contrast, the wall-effect lift decreases when the major axis is parallel to the flow, and the particle migrates back toward the wall. The focusing positions of the non-spherical particle are determined by the maximum diameter of the particle and the focusing trend of this non-spherical particle follows that of the sphere of rotational diameter equal to the maximum diameter of the particle [48]. For example, RBCs with a discoid shape that have a maximum diameter of $8\mu m$ can be treated as spheres with $8\mu m$ diameter.

Particle's deformability Deformable particles, including cells, liquid drops, vesicles and viscous capsules surrounded by elastic membranes can migrate laterally and the direction of the lift force is generally towards the channel center (Figure 10.3b) [1, 10, 31, 49, 62, 75, 114]. The deformable particles under shear flow take ellipsoidal shapes, which are elongated forms of original spherical shapes, which can lead to a net lift force. Unlike the shaped particles that take tumbling rotational motion, deformable particles can perform a tank-treading motion to keep the shape of the particle-fluid boundary, which

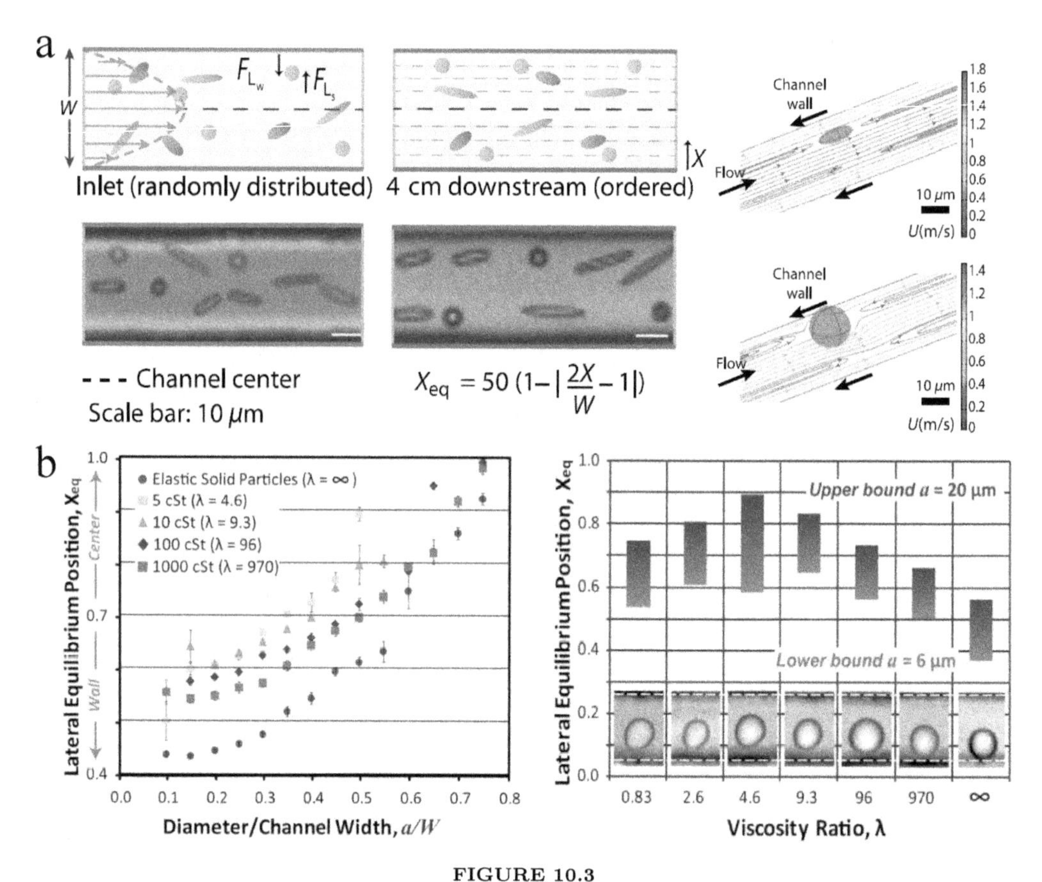

FIGURE 10.3
(a) Inertial focusing of different shaped particles in a straight microchannel [80]. (b) Focusing position changes with deformability of particle [49]. As the deformability of particle or droplet increases, the focusing position becomes closer to the channel center.

leads to a lift force regardless of Re. The cell-free layer in the blood flows can be explained by this lift force. In addition, lift force on a deformable particle was reported to arise by nonlinearity from matching of velocities and stresses at the interface of the deformable particles [75].

The deformed shape of the deformable particles strongly affects the magnitude of lateral drift velocity and lift force. For droplets, two dimensionless numbers, Weber number (We, the ratio of inertial stress and surface tension) and capillary number (Ca, the ratio of viscous stress and surface tension), allow characterizing the relative deformation of the droplet [85]. We and Ca are expressed by $We = \rho_f U_{max} a/\sigma$ and $Ca = \mu_f U_{max} a/\sigma$ where U_{max} and σ are the maximum fluid velocity and surface tension of the droplet. Generally, as We and Ca increases, deformability-induced lift force becomes stronger. In addition, the internal-to-external viscosity ratio, $\lambda_v = \mu_{in}/\mu_f$, is another dimensionless parameter to characterize deformability and drift of the droplet where μ_{in} is the viscosity of the fluid in the droplet [10, 31, 75, 85]. As the deformability of the droplet increases ($4.6 \leqslant \lambda_v \leqslant 970$), the focusing position shifts towards the channel center compared with elastic solid particles ($\lambda_v = \infty$) [49]. For lower viscosity ratio ($\lambda_v < 4.6$) on the other hand, the focusing position becomes closer to the channel wall since the droplet of the lower viscosity ratio slows down the surrounding flow between the droplet and the wall less effectively and the internal circulation affects the lift [85, 117]. The focusing positions of cells, due to their deformability tend to be closer to the channel centerline than the solid particles of the same size [49]. Even though the size of breast cancer cells with increased metastatic potential (modMCF-7 cells) and benign breast cancer cells (MCF-7 cells) is similar, the metastatic cells migrate closer to the channel center, which can allow the deformability-based separation.

10.2.2.2 Particle-particle and particle-fluid interaction

Particle-particle interaction In addition to the lateral migration and focusing of particles caused by the inertial lift forces, particles are longitudinally ordered. The randomly distributed particles are self-assembled with specific inter-particle spacing by the hydrodynamic particle-particle interactions. When the distance between two particles is small ($\lesssim 10a$), they feel particle-particle interaction which leads to oscillatory motion of particles until they achieve an organized state. The reversing streamlines which arise from the reflection of the disturbance flow off the channel wall are the key factor of particle-particle interaction. Matas et al. suggested the existence of reversing streamlines in a pipe [81] and Lee et al. provided the mechanism of particle ordering within a rectangular channel (Figure 10.4b) [65].

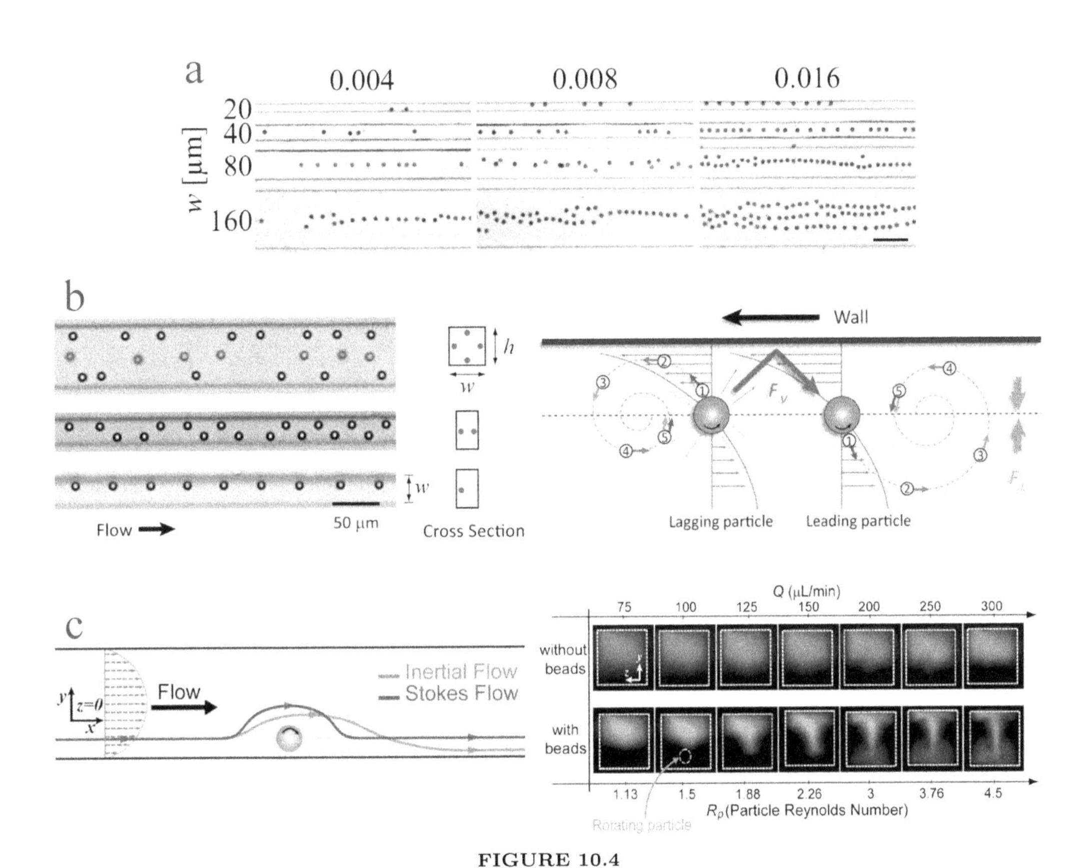

FIGURE 10.4
(a) Trains of particles formed by inertial ordering. Transition of ordering lines of focused particles can be observed in accordance with width of channel and bead volume fractions [47]. (b) Inertial ordering and self-assembly in various rectangular channels. With appropriate particle concentration, highly uniform inter-particle spacing can be found. The ordering mechanism involves repulsive viscous interaction (F_v), inertial lift forces (F_L) and parabolic velocity profile (gray arrows) [65]. (c) A rotating particle in finite-Re channel flows can lead to particle-induced convection. In contrast to the fore-aft-symmetric streamlines in Stokes flow, net flows around a particle exist in the case of inertial flow. Confocal images show the secondary flow induced by a rotating particle [3].

Once particles are ordered to focusing lines and form trains of particles, the concentration of particle suspension can be translated as a length fraction (λ) that is the number of the particle diameter per channel length. The minimal length fraction of focused particles typically does not exceed ~ 3. For λ beyond the minimum, particle trains can form additional ordering lines (Figure 10.4a) [47]. A high length fraction inevitably leads to defocusing of particles from equilibrium focusing positions, which deteriorate the efficiency of particle separation. The inertial microfluidic systems, therefore, require rather low concentration of particles, in the order of 0.1% volume fraction. For applications in hematology, whole blood must be diluted $\sim 100 - 1000$ times typically for tight focusing of cells. Separation of CTCs in spiral channels can be done at a higher concentration of blood because the RBCs are not required to be as tightly focused. Inertial focusing of CTC in whole blood was studied by particle tracking analysis (PTA) under the strong particle-particle interactions with blood cells [70]. Fluorescent-labeled PC-3 cells (prostate cancer cells) were spiked into whole blood (hematocrit, hct $= 45\%$) and their positions were measured at multiple vertical positions of the microchannel, which were reconstructed as 2-D intensity maps of particle frequency. They found the focusing positions of PC-3 cells near side walls and shifting towards the channel center as increasing RBC concentration and increasing cell size. Surprisingly, focusing positions of PC-3 cells change radically in the case of whole blood; PC-3 cells are focused near the short channel faces which are unstable focusing positions under other conditions. The exact mechanism of this radical focusing positions change is not understood at this point but the Newtonian nature of rheology of whole blood and the deformability of the cells are suggested to be affecting the focusing.

Particle-fluid interaction The reversing streamlines are generated by the channel confinement effect irrespective of *Re*. However, the inertial effect creates net secondary flows around rotating particles which are similar to Dean flows in curved channels. Amini et al. investigated particle-induced convection flow around a rotating particle with finite inertia in a rectangular channel [3]. Simulation results for comparison of fluid streamlines fore and aft of the particle show clear differences in the case of Stokes flow and finite-*Re* flow. Fluid streamlines are symmetric in Stokes flow as expected, whereas fluid streamlines are diverted towards the particle due to particle-induced convection in inertia flow (Figure 10.4c). The magnitude of convection is proportional to particle size ($\sim a^3$) and flow velocity ($\sim U^2$). The fluorescence dye from one side is transferred towards the other side due to the strong convection by the presence of the rotating particles (Figure 10.4c). This particle induced convection can lead to applications such as fluid switching and mixing around rigid particles. However, it may also affect adversarily in particle separation due to drag force induced by the secondary flows.

10.3 APPLICATIONS

10.3.1 Blood sample preparation with inertial microfluidics

10.3.1.1 Separation of CTCs and WBCs

CTC separation Isolation and analysis of CTCs have drawn much attention due to their importance in cancer research [73,86,97]. However, their scarcity in contrast to blood cells has made it challenging to collect or enumerate CTCs. While the concentration of RBCs and WBCs are 10^9 cells/ml and 10^6 cells/ml, respectively, the concentration of CTCs is extremely low ~ 100 cells/ml at most [83,97]. Therefore, important requirements for CTC separation techniques are high throughput and high yield, which could be met by inertial microfluidic separations. Similar to other CTC separation methods, inertial microfluidic separation is typically based on the size of cells. CTCs have size $\sim 14 - 30\mu m$ [11,52,71] while the size of WBCs is $\sim 7 - 15\mu m$ and, which are greater than other components such as RBCs ($\sim 6 - 8\mu m$) or platelets ($\sim 2 - 3\mu m$). These differences in sizes lead to different inertial focusing positions with a small overlap in the inertial microfluidic system, which could allow high efficiency and purity.

Among various inertial microfluidic device designs, CTC separations were mostly demonstrated with spiral microchannels due to their high flow rate and capability of handling less diluted blood. Sun et al. achieved label-free separation and enrichment of MCF-7 cells and HeLa cells using a double spiral microchannel with the concentration of 100 CTCs/million hematologic cells (Figure 10.5a) [111]. The double spiral microchannel of a very low aspect ratio ($H/W = 0.167$) and an alternation of flow direction through the S-turn at the center of channel design allowed the better focusing behavior of small blood cells and improvement of separation efficiency compared to the single spiral microchannel. In the low aspect ratio channel, shear rate of the channel height direction is much greater than the channel width direction, which pushes the particles and cells toward the z-direction equilibrium positions very quickly. The recovery rate for tumor cells was 88.5% and the purities of blood cells and tumor cells were 92.28% and 96.77%, respectively, with a throughput of 3.33×10^7 cells/min.

As discussed earlier, the requirement of a low concentration of particle for efficient focusing is the main limitation of the inertial microfluidic system, which inevitably leads to dilution of the blood sample or lysis of RBCs. For the first time, Hou et al. demonstrated isolation and retrieval of CTCs (MCF-7 cells) from a less diluted blood sample using the spiral microchannel [45]. The blood sample was diluted as $\sim 20 - 25\%$ hematocrit with the concentration of $\sim 10^5$ MCF-7 cells/ml. To enhance the separation efficiency, a 2-stage cascade spiral system was introduced and achieved the relatively high yield of $> 85\%$ and the enrichment ratios of $\sim 10^9$-fold over RBCs and $\sim 10^3$-fold over WBCs each. For demonstrating cell viability, the sorted cells from the separation channels were enumerated by fluorescent immune-staining and cultured, and successful culture of sorted MCF-7 cells in 96-well plate was achieved. The trapezoidal cross-sectional spiral microchannel has a merit of altering the Dean vortex core to achieve efficient separation [42]. Spiral channels with rectangular cross-section typically require a sheath flow for efficient particle dispersion, especially with high particle concentration. In contrast, spiral channels with trapezoidal cross-section have advantages in that single inlet operation is sufficient for separation and enrichment of cells. For example, Warkiani et al. introduced a spiral microchannel with a trapezoidal cross-section for a low concentration CTCs (~ 500 cells/7.5ml) separation with $2\times$ diluted blood sample [120]. The throughput and separation efficiency was as high as 1.7 ml/min and $\sim 80 - 90\%$, respectively. This device has merit in respect to cost with high resolution due to the fabrication using conventional micro-milling and PDMS casting. For the particular case, Kulasinghe et al. recently demonstrated enrichment of circulating head and neck tumor cells, which account for the seventh most common tumor type, using the spiral microfluidic channel for the first time with the yield of 60-70%, which widens the variety of the kinds of the CTCs separated [59].

Expansion-contraction microchannels were also used to separate CTCs from blood samples. The expansion-contraction microchannel has advantages in CTC capturing owing to its high selectivity and throughput. For instance, Sollier et al. achieved size-selective collection of CTCs in parallelized expansion-contraction microchannels (Figure 10.5c) [107]. This study has adopted a previous technique using micro-scale vortices and inertial focusing [49], and arrived at optimal operating conditions such as channel dimensions and flow rates for maximum trapping efficiency and purity. This technology using micro-vortices provides

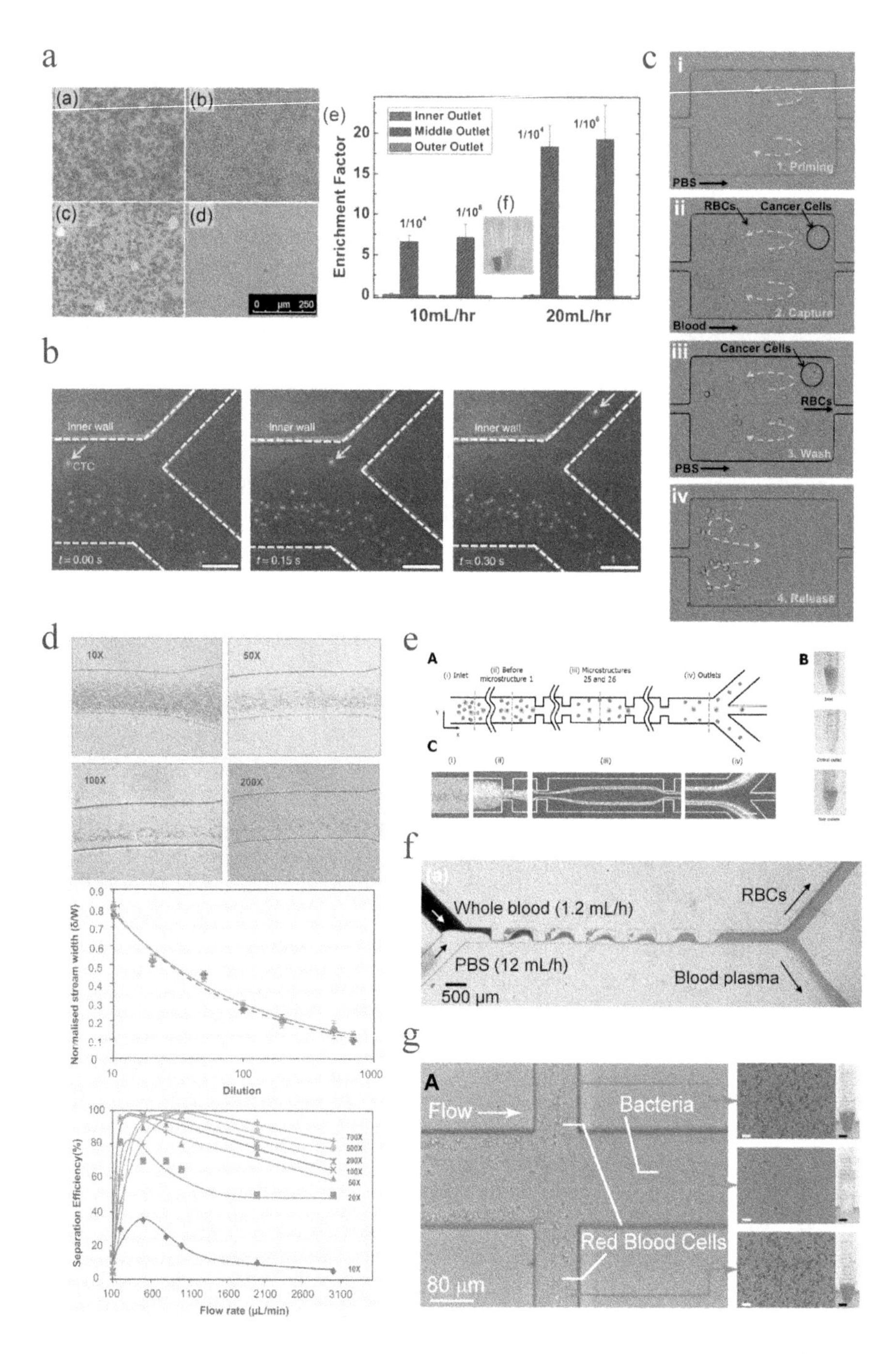

FIGURE 10.5

Blood separation using inertial microfluidics. (a) Separation of CTCs from diluted blood sample using double spiral microchannel [111]. (b) Isolation of CTCs from blood sample using 3-layered multiplexed spiral channel [121]. (c) Separation and enrichment of CTCs using size-selective CTCs trapping in micro-scale vortices [107]. (d) Separation of WBCs using spiral channel [87]. (e) Separation of WBCs using expansion-contraction channel [124].(f) Plasma extraction using expansion-contraction channel [63,64]. (g) Separation of E.coli K-12 bacteria and RBCs in massively parallel straight channel [74]. Smooth channel expansion near the outlet enhanced separation efficiency.

important advantages compared to other techniques in terms of sample processing time (20 min for 7.5 ml of whole blood), sample concentration (hundreds of cells in 20 ml of 10× diluted blood), applicability to various cancer types, cell integrity and purity (89% of purity and 3.5×10^4 of enrichment ratio).

Ozkumur et al. developed the CTC-iChip for CTC separation [90]. The CTC-iChip has both label-free and antibody-conjugated magnetic bead-based separation stages. Here, inertial microfluidics provided single stream focusing rather than size-based separation. First, the majority of RBCs were separated out from whole blood using DLD. Then the relatively larger WBCs and CTCs were inertially focused at the channel center with the serpentine channel. The focused stream of cells was flown under a magnetic field and CTCs were separated either by positive selection on CTCs based on EpCAM expression or depletion of WBCs based on CD45 and CD15.

Table 10.1 summarizes different typical approaches of separation of CTCs found in the literature of inertial separation.

TABLE 10.1
Separation of CTCs from a blood sample in inertial microfluidic systems.

Separation Cell Type	Dilution (Whole Blood=WB, Lysed Blood=LB)	Spiking	Yield	Purity	Enrichment ratio	Throughput	Ref
MCF-7, Hela	50× of WB	10^4 cells/ml	88.5%[a]	92.28% of Blood Cells 96.77% of Tumor Cells[a]	18.38[a]	0.33 ml/min (3.33×10^7 cells/min)	[111]
MCF-7	2× of WB	10^5 cells/ml	>85%[a]	N/A	10^9 and 10^3[a]	0.05 ml/min[a]	[45]
MCF-7 T24 MDA-MB-231	2× of WB	66 cells/ml	85% 80% 87%[a]	N/A	N/A	0.9 ml/min	[120]
MCF-7 T24 MDA-MB-231	Undiluted LB	133 cells/ml	84.9% 87.8% 83.7%[b]	N/A	N/A	0.75 ml/min (3-layered multiplexed channel)	[121]
MCF-7	100× of WB Undiluted LB	5×10^4 cells/ml 5×10^2 cells/ml	75.4%[a] 69.1%[a]	8.1%[a] N/A	81[b] N/A	0.4 ml/min	[46]
Head and Neck cancers (HCNs)	Undiluted LB	1-15 cells/ml	60-70%[a]	N/A	N/A	1.7 ml/min	[59]
MCF-7 Hela	45× of WB	2×10^3 cells/ml	23%[c] 10%[c]	N/A N/A	7.06[a] 5.53	7.5×10^6 cells/min (8 parallel channels)	[50]
MCF-7 (OVCAR-5, M395, PC-3,A549)	20× of WB	20-70 cells/ml	20.7%[c]	89%[a]	3.5×10^4[b]	0.375 ml/min[a] (8 parallel channels)	[107]
MDA-MB-231	20× of WB	1000 cells spiked in total volume	24%[c]	80%[a]	N/A	N/A	[54]

Yield
[a] The number of cancer cells collected off-chip/the number of injected cells
[b] The number of cancer cells counted on-chip/the number of injected cells
[c] The number of cancer cells captured/the number of injected cells (capture efficiency)
Purity
[a] The number of cancer cells from target outlet/the number of whole cells from target outlet
Enrichment ratio
[a] $(\text{Cancer cells/WBCs})_{outlet}/(\text{Cancer/WBCs})_{inlet}$
[b] $(\text{Cancer cells/RBCs})_{outlet}/(\text{Cancer/RBCs})_{inlet}$
Throughput
[a] Converted to flow rate of the equivalent amount of whole blood

WBC separation Similar with CTCs, WBCs are often the target of blood separation by various microfluidic techniques. Separation or removal of RBCs from blood samples is an essential preparation step before clinical and diagnostic tests, which involve isolation and analysis of WBCs [43]. Size-based separation including inertial microfluidics has great potential because inadvertent lysis of RBCs can adversely affect target cells and it is important to use methods that minimize the possibility of artificial alteration on the phenotypes of target cells [2]. Wu et al. achieved separation of WBCs using a spiral microchannel with a trapezoidal cross-section [123]. As mentioned in the CTCs section, a Dean vortex core in the spiral microchannel with trapezoidal cross-sections shifts and allows better separation results [42]. They optimized the design of the microchannel and demonstrated its ability in separation and recovering polymorphonuclear leukocytes (PMNs) and mononuclear leukocytes (MNLs) from a diluted blood sample of 1-2% hematocrit with high efficiency of $> 80\%$ [123]. Nivedita and Papautsky reported the optimized spiral microchannel device for blood cell separations (Figure 10.5d) [87]. The device showed high separation yield of $\sim 95\%$ and the throughput as high as 1×10^6 cells/min ($\geqslant 0.1$ hematocrit). Wu et al. achieved high throughput using parallelized straight channels with local microstructures [124]. As discussed earlier, spiral channel designs are not favorable for parallelization. Secondary flows arising from expansion-contraction lead to size-dependent focusing positions. While WBCs are dominated by inertial lift and maintain original focusing positions, RBCs are dragged away towards the side walls by secondary flows. The resultant purity and throughput of WBCs were 91% and 10.8 ml/min in the case of 72 radially arranged channels (Figure 10.5e). Table 10.2 summarizes these studies and the efficiency of inertial separation for WBCs.

TABLE 10.2
Separation of WBCs from a blood sample in inertial microfluidic systems.

Separation Cell Type	Dilution of Whole Blood	Yield[a]	Purity[a]	Throughput	Ref
WBC	10×	98.4%	N/A	0.8 ml/min	[123]
WBC	10×	84%	N/A	10 ml/min	[116]
WBC	500×	95%	N/A	1.8 ml/min (10^5 cells/min)	[87]
WBC	180×	89.7% (WBCs) 99.8% (RBCs)	91% (WBCs) 99.6% (RBCs)	10.8 ml/min (72 parallel channel)	[124]
WBC	10×	99%	N/A	1 ml/min	[99]

Yield
[a] The number of collected target cells from off-chip/the number of injected target cells
Purity
[a] The number of collected target cells/the number of collected total cells

10.3.1.2 Separation of plasma and bacteria

Plasma as a liquid component occupies approximately 55% of whole blood and contains various small molecules like proteins, metabolites and circulating nucleic acids (CNAs) [53]. Separation of blood plasma is routinely done for transfusion. Separated plasma may be further fractionated for the proteins contained in the plasma due to their important therapeutic uses [9]. In addition, the small molecules including proteins and CNA can be used as important biomarkers for various diseases [68, 95, 118]. Bacteria, although they are not considered as a component of blood, are also important in blood separation to diagnose sepsis [37]. The early detection of bacteria at low concentration is critical. Inertial focusing had been successfully applied to particle size typically larger than 5 μm in diameter in general. Focusing of platelet and bacteria is hard to realize due to their size. Therefore, separation of plasma and enrichment of small size components of the blood was achieved by removal of larger particles from blood (i.e., RBCs, WBCs and CTCs). Table 10.3 summarizes demonstrations of inertial separation for plasma and bacteria. The yield of plasma was calculated by the volumetric extraction of diluted plasma and the purity of plasma by the ratio of RBC rejection.

TABLE 10.3
Separation of plasma and bacteria from blood sample in inertial microfluidic systems.

Separation Cell Type	Dilution	Yield	Purity	Throughput	Ref
Plasma	Whole Blood	62.2%[a]	60%[a]	0.02 ml/min (10^6 cells/min)	[63]
Plasma	20×	18 – 25%[a]	N/A	0.4 ml/min	[77]
Plasma	20×	46%[a]	99.95%[a]	2.8 ml/min (7×10^5 cells/min) (8 parallel channels)	[132]
Plasma	20×	38.5%[a]	~100%[a]	0.7 ml/min	[127]
Plasma	40×	N/A	100%[a]	24 ml/min (16 parallel channels)	[98]
Plasma	20×	N/A	~99%[a]	1.5 ml/min (10^5 cell/min)	[101]
Bacteria	0.5% (v/v)	88%[b]	>88%[b]	8 ml/min (4×10^8 cell/min) (40 parallel channels)	[74]
Bacteria	0.05% (v/v)	62%[b]	99.87%[b]	0.015 ml/min (3.5×10^8 cell/min)	[125]

Yield
[a] Extracted plasma volume/the total volume of blood injected
[b] The number of bacteria collected off-chip/the number of injected cells
Purity
[a] 1– (the number of blood cells in the isolated plasma/the number of blood cells in the initial blood sample)
[b] The ratio between the number of recovered target cells to the total number of recovered cells

Separation of plasma mainly relies on the creation of a cell-free region in the microchannel to extract plasma. Marchalot et al. conducted plasma separation using a multi-step microfluidic channel [77]. The channel consisted of the upstream channel, contraction channel and a diverging channel including circulation zones. The cell-free layer close to the wall was induced by lateral migration of particles. Recirculation zones containing only plasma were created in the diverging channel and plasma was extracted from this zone. This simple straight channel yielded 18-25% plasma from the injected sample. Most of the devices separate particles using secondary flow and the separation cutoff value is modulated by controlling the force balance between inertial lift and Dean drag forces. Lee et al. used the expansion-contraction channel for plasma separation with high flow rate and achieved a high level of throughput ($\sim 10^8$ cells/min) (Figure 10.5f) [63]. Serpentine and spiral channels may enable high efficiency and high-throughput simultaneously with low dilution ratio. The connection of several devices in series allows an increase of separation efficiency and in parallel to increase throughput. Zhang et al. separated plasma using a serpentine device with eight parallel channels to achieve a massive throughput (7.0×10^8 cells/min) [132]. Rafeie et al. also constructed a high-throughput device consisting of eight parallel spiral channels with trapezoidal cross-sections, which were built by 3D printing technology [98]. This connected channel allows us to scale up the total flow rate (1.5 ml/min) and high-throughput. Robinson et al. fabricated a sequential spiral microfluidic device [101]. Blood cells that were not completely filtered at the first bifurcation could be filtered again at the second bifurcation. The separation efficiency was improved from 55% (single spiral channel) to ~ 99% (double spiral channel) for RBCs. Separation of bacteria was also done by removal of blood cells. Wu et al. suggested soft inertial force generated by a combination of asymmetrical sheath flow and channel geometry [125]. This method reduces a risk of stressing or damaging of cells using a protective sheath flow. Mach and Di Carlo fabricated a massively parallel blood filtration device for bacteria separation which facilitates 4.0×10^8 cells/min throughput (Figure 10.5g) [74].

10.3.2 Analysis of biological cells via inertial microfluidic system

10.3.2.1 Flow cytometry

There have been many studies to realize flow cytometry as an integrated microfluidic system [12, 96]. Microflow cytometry, or on-chip flow cytometry, requires microfluidic particle focusing capability and integrated optical or electrical sensors. Early studies for microflow cytometry adopted simple pinch flow as a means for hydrodynamic focusing of the particles. Inertial microfluidics drew much attention as a sample preparation technique for flow cytometry due to its capability of extremely high throughput and tight particle focusing without sheath flow. In order to focus and order the particles or cells effectively, various inertial microfluidic devices with straight and serpentine channels were designed and demonstrated the capability as the flow cytometry.

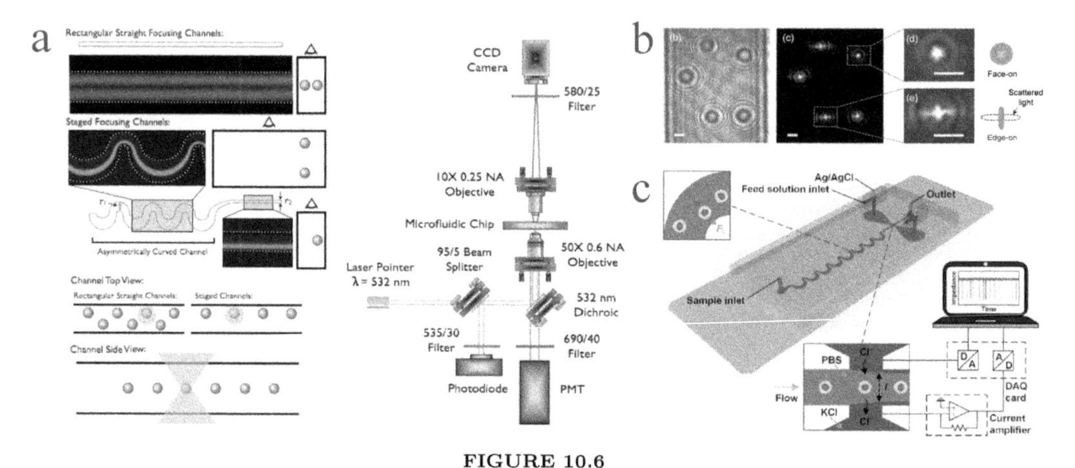

FIGURE 10.6
Micro-cytometry system combined with inertial microfluidics. (a) A staged channel consisting of curved and straight sections, which aligns particles in a single streamline with longitudinal spacing [88]. (b) A digital in-line holographic microscopy technique is applied to investigation of the inertial migration of RBCs and its relation to the cell orientation in low-viscosity and high shear rate microtube flows [15]. (c) High-throughput cell counting and discrimination are achieved by an impedance micro-cytometry integrated with inertial focusing and liquid electrode techniques [115].

Oakey et al. built microfluidic systems for flow cytometry with staged inertial microfluidic channels consisting of a serpentine channel and a rectangular channel, which allows the particles to be aligned with periodic longitudinal spacing in a single streamline [88]. The inertial ordering of particles is an additional advantage that guarantees the minimum spacing between particles and prevents particle overlap at the sensing region. In an asymmetric curved channel, Dean flow dragged the particles toward the inner edge of the focusing curve, and thus, the particles migrated toward one side of the channel prior to moving into the straight channel. In the rectangular channel, the particles focused into a single stream as shown in Figure 10.6a. In this study, the quality of focusing was evaluated by a standard flow cytometry coefficient of variation (cv), which was measured with a photomultiplier tube for single-channel fluorescence, and a silicon photodiode for scatter signal. The quality of the focusing stream was dependent on velocity and concentration of the particles. A minimum cv of 6% was observed at the flow rate of 100 μl/min with 0.1 w/v% (about 2×10^6 particles/ml). This value is close to a commercial system at $\sim$ 5% cv. The particles were focused more effectively with the high concentration since hydrodynamic interactions aided the particles to reach equilibrium positions faster [39]. With the optimal conditions, they achieved an analysis rate of 2.5×10^4 particles/s and a higher analysis rate could be possible by an increase of the fluid velocity without coincidences of the particles at the point of analysis.

The key issue in channel design is how to restrict the access of particles to multiple focusing positions and to achieve single stream focusing. Chung et al. presented novel inertial focusing platforms for single stream focusing [16, 17]. The Dean flows in a curving channel intrinsically perturb the entire flow field and the focusing positions dependent on the particle size [29, 60]. Secondary flows generated by expansion and contraction allow stable single stream focusing. Furthermore, the focused particle focusing close to the side wall can be an obstacle to measuring for flow cytometer due to the wall interface-induced scattering of the excitation beam [36]. Changing the beam path in the horizontal direction would complicate the device fabrication and interface with the optical system. Use of a low-aspect-ratio channel and implementing expansion-contraction in the height direction solved the issue [16]. Recently, we have developed a novel method to focus particles in a single stream by connecting channels with different cross-sectional shapes [56]. One of the focusing positions in the half-circular channel was made inaccessible for particle focusing by mapping of focusing positions and the corresponding basins of attraction of rectangular channel and triangular channel.

Unconventional imaging techniques other than laser optics were also applied for microflow cytometry. Goda et al. combined inertial microfluidics with the serial time-encoded amplified microscopy (STEAM) imaging technique to achieve high-speed imaging of particles moving as fast as 8 m/s [36]. Laser optics of flow cytometry can provide high throughput analysis of single cells. However, the acquired information is limited to cell size and fluorescent intensity of specific labeling. On the other hand, microscopic imaging with conventional high-speed cameras is limited by relatively long shutter speed ($\geqslant 1\mu$s) and data processing from image sensors. In addition, images of a microscopic object moving at high speed tend to blur and deteriorate single cells-based analysis. The STEAM camera can take blur-free images of fast-flowing particles and the acquired images are optoelectronically processed with an ultra-high throughput of 10^5 particles/s and shutter speed as fast as 27 ps. Choi and Lee introduced a digital in-line holographic microscopy to observe the inertial migration of RBCs in low viscosity and high shear rate microtube flows (Figure 10.6b) [15]. Depending on the viscosity of media and the applied shear rate, RBCs can show three different behaviors: Jeffery orbit, tank-treading motion, and aligned rotation [8, 38, 51, 82] (see Chapter 3). The aligned rotational motion is a spinning motion with symmetry axes aligned with the vorticity axis of the shear field, which can be observed with inertially ordered RBCs. Depending on the orientation of the RBCs, they showed the trajectories of the cells in edge-on and face-on orientations when the shear rate was high (Figure 10.6b). Additional viscoelastic lift force can align the rotational axis in a vertical direction that is more favorable for holographic measurements of RBC [35, 92]. There is increasing demand for integrating the optical component in microfluidic systems. However, such integrated optical components typically have limited sensing capability, which is not suitable for inertial microfluidic systems due to high flow speed.

An inertial microfluidic system for impedance cytometry has also been developed. As an alternative to expensive optical sensing components, an impedance measurement allows simple device fabrication and use of low-cost measurement electronics [128]. Tang et al. recently developed a high-throughput liquid electrode-based impedance microflow cytometer involving an inertial microfluidic system (Figure 10.6c) [115]. They introduced a liquid electrode composed of Ag/AgCl wire into the electrode chamber filled with flowing highly conductive electrolyte solution. The inertial focusing assisted in reducing the possibility of cell adhesions and made the particles and cells flow in a single streamline through the detection region. Based on the optimized system, they achieved a high detection throughput of $\sim$ 5000 cells/s and further investigated the size distributions of MCF-7 cells and WBCs, which resulted in a successful discrimination of the MCF-7 cells with WBCs.

10.3.2.2 Mechanical analysis

In addition to the size of cells, physical properties such as impedance deformability of the cell can provide important information. For example, cancerous cells are known to be more deformable than normal cells, which give an advantage to their ability to invade the secondary site during metastasis [20, 67, 69, 112]. Gossett et al. have developed a device with a unique combination of inertial focusing, hydrodynamic stretching, and automated image analysis for mechanical phenotyping of cells [40]. This method enables measurements of deformability of individual cells with a throughput of approximately 2000 cells/s. Serpentine channels were used to focus cells at the channel center. The focused cells move to the intersection near the outlet that is designed to effectively "squeeze" the cell (Figure 10.7a). The hydrodynamic stretching allows cell deformation in the flow direction and the deformed cell shape was imaged with a high-speed camera. High throughput of the inertial microfluidics provides a large set of data for statistical analysis. In this study, they have shown that mechanical cell deformability measurements helped to analyze populations of cells within the blood and pleural fluids.

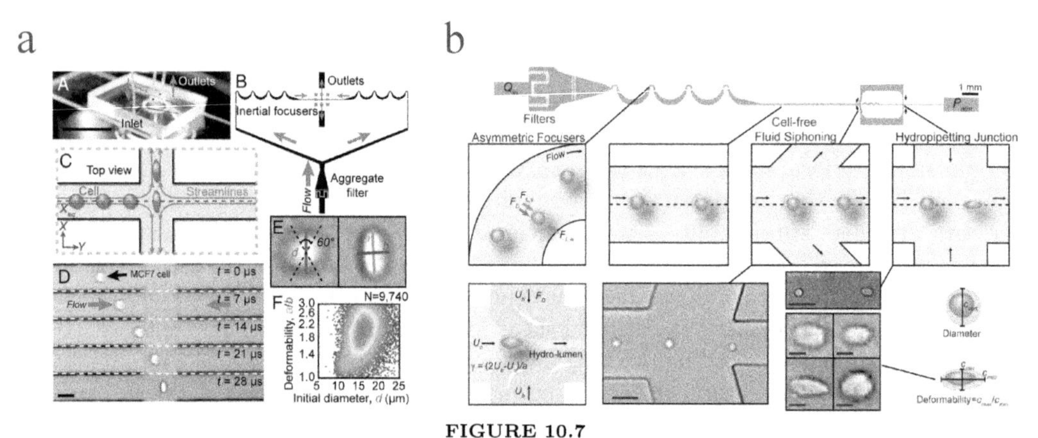

FIGURE 10.7
Mechanical analysis of cells using inertial microfluidic particle manipulation techniques. (a) Deformability cytometry enables high-throughput analysis of individual cell deformability using inertial focusing and cell stretcher [40]. High-speed images show morphology changes of cells at the extensional flow. Serpentine channels were used to focus cells at the channel center. (b) Inertial microfluidic device for hydropipetting [32].

The Hydrodynamic stretching was improved by hydrodynamic control of particle and flow velocity [40]. The previous realization hydrodynamic stretching had a limitation on the amount of the time that cells stay at the stagnation point of extensional flow (tens of microseconds). A method called "hydropipetting" increased throughput by over an order of magnitude by squeezing cells with perpendicular cross-flows (Figure 10.7b) [32]. By applying the sheath flow on to the cell, the cell was stretched in the direction of the flow. This new hydropipetting method could process up to 65000 cells/s Jurkat cells, MCF-7, and HeLa cells were used to analyze three different deformability profiles. In addition, increases in deformability were successfully measured in invasive cell models. Yet, both the hydropipetting and hydrodynamic stretching method have limitations with image processing time. A large number of stacked images were taken from the high-speed camera, which significantly increases the processing time as well as an enormous drop of actual throughput to $<\sim$ 10 cells/s. A novel research done by Deng et al. developed an inertial microfluidic cell stretcher (iMCS), which not only operates as a fully automated system, but also phenotypes a cell deformability in near real-time [25]. The method Deng et al. applied to measure maximum cell elongation was by colliding cells to the wall at the T-junction. Near real-time automated image analysis was done by developing a system control algorithm via LabVIEW and a self-written C++ routine. Through the novel cell stretcher platform (iMCS), it was able to achieve actual high-throughput ($\sim$ 450 cells/s) and high speed (10^5 frames/s) cell mechanical analysis, resulting in the high statistical significance of various cell lines. Cell deformability has a great potential to provide a physiologically meaningful metric to sort and analyze cells. Mechano-phenotyping with high-throughput deformability cytometry is expected to bring the statistical accuracy of traditional flow cytometric methods to label-free biophysical biomarkers, enabling applications in clinical diagnostics, stem cell characterization, and single-cell biophysics.

10.4 CONCLUSION AND PERSPECTIVES

Inertial microfluidics has drawn much attention not only for its interesting physics but also for diverse applications. Especially, the capability of label-free separation of microparticles with high throughput is well-suited for rare cell separation from blood, which was extensively studied and demonstrated with various device designs. Inertial microfluidics provides diverse methods for hydrodynamic manipulation of fluid flows and particle motion without external forces. However, the non-intuitive physics of fluid mechanics at finite Reynolds number flows led to initial difficulties in both understanding and applications of the inertial microfluidics. Currently, mechanisms of many inertial microfluidic phenomena have been revealed and they are providing guidelines for practical device designs. Despite the progress, there is still a largely unexplored territory in the field and reports of new findings do not seem to stop. For example, there were interesting focusing position shifts found in triangular channels where the direction of wall effect lift force can be adjusted [55,56]. The channel walls without mirror symmetry allow the methods to control the wall effect lift force from each wall, which leads to interesting focusing position changes. For applications for hematology, in particular, it is still imperative to have a better understanding of the effects of particle deformability, particle-particle interactions, and fluid rheology. For instance, inertial focusing using whole blood was studied by PTA analysis and showed puzzling results, which was speculated as results of strong particle-particle interactions with blood cells and the non-Newtonian nature of whole blood [70].

Inertial microfluidic particle manipulations have several limitations that may need some attention. High particle-particle concentrations lead to defocusing of particles and reduce the accuracy of particle manipulations. In the case of blood cell manipulations, this limitation leads to the necessity of blood dilution. Spiral channels can circumvent this limitation by allowing relatively small RBCs to be dominated by Dean drag force. Otherwise, RBCs need to be lysed or separated with an other technique before inertial microfluidic manipulations [90]. The more significant limitation could be the limitations on the particle size that can be manipulated. Due to strong size dependency of inertial lift forces, it is difficult to focus small particles in inertial microfluidic systems. Minimum particle size applicable to inertial focusing has been recognized as $\sim 5\mu m$ in diameter. Wang et al. recently showed inertial separation of a sub-micron size particle using high flow rate and serpentine channels [119]. The high flow rate and high *Re* were achieved using rigid polymeric microfluidics. Particles as small as $0.92\mu m$ could be focused into a narrow stream and separation of cyanobacteria and $2\mu m$ particles were demonstrated. Although it has not been shown, the device would be promising for applications in the separation of bacteria or platelets from the blood.

In addition to the efforts to understand the physics better and to overcome limitations, broadening the application area is also an important future direction. Inertial microfluidics has a high potential for meeting medical and clinical needs. The liquid biopsy requires fast, high purity separation of diverse target analytes that are often scarce in body fluids. Integration of sample preparation and downstream analysis in a single microfluidic chip will enable "sample-in-answer-out" point-of-care devices. Fast diagnosis of bacterial and fungal infection of blood is challenging due to the low concentration and small size of the bacteria and fungus. The capability of inertial microfluidic devices has been and will further be improved and we expect inertial microfluidics has a high potential in providing solutions to not only the issues aforementioned but also other unexplored applications in hematology.

Bibliography

[1] M. Abkarian and A. Viallat. Dynamics of vesicles in a wall-bounded shear flow. *Biophysical Journal*, 89:1055–1066, 2005.

[2] W. A. Al-Soud and P. Rådström. Purification and characterization of PCR-inhibitory components in blood cells. *Journal of Clinical Microbiology*, 39:485–493, 2001.

[3] H. Amini, Sollier E., Weaver W.M., and D. Di Carlo. Intrinsic particle-induced lateral transport in microchannels. *Proceedings of the National Academy of Sciences of the United States of America*, 109:11593–11598, 2012.

[4] H. Amini, Lee W., and D. Di Carlo. Inertial microfluidic physics. *Lab-on-a-Chip*, 14:2739–2761, 2014.

[5] E. S. Asmolov. The inertial lift on a spherical particle in a plane Poiseuille flow at large channel Reynolds number. *Journal of Fluid Mechanics*, 381:63–87, 1999.

[6] S. Berger, Talbot L., and Yao L. Flow in curved pipes. *Annual Review of Fluid Mechanics*, 15:461–512, 1983.

[7] A. A. S. Bhagat, Kuntaegowdanahalli S.S., and Papautsky I. Continuous particle separation in spiral microchannels using dean flows and differential migration. *Lab on a Chip*, 8:1906–1914, 2008.

[8] M. Bitbol. Red blood cell orientation in orbit C= 0. *Biophysical Journal*, 49:1055–1068, 1986.

[9] T. Brodniewiczproba. Human plasma fractionation and the impact of new technologies on the use and quality of plasma-derived products. *Blood Reviews*, 5:245–257, 1991.

[10] P.-H. Chan and L. Leal. The motion of a deformable drop in a second-order fluid. *Journal of Fluid Mechanics*, 92:131–170, 1979.

[11] Y. Chen, P. Li, P.-H. Huang, Y. Xie, J. D. Mai, L. Wang, N.-T. Nguyen, and T. J. Huang. Rare cell isolation and analysis in microfluidics. *Lab on a Chip*, 14:626–645, 2014.

[12] S. H. Cho, J. M. Godin, C.-H. Chen, W. Qiao, H. Lee, and Y.-H. Lo. Recent advancements in optofluidic flow cytometer. *Biomicrofluidics*, 4:043001, 2010.

[13] S. Choi, S. Song, C. Choi, and J.-K. Park. Continuous blood cell separation by hydrophoretic filtration. *Lab on a Chip*, 7:1532–1538, 2007.

[14] S. Choi, S. Song, C. Choi, and J.-K. Park. Microfluidic self-sorting of mammalian cells to achieve cell cycle synchrony by hydrophoresis. *Analytical Chemistry*, 81:1964–1968, 2009.

[15] Y.-S. Choi and S.-J. Lee. Inertial migration of erythrocytes in low-viscosity and high-shear rate microtube flows: Application of simple digital in-line holographic microscopy. *Journal of Biomechanics*, 45:2706–2709, 2012.

[16] A. J. Chung, D. R. Gossett, and D. Di Carlo. Three dimensional, sheathless, and high-throughput microparticle inertial focusing through geometry-induced secondary flows. *Small*, 9:685–690, 2013.

[17] A. J. Chung, D. Pulido, J. C. Oka, H. Amini, M. Masaeli, and D. Di Carlo. Microstructure-induced helical vortices allow single-stream and long-term inertial focusing. *Lab on a Chip*, 13:2942–2949, 2013.

[18] A. T. Ciftlik, M. Ettori, and M. A. Gijs. High throughput-per-footprint inertial focusing. *Small*, 9:2764–2773, 2013.

[19] D. C. Colter, I. Sekiya, and D. J. Prockop. Identification of a subpopulation of rapidly self-renewing and multipotential adult stem cells in colonies of human marrow stromal cells. *Proceedings of the National Academy of Sciences*, 98:7841–7845, 2001.

[20] S. E. Cross, Y.-S. Jin, J. Rao, and J. K. Gimzewski. Nanomechanical analysis of cells from cancer patients. *Nature Nanotechnology*, 2:780–783, 2007.

[21] E. M. Darling, M. Topel, S. Zauscher, T. P. Vail, and F. Guilak. Viscoelastic properties of human mesenchymally-derived stem cells and primary osteoblasts, chondrocytes, and adipocytes. *Journal of Biomechanics*, 41:454–464, 2008.

[22] C. S. De Paiva, S. C. Pflugfelder, and D. Q. Li. Cell size correlates with phenotype and proliferative capacity in human corneal epithelial cells. *Stem Cells*, 24:368–375, 2006.

[23] W. Dean. XVI. Note on the motion of fluid in a curved pipe. *The London, Edinburgh, and Dublin Philosophical Magazine and Journal of Science*, 2:208–223, 1927.

[24] W. Dean. LXXII. The stream-line motion of fluid in a curved pipe (Second paper). *The London, Edinburgh, and Dublin Philosophical Magazine and Journal of Science*, 5:673–695, 1928.

[25] S. Deng, Y., P. Davis, F. Yang, K. S. Paulsen, M. Kumar, R. Sinnott DeVaux, X. Wang, and D. S. Conklin. Inertial Microfluidic Cell Stretcher (iMCS): Fully automated, high-throughput, and near real-time cell mechanotyping. *Small*, 13:1700705, 2017.

[26] M. Dhar, J. Wong, A. Karimi, J. Che, C. Renier, M. Matsumoto, M. Triboulet, E. B. Garon, J. W. Goldman, and M. B. Rettig. High efficiency vortex trapping of circulating tumor cells. *Biomicrofluidics*, 9:064116, 2015.

[27] D. Di Carlo. Inertial microfluidics. *Lab on a Chip*, 9:3038–3046, 2009.

[28] D. Di Carlo, J. F. Edd, K. J. Humphry, H. A. Stone, and M. Toner. Particle segregation and dynamics in confined flows. *Physical Review Letters*, 102:094503, 2009.

[29] D. Di Carlo, J. F. Edd, D. Irimia, R. G. Tompkins, and M. Toner. Equilibrium separation and filtration of particles using differential inertial focusing. *Analytical Chemistry*, 80:2204–2211, 2008.

[30] D. Di Carlo, D. Irimia, R. G. Tompkins, and M. Toner. Continuous inertial focusing, ordering, and separation of particles in microchannels. *Proceedings of the National Academy of Sciences*, 104:18892–18897, 2007.

[31] S. K. Doddi and P. Bagchi. Lateral migration of a capsule in a plane Poiseuille flow in a channel. *International Journal of Multiphase Flow*, 34:966–986, 2008.

[32] J. S. Dudani, D. R. Gossett, T. Henry, and D. Di Carlo. Pinched-flow hydrodynamic stretching of single-cells. *Lab on a Chip*, 13:3728–3734, 2013.

[33] E. Furlani. Magnetophoretic separation of blood cells at the microscale. *Journal of Physics D: Applied Physics*, 40:1313, 2007.

[34] P. R. Gascoyne and J. Vykoukal. Particle separation by dielectrophoresis. *Electrophoresis*, 23:1973, 2002.

[35] T. Go, H. Byeon, and S. J. Lee. Focusing and alignment of erythrocytes in a viscoelastic medium. *Scientific Reports*, 7:41162, 2017.

[36] K. Goda, A. Ayazi, D. R. Gossett, J. Sadasivam, C. K. Lonappan, E. Sollier, A. M. Fard, S. C. Hur, J. Adam, and C. Murray. High-throughput single-microparticle imaging flow analyzer. *Proceedings of the National Academy of Sciences*, 109:11630–11635, 2012.

[37] M. Goldman and M. A. Blajchman. Blood product-associated bacterial sepsis. *Transfusion Medicine Reviews*, 5:73–83, 1991.

[38] H. Goldsmith, J. Marlow, and F. MacIntosh. Flow behaviour of erythrocytes-I. Rotation and deformation in dilute suspensions. *Proceedings of the Royal Society Series B-Biological Sciences*, 182:351–384, 1972.

[39] D. R. Gossett and D. D. Carlo. Particle focusing mechanisms in curving confined flows. *Analytical Chemistry*, 81:8459–8465, 2009.

[40] D. R. Gossett, T. Henry, S. A. Lee, Y. Ying, A. G. Lindgren, O. O. Yang, J. Rao, A. T. Clark, and D. Di Carlo. Hydrodynamic stretching of single cells for large population mechanical phenotyping. *Proceedings of the National Academy of Sciences*, 109:7630–7635, 2012.

[41] D. R. Gossett, H. T. K. Tse, J. S. Dudani, K. Goda, T. A. Woods, S. W. Graves, and D. Di Carlo. Inertial manipulation and transfer of microparticles across laminar fluid streams. *Small*, 8:2757–2764, 2012.

[42] G. Guan, L. Wu, A. A. Bhagat, Z. Li, P. C. Chen, S. Chao, C. J. Ong, and J. Han. Spiral microchannel with rectangular and trapezoidal cross-sections for size based particle separation. *Scientific Reports*, 3:1475, 2013.

[43] W. G. Guder, S. Narayanan, H. Wisser, and B. Zawta. *Diagnostic Samples: From the Patient to the Laboratory: The Impact of Preanalytical Variables on the Quality of Laboratory Results*, 4th, Updated Edition. Wiley-Blackwell, NY, USA, 2008.

[44] B. P. Ho and L. G. Leal. Inertial migration of rigid spheres in 2-dimensional unidirectional. *Journal of Fluid Mechanics*, 65:365–400, 1974.

[45] H. W. Hou, M. E. Warkiani, B. L. Khoo, Z. R. Li, R. A. Soo, D. S.-W. Tan, W.-T. Lim, J. Han, A. A. S. Bhagat, and C. T. Lim. Isolation and retrieval of circulating tumor cells using centrifugal forces. *Scientific Reports*, 3:1259, 2013.

[46] D. Huang, X. Shi, Y. Qian, W. L. Tang, L. B. Liu, N. Xiang, and Z. H. Ni. Rapid separation of human breast cancer cells from blood using a simple spiral channel device. *Analytical Methods*, 8:5940–5948, 2016.

[47] K. J. Humphry, P. M. Kulkarni, D. A. Weitz, J. F. Morris, and H. A. Stone. Axial and lateral particle ordering in finite Reynolds number channel flows. *Physics of Fluids*, 22:081703, 2010.

[48] S. C. Hur, S.-E. Choi, S. Kwon, and D. Di Carlo. Inertial focusing of non-spherical microparticles. *Applied Physics Letters*, 99:044101, 2011.

[49] S. C. Hur, N. K. Henderson-MacLennan, E. R. McCabe, and D. Di Carlo. Deformability-based cell classification and enrichment using inertial microfluidics. *Lab on a Chip*, 11:912–920, 2011.

[50] S. C. Hur, A. J. Mach, and D. Di Carlo. High-throughput size-based rare cell enrichment using microscale vortices. *Biomicrofluidics*, 5:022206, 2011.

[51] G.B. Jeffery. The motion of ellipsoidal particles immersed in a viscous fluid. *Proceedings of the Royal Society of London A: Mathematical, Physical and Engineering Sciences*, 102:161–179, 1922.

[52] C. Jin, S. M. McFaul, S. P. Duffy, X. Deng, P. Tavassoli, P. C. Black, and H. Ma. Technologies for label-free separation of circulating tumor cells: from historical foundations to recent developments. *Lab on a Chip*, 14:32–44, 2014.

[53] M. Kersaudy-Kerhoas and E. Sollier. Micro-scale blood plasma separation: from acoustophoresis to egg-beaters. *Lab on a Chip*, 13:3323–3346, 2013.

[54] R. Khojah, R. Stoutamore, and D. Di Carlo. Size-tunable microvortex capture of rare cells. *Lab on a Chip*, 17:2542–2549, 2017.

[55] J. Kim, J. Lee, T. Je, E. Jeon, and W. Lee. Size-dependent inertial focusing position shift and particle separations in triangular microchannels. *Analytical Chemistry*, 90:1827–1835, 2018.

[56] J. Kim, J. Lee, C. Wu, S. Nam, D. Di Carlo, and W. Lee. Inertial focusing in non-rectangular cross-section microchannels and manipulation of accessible focusing positions. *Lab on a Chip*, 16:992–1001, 2016.

[57] Y. W. Kim and J.Y. Yoo. Axisymmetric flow focusing of particles in a single microchannel. *Lab on a Chip*, 9:1043–1045, 2009.

[58] D. L. Koch and R. J. Hill. Inertial effects in suspension and porous-media flows. *Annual Review of Fluid Mechanics*, 33:619–647, 2001.

[59] A. Kulasinghe, T. H. P. Tran, T. Blick, K. O'Byrne, E. W. Thompson, M. E. Warkiani, C. Nelson, L. Kenny, and C. Punyadeera. Enrichment of circulating head and neck tumour cells using spiral microfluidic technology. *Scientific Reports*, 7:42517, 2017.

[60] S. S. Kuntaegowdanahalli, A. A. S. Bhagat, G. Kumar, and I. Papautsky. Inertial microfluidics for continuous particle separation in spiral microchannels. *Lab on a Chip*, 9:2973–2980, 2009.

[61] T. Laurell, F. Petersson, and A. Nilsson. Chip integrated strategies for acoustic separation and manipulation of cells and particles. *Chemical Society Reviews*, 36:492–506, 2007.

[62] L.G. Leal. Particle motions in a viscous fluid. *Annual Review of Fluid Mechanics*, 12:435–476, 1980.

[63] M.G. Lee, S. Choi, H.J. Kim, H.K. Lim, J.H. Kim, N. Huh, and J.K. Park. Inertial blood plasma separation in a contraction-expansion array microchannel. *Applied Physics Letters*, 98:253702, 2011.

[64] M.G. Lee, S. Choi, and J.K. Park. Inertial separation in a contraction-expansion array microchannel. *Journal of Chromatography A*, 1218:4138–4143, 2011.

[65] W. Lee, H. Amini, H. A. Stone, and D. Di Carlo. Dynamic self-assembly and control of microfluidic particle crystals. *Proceedings of the National Academy of Sciences of the United States of America*, 107:22413–22418, 2010.

[66] W. Lee, P. Tseng, and D. Di Carlo. *Microtechnology for Cell Manipulation and Sorting.* Springer International Publishing, 2017.

[67] M. Lekka, P. Laidler, D. Gil, J. Lekki, Z. Stachura, and A. Hrynkiewicz. Elasticity of normal and cancerous human bladder cells studied by scanning force microscopy. *European Biophysics Journal*, 28:312–316, 1999.

[68] S. Leon, B. Shapiro, D. Sklaroff, and M. Yaros. Free DNA in the serum of cancer patients and the effect of therapy. *Cancer Research*, 37:646–650, 1977.

[69] Q. Li, G. Lee, C. Ong, and C.T. Lim. AFM indentation study of breast cancer cells. *Biochemical and Biophysical Research Communications*, 374:609–613, 2008.

[70] E.J. Lim, T.J. Ober, Edd J.F., G.H. McKinley, and M. Toner. Visualization of microscale particle focusing in diluted and whole blood using particle trajectory analysis. *Lab on a Chip*, 12:2199–2210, 2012.

[71] H.K. Lin, S. Zheng, A.J. Williams, M. Balic, S. Groshen, H.I. Scher, M. Fleisher, W. Stadler, R.H. Datar, and Y.-C. Tai. Portable filter-based microdevice for detection and characterization of circulating tumor cells. *Clinical Cancer Research*, 16:5011–5018, 2010.

[72] C. Liu, G. Hu, X. Jiang, and J. Sun. Inertial focusing of spherical particles in rectangular microchannels over a wide range of Reynolds numbers. *Lab on a Chip*, 15:1168–1177, 2015.

[73] M.C. Liu, P.G. Shields, R.D. Warren, P. Cohen, M. Wilkinson, Y.L. Ottaviano, S.B. Rao, J. Eng-Wong, F. Seillier-Moiseiwitsch, and A.-M. Noone. Circulating tumor cells: a useful predictor of treatment efficacy in metastatic breast cancer. *Journal of Clinical Oncology*, 27:5153–5159, 2009.

[74] A.J. Mach and D. Di Carlo. Continuous scalable blood filtration device using inertial microfluidics. *Biotechnology and Bioengineering*, 107:302–311, 2010.

[75] J. Magnaudet, S. Takagi, and D. Legendre. Drag, deformation and lateral migration of a buoyant drop moving near a wall. *Journal of Fluid Mechanics*, 476:115–157, 2003.

[76] X. Mao, J. R. Waldeisen, and T. J. Huang. "Microfluidic drifting": implementing three-dimensional hydrodynamic focusing with a single-layer planar microfluidic device. *Lab on a Chip*, 7:1260–1262, 2007.

[77] J. Marchalot, Y. Fouillet, and J.L. Achard. Multi-step microfluidic system for blood plasma separation: architecture and separation efficiency. *Microfluidics and Nanofluidics*, 17:167–180, 2014.

[78] J. M. Martel and M. Toner. Particle focusing in curved microfluidic channels. *Scientific Reports*, 3:3340, 2013.

[79] J.M. Martel and M. Toner. Inertial focusing in microfluidics. *Annual Review of Biomedical Engineering*, 16:371–396, 2014.

[80] M. Masaeli, E. Sollier, H. Amini, W. Mao, K. Camacho, N. Doshi, S. Mitragotri, A. Alexeev, and D. Di Carlo. Continuous inertial focusing and separation of particles by shape. *Physical Review X*, 2:031017, 2012.

[81] J.P. Matas, V. Glezer, E. Guazzelli, and J.F. Morris. Trains of particles in finite-Reynolds-number pipe flow. *Physics of Fluids*, 16:4192–4195, 2004.

[82] S. Mendez and M. Abkarian. In-plane elasticity controls the full dynamics of red blood cells in shear flow. *Phys. Rev Fluids*, 3:101101(R), 2018.

[83] M.C. Miller, G.V. Doyle, and L.W. Terstappen. Significance of circulating tumor cells detected by the CellSearch system in patients with metastatic breast colorectal and prostate cancer. *Journal of Oncology*, 2010:617421, 2010.

[84] H. Moffatt. Viscous and resistive eddies near a sharp corner. *Journal of Fluid Mechanics*, 18:1–18, 1964.

[85] S. Mortazavi and G. Tryggvason. A numerical study of the motion of drops in Poiseuille flow. Part 1. Lateral migration of one drop. *Journal of Fluid Mechanics*, 411:1–18, 2000.

[86] S. Nagrath, L. V. Sequist, S. Maheswaran, D. W. Bell, D. Irimia, L. Ulkus, M. R. Smith, E. L. Kwak, S. Digumarthy, and A. Muzikansky. Isolation of rare circulating tumour cells in cancer patients by microchip technology. *Nature*, 450:1235–1239, 2007.

[87] N. Nivedita and I. Papautsky. Continuous separation of blood cells in spiral microfluidic devices. *Biomicrofluidics*, 7:054101, 2013.

[88] J. Oakey, R. W. Applegate Jr, E. Arellano, D. Di Carlo, S. W. Graves, and M. Toner. Particle focusing in staged inertial microfluidic devices for flow cytometry. *Analytical Chemistry*, 82:3862–3867, 2010.

[89] M. Ozkan, M. Wang, C. Ozkan, R. Flynn, and S. Esener. Optical manipulation of objects and biological cells in microfluidic devices. *Biomedical Microdevices*, 5:61–67, 2003.

[90] E. Ozkumur, A. M. Shah, J. C. Ciciliano, B. L. Emmink, D. T. Miyamoto, E. Brachtel, M. Yu, P.-i. Chen, B. Morgan, and J. Trautwein. Inertial focusing for tumor antigen-dependent and -independent sorting of rare circulating tumor cells. *Science Translational Medicine*, 5:179ra47, 2013.

[91] N. Pamme and C. Wilhelm. Continuous sorting of magnetic cells via on-chip free-flow magnetophoresis. *Lab on a Chip*, 6:974–980, 2006.

[92] H. Park, S.-H. Hong, K. Kim, S.-H. Cho, W.-J. Lee, Y. Kim, S.-E. Lee, and Y. Park. Characterizations of individual mouse red blood cells parasitized by Babesia microti using 3-D holographic microscopy. *Scientific Reports*, 5:10827, 2015.

[93] J.-S. Park and H.-I. Jung. Multiorifice flow fractionation: continuous size-based separation of microspheres using a series of contraction/expansion microchannels. *Analytical Chemistry*, 81:8280–8288, 2009.

[94] J.-S. Park and H.-I. Song, S.-H.and Jung. Continuous focusing of microparticles using inertial lift force and vorticity via multi-orifice microfluidic channels. *Lab on a Chip*, 9:939–948, 2008.

[95] S. Patyar, R. Joshi, D. P. Byrav, A. Prakash, B. Medhi, and B. Das. Bacteria in cancer therapy: a novel experimental strategy. *Journal of Biomedical Science*, 17:21, 2010.

[96] M. E. Piyasena and S. W. Graves. The intersection of flow cytometry with microfluidics and microfabrication. *Lab on a Chip*, 14:1044–1059, 2014.

[97] V. Plaks, C. D. Koopman, and Z. Werb. Circulating tumor cells. *Science*, 341:1186–1188, 2013.

[98] M. Rafeie, J. Zhang, M. Asadnia, W. H. Li, and M. E. Warkiani. Multiplexing slanted spiral microchannels for ultra-fast blood plasma separation. *Lab on a Chip*, 16:2791–2802, 2016.

[99] H. Ramachandraiah, H. A. Svahn, and A. Russom. Inertial microfluidics combined with selective cell lysis for high throughput separation of nucleated cells from whole blood. *RSC Advances*, 7:29505–29514, 2017.

[100] J. M. Rhea and R. J. Molinaro. Cancer biomarkers: surviving the journey from bench to bedside. *MLO: Medical Laboratory Observer*, 43:10, 2011.

[101] M. Robinson, H. Marks, T. Hinsdale, K. Maitland, and G. Cote. Rapid isolation of blood plasma using a cascaded inertial microfluidic device. *Biomicrofluidics*, 11:024109, 2017.

[102] S. Rubinow and Keller J. B. The transverse force on a spinning sphere moving in a viscous fluid. *Journal of Fluid Mechanics*, 11:447–459, 1961.

[103] P. Saffman. The lift on a small sphere in a slow shear flow. *Journal of Fluid Mechanics*, 22:385–400, 1965.

[104] G. Segre and Silberberg A. Radial particle displacements in poiseuille flow of suspensions. *Nature*, 189:209–210, 1961.

[105] J. Seo, M. H. Lean, and A. Kole. Membraneless microseparation by asymmetry in curvilinear laminar flows. *Journal of Chromatography A*, 1162:126–131, 2007.

[106] J. Shi, H. Huang, Z. Stratton, Y. Huang, and T. J. Huang. Continuous particle separation in a microfluidic channel via standing surface acoustic waves (SSAW). *Lab on a Chip*, 9:3354–3359, 2009.

[107] E. Sollier, D. E. Go, J. Che, D. R. Gossett, S. O'Byrne, W. M. Weaver, N. Kummer, M. Rettig, J. Goldman, and N. Nickols. Size-selective collection of circulating tumor cells using Vortex technology. *Lab on a Chip*, 14:63–77, 2009.

[108] T. M. Squires and S. R. Quake. Microfluidics: Fluid physics at the nanoliter scale. *Reviews of Modern Physics*, 77:977–1026, 2005.

[109] G. Subramanian and D. Koch. Centrifugal forces alter streamline topology and greatly enhance the rate of heat and mass transfer from neutrally buoyant particles to a shear flow. *Physical Review Letters*, 96:134503, 2006.

[110] A. P. Sudarsan and V. M. Ugaz. Multivortex micromixing. *Proceedings of the National Academy of Sciences*, 103:7228–7233, 2006.

[111] J. Sun, M. Li, C. Liu, Y. Zhang, D. Liu, W. Liu, G. Hu, and X. Jiang. Double spiral microchannel for label-free tumor cell separation and enrichment. *Lab on a Chip*, 12:3952–3960, 2012.

[112] S. Suresh. Biomechanics and biophysics of cancer cells. *Acta Materialia*, 55:3989–4014, 2007.

[113] J. Takagi, M. Yamada, M. Yasuda, and M. Seki. Continuous particle separation in a microchannel having asymmetrically arranged multiple branches. *Lab on a Chip*, 5:778–784, 2005.

[114] C. K. Tam and W. A. Hyman. Transverse motion of an elastic sphere in a shear field. *Journal of Fluid Mechanics*, 59:177–185, 1973.

[115] W. Tang, D. Tang, Z. Ni, N. Xiang, and H. Yi. Microfluidic impedance cytometer with inertial focusing and liquid electrodes for high-throughput cell counting and discrimination. *Analytical Chemistry*, 89:3154–3161, 2017.

[116] H. C. Tseng, R. G. Wu, H. Y. Chang, and F. G. Tseng. High-throughput white blood cells (leukocytes) separation and enrichment from whole blood by hydrodynamic and inertial force. *IEEE 25th International Conference on Micro Electro Mechanical Systems (MEMS)*, 2012.

[117] W. S. Uijttewaal, E. J. Nijhof, and R. M. Heethaar. Droplet migration, deformation, and orientation in the presence of a plane wall: A numerical study compared with analytical theories. *Physics of Fluids A: Fluid Dynamics*, 5:819–825, 1993.

[118] J. Wagner. Free DNA-new potential analyte in clinical laboratory diagnostics? *Biochemia Medica*, 22:24–38, 2012.

[119] L. Wang and D. S. Dandy. High-throughput inertial focusing of micrometer- and sub-micrometer-sized particles separation. *Advanced Science*, 4:1700153, 2017.

[120] M. E. Warkiani, G. Guan, K. B. Luan, W. C. Lee, A. A. S. Bhagat, P. K. Chaudhuri, D. S.-W. Tan, W. T. Lim, S. C. Lee, and P. C. Chen. Slanted spiral microfluidics for the ultra-fast, label-free isolation of circulating tumor cells. *Lab on a Chip*, 14:128–137, 2014.

[121] M. E. Warkiani, B. L. Khoo, L. D. Wu, A. K. P. Tay, A. A. S. Bhagat, J. Han, and C. T. Lim. Ultra-fast, label-free isolation of circulating tumor cells from blood using spiral microfluidics. *Nature Protocols*, 11:134–148, 2016.

[122] D. Wu, J. Qin, and B. Lin. Electrophoretic separations on microfluidic chips. *Journal of Chromatography A*, 1184:542–559, 2018.

[123] L. Wu, G. Guan, H. W. Hou, A. A. S. Bhagat, and J. Han. Separation of leukocytes from blood using spiral channel with trapezoid cross-section. *Analytical chemistry*, 84:9324–9331, 2012.

[124] Z. Wu, Y. Chen, M. Wang, and A. J. Chung. Continuous inertial microparticle and blood cell separation in straight channels with local microstructures. *Lab on a Chip*, 16:532–542, 2016.

[125] Z. Wu, B. Willing, J. Bjerketorp, J. K. Jansson, and K. Hjort. Soft inertial microfluidics for high throughput separation of bacteria from human blood cells. *Lab on a Chip*, 9:1193–1199, 2009.

[126] N. Xiang, K. Chen, D. Sun, H. Wang, and Z. Ni. Quantitative characterization of the focusing process and dynamic behavior of differently sized microparticles in a spiral microchannel. *Microfluidics and Nanofluidics*, 14:89–99, 2013.

[127] N. Xiang and Z. Ni. High-throughput blood cell focusing and plasma isolation using spiral inertial microfluidic devices. *Biomedical Microdevices*, 17:110, 2015.

[128] Y. Xu, X. Xie, Y. Duan, L. Wang, Z. Cheng, and J. Cheng. A review of impedance measurements of whole cells. *Biosensors and Bioelectronics*, 77:824–836, 2016.

[129] M. Yamada, M. Nakashima, and M. Seki. Pinched flow fractionation: continuous size separation of particles utilizing a laminar flow profile in a pinched microchannel. *Analytical Chemistry*, 76:5465–5471, 2004.

[130] Z. T. F. Yu, K. M. Aw Yong, and J. Fu. Microfluidic blood cell sorting: now and beyond. *Small*, 10:1687–1703, 2014.

[131] J. Zhang, W. Li, M. Li, G. Alici, and N.-T. Nguyen. Particle inertial focusing and its mechanism in a serpentine microchannel. *Microfluidics and Nanofluidics*, 17:305–316, 2014.

[132] J. Zhang, S Yan, W. Li, G. Alici, and N.-T. Nguyen. High throughput extraction of plasma using a secondary flow-aided inertial microfluidic device. *RSC Advances*, 4:33149–33159, 2014.

[133] J. Zhang, S Yan, R. Sluyter, W. Li, G. Alici, and N.-T. Nguyen. Inertial particle separation by differential equilibrium positions in a symmetrical serpentine microchannel. *Scientific Reports*, 4:4527, 2014.

[134] J. Zhang, S. Yan, D. Yuan, G. Alici, N.-T. Nguyen, M. E. Warkiani, and W. Li. Fundamentals and applications of inertial microfluidics: a review. *Lab on a Chip*, 16:10–34, 2016.

[135] J. Zhou, P. V. Giridhar, S. Kasper, and I. Papautsky. Modulation of aspect ratio for complete separation in an inertial microfluidic channel. *Lab on a Chip*, 13:1919–1929, 2013.

CHAPTER 11

Microfluidic biotechnologies for hematology: separation, disease detection and diagnosis

Kuan Jiang

Mechanobiology Institute, National University of Singapore, Singapore

Chwee Teck Lim

Department of biomedical engineering, National University of Singapore, Singapore

CONTENTS

BLOOD is one of the most collected bodily fluids that can yield tremendous diagnostic information with regard to the health and well-being of a person. With the recent advancement in micro- and nanotechnologies, blood-based diagnosis has become more specific, accurate, faster and more readily available. Microfluidics is one such technology that has been extensively explored for use in hematology and related diseases since they have the ability to handle small volumes of liquid and effectively perform separation, sorting and analysis of blood cells and molecular components. In this chapter, after a brief introduction on a few general physical principles applied in microfluidic platforms as well as their application and performance in blood component separation, we will examine two types of diseases where diseased cells are manifested in the blood circulatory system and as such, present themselves as excellent circulating biomarkers for diagnosis: malaria and cancer. We will cite and discuss examples of microfluidic platforms that have been developed for malaria detection, and for liquid biopsy with the aim to better detect and diagnose cancer. Through such examples, we hope to highlight the important roles such microfluidic-based systems can play in health care and on the challenges and perspectives for the future development of the field.

11.1 INTRODUCTION

Blood is a critical bodily fluid that facilitates the function of the human body. Its primary roles are to exchange nutrients, remove metabolic wastes and transmit molecular signals to promote tissue regeneration and maintain homeostasis. Cellular and molecular changes in blood components can reflect the health condition of our body. For example, blood cell count is one of the most common blood tests performed in clinics. The numbers of red blood cells (RBCs), white blood cells (WBCs) and platelets can indicate the status of the human hematopoietic system, immune system, and blood coagulation function. The concentration of small molecules such as glucose, lipids, cholesterol and large molecules like alanine aminotransferase (ALT) are good functional indicators of specific organs and reflect the health status of a person. Besides testing on common blood components, probing the presence of disease-related biomarkers, such as hepatitis B virus surface antigen (HBsAg) in hepatitis B, can assist in disease diagnosis.

There are two trends in the development of blood-related tests. On one hand, industries are making efforts to make existing tests more sensitive and specific, reduce the cost of them as well as simplify the operating procedures. On the other hand, researchers have been identifying new biomarkers as well as promoting new technologies for expanding the applications of blood tests. One of the recent achievements is the screening of circulating biomarkers of tumors for cancer diagnosis, prognosis, and treatment. The emergence of microfluidic technologies in biomedical applications has contributed significantly to promote the progression of this kind of blood-based clinical testing. As the name of the technique suggests, microfluidic platforms are usually small in size, allow processing of a smaller amount of samples, require fewer reagents and as such, can significantly reduce the cost of each test. As blood is a complex mixture of different blood components, it is a big challenge to extract the desired biological information or to obtain one pure fraction of blood component for downstream application. Thus, blood components separation and blood cell sorting have been a hot research area due to its significant impact on blood based diagnosis and blood related biomedical and physiological studies.

In the following sections, we will first give a brief introduction to microfluidic technology. Following that, we will introduce the principles of several representative microfluidic systems as well as their latest applications and performance in blood cell separation. Next, we will discuss specific applications of microfluidics in two types of diseases: malaria and cancer. We will end with what the current challenges are and offer our future perspectives on microfluidics on healthcare.

11.2 MICROFLUIDIC TECHNOLOGY

Microfluidic technology essentially involves a device or system that deals with fluid confined in micro-sized channels. Unlike other bulk processing methods (e.g., centrifugation), a small volume of liquid can be precisely controlled, processed and analyzed inside microchannels to achieve high spatial-temporal resolution and high accessibility of the fluid components. Such devices are usually small in footprint even when multiple microfluidic modules with different functions are integrated together. Capillary forces can dominate in a micron-domain making fully automated self-driven devices possible in microfluidic platforms.

In the field of blood-based medical screening, microfluidic-based platforms usually have one or more of the following features:

i Small and portable;

ii Self-driven and automated;

iii Multi to single-cell resolution;

iv High sensitivity;

v Low fabrication cost;

vi Small sample volume; and

vii Easy to operate.

These unique features of microfluidic platforms address most of the issues in current clinical blood tests including:

i Being available to resource-scarce settings;

ii Enabling point-of-care testing;

iii Reducing blood sample volume required for testing; and

iv Assaying cells down to single-cell resolution in a high throughput manner.

As such, the application of microfluidics for such clinical use can now be better implemented. Nevertheless, factors influencing the adoption of microfluidics will differ from case to case depending on their specific applications. In the following sections, we will discuss the basic concepts in handling and separating blood components and subsequently discuss how and why microfluidics has become more and more important in hematology by focusing on applications for two sadly notorious diseases which are malaria and cancer.

11.3 BLOOD COMPONENTS SEPARATION

11.3.1 Introduction

There are four major components of blood: plasma, RBC, platelet and leukocyte. Leukocytes can be further categorized into five subtypes, and they are neutrophil, lymphocyte, monocyte, basophil and eosinophil, based on their different roles in the immune system. These cells can also be classified into even smaller groups like T lymphocytes and B lymphocytes. Here, we focus on the separation of the major blood components: plasma, platelets and leukocytes.

Traditional methods for blood component separation rely mostly on the difference in physical properties between different blood cell fractions. Density based centrifugation is the mainstream method for the preparation of blood components for clinical blood transfusion and analysis. As for the sorting of blood cells, antibody-based recognition of individual cell types followed by a magnetic or electric based separation method have been extensively studied over the past 20 years. The density based bulk processing methods are labor intensive involving multiple steps and require a long processing time using expensive centrifuge machines. They also suffer from low separation resolution and require a large amount of blood for processing. Magnetic activated cell sorting (MACS) and fluorescence activated cell sorting (FACS) are promising technologies that can give high separation specificity [3, 37]. However, the expensive antibodies used in both methods make the technology costly, while the labelling process can impede downstream molecular analysis.

Microfluidics is an emerging field that uses miniaturized platforms, usually at the micrometer scale, to process a small volume of sample. The biggest virtue of this type of system is its portable size and ability to handle blood at the microliter level. This makes the overall process more controllable, and also provides higher separation efficiency and resolution. Additionally, it is easier to automate the separation process thus making the entire process less labor intensive. In the following sections, we will introduce the principles of several representative microfluidic systems as well as their latest applications and performance in blood cell separation.

11.3.2 General principle

Every cell type in blood is unique and has its specific function. Correspondingly, they exhibit very different physical properties. Size, deformability and density difference can all result in different behavior of blood cells under different applied force fields. Also, the specific surface markers expressed on different cell types make it possible to mark cell populations with different antibodies and achieve separation through active selection or capture. In this section, we will introduce the physical principles that can be employed to target and separate one or more blood components from whole blood. In general, magnetic, electric and acoustic forces can be used to actively select cells based on their physical or molecular differences. Hydrodynamic force, filtration and density gradient can passively separate cells with different physical parameters. For active selection, it can be further divided into two categories: label-based and label-free methods depending on whether or not an antibody is used to tag different cell types.

11.3.2.1 Filtration based on size and deformability

The most obvious physical difference between different blood cells is their size and morphology. RBCs have a unique biconcave shape with a diameter of about 7 μm and height of around 2 μm. The platelet is the smallest cell component in blood having a diameter of around 2.5 μm. Leukocytes exhibit a size range from 6 μm to 12 μm, but their different nuclear to cytoplasm ratio not only results in different deformability as compared to the RBC and platelet, but also a subtle difference among different leukocyte subtypes. Owing to these observed differences in size and deformability, one of the most straightforward ways of selecting one blood cell population over the other is through the retention of cells with specifically designed constrictions. A filtration system can be easily incorporated into microfluidic channels and multiples of such systems have been developed over the past few years and efforts have now been focused on increasing the throughput and solving the clogging issue inside the small channel. Figure 11.1(A) shows a microfluidic ratchets array developed recently for size and deformability based separation of each blood cell fraction, which is representative for technologies in this category [42]. This chip explores the deformability difference among all components of human blood. While blood cells are processed in the chip, the smallest and most deformable cells go to the upper side of the assay while larger and stiffer WBCs and rarely occurring circulating tumor cells stay in the bottom layer. The chip has the potential to achieve a full fractionation of blood in one chip. However, to guarantee a complete separation and migration of different components, the time required for processing is longer than the other passive technologies (1 hour to process 1 ml whole blood), which impeded the application of the chip.

11.3.2.2 Inertial focusing in fluid channel

When a cell is flowing in a microfluidic channel, there will be hydrodynamic forces arising from different sources being exerted on cells through the fluid field. In a straight channel, lift force induced by the parabolic velocity profile of the fluid flow is dominant. In laminar flow, due to the friction of the channel wall, the fluid in the central of the channel will flow faster than the fluid near the wall. The velocity gradient results in a shear force in the cell body along the flow direction. The net effect of the fluid shear pushes the cell away from the central line, which is termed shear-induced lift force. On the other hand, there is a "wall effect" that tends to push the cells back to the central line, which is termed wall-induced lift force. These two forces are both dependent on cell size but with different orders of correlation. Thus, in properly designed channels, there will be equilibrium positions on the cross-section of the fluid channel for cells with different sizes, so that separation is achieved [10]. In a curved channel, additionally, there will be a centrifugal force arising from the channel curvature acting on the cells. The centrifugal force forms two counter rotating vortices as a secondary flow in the flow cross section area which is termed Dean Vortices, exerting a Dean Drag force to act on the cells. The interplay between the lift force and Dean Drag force causes cells to either migrate with the Dean cycle or be focused when they reach their focusing position, depending on their size range. By designing the length, curvature and dimensions of the channel, cells with specific size range can be selected from blood. Figure 11.1(B) shows the focusing position of particles in straight rectangular and circular channels as well as the flow profile in a curved rectangular channel [43].

11.3.2.3 Hydrodynamic focusing

Besides the inertial forces, microstructures introduced in microfluidic channels can be designed to direct cells with different sizes into different flow streams. Hydrodynamic deterministic lateral displacement is a phenomenon where particles of different sizes are directed to different positions of the channel by an array of micropillars as shown in Figure 11.2(A) [2]. Each row of the posts is laterally shifted by a small distance, which creates separate flow laminae that can be controlled by the distance between the posts and length of lateral displacement. Different flow streamlines will be formed and if the center of a particle is outside the first streamline, it will migrate to the second streamline after passing through a post [26]. This mechanism can continuously migrate particles larger than a critical size to the side of the flow channel and this principle has successfully been applied to form various microfluidic cell separation systems.

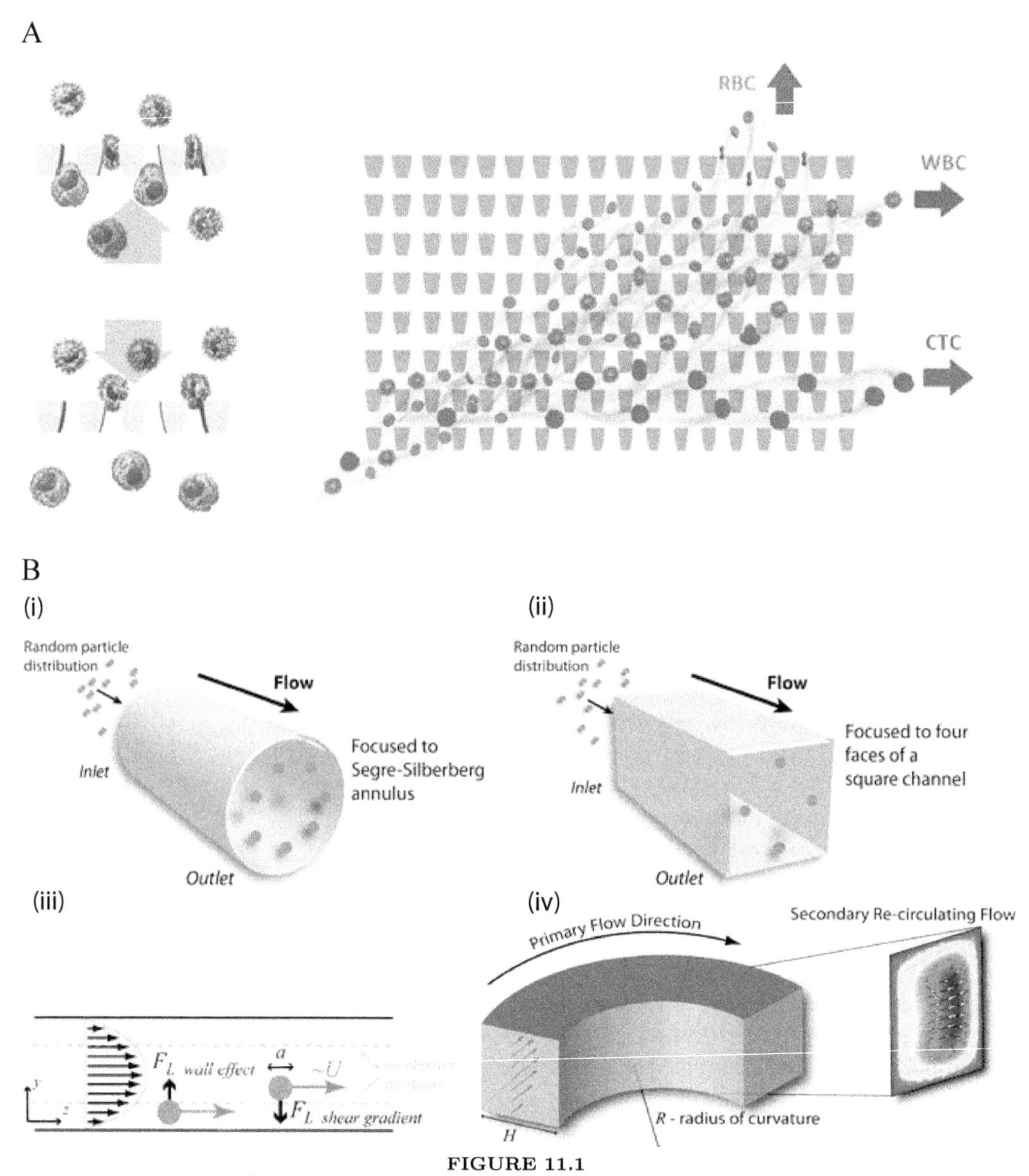

FIGURE 11.1
Passive sorting methods. (A) A deformability based ratchets array with oscillation perpendicular to the blood flow direction. More deformable components tend to migrate to the upper part of the device at the end of the channel. Reprinted from [42], Copyright 2016, with permission from John Wiley and Sons. (B) Principle of inertial focusing. The balance of fluid lift force causes particles to align along the channel wall and the position is dictated by the size of particles. (a) and (b) show the focusing positions in circular and rectangular channels, respectively. (c) shows the force contributing to the focusing. (d) shows the flow profile in the cross-sectional area of a curved channel. Reprinted from [10], Copyright 2009, with permission from Royal Society of Chemistry.

Figure 11.2(B) shows a biomimetic microfluidic system designed to separate WBCs. The channel targets the RBC depletion layer near the channel side wall whereas WBCs will undergo lateral migration [49]. Concentrated WBCs can be harvested from the small side outlet of the device. When targeting to separate the liquid phase from the solid phase of blood, there are several similar hemodynamic behaviors that can be utilized to form cell focusing or cell depletion layers in blood flow. Technologies based on hemodynamics and hydrodynamics will be discussed in more detail in the plasma separation section.

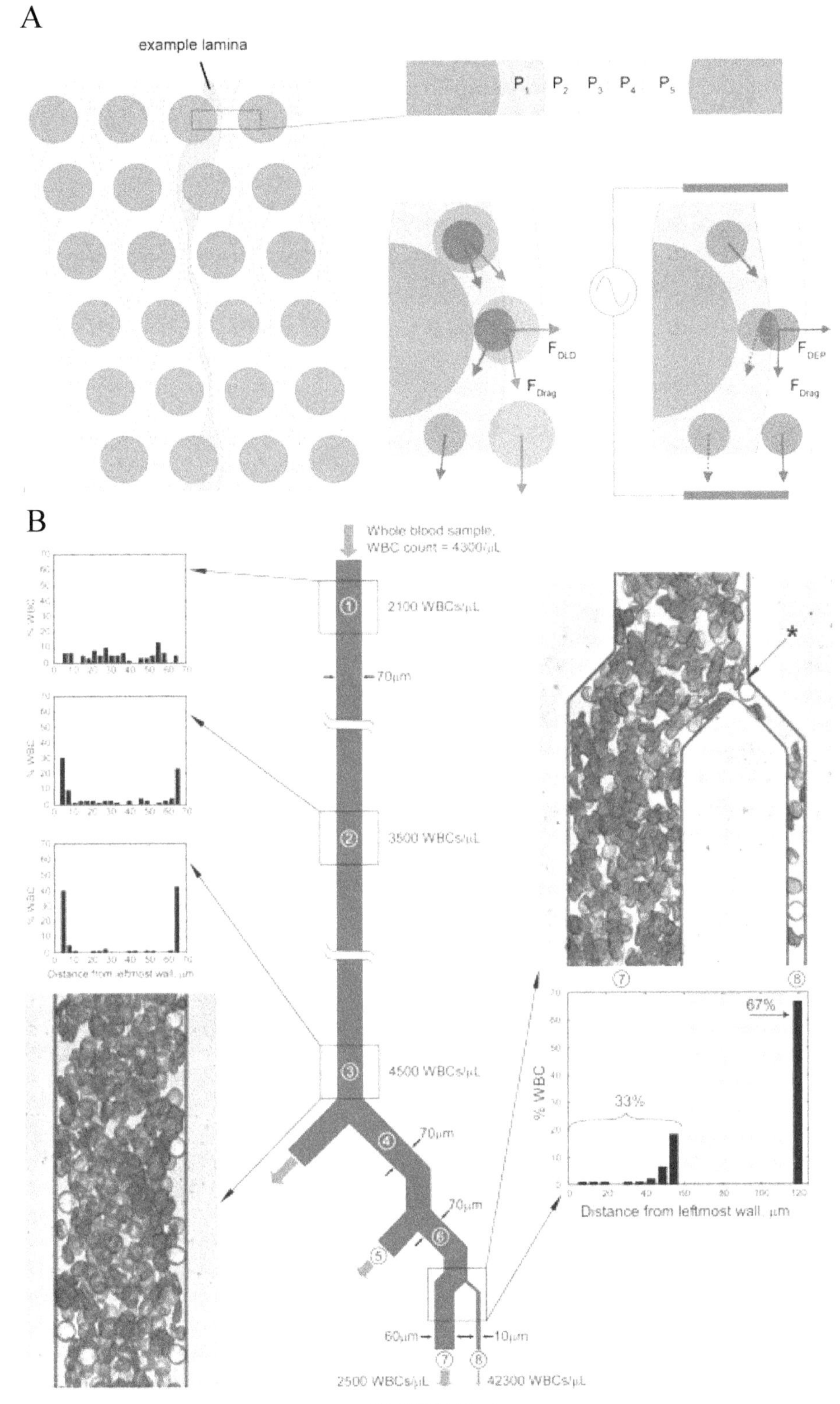

FIGURE 11.2
Hydrodynamic effects for particle separation. (A) Deterministic lateral displacement for cell separation by size. The pillars array segments the fluid flow to different partitions where smaller cells migrate in the array closer to the center while larger cells are directed to the corner. Reprinted from [2], Copyright 2009, with permission from Royal Society of Chemistry. (B) A biomimetic device based on hemodynamics for leukocyte separation. RBCs tend to migrate to the central axis and WBCs can be collected from the side of channel. Reprinted from [49], Copyright 2005, American Chemical Society.

11.3.2.4 Dielectrophoresis

Cell membrane acts like an insulator that isolates cytoplasm from the extracellular environment. In addition, there are charged molecules and ions distributing inside and outside of the cell membrane. As a result, when a cell is subjected to an electric field, an electric dipole can be induced in the cell. By adjusting the direction and strength of an external electrical field, it is possible to exert a dielectrophoretic force on the cell of a safe level. In a non-uniform electric field, the induced electric force is dependent on the cell volume, density and other physical parameters. As cells also experience gravitational force under flow, by adjusting the electric field, they can be focused at different positions of a microfluidic channel according to their size and density difference. Figure 11.3(A) shows a representative microfluidic device designed based on dielectrophoresis to separate blood cells [16].

11.3.2.5 Acoustophoresis

Acoustic force is another active force that has been extensively studied for particle separation. The acoustic primary radiation force (PRF) acting on flowing particles is dependent on the particle size and density. The level of force can be controlled through the density of flow medium so that cells with different sizes migrate to different positions of the channel during free flow. Figure 11.3(B) shows schematically how acoustophoresis happens and this principle has been developed to separate RBCs, WBCs and platelets from whole blood in several microfluidic-based systems [43].

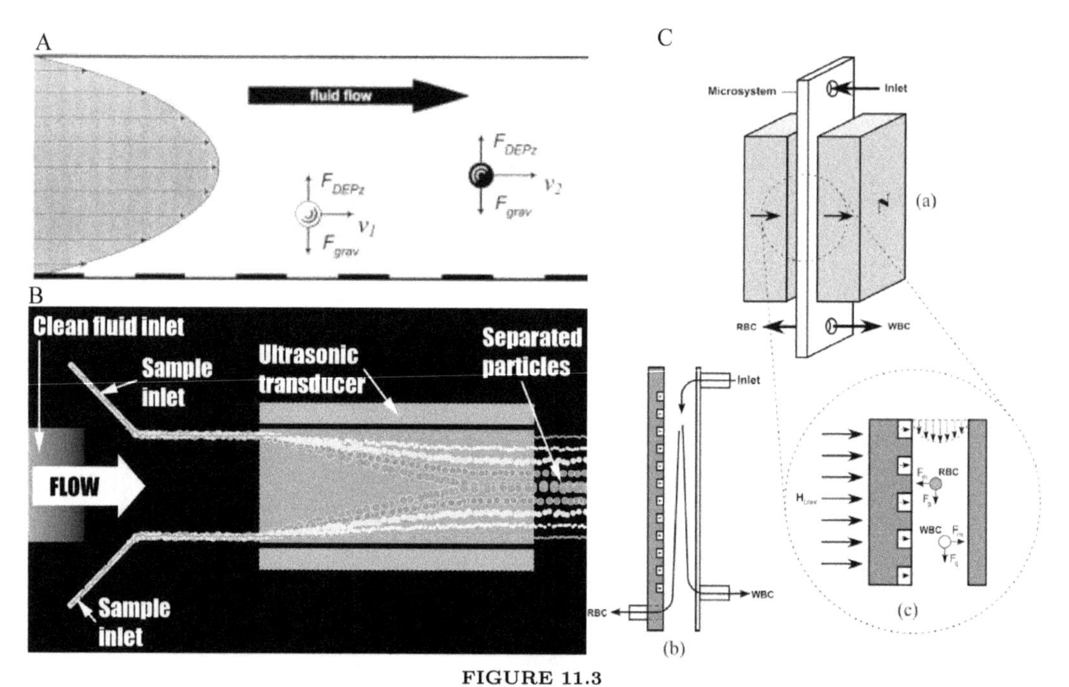

FIGURE 11.3
Principles of active separation methods. (A) Cell sorting based on dielectrophoretic force. The dependence of DEP force and gravitational force on size is different resulting in different balance positions of cell components with different size. Reprint from [16], copyright 2002, with permission from John Wiley and Sons. (B) Particle sorting based on ultrasonic radial force. Particles with different size will be focused in different positions of the channel. Reprint from [43], copyright 2007, with permission from American Chemical Society. (C) Blood cell sorting based on magnetophoretic force. RBC and WBC have different magnetic susceptibility resulting in an opposite magnetic force acting on them. This device is also gravity driven where RBCs and WBCs are collected from different sides of the channel. Reprinted from [15], Copyright 2007, with permission from IOP Publishing.

11.3.2.6 Magnetophoresis

Besides electric and acoustic fields, a magnetic field also has different effects on different blood cell types. Magnetic susceptibility of different cell types has been explored for cell separation. For example, RBCs and WBCs have different cellular content which results in opposing magnetic forces acting on them. Generally, deoxygenated RBCs tend to be paramagnetic while WBCs are diamagnetic and this can be utilized to separate WBCs and RBCs as shown in Figure 11.3(C) [15].

11.3.2.7 Discussion

RBCs are the most abundant cell type in blood with about 5 billion in one milliliter of whole blood. To achieve separation of each blood component, the first question is how RBCs will react differently from the desired cell type in the given force field or constriction array, and whether the difference is good enough to take them apart and remove most of the RBCs. The size, deformability and morphology can be considered unique for RBCs compared to other cell components in blood. Some overlapping of physical properties, such as cell size, happens only to a small portion of other blood cell types due to cell heterogeneity. The difficulties faced are 1) RBC number is extremely high and dominant in whole blood, which makes it hard to be reduced to an acceptable level, and 2) due to the high number of RBCs, whole blood is a non-Newtonian fluid and cell-cell interaction in the fluid system makes the behavior of blood cells in the force field designed for cell separation much more unpredictable and complicated. A complete separation is not realistic with current technologies but we have witnessed an increasing amount of satisfying performance as the microsystem technologies mature.

11.3.3 Plasma separation

Plasma is the liquid phase of blood that contains abundant molecular information about the human body. Currently, the most effective way of removing all the cellular components from whole blood is by using the density-based centrifuge. However, for diagnosis purposes, a bulk centrifuge machine is not always available at all sites and the use of a centrifuge requires trained personnel, which is both time and labor consuming. Microfluidic devices have achieved great success in the field of plasma separation. In general, the techniques mainly used the differences in physical properties of different blood components. Gravity based sedimentation, size based filtration, lateral displacement and hydrodynamic focusing have all been explored in the past decade.

When blood is flowing in microfluidic channels, it is subjected to hydrodynamic forces that tend to induce an axial migration of RBCs to the central line of the flow, which is termed "plasma skimming" or also called Fahraeus effect. This migration tendency creates a cell-free layer along the channel wall. By optimizing flow rate, dimensions and geometry of the fluid channel, it is possible to get very pure plasma from whole blood. Jaeggi et al. designed a microchannel having vertical branches along the channel to extract plasma using this effect. They achieved a 92% removal of blood cells at a flow rate of 5 ml/min [28]. Bifurcation law or Zweifach-Fung effect is another phenomenon where blood cells tend to move into a larger opening channel of the branching point. A device based on this effect has demonstrated a near 100% pure plasma separation. However, the flow rate in the device is only 3-4 μl/min [58]. Prabhakar et al. designed a microfluidic system that integrates several hydrodynamic effects. As shown in Figure 11.4(A), the principles involve the previously discussed bifurcation law, Fahraeus effect, centrifugal action and constriction

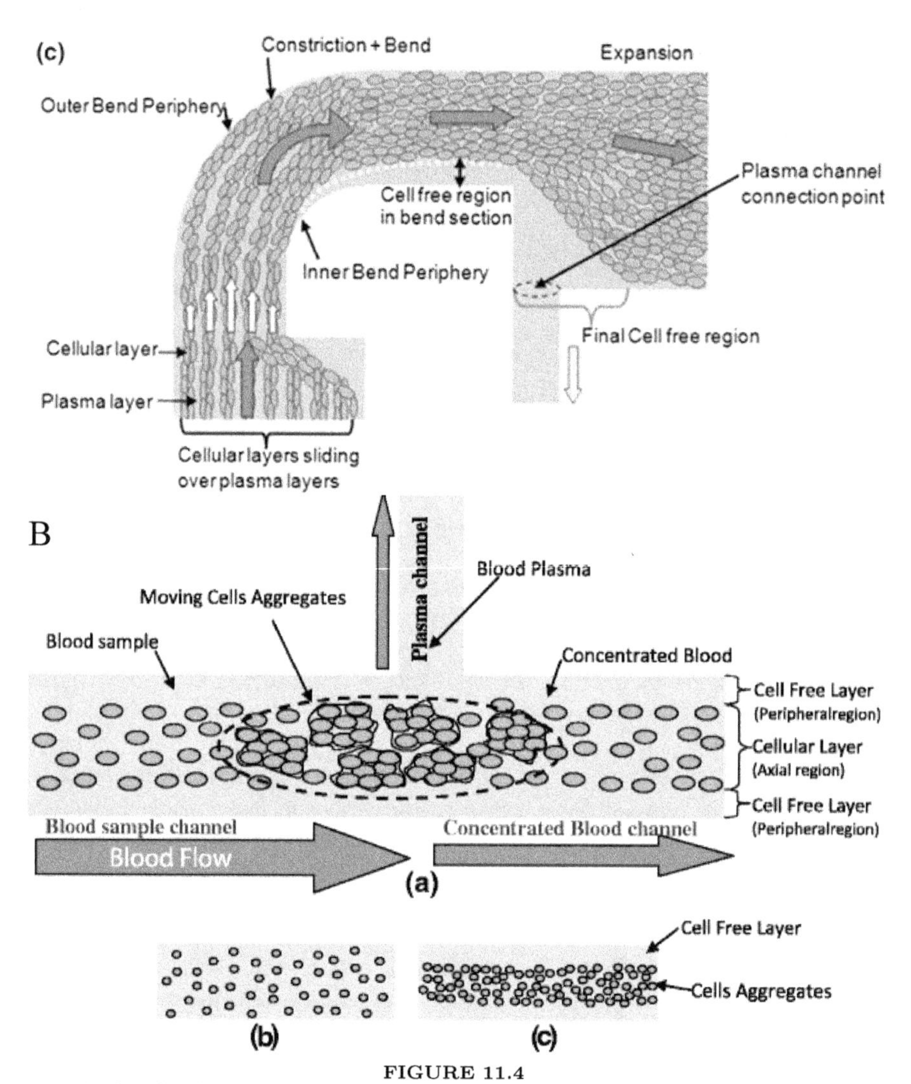

FIGURE 11.4
Plasma separation based on hemodynamic and hydrodynamic effect. (A) A microfluidic device employing multiple cell margination effect for creating cell free region and collecting plasma from whole blood. Reprinted from [45], Copyright 2015, with permission from Springer. (B) A T-shape channel for plasma collection using the cell free layer created by axial migration of RBCs. Reprinted from [53], Copyright 2013, with permission from Springer.

expansion. They achieved a nearly 100% separation efficiency for 20% hematocrit and 80% efficiency for undiluted whole blood. Meanwhile, the throughput is enhanced to around 0.5 ml/min [45].

Besides the hydrodynamic focusing of blood cells in a microfluidic channel, Dean drag force induced in a curved channel and inertial force can also interplay to facilitate the separation of the cells. Rafeie et al. developed a high-throughput multiplexing slanted spiral chip to separate plasma from diluted human whole blood. The chip has a dimension specifically designed to focus all cell components in blood so that blood plasma can be extracted from the opposite site in the channel. By connecting individual chips in parallel to form a microfluidics system, the platform is able to achieve a nearly 100% removal of RBC at HCT value below 1% with a maximized overall flow rate of ~530 μl whole blood every minute [46].

Active energy like an electrical field and acoustic field can also be used to migrate and remove blood cells and derive pure plasma. Ultrasonic standing waves can generate a radiation force that migrates blood cells into the pressure nodes of an acoustic force field. By designing a microfluidic channel matching the half wavelength of the ultrasonic standing wave, cells can be focused to the pressure nodes located in the center of the microfluidic channel. This focused cell phase can be removed in the middle outlet of the chip and plasma can be collected in the side outlets through a trifurcated outlet design. Lenshof et al. developed an acoustic based plasmapheresis chip for detection of prostate specific antigen based on this principle and Figure 11.5 schematically shows how this chip works. They achieved a more than 99% removal of RBCs from whole blood at a flow rate of 80 μl/min and the residual RBCs in the separated plasma are below the threshold recommended for plasma analysis by the council of Europe [35].

As blood cells are polarizable, they will experience a dielectrophoretic force in a non-uniform electric field and migrate towards the weaker electric field site. Cell free plasma can then be collected from the opposite side. Nakashima et al. developed a capillary-driven microfluidic chip that is able to extract 300 nl plasma from 5 μl blood with a 97% cell removal [38].

Although these techniques can hardly generate 100% cell free plasma, their capability in reducing blood cells at a factor of more than 1000 times has successfully demonstrated their potential application in quick clinical diagnosis using plasma. These simple miniaturized systems can be further integrated with downstream molecular assay to efficiently obtain useful information from blood plasma.

11.3.4 Separation of platelets

Platelets are the smallest cellular component in blood. They play an important role in stopping bleeding and enabling wound sealing. Separation of viable platelets is crucial for platelet transfusion in treatment of patients with acute bleeding or clotting malfunction. Plateletpheresis is widely adopted in clinics to extract only platelets from donor's whole blood and circulate the other blood components back to the donor. This process usually takes around an hour to collect enough platelets for one unit (around 3 10^{11}) of platelet transfusion. However, in physiological research and clinical diagnosis, fast separation of platelets for downstream analysis is always desired. Additionally, the centrifuge based plateletpheresis has been shown to cause damage or activation of platelets which is undesirable for their downstream application [5]. Several microfluidic systems have also been developed to perform on-chip plateletpheresis. Due to the significantly smaller size of platelets as compared to other blood cells, the separation efficiency is usually high and the separation process is relatively simpler as compared to the separation of other blood cells.

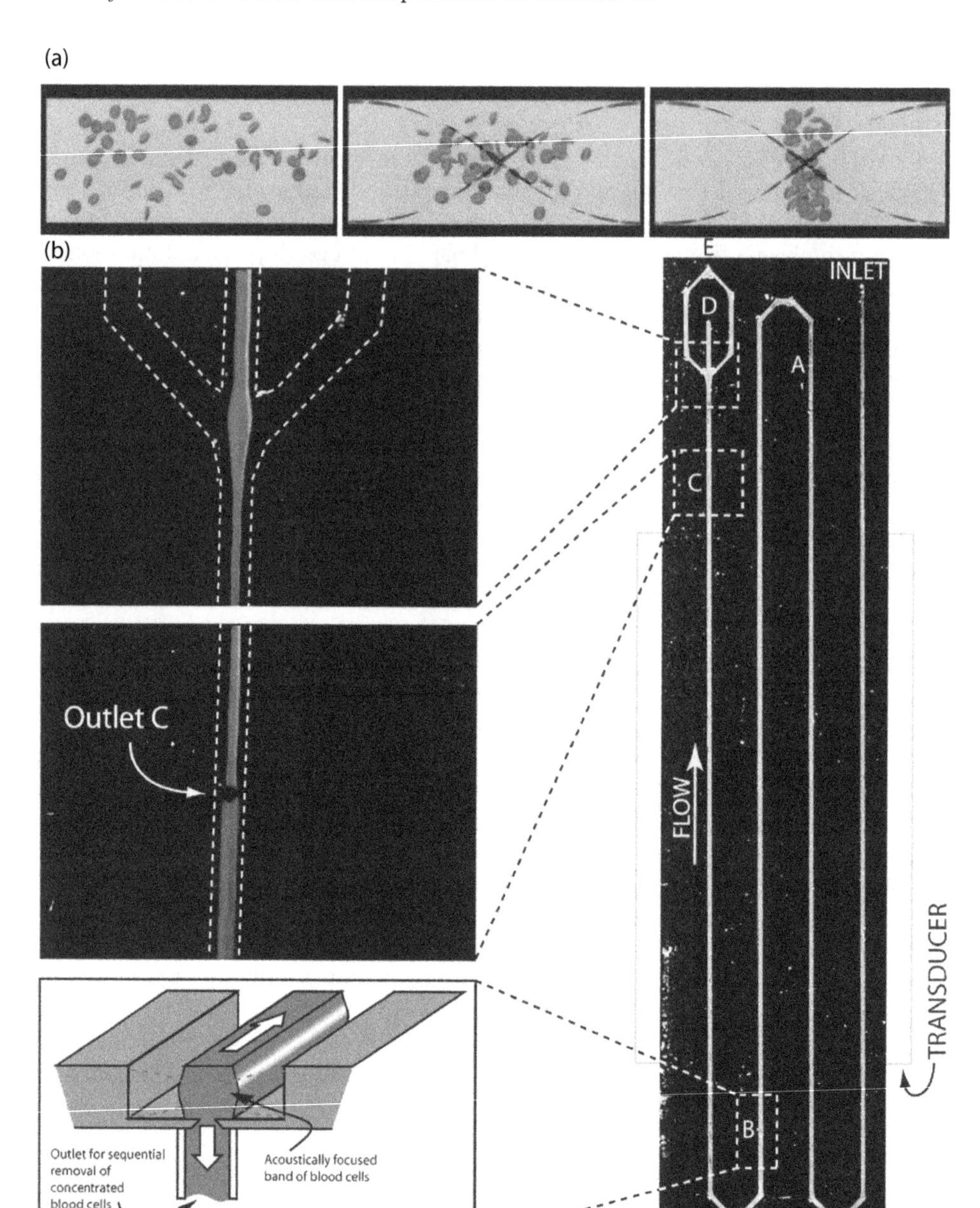

FIGURE 11.5
An acoustopheresis based plasma separation device. (A) shows the focusing of RBCs to the pressure node of acoustic radial force. The focused RBCs form a stream line in the microfluidic channel and there are outlets along the central part of the channel for continual removal of RBCs. (B) shows how RBCs are sequentially removed and plasma being collected from two side outlets. Reprinted from [35], Copyright 2009, with permission from American Chemical Society.

The small size of platelets results in a smaller force acting on them when subjected to both acoustic and dielectric fields. The separation purity based on these two techniques has been shown to be higher than 90% with more than 90% percent depletion of RBCs as well. However, to allow sufficient migration time of the cells, active energy based separation has resulted in a low overall throughput. Recently, Chen et al. developed an acoustic based chip that was able to separate platelets from undiluted whole blood at 10 ml/min.

(Figure 11.6(B), [7]) It introduces a buffer phase flowing with the whole blood phase and the pressure node is designed to be in the buffer phase. As WBCs and RBCs have larger volumes than platelets, they experience larger acoustic forces than platelets. When flowing in the acoustic force field, the larger WBCs and RBCs can be effectively removed from the whole blood phase to buffer phase, leaving platelets flowing with the plasma phase. This technique achieves more than 80% for platelet separation and RBC removal. From their characterization of the separated platelets, the method has also been proved to produce high integrity platelets as compared to conventional methods. Pommer et al. built a similar microfluidic device based on dielectrophoresis which can separate 95% pure platelets from diluted blood with minimal activation as shown in Figure 9.6(A). However, the chip can only process several microliters of whole blood in a minute [44]. Besides active field assisted separation, Choi et al. developed a hydrophoretic chip that is able to process millions of cells per second and can separate platelets from other blood cells at a purity of around 76.8% [9]. The chip employs convective vortices, steric hindrance and rotational flow generated by obstacles inside the channel as shown in Figure 11.6(C) to achieve cell ordering according to their size.

11.3.5 Separation of leukocytes

White blood cells in blood form the secondary barrier of the human immune system. Different kinds of WBCs play different roles during the process of immune response. There are two major applications for WBC separation in biomedicine. One is to get rid of WBCs in whole blood for safer blood transfusion, and another is to separate specific WBC subtypes for studying immune reaction, immune cell biology and physiology. As discussed in the previous sections, the major difficulty faced by white cell separation is the high RBC content in blood and the similarity of these two populations in physical properties. Deriving pure enough WBCs will have many downstream applications, including the study of immune cell biology and purification of lymphocytes for immune therapy. Providing viable primary white cells for cell and molecular study will help to promote people's understanding of immune cell function and further benefit development of immune cell related clinical applications.

Traditionally, centrifuging whole blood through a medium with density gradient can reliably separate blood cells into different density segments. The shortage of this method, similar to the separation of other components through density centrifugation, is the bulky expensive equipment and demand of time and labor. Since different WBCs have unique surface markers. FACS and MACS have also been applied in WBC separation and achieved great success. However, the antibody labelling process in these techniques can interrupt the downstream molecular study and analysis of these cells, which hinder their application for clinicians and biomedicine researchers. Multiple groups have tried to use solely physical measures to extract WBCs from blood in microfluidic platforms, and the progression in this field has been very rapid in recent years.

WBCs are considered slightly larger and less deformable as compared to red cells. Filtration of whole blood is a simple and straightforward way of achieving separation of WBCs. There are four basic forms of these microfiltration systems for blood cell separation: weir-based, membrane-based, pillar-based and cross-flow [20]. The critical size identified for efficient white cell separation is 3.5 μm. The disadvantage of using filters to separate cells is the clogging of cells which will directly compromise the purity of separated cells. Additionally, once clogging happens in the microchannel, the accumulation of cells is rapid and can disrupt separation. Due to extremely high cell density in blood, filtration-based systems are sometimes considered non-stable and non-reliable in separating blood cells, but membrane

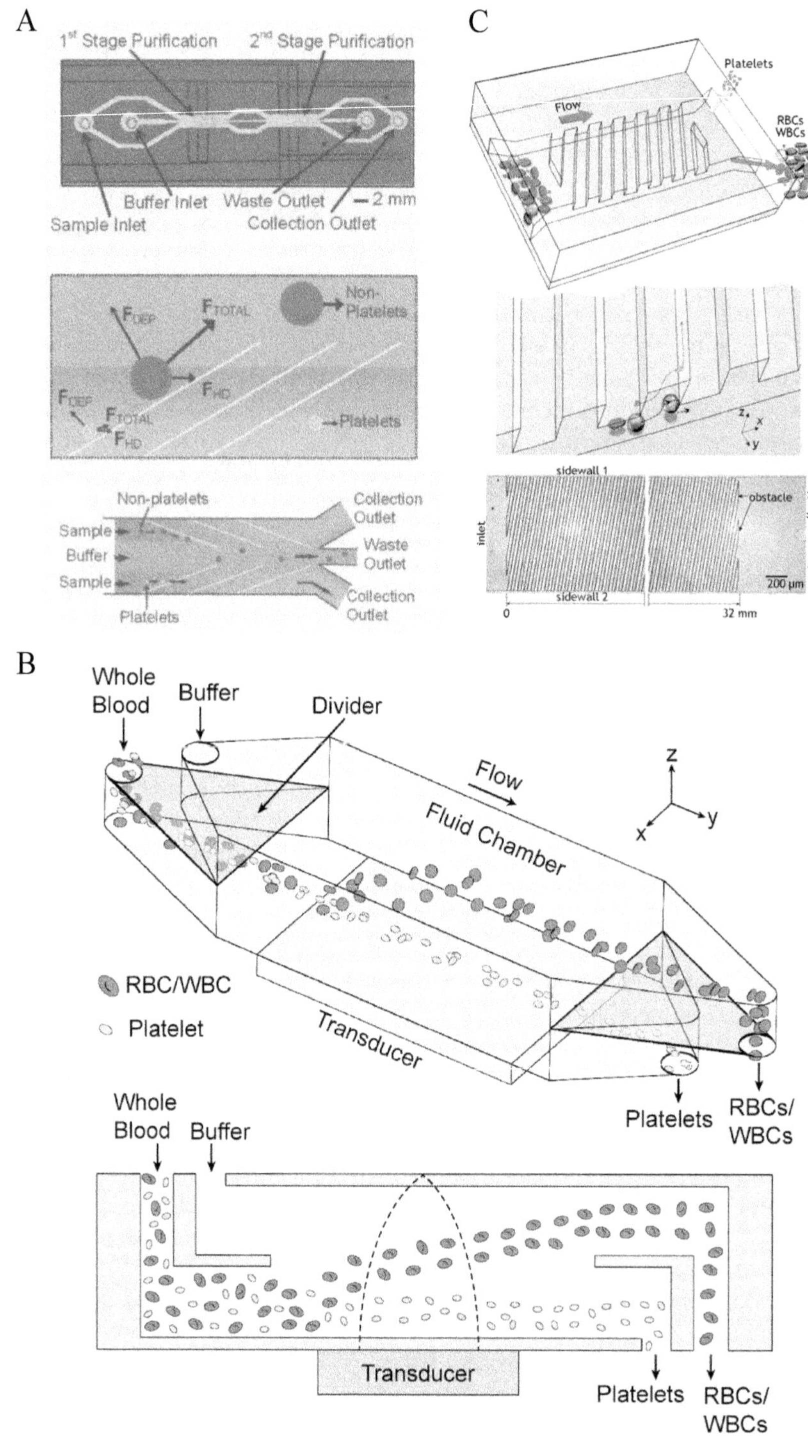

FIGURE 11.6
Microfluidic platforms for platelet separation. (A) Dielectrophoresis based platform for blood cell removal and platelets collection. Since platelets are significantly smaller than the other cell components in blood, the DEP force acting on them is much smaller compared to the force on other components. Platelets maintain in the side position while the larger cells are being deflected closer to the central region where they are discarded as waste. Reprinted from [44], Copyright 2008, with permission from John Wiley and Sons. (B) An acoustophoresis based platform for platelet separation. Larger blood cells experience higher acoustophoretic force and migrate to the upper part of the channel while platelets maintain the position in the lower half and are collected. Reprinted from [7], Copyright 2016, with permission from Royal Society of Chemistry. (C) A hydrophoresis microfluidic device with anisotropic obstacles. Larger cells will remain near the sidewall while smaller cells being unfocused migrate to the upper side. Platelets can thus be separated from other blood cells. Reprinted from [9], Copyright 2011, with permission from Royal Society of Chemistry.

based WBCs depletion has demonstrated its capability for pre-removal of white cells for clinical blood transfusion.

Deterministic lateral displacement, which is also known as bump array, is a phenomenon where particles bump to different a streamline when encountering a pillars array with predefined distance between each pillar. As discussed previously, the position of each cell coming out from the pillars array is purely determined by the size of the cell. WBCs are considered larger than RBCs, so efficient separation of these two cell types can be achieved by identifying a suitable cutoff size. It has been explored as early as a decade ago to use DLD array for WBC separation from whole blood. However the overall throughput is only a few μl/min. Kim et al. developed a deterministic migration based device for WBC separation [30]. The chip is capable of processing 1 ml of whole blood in 7 min, which is a considerable progression as compared to other platforms in this category. The separation efficiency can be up to 97.6% and functional analysis proved the separated T lymphocytes to be viable and functional.

Leukocytes are CD45 expressed cells and each different fraction of WBCs has its own specific surface markers. Techniques targeting these biomarkers can achieve high specificity separation of the desired population of white cells from whole blood. The general method involves employing antibody-attached magnetic beads to label the target WBCs. By applying an external magnetic field, the immunomagnetically labeled cells can be pulled out from the background cells. These MAC based separation techniques have been extensively studied and this concept has been well proven to be effective in retrieving target cells. Inglis et al. developed a silicon substrate chip with nickel magnetic stripes [27]. Leukocytes are labelled with CD45 magnetic beads and their flow direction can be altered by the slanted oriented magnetic stripes. This device achieved a high leukocyte selectivity, however, there is a $sim50\%$ cell loss because of cell loss in the device or non-sufficient separation. Moreover, due to the complexity of whole blood, direct labeling and separation of white cells from unprocessed whole blood require a long processing time and the efficiency is always compromised by the large number of RBCs. Besides, the magnetic labelling process could possibly damage the cells inadvertently.

Besides magnetic assisted separation, WBCs also exhibit a significantly lower density as compared to RBCs. This physical difference results in a different migration pattern when the cells are subjected to dielectrophoretic force. Han et al. developed a microfluidic based DEP separator [19]. The degree of deflection is different for RBCs and WBCs in this electric field resulting in the separation of two cell populations. The device can continuously separate WBCs and RBCs from whole blood at a volumetric flow rate of about 50 μl/hour. The separation efficiency is 87% for RBCs and 92.1% for WBCs.

Hou et al. developed a spiral chip that is able to separate neutrophils, lymphocytes and monocytes into different outlets from lysed blood using inertial focusing [24]. Figure 11.7(C) schematically shows the working principle of the chip. They successfully used the separated neutrophils to study the alteration of cell behavior in type II diabetes patients. The chip can process a small volume of blood $\sim$100 μl at an optimized flow rate of 130 μl/min. The separated neutrophils have a more than 90% purity which is suitable for downstream investigation of cell and molecular biology of the cells. The problem with inertial focusing separation is the requirement for RBC lysis (otherwise the separation purity will be greatly compromised) and non-specificity due to overlapping of cell sizes.

11.3.6 Summary

Taking blood apart has been a hot topic in the field of biomedical engineering for the past two decades. Traditional bulk separation based on centrifugation cannot fullfil the re-

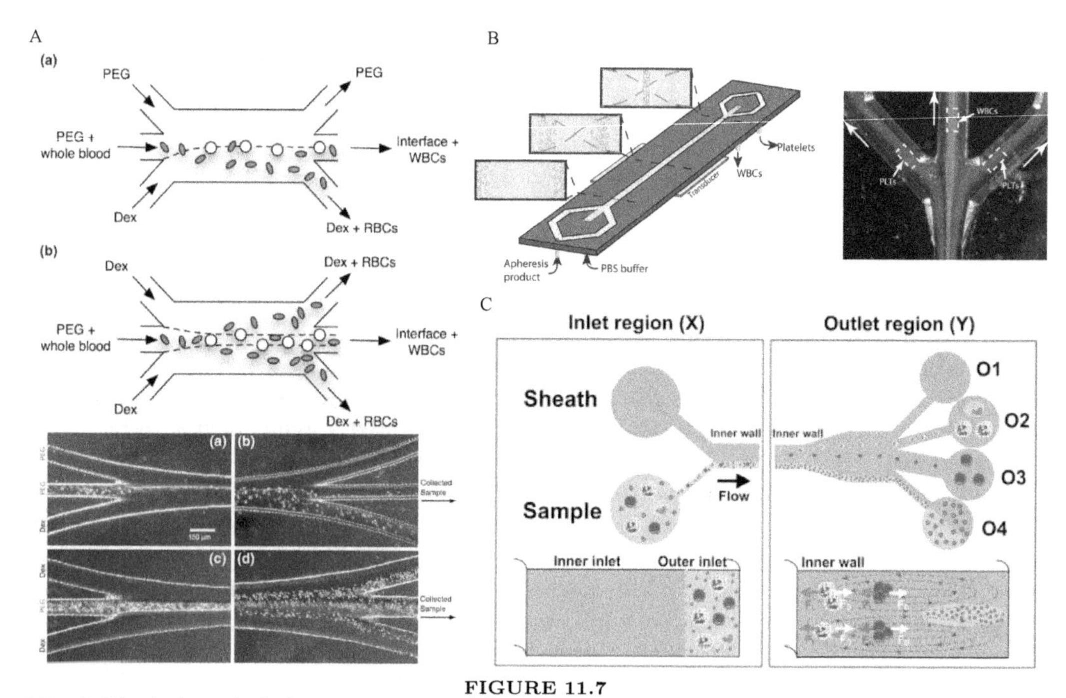

FIGURE 11.7
Microfluidic platforms for leukocytes separation. (A) Aqueous two phase system for leukocyte separation. Because of the difference in surface energy, WBCs will tend to maintain a position at the interface of PEG-Dex while RBCs will migrate to the Dex region. Reprinted from [50], Copyright 2009, with permission from Springer. (B) Acoustophoresis for white cell separation. Larger WBCs migrate and focus to the central of the channel and can be collected from the middle outlet. Reprinted from [11], Copyright 2011 Dykes et al., PloS One. (C) A spiral microfluidic chip for separation of neutrophils from lysed blood. As neutrophils are the largest white cell population, they are focused to the stream line closest to the inner wall of the spiral channel. Reprinted from [24], Copyright 2016 Nature Publishing Group.

quirement for quick and simple diagnosis from blood. Microfluidic platforms have attracted great interest, because of their ease in handling, low operating cost, and portability for blood component separation. There are many promising results being witnessed using these platforms. Active energy and passive based separations both showed their strengths as well as weaknesses in addressing the issue. Owing to the nature of the ease and convenience in integration of engineering components, it is expected that more integrative microfluidic systems will be developed to assemble microfluidic components with different physical principles and their unique advantage in achieving more robust and precise separation of blood components. Also, it is notable that microfluidic platforms have been used to study the physiological functions of blood components and these two different areas have the potential to be integrated together to achieve a more automatic research platform in this area. It is promising that with the development of these techniques, researchers will be able to retrieve more biologically useful information from human blood and translate the findings for eventual clinical utility.

11.4 MICROFLUIDIC APPLICATIONS IN MALARIA

Malaria is an infectious disease caused by the Plasmodium group parasites and transmitted commonly by the female Anopheles mosquitoes. The malarial parasite is capable of invading a RBC where they will multiply and then break out to continue to infect other RBCs within a span of two days. The symptoms include fever, tiredness, and vomiting which are similar to other febrile diseases. It has been estimated by the World Health Organization that in 2016 that there were 216 million malaria cases and 445,000 deaths around the world, the majority of which were African children.

As the pathology of Plasmodium parasites mainly occurs in the circulatory system, the diagnosis of malaria usually involves a blood smear to detect infected RBCs or parasites. In fact, the gold-standard diagnosis for malaria comprises a microscopy-based Giemsa smear method in which blood smear on a glass slide is Giemsa stained and checked under an optical microscope. This clinical test is labor-intensive requiring trained personnel to perform the examination, and sensitivity can vary due to human error. Another widely used diagnostic method is antigen-based rapid diagnostic tests (RDTs) which detect pathogen-specific antigens with the lowest detection limit of 100 parasites/μl. This fast diagnostic method is easy to use, and the results are straightforward. However, it cannot be used to evaluate the disease status or monitor disease progression.

For guiding the technology development for malaria detection trials, the malERA consultative group proposed a target product profile for malaria diagnostics in 2011 (The malERA Consultative Group on Diagnoses, 2011). The case management requires 95% sensitivity, 90% specificity with an analytical sensitivity of 100-200 parasites/μl. As for disease screening and surveillance, the standard has been raised to 95% sensitivity, 99% specificity with analytical sensitivity at 20 parasites/μl. Additionally, the test should require no professional personnel, use only a finger prick of blood with a commercially available price less than 1 USD. Up until 2018, no single commercialized technology can meet all the standards listed above, but microfluidic-based technologies have shown great potential in tackling this problem.

11.4.1 Microfluidics in pre-processing blood for clinical tests

The malaria detection methods that are currently widely used in clinical trials have their unique features that allow successful commercialization and clinical applications. The gold-standard method, although it is labor-intensive, has precise quantitative readouts for disease monitoring. The RDTs, on the other hand, are cheap and easy to use making it an ideal POC product. They are good but not perfect, which leaves room for researchers to keep improving them. There are many research efforts aimed at overcoming the disadvantages associated with existing technologies. Horning et al. developed a paper-based microfluidic cartridge to simplify the sample preparation and result readout procedures of the gold-standard method. They used dyeing paper to stain blood samples with acridine orange (AO) fluorescence dye which is a substitute for Giemsa stain. After blood is driven through the dyeing paper and stained by the dye, it enters a cartridge composing a coverslip for microscopy imaging [21]. Figure 11.8A schematically shows the workflow of the device. In their study, they also designed an imaging processing algorithm to automate the counting of RBCs and iRBCs. The device shortens the sample processing procedure to as short as 1 minute and requires no further operation from test operators after sample loading. This study shows the capability of microfluidic systems in automating and simplifying diagnosis processes. Similarly, Bauer et al. developed a magnetic-based biomarker enrichment device that accommodates to currently available RDTs which added only 0.25 USD and 3 minutes to the tests but can increase the sensitivity to single digit parasitemia [1].

As discussed before, detection sensitivity is one of the most important parameters in evaluating the performance of malaria detection devices. To avoid complicated processing procedures and eliminate human error in performing the test, quantifiable molecular readings are more favorable in devices intended for disease screening and monitoring. Using PCR to directly probe the presence of parasite nucleic acid is one of the preferred methods. However, the specificity and sensitivity of PCR-based methods suffered from the high

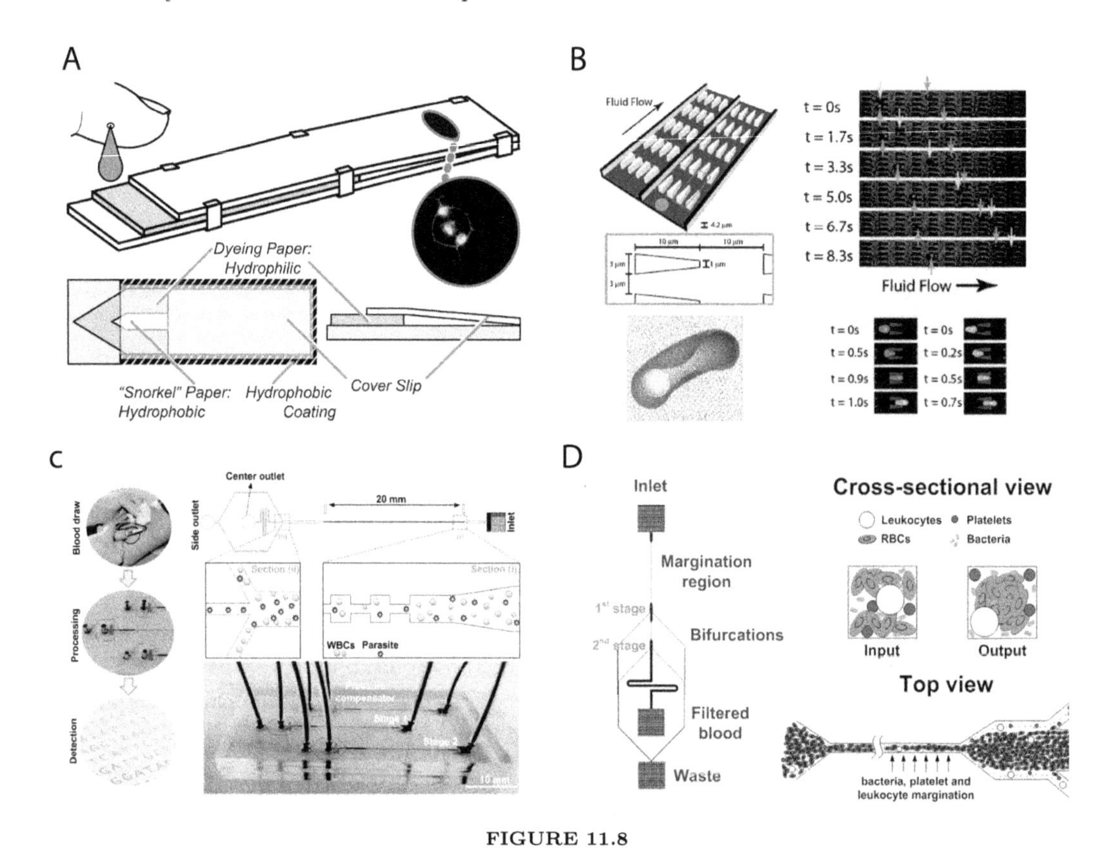

FIGURE 11.8
Microfluidic applications in malaria. (A) A microfluidic cartridge device for automatic dyeing of blood samples for microscopy imaging. Reprinted from [21] (B) A microfluidic deformability device that detects infected red blood cells from their transit time on the device. Reprinted from [4] (C) An inertial microfluidic device that concentrates malaria parasites for downstream PCR analysis. Reprinted from [56] (D) A cell margination device for parasites and infected cell removal of blood samples from patients with sepsis. Reprinted from [22].

background DNA from other nucleated cells in the blood. Warkiani et al. developed a microfluidic system based on inertial focusing to concentrate parasites presented in blood before proceeding to qPCR screening, which is schematically shown in Figure 11.8C. The sensitivity of the system can go down to 2-10 parasites per μl [56]. Although this device still involves multiple steps like blood lysis and relies on PCR machines, its working principle is simple, and the sensitivity is extremely high. Therefore, it can potentially be integrated into other diagnostic systems as a pre-concentration component. The efforts discussed above have shown the capability of microfluidic technologies in optimizing current malaria diagnostic tests to simplify the procedures or improve test results. We can see the potential of microfluidic-based platforms in becoming an indispensable component for the next generation malaria detection assays from these examples.

11.4.2 Microfluidics for malaria detection

Besides being used as a pre-processing component, people have also tried to develop microfluidic systems directly for disease detection. In malaria patients' blood, RBCs infected with Plasmodium falciparum parasites have altered mechanical properties making these cells stiffer than the uninfected RBCs. Increasing evidence is showing that cell mechanical properties can be used as a reliable biomarker for health and disease [34]. There are many efforts in recent years trying to do disease detection from a cell mechanics perspective.

However, most of the research either focused only on the population behavior, for which is difficult to capture the differences as iRBCs are rare as compared to blood cells, or encountered problems like low throughput due to a large number of blood cells that need to be observed or processed. Bow et al. developed a microfluidic deformability flow cytometer to quantitatively measure the mechanical properties of individual RBCs from an infected blood sample with high throughput [4]. Figure 11.8B shows the workflow of the device. In their experiments, they observed a standalone population in the bimodal plot with both thiazole orange fluorescence intensity and deformability of cells, which represents infected RBCs. The screening can be done conveniently with computer-assisted automatic analysis. This work demonstrated the potential of combining both biomolecular and biomechanical markers in one microfluidic platform to perform an infected cell count and quantitatively study the mechanical alternations in disease cells with high precision, high throughput and full automation. This platform provides researchers with a tool to explore more deeply the mechanical and pathological changes during parasite infection, which in turn might give a new prognostic marker for clinicians.

11.4.3 Summary

Apart from the microfluidic technologies discussed above, there are many more relevant platforms that are trying to tackle this issue from different perspectives. Kumar et al. used an antibody-coated electrode to capture iRBCs [32]. In their work, they showed the concentration of parasites presented in blood is proportional to the electrochemical readouts in the range from 102 cells -108 parasites/ml, which can potentially be utilized to monitor disease progression. Technologies employing different physical principles have been investigated widely in the detection and diagnosis of malaria in the past decade.

We used malaria detection as an example here to exhibit the changes brought by microfluidic platforms in infectious disease diagnosis. There are also efforts trying to develop novel intervention methods for this infectious disease. Hou et al. showed a cell margination-based chip for eliminating infected blood cells and parasites from blood of patients with sepsis which could potentially be used as a form of therapy for sepsis [22]. Figure 11.8D shows the working principle of the device. This work demonstrates the possibility of using microfluidic-based systems not only in infectious disease detection but also in disease treatment.

There are broader applications of microfluidics in other infectious diseases also, for example, hepatitis B and HIV. Though the application backgrounds of these technologies are different, they share great similarities in general principles. Kaminska et al. fabricated a microfluidic chip with anti-hepatitis B virus antigen (HBsAg) antibody and Au-Ag coated GaN substrate. The device is capable of detecting HBsAg with a detection limit as low as 0.01 IU/ml [29]. This device shares common characteristics with the immune-based malaria detection methods. Recently, there has also been great progress made in the field of fast and convenient HIV detection devices intended for resource-poor settings [36]. Point-of-care testing and novel treatment methods of infectious diseases are important applications of microfluidic technology, and there are also several up-to-date and comprehensive review papers describing these methods if readers want to read further on this topic [8,31,47,51].

11.5 CANCER DIAGNOSIS

Cancer is one of the most fatal diseases, causing human death worldwide. Early detection of cancer is crucial for disease management and good disease prognosis. Traditional methods for cancer diagnosis usually involve tumor biopsy where physicians take a piece of tissue through needle aspiration from the suspected tumor region of a patient for histological examination. This process is invasive and involves highly trained personnel. More importantly, it is nearly impossible to do biopsy frequently for disease monitoring and treatment assessment. Thus, a reliable non-invasive method that can reflect the disease status of a cancer patient is an urgent need in clinics. There is an emergent field in cancer detection, namely liquid biopsy that is receiving increasing attention among researchers and clinicians in recent years. As blood also metabolically interacts with tumor tissue, it also carries information that could potentially serve as biomarkers for cancer diagnosis. Liquid biopsy is proposed based on the idea of detecting cancer and monitoring cancer progression based on circulating biomarkers present in human blood. We should take note here that there are mature cancer-associated molecules that have an elevated level in different cancer cases such as CA15-3 in breast cancer, tumor M2-PK, etc. These traditional tumor markers although reflecting the presence of a tumor can hardly give information for early diagnosis, disease prognosis, and disease status. There are three major categories of novel biomarkers that have been established in the past decades that possess a greater potential in cancer diagnosis: circulating tumor cells (CTCs) circulating tumor DNAs (ctDNAs) and tumor-secreted circulating exosomes. A common characteristic of these blood-present tumor markers is their extremely low abundance with high background noise. For example, the number of CTCs can be as low as several cells per ml of blood, which requires a highly sensitive assay to detect them.

11.5.1 Microfluidics for CTC detection

CTCs have been well studied among these three biomarkers. The occurrence of CTCs in the bloodstream has been associated with the blood-borne metastasis of a tumor, and it has been found that CTC count in cancer patients' blood is strongly correlated with disease prognosis. So far there is only one US FDA approved medical device for CTC enumeration in clinics: the CellSearch system. This system requires 7.5 ml of patient blood and gives CTC count based on immunomagnetic labeling of CTCs for EpCAM markers. Although this is the method currently used in the clinics for CTC detection, there are issues related to this approach. Firstly, it requires an expensive machine to perform magnetic-based sorting and using antibodies for labeling. More importantly, the heterogeneity of CTCs and the downregulation of EpCAM during epithelial-mesenchymal transition means a number of CTCs may not be detected and captured.

Due to the unique capability of precisely manipulating and handling low sample volume, microfluidic technologies have been widely employed in CTC related biomedical applications to achieve unprecedented sensitivity and functionalization. Ozkumur et al. developed an integrated microfluidic system based on both hydrodynamic and magnetic-activated cell sorting (MACS) [41]. They designed a deterministic lateral displacement (DLD) array and an inertial focusing component before magnetic separation as shown in Figure 11.9A. The DLD array works to concentrate nucleated cells and reduce RBC concentration for downstream separation. Inertial focusing defined the position of the cells for separation making the migration of target cells less influenced by cell-cell interaction and effectively avoids the random migration of unwanted cells. In their device, they used a negative selection of CTCs, where leukocytes are labeled and removed instead of a positive selection of CTCs, and they successfully derived both CK and EpCAM negative tumor cells. This device can process whole blood at a flow rate of 107 10^7 cells/s with a purity higher than 0.1%, which is considered high enough for enumeration and further molecular analysis of CTCs [13].

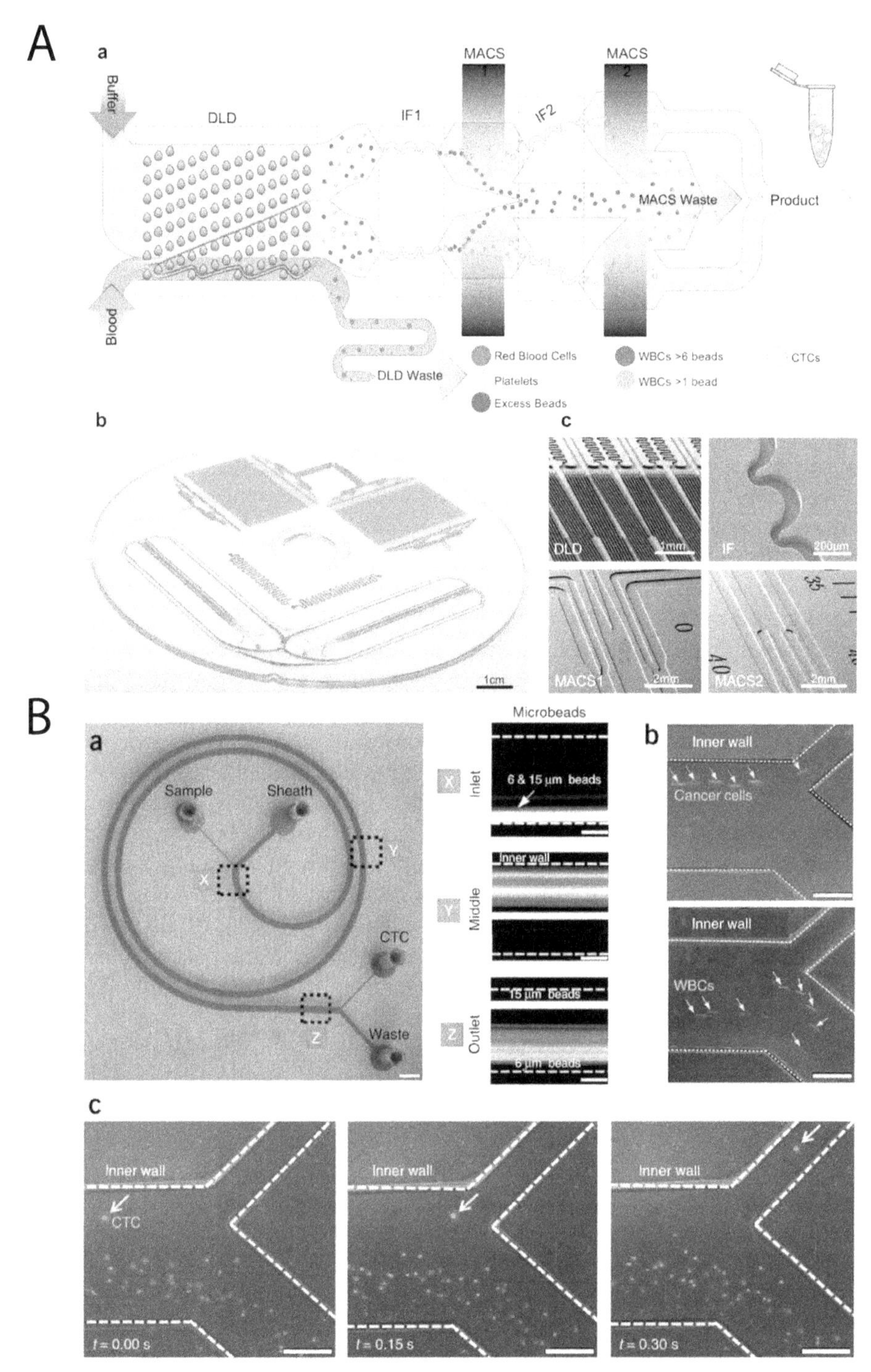

FIGURE 11.9

Microfluidic devices for retrieving circulating tumor cells. (A) A microfluidic device employs centrifugal force to focus large CTCs and collect them from the inner outlet. Reprinted from [56]. (B) CTC-iChip which separated CTC using magnetic sorting. The CTCs are first enriched through the DLD arrays and then focused with inertial focusing. Following that, the pre-labeled CTCs can be separated from WBCs with magnetic force. Reprinted from [13].

Besides the antibody-labeling and active selection methods based on external force fields, passive focusing and separation of CTCs have also attained great progress. Hou et al. developed a microfluidic system based on inertial focusing that can select CTCs that are larger than WBCs [25]. Figure 11.9B shows the working principle of the device. This simple working principle and ease of use of the chip allow more research and clinical investigation on patient-derived CTCs. This separation system has been successfully commercialized by Clearbridge Biomedics as the ClearCell FX system. Yeo et al. did further work based on the ClearCell FX system [60]. They developed a single cell capture chip where they established an operating procedure from receiving patient's blood samples to isolating single CTCs for downstream molecular profiling. This complete single CTC solution provides a helpful tool for researchers to decipher the genetic information hidden in patient CTCs. These related technologies have also been documented in some recent reviews [54, 57].

11.5.2 Microfluidics for ctDNA and exosome detection

Exosomes are extracellular vesicles released by cells, and there is increasing evidence showing their importance in intercellular communication. In the context of cancer, tumor-derived exosomes are shown to alter tumor microenvironment and potentially correlate with the progression of cancer. Similar to CTCs, exosomes possess the potential as a cancer early diagnosis marker as well as promote the understanding of the tumor progression mechanism. The detection of exosomes is generally more difficult than CTCs, as exosomes are much smaller with sizes down to tens of nanometers. Additionally, there are other extracellular vesicles presented in blood resulting in a high background noise separation. One of the most established methods in exosome separation is ultracentrifuge based techniques. However, the protocol usually involves a multistep time-consuming centrifuge process making the procedures complicated and labor intensive. There are a few microfluidic-based methods developed for exosome separation in the past few years. The technology can be typically categorized into antibody-based methods and passive on-chip centrifuge based methods. Some recent advancements also allow the separation and molecular analysis of exosomes to be performed on the same microfluidic platform, which significantly shortens the time for analysis bringing it one step closer to clinical applications. Fang et al. used an immune-based capturing system to quantify circulating exosomes from patient blood for breast cancer diagnosis and molecular classification as shown in Figure 11.10A. [14]. Yeo et al. on the other hand developed an on-chip centrifugation device for passive concentration of exosomes and the schematics of the device is shown in Figure 11.10B [59]. Shao et al. developed a system where the exosomes from tumor cells are immuno-captured on a chip by magnetic nanoparticles conjugated with antibodies of interest like the CD63 antibody as shown in Figure 11.10C. Following capture, they analyzed the protein profile of glioblastoma cells from patient samples, which demonstrated the real-time disease monitoring capability of the system [17]. These microfluidic isolation methods and quantification of exosomes are emerging as having great potential for cancer diagnosis and further clinical validation is necessary for eventual utility in the clinics.

Cell-free tumor DNAs are known to circulate in the bloodstream of patients. In the past decade, there is emerging use of ctDNAs for cancer detection and management, and there is a recent report suggesting that ctDNAs can serve as a routine tool for lung cancer patient management [55]. Quantitative real-time PCR is one of the most suitable technologies used in detecting these circulating genetic materials from tumor cells. However, the low abundance of ctDNAs, as well as nucleic acids from other normal cells, formed a high detection barrier to reliably detect these ctDNAs. Recent advancements in microfluidic nucleic acid enrichment platforms have provided powerful tools for researchers to conduct high

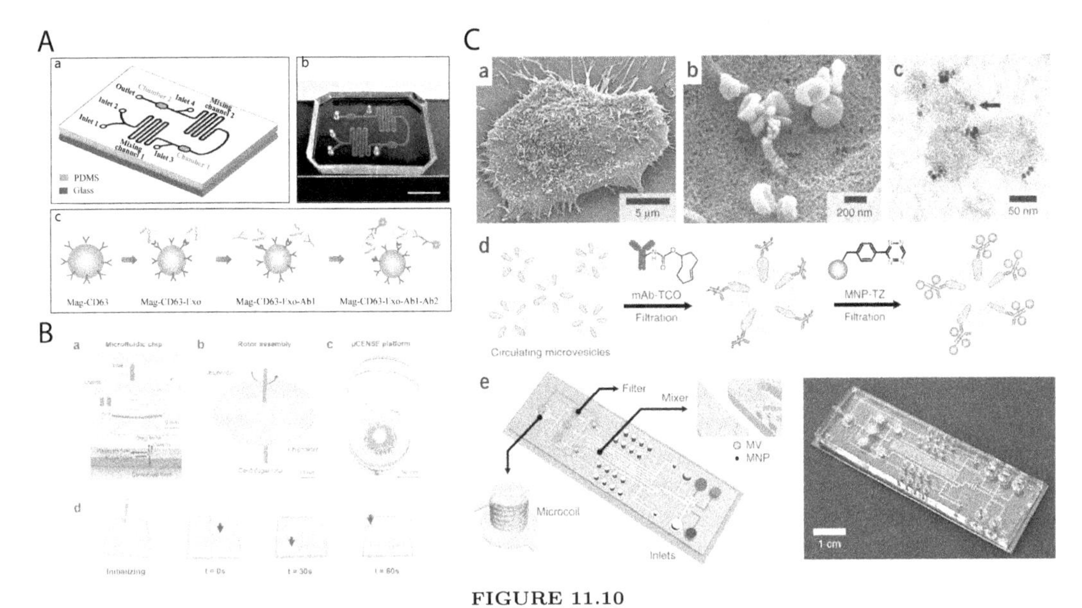

FIGURE 11.10
Microfluidic chips for exosomes separation. (A) An immune-based chip that captures exosomes through anti-CD63 antibodies. Reprinted from [14]. (B) A centrifugal chip that separates exosomes with on-chip centrifugation. Reprinted from [59]. (C) A microfluidic chip developed for exosome capturing and profiling on-chip. Reprinted from [17]

precision analysis on ctDNAs in patient plasma [12, 18]. There is potential where ctDNAs can be detected and analyzed conveniently all on a single microfluidic platform for better cancer detection and prognosis. Moreover, analyzing cancer genetics from ctDNAs might provide people a better understanding of patient-specific tumor phenotypes and thus make personalized treatment possible.

11.5.3 Cancer detection based on cell mechanics

Besides all these cancer-related cellular and molecular biomarkers, cell physical properties have also emerged as a potential disease marker in recent years. We discussed in the malaria section that the iRBCs are found to exhibit altered mechanical properties as compared to the uninfected cells and this mechanical property difference can be quantified by the velocities of cells transiting through microfluidic constrictions. The disease-related change of cell stiffness and other biophysical parameters have also been found in cancer cells and have attracted increasing research interest. Hou et al. compared the entry velocity of normal breast epithelial cells and breast cancer cells into a small constriction in microfluidic devices [23]. They found that the entry velocity of cancerous cells into a microfluidic constriction was significantly higher than normal cells. These results indicate that breast cancer cells appear to be softer when compared with their healthy counterparts. Similar results in many other types of cancer cells have also been reported using different techniques such as atomic force microscopy, optical tweezers, and micropipette aspiration to cite just a few examples [39]. However, compared to these techniques, microfluidic-based platforms can enable high throughput screening of mechanical properties of cells. Although microfluidics-based mechanical characterization has been thought to be semi-quantitative, with inputs from physical modeling and simulation, we can now obtain more precise and quantitative results based on the transit behavior of cells through microchannels [33, 40]. Recently there are also researchers trying to separate and measure mechanical properties of patient CTCs using integrated microfluidic platforms, which demonstrates the possibility to use altered cell mechanics as a biomarker for disease screening [6]. In general, these studies show that that the altered cellular biomechanical properties can become one of the more effective biomarkers for better detection and diagnosis of cancer.

11.6 CONCLUSIONS AND FUTURE OUTLOOK

In this chapter, we listed the key features of microfluidic platforms and discussed in detail how these features can help to improve disease detection and diagnosis. Apart from the examples we listed here, microfluidic technologies have also achieved great success in areas such as blood fractionation, blood cell phenotyping, and immune cell assays [48, 52, 61].

Microfluidics-based technologies have multiple advantages allowing them to competently perform clinical tests. The portable size of the platforms is highly favorable, especially for point-of-care applications. The highly ordered fluid flow in microchannels allows multiple modules to be integrated together and therefore enable a multi-step process to occur all on the same chip. Such platforms usually need only a few pre-processing steps before loading the clinical samples into the chip. Furthermore, the operations can also be automated on a chip and this can save labor as well as require minimal training in using it. Last but not the least, the high precision of liquid manipulation can enable tests to be done at the single cell or even single molecule resolution.

One noted trend in this field is that there are more and more integrated systems that combine different microfluidic units to perform multiple processes on a chip. For example, CTC-iChip has three different units that concentrate, focus and finally separate the CTCs [41]. Pre-enrichment of blood sample before expanding CTCs on a chip is also a strategy to optimize the culture of CTCs. There is no single all-purpose system, but we can expand their applications through integration. One of the real challenges in microfluidic technology is to bring such platforms from the bench to the bedside and market. Firstly, such devices need to be robustly tested and validated with clinical samples. Next, they also need to be suitable for large-scale manufacturing and at a very low cost. The best microfluidic systems may not be the ones with complex channel arrangements and many components, but those that can address a specific need or problem with as simple a design as possible. We believe through joint efforts and interdisciplinary collaboration among engineers, life scientists and clinicians, more robust, highly scalable and high-performance microfluidic systems can be realized.

Conflict of Interest: C.T.L. is co-founder and has financial interest with Clearbridge Biomedics Pte Ltd.

Bibliography

[1] W.S. Bauer, D.W. Kimmel, N.M. Adams, L.E. Gibson, T.F. Scherr, K.A. Richardson, J.A. Conrad, H.K. Matakala, F.R. Haselton, and D.W. Wright. Magnetically-enabled biomarker extraction and delivery system: Towards integrated ASSURED diagnostic tools. *Analyst*, 142(9):1569–1580, 2017.

[2] J.P. Beech, P. Jonsson, and J.O. Tegenfeldt. Tipping the balance of deterministic lateral displacement devices using dielectrophoresis. *Lab on a Chip*, 9(18):2698–2706, 2009.

[3] W. A. Bonner, H.R. Hulett, R.G. Sweet, and L. A. Herzenberg. Fluorescence activated cell sorting. *Review of Scientific Instruments*, 43(3):404–409, 1972.

[4] H. Bow, I.V. Pivkin, M. Diez-Silva, S.J. Goldfless, M. Dao, J.C. Niles, S. Suresh, and J. Han. A microfabricated deformability-based flow cytometer with application to malaria. *Lab on a Chip*, 11:1065–1073, 2011.

[5] L.F. Brass, T.J. Stalker, L. Zhu, and D.S. Woulfe. *Platelets*, Chapter 16: Signal Transduction during Platelet Plug Formation. Academic Press, 2012.

[6] J. Che, V. Yu, E.B. Garon, J.W. Goldman, and D. Di Carlo. Biophysical isolation and identification of circulating tumor cells. *Lab on a Chip*, 17(8):1452–1461, 2017.

[7] Y. Chen, M. Wu, L. Ren, J. Liu, P.H. Whitley, L. Wang, and T.J. Huang. High-throughput acoustic separation of platelets from whole blood. *Lab on a Chip*, 16(18):3466–3472, 2016.

[8] C.D. Chin, T. Laksanasopin, Y. K. Cheung, D. Steinmiller, V. Linder, H. Parsa, J. Wang, H. Moore, R. Rouse, G. Umviligihozo, E. Karita, L. Mwambarangwe, S.L. Braunstein, J. van de Wijgert, R. Sahabo, J.E. Justman, W. El-Sadr, and S.K. Sia. Microfluidics-based diagnostics of infectious diseases in the developing world. *Nature Medicine*, 17(8):1015–1019, 2011.

[9] S. Choi, T. Ku, S. Song, C. Choi, and J.-K. Park. Hydrophoretic high-throughput selection of platelets in physiological shear-stress range. *Lab on a Chip*, 11:413–418, 2007.

[10] D. Di Carlo. Inertial microfluidics. *Lab on a Chip*, 9:3038–3046, 2009.

[11] J. Dykes, A. Lenshof, I.-B. Astrand-Grundstrom, T. Laurell, and T. Scheding. Efficient removal of platelets from peripheral blood progenitor cell products using a novel microchip based acoustophoretic platform. *PLoS One*, 6(8):e23074, 2011.

[12] A. Egatz-Gomez, C. Wang, F. Klacsmann, Z. Pan, S. Marczak, Y. Wang, G. Sun, S. Senapati, and H.-C. Chang. Future microfluidic and nanofluidic modular platforms for nucleic acid liquid biopsy in precision medicine. *Biomicrofluidics*, 10:032902, 2016.

[13] F. Fachin, P. Spuhler, J.M. Martel-Foley, J.F. Edd, T.A. Barber, J. Walsh, M. Karabacak, V. Pai, M. Yu, K. Smith, H. Hwang, J. Yang, S. Shah, R. Yarmush, L.V. Sequist, S.L. Stott, S. Maheswaran, D.A. Haber, R. Kapur, and M. Toner. Monolithic chip for high-throughput blood cell depletion to sort rare circulating tumor cells. *Scientific Reports*, 7:10936, 2017.

[14] S. Fang, H. Tian, X. Li, D. Jin, X. Li, J. Kong, C. Yang, X. Yang, Y. Lu, Y. Luo, B. Lin, W. Niu, and T. Liu. Clinical application of a microfluidic chip for immunocapture and quantification of circulating exosomes to assist breast cancer diagnosis and molecular classification. *PLoS ONE*, 12(4):7–9, 2017.

[15] E. Furlani. Magnetophoretic separation of blood cells at the microscale. *Journal of Physics D: Applied Physics*, 40:1313, 2007.

[16] P. R. Gascoyne and J. Vykoukal. Particle separation by dielectrophoresis. *Electrophoresis*, 23:1973, 2002.

[17] Shao H., J. Chung, L. Balaj, A. Charest, D.D. Bigner, B.S. Carter, F.H. Hochberg, X.O. Breakefield, R. Weissleder, and H. Lee. Protein typing of circulating microvesicles allows real-time monitoring of glioblastoma therapy. *Nature Medicine*, 18:1835–1840, 2012.

[18] J. Han, X. Wang and Y. Sun. Circulating tumor DNA as biomarkers for cancer detection. *Genomics, Proteomics and Bioinformatics*, 15:59–72, 2017.

[19] K-H. Han and A.B. Frazier. Lateral-driven continuous dielectrophoretic microseparators for blood cells suspended in a highly conductive medium. *Lab on a Chip*, 8:1079, 2007.

[20] J. Hong Miao, V. Samper, Y. Chen, C.K. Heng, T.M. Lim, and L. Yobas. Silicon-based microfilters for whole blood cell separation. *Biomedical Microdevices*, 10(2):251–257, 2008.

[21] M.P. Horning, C.B. Delahunt, S.R. Singh, S.H. Garing, and K.P. Nichols. A paper microfluidic cartridge for automated staining of malaria parasites with an optically transparent microscopy window. *Lab on a Chip*, 14(12):2040–2046, 2014.

[22] H.W. Hou, H.Y. Gan, A.A.S. Bhagat, L.D. Li, C.T. Lim, and J. Han. A microfluidics approach towards high-throughput pathogen removal from blood using margination. *Biomicrofluidics*, 6(2):024115–1–13, 2012.

[23] H.W. Hou, Q.S. Li, G.Y.H. Lee, A.P. Kumar, C.N. Ong, and C.T. Lim. Deformability study of breast cancer cells using microfluidics. *Biomedical Microdevices*, 11(3):557–564, 2009.

[24] H.W. Hou, C. Petchakup, H.M. Tay, Z.Y. Tam, R. Dalan, D.E.K. Chew, H. Li, and B.O. Boehm. Rapid and label-free microfluidic neutrophil purification and phenotyping in diabetes mellitus. *Scientific Reports*, 6:29410, 2016.

[25] H.W. Hou, M.E. Warkiani, B.L. Khoo, Z.R. Li, R.A. Soo, D.S. Tan, W-T. Lim, J. Han, A.A.S. Bhagat, and C.T. Lim. Isolation and retrieval of circulating tumor cells using centrifugal forces. *Scientific Reports*, 3(1259):1–8, 2013.

[26] L.R. Huang, E.C. Cox, R.H. Austin, and J.C. Sturm. Continuous particle separation through deterministic lateral displacement. *Science*, 304(5673):987–990, 2004.

[27] D.W. Inglis, R. Riehn, R.H. Austin, and J.C. Sturm. Continuous microfluidic immunomagnetic cell separation. *Applied Physics Letters*, 85(21):5093–5095, 2004.

[28] R.D. Jaeggi, R. Sandoz, and C.S. Effenhauser. Microfluidic depletion of red blood cells from whole blood in high-aspect ratio microchannels. *Microfluidics and Nanofluidics*, 3(1):47–53, 2007.

[29] A. Kaminska, E. Witkowska, K. Winkler, I. Dziecielewski, J.L. Weyher, and J. Waluk. Detection of Hepatitis B virus antigen from human blood: SERS immunoassay in a microfluidic system. *Biosensors and Bioelectronics*, 66:461–467, 2014.

[30] B. Kim, Y.J. Choi, H. Seo, E.C. Shin, and S. Choi. Deterministic migration-based separation of white blood cells. *Small*, 12(37):5159–5168, 2016.

[31] N. Kolluri, C.M. Klapperich, and M. Cabodi. Towards lab-on-a-chip diagnostics for malaria elimination. *Lab on a Chip*, 18(1):75–94, 2017.

[32] B. Kumar, V. Bhalla, R.P. Singh Bhadoriya, and C.R. Suri. Label-free electrochemical detection of malaria-infected red blood cells. *RSC Advances*, 6(79):75862–75869, 2016.

[33] J.R. Lange, J. Steinwachs, T. Kolb, L.A. Lautscham, I. Harder, G. Whyte, and B. Fabry. Microconstriction arrays for high throughput quantitative measurements of cell mechanical properties. *Biophysical Journal*, 109(1):26–34, 2015.

[34] G.Y.H. Lee and C.T. Lim. Biomechanics approaches to studying human diseases. *Trends in Biotechnology*, 25(3):111–118, 2007.

[35] A. Lenshof, A. Ahmad-Tajudin, K. Jarås, A-M. Sward-Nilsson, L. Åberg, G. Marko-Varga, J. Malm, H. Lilja, and T. Laurell. Acoustic whole blood plasmapheresis chip for prostate specific antigen microarray diagnostics. *Analytical Chemistry*, 81(15):6030–6037, 2009.

[36] M. Mauk, J. Song, H.H. Bau, R. Gross, F.D. Bushman, R.G. Collman, and C. Liu. Miniaturized devices for point of care molecular detection of HIV. *Lab on a Chip*, 17(3):382–394, 2017.

[37] S Miltenyi, W. Muller, W. Weichel, and A. Radbruch. High gradient magnetic cell separation. *Cytometry*, 11(2):231–238, 1990.

[38] Y. Nakashima, S. Hata, and T. Yasuda. Blood plasma separation and extraction from a minute amount of blood using dielectrophoretic and capillary forces. *Sensors and Actuators, B: Chemical*, 145(1):561–569, 2010.

[39] Y. Nematbakhsh and C.T. Lim. Cell biomechanics and its applications in human disease diagnosis. *Acta Mechanica Sinica*, 31(2):268–273, 2015.

[40] K.D. Nyberg, M.B. Scott, S.L. Bruce, A.B. Gopinath, D. Bikos, T.G. Mason, J.W. Kim, H.S. Choi, and A.C. Rowat. The physical origins of transit time measurements for rapid, single cell mechanotyping. *Lab on a Chip*, 16(17):3330–3339, 2016.

[41] E. Ozkumur, A. M. Shah, J. C. Ciciliano, B. L. Emmink, D. T. Miyamoto, E. Brachtel, M. Yu, P.-i. Chen, B. Morgan, and J. Trautwein. Inertial focusing for tumor antigen-dependent and -independent sorting of rare circulating tumor cells. *Science Translational Medicine*, 5:179ra47, 2013.

[42] E.S. Park, C. Jin, Q. Guo, R.R. Ang, S.P. Duffy, K. Matthews, A. Azad, H. Abdi, T. Todenhofer, J. Bazov, K.N. Chi, P.C. Black, and H. Ma. Continuous flow deformability-based separation of circulating tumor cells using microfluidic ratchets. *Small*, 12:1909–19, 2016.

[43] F. Petersson, L. Åberg, A-M. Sward-Nilsson, and T. Laurell. Free flow acoustophoresis: Microfluidic-based mode of particle and cell separation. *Analytical Chemistry*, 79:5117–5123, 2007.

[44] M.S. Pommer, Y. Zhang, N. Keerthi, D. Chen, J.A. Thomson, C.D. Meinhart, and H.T. Soh. Dielectrophoretic separation of platelets from diluted whole blood in microfluidic channels. *Electrophoresis*, 29:1213–1218, 2008.

[45] A. Prabhakar, Y.V. Bala Varun Kumar, S. Tripathi, and A. Agrawal. A novel, compact and efficient microchannel arrangement with multiple hydrodynamic effects for blood plasma separation. *Microfluidics and Nanofluidics*, 18:995–1006, 2015.

[46] M. Rafeie, J. Zhang, M. Asadnia, W. H. Li, and M. E. Warkiani. Multiplexing slanted spiral microchannels for ultra-fast blood plasma separation. *Lab on a Chip*, 16:2791–2802, 2016.

[47] C. Rivet, H. Lee, A. Hirsch, S. Hamilton, and H. Lu. Microfluidics for medical diagnostics and biosensors. *Chemical Engineering Science*, 66:1490–1507, 2011.

[48] N. Shao and L. Qin. Biochips-new platforms for cell-based immunological assays. *Small Methods*, 2:1700254, 2017.

[49] S.S. Shevkoplyas, T. Yoshida, L.L. Munn, and M.W. Bitensky. Biomimetic autoseparation of leukocytes from whole blood in a microfluidic device. *Analytical Chemistry*, 77:933–937, 2005.

[50] J.R. SooHoo and G.M. Walker. Microfluidic aqueous two phase system for leukocyte concentration from whole blood. *Biomedical Microdevices*, 11(2):323–329, 2009.

[51] A. Tay, A. Pavesi, S.R. Yazdi, C.T. Lim, and M.E. Warkiani. Advances in microfluidics in combating infectious diseases. *Biotechnology Advances*, 34:404–421, 2016.

[52] N. Toepfner, C. Herold, O. Otto, P. Rosendahl, A. Jacobi, M. Krater, J. Stachele, L. Menschner, M. Herbig, L. Ciuffreda, L. Ranford-Cartwright, M. Grzybek, U. Coskun, E. Reithuber, G. Garriss, P. Mellroth, B. Henriques-Normark, N. Tregay, M. Suttorp, M. Bornhauser, E.R. Chilvers, R. Berner, and J. Guck. Detection of human disease conditions by single-cell morpho-rheological phenotyping of blood. *eLife*, 7:1–22, 2018.

[53] S. Tripathi, A. Prabhakar, N. Kumar, S.G. Singh, and A. Agrawal. Blood plasma separation in elevated dimension T-shaped microchannel. *Biomedical Microdevices*, 15:415–425, 2013.

[54] M. Umer, R. Vaidyanathan, N.T. Nguyen, and M.J.A. Shiddiky. Circulating tumor microemboli: Progress in molecular understanding and enrichment technologies. *Biotechnology Advances*, 36:1367–1389, 2018.

[55] J.A. Vendrell, F.T. Mau-Them, B. Béganton, S. Godreuil, P. Coopman, and J. Solassol. Circulating cell free tumor DNA detection as a routine tool for lung cancer patient management. *International Journal of Molecular Sciences*, 18:E264, 2017.

[56] M.E. Warkiani, A.K.P. Tay, B.L. Khoo, X. Xiaofeng, J. Han, and C.T. Lim. Malaria detection using inertial microfluidics. *Lab on a Chip*, 15:1101–1109, 2015.

[57] J. Wu, Q. Chen, and J.-M. Lin. Microfluidic technologies in cell isolation and analysis for biomedical applications. *The Analyst*, 142:421–441, 2017.

[58] S. Yang, A. Undar, and J.D. Zahn. A microfluidic device for continuous, real time blood plasma separation. *Lab on a Chip*, 6:871–880, 2006.

[59] Kenry Yeo, J.C., Z. Zhao, P. Zhang, Z. Wang, and C.T. Lim. Label-free extraction of extracellular vesicles using centrifugal microfluidics. *Biomicrofluidics*, 12:024103, 2018.

[60] T. Yeo, S.J. Tan, C.L. Lim, D.P.X. Lau, Y.W. Chua, S.S. Krisna, G. Lyer, G.S. Tan, T. Kiat Hon Lim, D.S.W. Tan, W-T. Lim, and C.T. Lim. Microfluidic enrichment for the single cell analysis of circulating tumor cells. *Scientific Reports*, 6:22076, 2016.

[61] Z.T.F. Yu, K.M. Aw Yong, and J. Fu. Microfluidic blood cell sorting: Now and beyond. *Small*, 10:1687–1703, 2014.

CHAPTER 12

Blood suspensions in animals

Ursula Windberger

Medical University of Vienna, Vienna, Austria

CONTENTS

THE MAIN DUTY OF RED BLOOD CELLS (RBCs) in all vertebrates is the transport of oxygen to fulfill the tissue demand, and in parallel to remove CO2 as the metabolic end product. Even though some cellular structures and properties might be explained as a response to external circumstances or as an attempt to make the blood suspension more fluid, the question still arises regarding why nature did develop so many forms of red cells to fulfill the identical task. Our current knowledge on structure-function relationships by including animals into our understanding of RBCs is still not more than the tip of an iceberg. Due to the huge number of vertebrate species on earth and the limited number of researchers in the field of veterinary hematology and hemorheology we would always be in the dilemma of omitting knowledge. In the inaugural number of Comparative Hematology International one can read: "Although there are many thousands of hematologists working on the blood of one species, Homo sapiens, the number concerned with the hundreds of thousands of other species is several orders of magnitude less " [92]. In the context of research on cardiovascular dynamics many decades ago, hemorheology became a matter of intensive research as well. Fascinating in-vivo studies were performed that are still the basis of our understanding of microvascular flow [46, 72, 74, 132–135, 208, 209, 214]. Although the animals served as a model for this basic research, the outcomes reflect also the physiological circumstance in the entire species. This can have a clinical relevance in veterinary medicine and in current translational research. In parallel to these experiments, in-vitro studies were performed on domestic animal blood [75, 76, 229, 237, 238], but blood was also withdrawn from exotic species like crocodiles [263], Antarctic birds [87], icefish [263], seals [149, 267], frogs and toads [32, 96, 121, 176, 254] - to map only a few comparative studies from these earlier days. Meanwhile, not only have species been added to this list, but also the methods have been refined or new methods have been developed. For instance, the assembly of the RBC cytoskeleton network can be nicely seen by super-resolution microscopy [177], and sophisticated in-vivo imaging methods have been developed for rodents. This chapter gives an overview of the available literature concerning animal hemorheology. It starts with a description of animal blood and continues with an overview of the differences of geometrical factors and molecular structures of mammalian RBCs. Subsequently the mechanical properties that arise from these parameters and the bulk behavior of the suspension are explained. At the end, specific animal species are described in more detail and an outlook is provided.

12.1 BLOOD OF INVERTEBRATE ANIMALS

Where do we start when we speak about animal blood? Blood is not restricted to vertebrates. Invertebrates can also own a flowing liquid that moves in a cardiovascular system. Even cells containing respiratory pigments with metal ions (Fe^{2+}, Cu^{2+}) at the active sites can circulate within the body fluids. Only in a small number of species do these RBCs pass narrow capillary systems (among them the phoronids and non-vertebrate chordata), depending on the status of the development of the cardiovascular system. A circulatory system had to evolve with the rise in body size, when the diffusion of oxygen and nutrients alone could no longer fulfill the tissue demand. Pumping organs and conduits developed, but "real" vessels with endothelial lining developed in vertebrates. The pumping organs of annelids and arthropods moves fluid in well-defined spaces [154], like vessels or plexi that are lined by extracellular matrix or by the basal cell surface of the adjacent tissue. Open cardiovascular systems evolved to supply organs in series, but the more developed closed circuits supply organs in parallel, so that all organs receive the identical amount of oxygen and nutrients.

Pigments for oxygen transport circulate mainly in solution (e.g., hemocyanin (blue color in arthropods, mollusks), hemerythrine (pink color in annelids, brachiopods), chlorocruorin (green color in some marine species)), but cells are also used for their transport. A comprehensive survey on invertebrate oxygen carriers including also plant hemoglobins can be found in a textbook, previously edited by Ch. P. Mangum [141]. It appears - despite the fact that the knowledge on this issue is still limited by the huge number of species that must be systematically investigated - that no simple explanation is viable to help us understand the nature of blood and RBCs in the invertebrates. Tissue heme proteins are present in lower invertebrate animals, and RBCs developed together with the circulation, but circulatory systems and "blood" are so diverse that no simple conclusion can be drawn. Generally, circulating hemocytes are nucleated and their function is not restricted to oxygen transport but includes phagocytosis/pinocytosis, blood clotting, wound healing, and metabolic processes, as well [239]. Surprisingly, it appears that oxygen transport plays an underlying role. Although RBC hemoglobins of non-vertebrates can store and transfer oxygen, only the pigment of vertebrate RBCs shows the typical respiratory properties that we typically attribute to hemoglobin [142].

12.2 BLOOD OF VERTEBRATE ANIMALS: SPECIES DIFFERENCES IN RBC SIZE AND SHAPE

Vertebrates possess circulating red blood cells full of hemoglobin, designed for a passage through narrow tubes. The red cells have lost the ability of phagocytosis, while their exclusive task became gas exchange in every capillary system in a body. It is mandatory that they remain in the circulatory system. Therefore cell properties developed to ensure that RBCs do not adhere to vessel walls like white blood cells and that RBCs preferentially flow in the center of a vessel.

The benefit of enclosing respiratory pigments by a membrane was explored in the late sixties of the last century [53, 212]. Compared to a circulating hemoglobin solution, the presence of blood cells containing the identical amount of hemoglobin generated a non-Newtonian behavior of blood through an increase of blood viscosity at low shear rates. This appears as an undesired effect at first sight, but the authors explained that the preservation of the oncotic pressure, as well as the proximity of hemoglobin to enzymes associated with gas exchange would be the developmental benefit. Schmidt-Nielsen [212] saw the advantage in the bolus flow of a suspension in comparison to a laminar flow of a hemoglobin solution. When hemoglobin is packed into cells, there will be no stagnation layer of solute pigments along the endothelium, and all the respiratory pigment can be transported and used for gas exchange.

The cell membrane of vertebrate RBCs is composed of a phospholipid bilayer, supported by an underlying 2D-cytoskeleton network. Linker proteins connect the triangular spectrin network to the membrane proteins in the bilayer. This general RBC structure is well known from the human red cell and maintained in the vertebrates. But despite this general composition, the cell geometry can be very diverse among the mammals (see Figure 12.1).

When a nucleus or a tubulin system is present, the geometry is further different [168]. For instance, fish RBCs are ellipsoidal and bulge in the region of their nucleus, whereas the enucleated mature mammalian RBC is round with biconcave lateral faces depending on the magnitude of cell surface per enclosed volume (Figure 12.1). But exceptions exist: some fish RBCs do not contain a nucleus [276], and camelid RBCs - although they are mammalian red cells - are not circular, but flat and ellipsoidal.

In the biconcave mammalian red cells, the surface to volume ratio determines the final shape [6, 48], and an index was developed to correlate cell volume, surface area, and cell roundness. This index predicts that small cells (e.g. from sheep, goat) are more spherical than human RBCs [238]. Generally, a small RBC size is associated with an increase in cell roundness since the architecture of the triangular spectrin network inhibits deflated biconcave shapes in cells with small diameter (Figure 12.1). In addition, the leaflets can influence the shape on the basis of their relative magnitude [160], an issue that might further explain the poikilocytosis in goat [120] or the surface roughness, e.g. of carp RBCs [234]. But despite the diversity of red cell size and sphericity, all mammalian blood suspensions tested so far present a shear-thinning behavior with a species-specific magnitude.

If RBC size is small, the number of circulating red cells must increase to provide the sufficient amount of hemoglobin for the blood gas transport. Blood of species with low RBC volume like goat and sheep (15-40 fL; $8 - 14 \times 10^9$ cells L^{-1}) contains more RBCs per mL than human (about 90 fL; $4 - 6 \times 10^9$ cells L^{-1}). There is no relationship between RBC

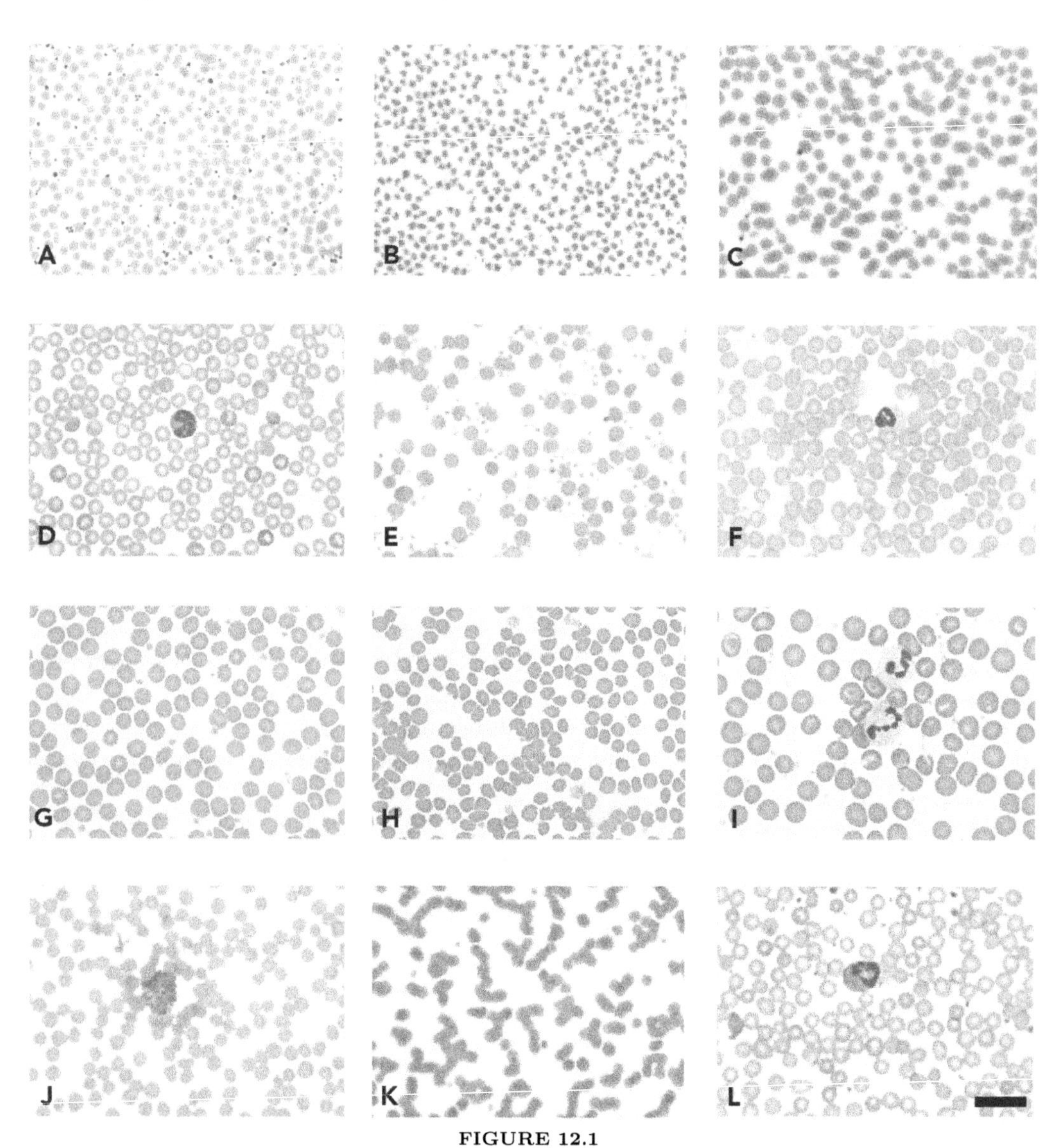

FIGURE 12.1
Photomicrographs of erythrocytes from mammalian species. A: sheep (Ovis gmelini aries), B: goat (Capra aegagrus hircus), C: cow (Bos primigenius taurus), D: mouse (Mus musculus), E: rat (Rattus rattus), F: hamster (Cricetus cricetus), G: dog (Canis lupus familiaris), H: cat (Felis silvestris catus), I: elephant (Loxodonta africana), K: horse (Equus ferus caballus), L: pig (Sus scrofa domesticus), M: rabbit (Oryctolagus cuniculus). 1000x magnification in oil immersion. Romanowsky (D-K, M) or Wright (A, B, C, L) staining. By courtesy of Dr. Furman & Dr. Leidinger (Laboratory In-Vitro, Vienna, Austria), and Dr. Schwendenwein (Clinical Pathology Platform, Veterinary University Vienna, Austria). Horizontal bar: 20 μm.

size and the size of the animal, e.g. large animals like horses have smaller RBCs (45 fL) than smaller species like rabbits (62 fL). Throughout the mammals, the human RBC is a large red cell. The farm and companion animals have generally much smaller RBCs. Only seals, dolphins, and whales have larger cells because they need as much oxygen as possible during a dive [43, 149, 269]. Therefore their RBCs are large and inflated to carry a high amount of hemoglobin. The RBC of the African elephant is also larger than the human cell, a finding that cannot be associated to its way of life so far. A report from the middle of the 19th century [88] has provided the RBC dimensions together with the size of its nucleus

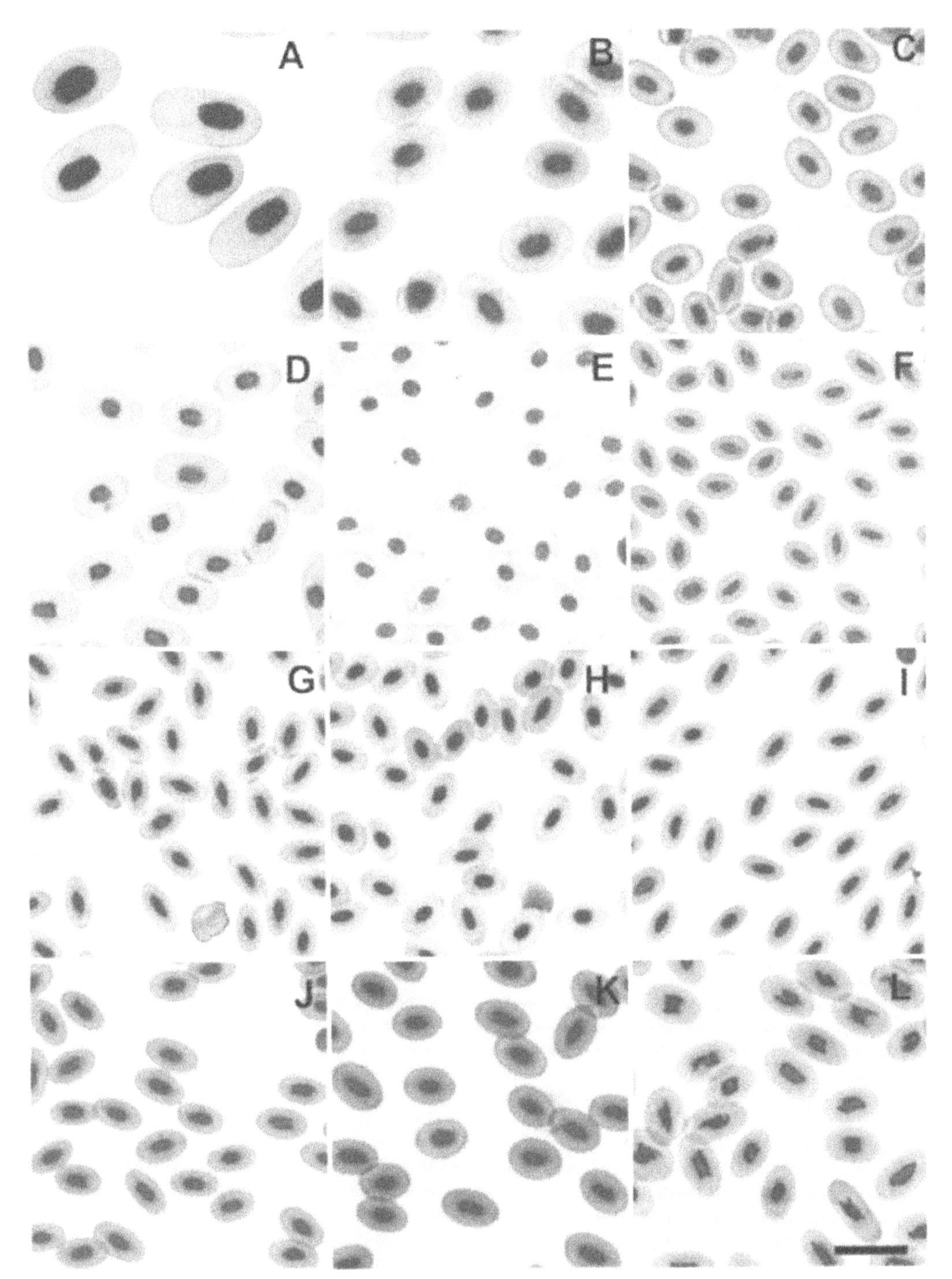

FIGURE 12.2
Photomicrographs of erythrocytes of some amphibian and reptile species. A. Ommatotriton vittatus, B: Pelophylax caralitanus, C: Pelophylax caucasus, D: Emys orbicularis, E: Testudo graeca, F: Ophisops elegans, G: Mesalina brevirostris, H: Anatololacerta danfordi, I: Lacerta trilineata, J: Leptotyphlops macrorhynchus, K: Hemorrhois ravergieri, L: Montivipera xanthina. Horizontal bar: 20 μm. From [10] with permission.

(if present) of a huge number of species from all vertebrate classes. The largest cells were not present in fishes, but in species of the amphibian class (see Figures 12.2 and 12.3). The largest RBC was the elliptical cell from Siren lacertina (58 × 32μm), whereas the smallest cell was the round cell of Moschus javanicus (2.1 μm). Almost 100 years later, other authors reported that the largest RBC is found in Amphiuma tridactilum (a species of aquatic salamander) [245], see also the review of [10]. To know more about hematological values of animals the reader is invited to look into the related textbooks [68, 169]. A survey on the hypothesis regarding why the size of vertebrate erythrocytes spans such a large range can be found in the introduction section of a recent thesis [70].

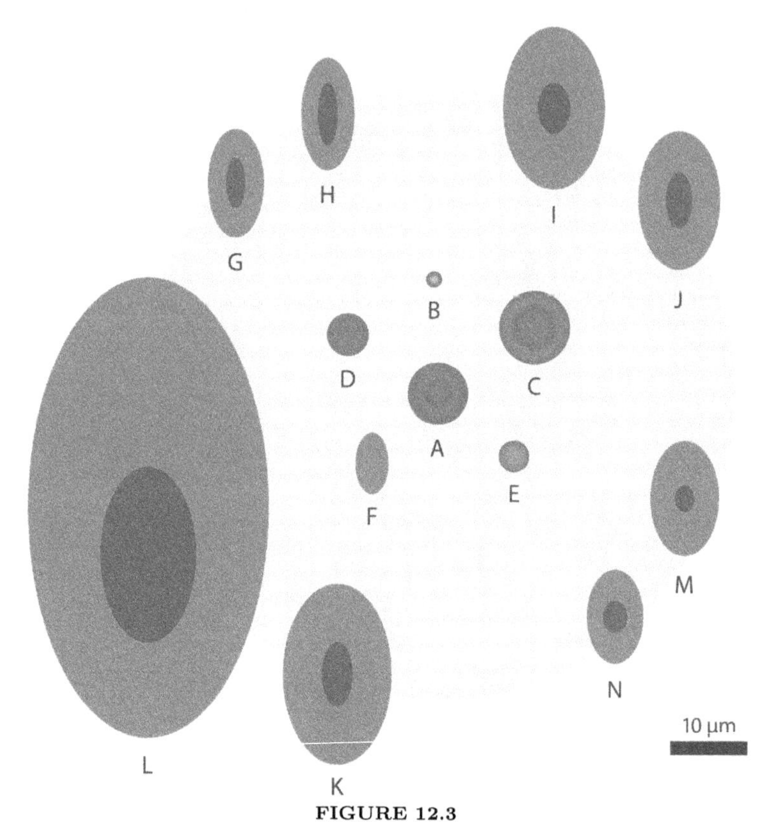

FIGURE 12.3
Schematic drawing of RBCs from vertebrates. A: Homo, B. Moschus javanicus, C: Elaphas indicus, D: Equus caballus, E: Capra hircus, F: Camelus dromedarius, G: Vultur auricularis, H: Pelicanus onocrotalis, I: Chelidonia mydas, J: Python tigris, K: Rana temporaria, L: Siren lacertina, M: Anguilla vulgaris, N: Cyprinus carpio. Dimensions of cells and cell nucleus were taken from [88]. The largest and the smallest RBC from each vertebrate class were selected.

Shapes of RBCs that deviate from round forms obviously need additional intracellular structures [113] underneath the membrane cytoskeleton to stabilise this geometry and to prevent buckles on the cell surface especially when the lateral face is large. RBC length correlated positively with the number of microtubules in a wide range of species [82]. This reinforcing bundle of microtubules is termed the marginal band [168]. (Remark: proteomics showed the presence of tubulin isoforms in human erythrocytes as well [7]). Both the presence of nucleus and tubulin filaments stabilises the cell against deformation, either locally or globally. It appears that the cell stiffness can differ locally, e.g. a RBC's apex can indent the basal part of another RBC (Figure 12.4).

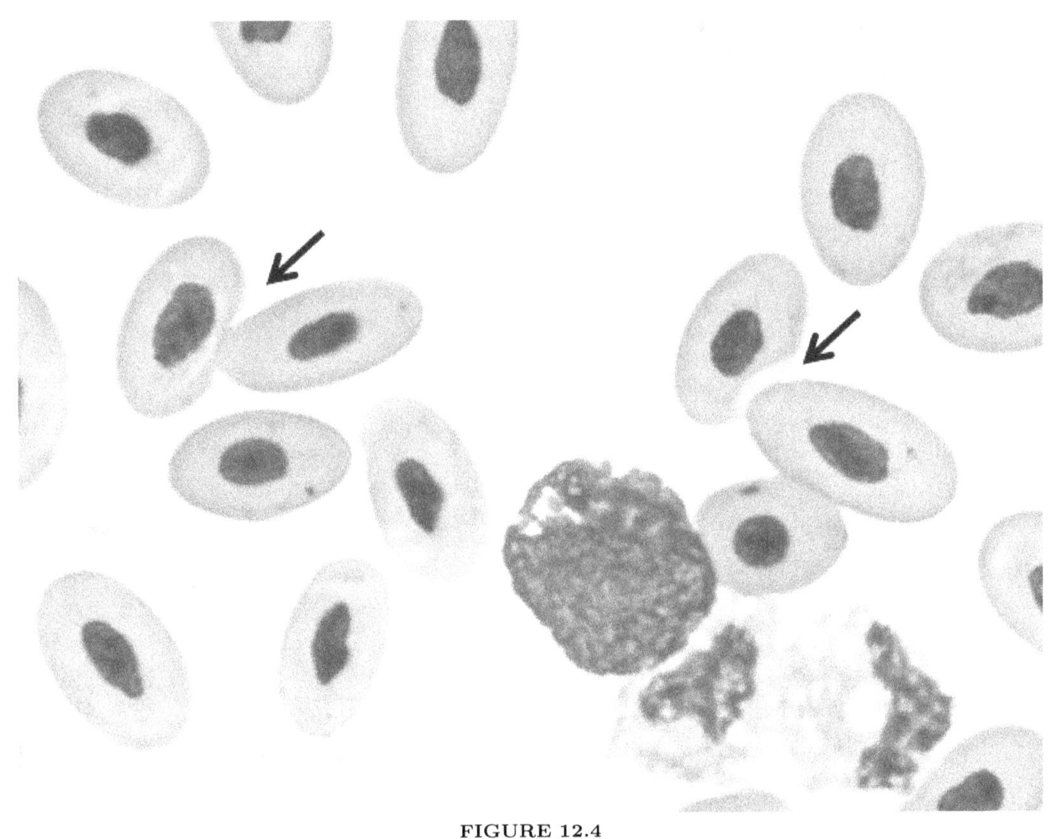

FIGURE 12.4
Example of local deformability change of nucleated RBCs. The two arrows indicate that red cells of tiger python can indent their neighbors by their poles (1000x magnification). Courtesy of Dr. Furman and Dr. Leidinger (laboratory In-Vitro, Austria)

The higher stiffness together with the ellipsoidal shape should favour an orientation in a blood vessel that is different from human RBCs. It is unlikely that nucleated RBCs have parachute shapes but needs to be proven. In straight glass capillaries, other deformations have been observed to reduce red cell width: the elongated RBCs of the Peking duck (Anas platyrhynchos) displace their hemoglobin solution into the leading part of the cell whereas the membrane at the rear part kinks next to the nucleus [75]. Gaehtgens and co-workers also observed that duck red cells were excluded from flowing through such glass capillaries with diameters between 5 and 11 μm. Increasing the feeding hematocrit (concentration of RBCs, noted HCT) could not augment the HCT outflow [76]. On the contrary, the relative discharge hematocrit decreased when more cells were fed into the system, which the authors associated with unsorted cell crowding in front of the inlet. The authors went even further and suggested that "clustering" with subsequent formation of a cell-free layer might be a requisite of deformable RBCs (as for mammalian RBCs) only. It is interesting to note that the chicken RBC membrane itself is not that "stiff", although fortified by tubulin [131, 136]. Single cell spectroscopy studies recently showed that indentation beneath the nucleus yielded low apparent Young's modulus values in the range of 10^2 Pa at the physiological body temperature of chicken (personal observation). In contrast to poultry, fish RBCs are more as shown by filtration tests [165]. The authors concluded that a slow travel of fish RBCs through the capillaries would give more time for O_2 delivery.

Shape changes resulting from an increase or decrease of the cytoplasmic volume include changes in pH [162], ion content [73, 151, 194, 258], osmolality [49, 152], and ATP content [161]. The RBC volume regulates the amount of hemoglobin that can be carried per cell (MCH) to keep the mean corpuscular hemoglobin concentration (MCHC) reasonable (exceptions exist). A significant rise in cytoplasmic viscosity occurs when the hemoglobin concentration exceeds a certain level [198, 264]. This cannot be avoided in nucleated RBCs. Avian RBCs show an approximately 20% increase in MCHC due to the presence of a nucleus.

12.3 SPECIES DIFFERENCES IN THE MOLECULAR STRUCTURE OF RBC MEMBRANES

Species-specific differences were found in regard to the protein composition of the RBC membrane [84, 100, 111, 126, 144, 179, 216, 217, 231, 241, 266] and to the lipid composition of the bilayer [63, 108, 119, 166, 175, 183, 190, 196, 233, 240, 252]. Specific studies on animal RBC bilayers have been undertaken in terms of the binding and membrane transfer rates of unsaturated fatty acids in sheep [35], and the influence of fatty diets in rat, rabbit [93, 242], and mouse [55].

12.3.1 Membrane proteins

In one of the first comparative RBC gel electrophoresis studies, the author stated that general proteins like spectrin, ankyrin, band 3, band 4.1, and actin are preserved in all RBCs, although it can be read from the images that they appear at least quantitatively different [126]. Later, researchers pointed out the differences between species, e.g. quantitative differences in band 3 and band 4.1 in mouse versus pig RBCs were indicated [266]. Guerra-Shinohara reported that band 4.2 protein was absent in horse, guinea pig and two rodents. Baskurt and co-workers associated the lack of band 4.2 protein with the large number of echinocytes in the horse [16]. Matei [144] provided a comprehensive view over a range of eight mammalian species including human. The author measured the highest concentration of band 1+2 (spectrins) and band 3 protein in human, and the highest concentration of band 2.1 (ankyrin) in sheep (Figure 12.5). Pasini [179] suggested differences in the molecular weight and glycosylation of band 3 protein among the animals; the protein has a higher molecular weight in cow, guinea pig and mouse RBCs, and a lower molecular weight in sheep RBC. Sharma and coworkers [216, 217] analysed the bands by mass spectrometry and found quantitative and qualitative differences in three ruminant species and the pig compared to human. The authors also identified heat shock proteins in animal membranes (identified as band G1, G2, C1 proteins). It was found that goat and buffalo membranes possess a higher number of glycoproteins compared to human [216] and that in camelids the rotational and lateral mobility of band 3 protein is low, whereas the connection of the anion exchanger to ankyrin is tight and close [147]. Antibodies to the cytoplasmic domain of the human band 3 did not cross-react with chicken band 3 protein [139], indicating significant differences in the functional residues of the protein. It was shown that the oxygenation-driven binding of hemoglobin to band 3 protein occurs in species of all vertebrate classes [257]. The authors suggested that this interaction would therefore be ubiquitous in vertebrates. However, the cdB3-Hb interaction showed quantitative differences: it was lowest in the tested fish and highest in human. The influence of oxygenation with the mechanical performance of the membrane or ion fluxes across the membrane is therefore species-specific. Another protein interaction (band 3 - ankyrin) that is also driven by the available oxygen in human RBCs [226] has not been investigated comparatively.

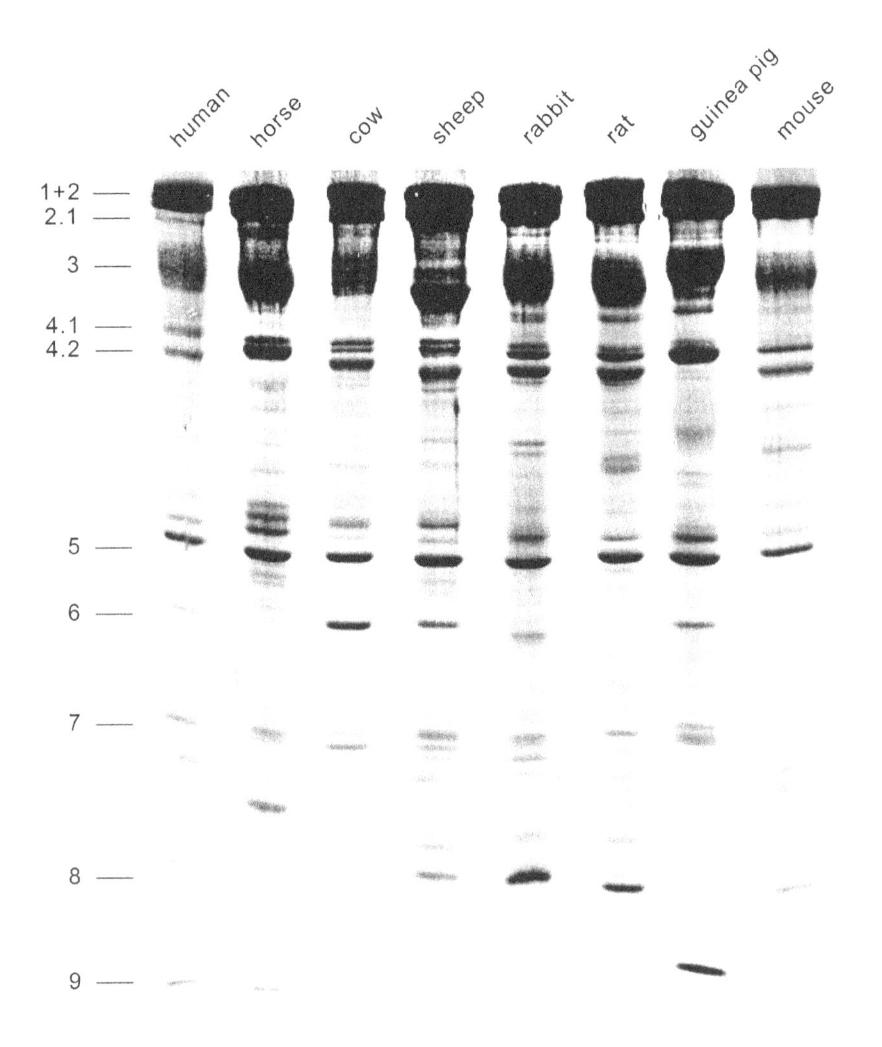

FIGURE 12.5
Electrophoretic pattern of membrane polypeptides from RBC membranes of eight mammalian species. Adapted from Figure 1 in comparative studies of the protein composition of red blood cell membranes from eight mammalian species/H. Matei , L. Frentescu , Gh. Benga, Comparative studies of the protein composition of red blood cell membranes from eight mammalian species, Journal of Cellular and Molecular Medicine, 4(4) 2007. Copyright (c) [2000] [copyright owner as specified in the Journal] https://onlinelibrary.wiley.com/doi/abs/10.1111/j.1582-4934.2000.tb00126.x

In summary, the important proteins involved in cytoskeleton build-up and connection to the bilayer are maintained in all species (Figure 12.5), but differences exist with respect to the molecular weight of the proteins [110,192,266] and perhaps with respect to the sequence of the residues, thus leading to different functions [168].

12.3.2 Membrane lipids

The bilayer is an area-incompressible fluid membrane and contributes to the viscosity and the bending modulus of the membrane. The fatty acid composition of the phospholipids in the bilayer and the cytoplasmic hemoglobin concentration vary among species and is influenced by external factors (see Section 12.7). The ability of a RBC membrane to dissipate viscous forces, which affects cell flow behaviors, therefore differs as well . But the influence of the membrane lipid structure on RBC deformation is limited. For instance, the quantity of unsaturated fatty acids in the RBC membrane of sheep is twofold higher than in pig and horse [183], but sheep shows the lowest RBC deformation among a series of species [182] (see also Figure 12.7). The importance of the bilayer composition may gain more relevance in terms of mechanical issues when the membrane loses tethers, or when microparticles are formed during disease [172] or mechanical trauma.

12.4 SPECIES DIFFERENCES IN THE INTRINSIC PROPERTIES OF RBCS

12.4.1 RBC aggregability

RBC aggregability seems to be reserved only for a relative small number of species [265,271]. The presence of a nucleus in red cells of non-mammalian species is an obstacle for rouleaux formation, but great differences exist even among mammals. When aggregation exists, it occurs only at low shear forces and can be detected by measuring the shear modulus at low shear stress. The yield stress of blood can be obtained by fitting values of low shear viscosity to the Casson equation. This approach was used recently to quantify RBC aggregation. Good correlation to aggregation indices was achieved [124]. But the yield stress can also be measured in oscillatory tests. Figure 12.6 shows the variation of the storage viscoelastic modulus (G') of blood with the shear stress (see Chapter 1 for definition). It is clearly species-specific at low stresses. Frequency (1.5Hz), HCT (40%), temperature (37°C), and gap width (1 mm) were kept constant in these tests. The figure also shows that although G'-values can be measured at stresses below 200 mPa in sheep, horse, pig, and rat, a correct linear viscoelastic range is difficult to identify on an individual basis, even in blood samples with high RBC aggregability like in horse.[224, 274].

It is easier to identify the abrupt change of viscoelasticity at stresses between 5 and 200 mPa. If the hematocrit and the plasma fibrinogen concentration rise, not only G'-values, but also the yield point increases. RBC aggregates govern low shear blood viscosity and yielding. It is therefore easily explained that the hyper-aggregating horse blood needs a higher transition stress to change from a viscoelastic fluid to a viscous fluid than the non-aggregating sheep blood at equal condition [274]. But the presence of RBC aggregates cannot be the only cause for yielding, since the non-aggregating rat blood shows significant G'-values within a stress range between 1 and 20 mPa (Table 12.1 at the end of the chapter and Figure 12.6).

Why nature has only generated a limited number of species that have the ability to form rouleaux is not known.

Animal size is not a trigger factor, as RBCs of healthy large animals can strongly aggregate (horses) or not (cows). For small animals with a body weight of less than 500 g, no species has so far shown RBC aggregation. Hypothetically, the length of the vessel between two bifurcations could be too short to allow axial migration. In the spinotrapezius muscle of rat, the length of venular segments between branch points has been plotted against their diameter [30]. Dextran has been added to rat blood to allow RBC aggregation. It was found out that aggregation increases the rate of axial migration for RBCs traveling near the vessel wall from less than 1% to 1.27%. This means that during a longitudinal transit of 100 μm the RBC would migrate less than 1 μm (native rat blood without aggregation) or 1.27 μm (rat blood with aggregation). Although the formation of a cell-free layer width scales with

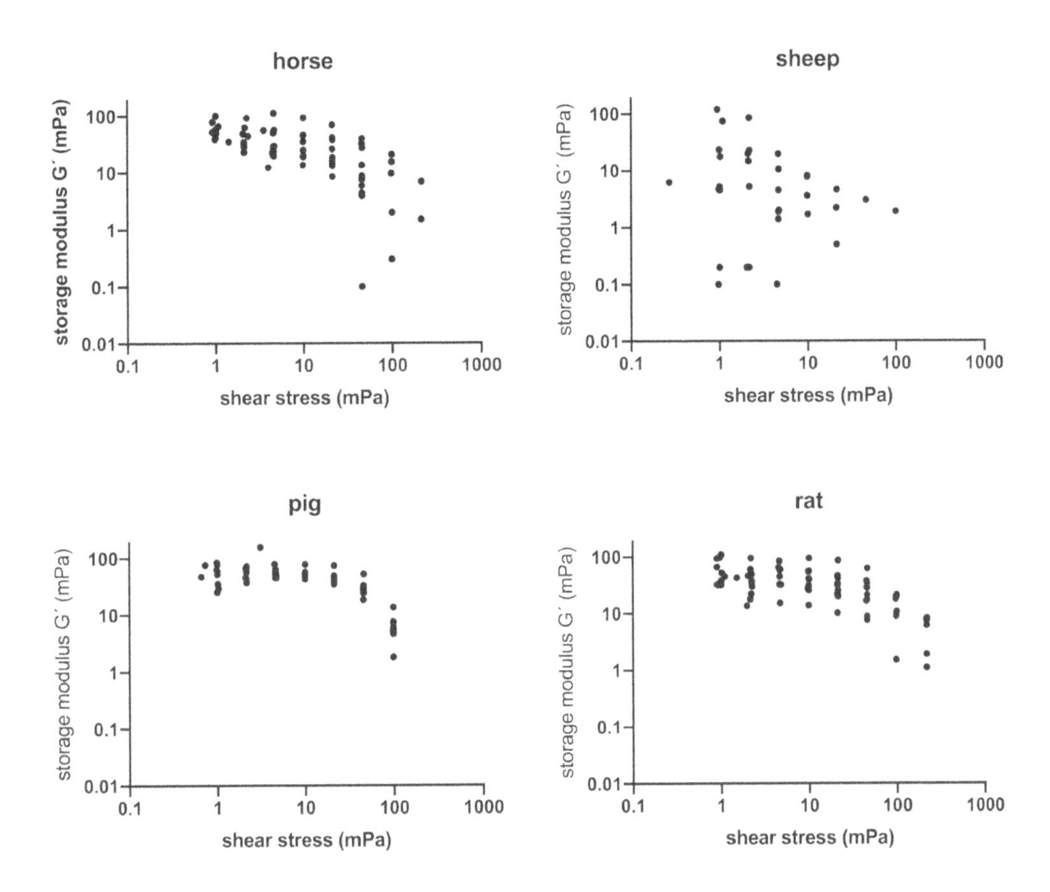

FIGURE 12.6
Stress sweep tests of horse (a), sheep (b), pig (c), and rat (d) blood suspensions. Blood was hematocrit adjusted (40%) and tested at 37°C. Scatterplots of storage modulus (G′, in mPa) at 1.5Hz. G′ was nil in dromedary camel blood.

RBC aggregability, it also needs time, and frequent branching of the venular network attenuates its development. This effect can be amplified at high heart rate. This could make RBC aggregability unnecessary. However in-vivo microscopy showed that some organisation of cells in the flow had developed in rat venules despite the lack of red cell aggregability [29]. It could simply be the effect of wall lift.

In the group of large animals, in equidae (horse, donkey, zebra), but also in some seals [43] there is pronounced RBC aggregation, whereas domestic ruminants (cattle, sheep, goat) do not show any [22,68,171,270]. But when fibrinogen is added to cow blood, RBC aggregation indices increase significantly [107], suggesting that RBC aggregation can rise when inflammatory processes affect the animals. RBC aggregation is therefore not completely impossible in cattle, sheep, and goat RBCs, but rather depends on the amount of pro-aggregating proteins. It was proposed that rouleaux formation is associated with a biconcave red cell shape [20], which makes aggregation difficult for the small red cells of sheep and goat. But cattle and horse RBC volumes are comparable. The absence of rouleaux is even preserved in cattle strains that live in different environments. For instance, the South African Nguni cattle in

Gauteng show comparable aggregation-values with the Simmental breed in Austria [207]. RBC aggregation was associated with the role of edema prevention by keeping the capillary pressure low at high blood flow velocity [20]. This is important when animals exercise. Popel [186] suggested that the athletic phenotype of a species could serve as a trigger for the RBC aggregation phenotype, and this approach is still discussed. This would suggest that domestic cattle, sheep and goats belong to the sedentary animal group, and movement of animals during pasture management should be undertaken with care. Interestingly, other animals of the bos genus like bison and yak show little higher whole blood viscosity (WBV) values at low strains and a higher shear thinning than cattle or sheep [105]. Bison and yak blood should thus have some degree of RBC aggregation. This might indicate a selection out of a benefit against a stressor (a predator or an environmental stress) in bison and yak, or the loss of a former property could be a matter of inbreeding. Reduced heterozygosis is a common result of domestication. Generally, this results in reduced relative fitness of the animal, which - in the case of RBC aggregability - might have resulted in a loss of this function. One study supports this: northern elephant seals that are bred in aquaria show reduced low shear viscosity compared to free-ranging members [267]. But since viscosity depends on many factors, the conclusion remains weak.

12.4.2 RBC deformability

Erythrocytes must survive for months in a high-shear environment while they constantly circulate through the vascular system. RBCs must therefore deform in the capillaries, irrespective of their membrane stiffness or size. Although it can well be assumed that the capillary diameter and the intra-cardiac pressure at volume ejection are comparable in most mammalian species, mammalian RBCs are various in size and response to mechanical stress [167,218]. Several techniques have been used to probe animal RBC deformability: ektacytometry, filtration tests, pipette aspiration tests, and AFM. Other methods like optical and magnetic tweezers are appropriate as well [200, 284].

In laser diffractometry the varying diameter and shape of the cells make species comparisons difficult. A method to compare cells of different sizes was developed [18]. Data are shown in Figure 12.7 and Table 12.2. Sheep has the lowest elongation indices among a wide range of animals, which indicates the poorest cell deformability among the listed animals. However, the shear stress for half-maximal deformation ($SS_{1/2}$) is comparable to many other species on this list, which shows that the red cell starts to deform at lower stresses than e.g. horse or cattle RBCs. Therefore the sheep RBC membrane is not as stiff as postulated from the maximum elongation only. It cannot be neglected that this method, although very practical in its use, can display membrane stiffness only by indirect means. Micropipette aspiration techniques are also tricky because the cytoplasmic viscosity, very sensitive to temperature fluctuations, affects the finding [255]. The shear modulus of elasticity of the membrane, filterability, and erythrocyte elongation have been tested by different techniques and there is now a consensus that erythrocyte deformability is species-specific. Waugh and coworkers [254] tested RBCs from vertebrates of all classes by pipette aspiration. They showed that membranes of nucleated RBCs had larger stiffness than mammal RBC membranes. Interestingly, toadfish RBCs showed a lower stiffness than turtle and snake RBCs, the values being more comparable to turkey. Years later the authors confirmed their previous findings and included new species [253]. Llama RBCs showed high rigidity, which was even comparable to turtle and snake RBCs, whereas bullfrog, turkey and toadfish showed lower values, lying in-between mammals and turtle. Automated filtration methods were used to measure RBC capillary entry and transit times [8, 14, 64, 86, 109] in animals. These studies showed that the large fish RBCs needed the largest pore diameter (9 μm), whereas frog and turtle cells could pass through 4 μm pores. Mammalian cells could be filtered through 3 μm pores [48]. When test conditions are modified, RBCs respond species-specifically. For instance, hypoxia had a stronger impact on RBC deformability in rat, rabbit, and cat compared to dog [89], and the sensitivity of RBCs to glutaraldehyde hardening is different among guinea pig, dog, rabbit, rat, mouse, and sheep [14].

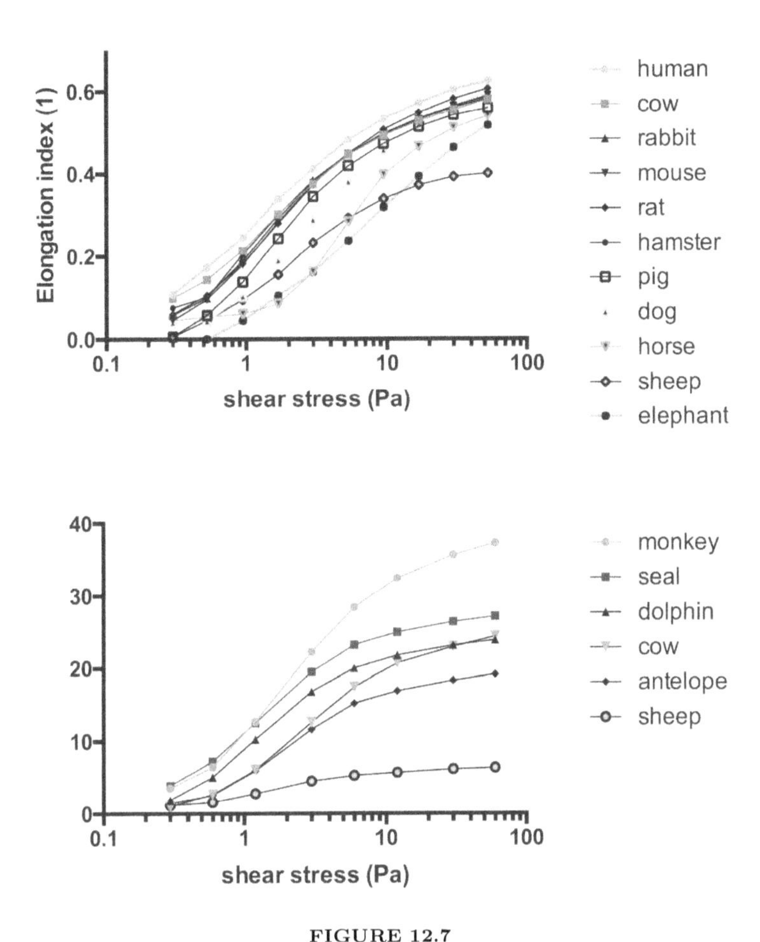

FIGURE 12.7
RBC elongation of several mammalian species. Rotational ektacytometry was performed with (a) LORRCA (Mechatronics, The Netherlands) and (b) Rheodyn SSD (Myrenne, Germany). The small animal species (mouse, rat, hamster, rabbit) show similar EI-SS curves [18].

The easiest way to test RBC deformation is by rotational ektacytometry; therefore most animal values are available from this method and the widest number of species can be used for comparison [98, 116, 150, 182, 220, 269, 273]. Mammals can be ranked according to the maximum RBC deformation (EI_{max}) as follows: marine animals (sea lion, sea elephant) and vervet monkey have good RBC deformation, followed by a group comparable to human (rat, rabbit, hamster, mouse, rhesus monkey and baboon), further followed by a group with less RBC deformation than human (pig, dog, horse, African elephant), and a group with poor red cell deformation (sheep, goat). Figure 12.7 shows the EI/shear stress dependency of several species tested in our laboratory - please note that different devices were used and that the elongation indices cannot be directly compared between the two methods. Table 12.2 shows the $SS_{1/2}$ and EI_{max} values of these measurements. As already mentioned, the applied stress to result in half of the maximum elongation can draw a different species ranking. By using $SS_{1/2}$, mammals can be ranked as follows: vervet monkey, seal, dolphin, Chinese hamster, and rabbit RBCs show the best deformability, and this group is followed closely by sheep, rat, mouse, Syrian hamster, pig, and antelope. Cattle and dog show lower values, and horse and African elephant RBCs show the worst RBC deformability among the animals on the list. In summary, sheep and dog change their places in the two rankings. It is noticeable that many small animals species like rats, mice, hamsters, and the rabbit show good RBC deformability. Ektacytometry is not the appropriate method to explore ellipsoidal RBCs (camel, chicken, goose, penguin) [269]. Such RBCs do not align, but tumble in the Couette flow.

12.5 SPECIES DIFFERENCES IN THE MACROSCOPICAL BEHAVIOR OF ANIMAL WHOLE BLOOD

Parameters that contribute to whole blood viscosity are the RBC fraction (hematocrit), the RBC size, shape, and mechanical properties (including the cytoplasmic viscosity), and the viscosity of the suspending medium (plasma viscosity). If aggregates are present, they generate a yield-type behavior of the suspension [140], and the yield stress is shifted to lower or higher values. It is logical that flow curves appear species-specific due to the species-specificity of these properties. Microvascular network models that use these parameters were previously reviewed [41, 170, 189]. It is difficult to interpret the viscosity spectrum if kinetic effects are present in parallel. This is especially the case in equine blood due to the hyperaggregating RBCs that induce structural inhomogeneity in the blood. For more details concerning the problems that can arise during the measurement of horse blood see Section 6 of this chapter.

Viscosity scales exponentially with hematocrit, but hematocrit dependency (the differences are more visible at low shear rates) is higher if RBC aggregation is present. Viscosity also scales with temperature through a species-specific power law [43, 247, 268], a finding that has been brought into the context with the specific environment in which the animal lives. For instance, a reduced temperature dependency of blood viscosity was observed in the dromedary camel. Dehydrated camels are heterothermic [95] and save body water by not losing heat through evaporation. Within the physiological nychthemeral amplitude of body temperature - which can be even 5°C - $WBV_{1000s^{-1}}$ is expected to vary only by 0.3 mPa.s (compared to 0.9 mPa.s in human controls). The potential benefit for the animal is described in the next part. The RBC membrane viscosity also depends on temperature (tested by micropipette aspiration in animals [97] and human [255]).

In a mechanical context, mammalian RBCs are soft deformable capsules, similar to fluid drops made of a viscous liquid that is surrounded by an elastic membrane. Another important parameter is the cell surface charge, which can mediate repulsion between cells [69]. Decades ago Abramson [5] quantified the net RBC surface charge in 11 species. He found out that pig and rabbit showed the lowest, and human and dog the highest charge (15 billions of electrons on each human RBC surface, showing the "valence" of each cell). In a second step he normalised the electron quantity to the RBC surface area and showed that about 1% of the RBC surface was charged in each species. Sialic acid groups on membrane glycoproteins are the main carrier of this charge. Eylar [65] quantified the sialic acid molecules on RBC membranes in 6 species, and confirmed the previous information of Abramson. Pig RBCS were very different from human ones, with human showing the higher amount of sialic acid by a factor of 4. This resulted in a surface charge that was more than doubled in human red cells. The surface charge can be also calculated out of the electrophoretic mobility [15, 148, 213]. It was found that the electrophoretic mobility of RBCs in buffer was similar for rat, human, and horse, followed by other species like guinea pig and rabbit. Surface charge can be estimated from the adhesion of a cell to a cantilever, too. We found out that the horse showed the highest adhesion, closely followed by camel. Low adhesion was found for human RBCs in this setup [275]. Note that surface charge does not predict aggregability. Indeed, in a hydrophilic environment the surface charge on the cells will always be shielded by the surrounding molecules. Recently, it was shown that dextran and albumin could coat RBCs [125].

In a wide range of animals, species differences in WBV were shown decades ago [48, 105]. In nine mammalian species, it was shown [270] that blood viscosity (normalised by hematocrit) is rather uniform at high shear forces (rat excepted), while species differences exist at low shear forces with rat and equine blood showing the highest, and domestic ruminants the lowest viscosity values. WBV at low shear rates is frequently used to estimate RBC aggregation. However, this is not true for every circumstance. For example, in rats, the aggregation of RBCs measured by light transmission is minimal, but in rats, it is higher than in horses in standardised HCT samples (Figure 12.8. This is surprising since equine RBCs are famous for their pronounced RBC aggregability, which should by such theory result in the highest WBV among the species. Horse blood is thixotropic so its viscosity decreases with time after the stress has been imposed [270]. Rat blood did not show such a reduction of WBV with time. Therefore, the estimation of RBC aggregation out of a flow curve can be misleading, if the factors that stabilise or destabilise a suspension are not taken into account.

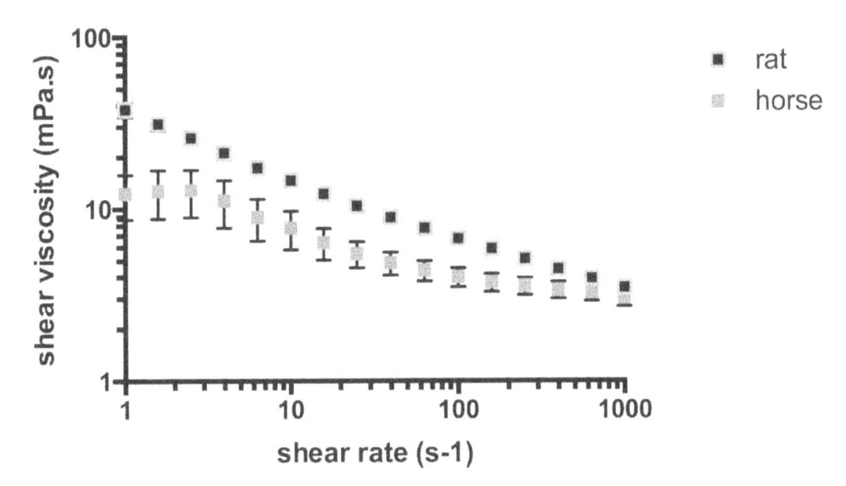

FIGURE 12.8
Whole blood viscosity of (a) horse (warmblood) versus (b) rat (wildtype inbred strain) at 40% HCT, tested at 37°C. Although rat does not show RBC aggregation, blood viscosity is higher than in horse at shear rates between 3 and 100s-1, although horse is well known as a species with profound RBC aggregability, whereas rat does not exhibit RBC aggregation.

The storage and loss moduli of animal blood at distinct HCT increments has also been compared from small amplitude oscillatory tests [274]. As expected, G' - G'' (loss modulus) - spectra of whole blood ($0.1 - 10$ rad.s^{-1}) in linear regime differed in species that show the highest differences in RBC aggregability. Difficulties in the identification of a linear viscoelastic range is described in Section 4 of this chapter.

G' and G'' are increasing functions of frequency in human, horse, sheep, pig, and rat. At HCT higher than 80%, blood becomes gelly and RBCs form a percolating network

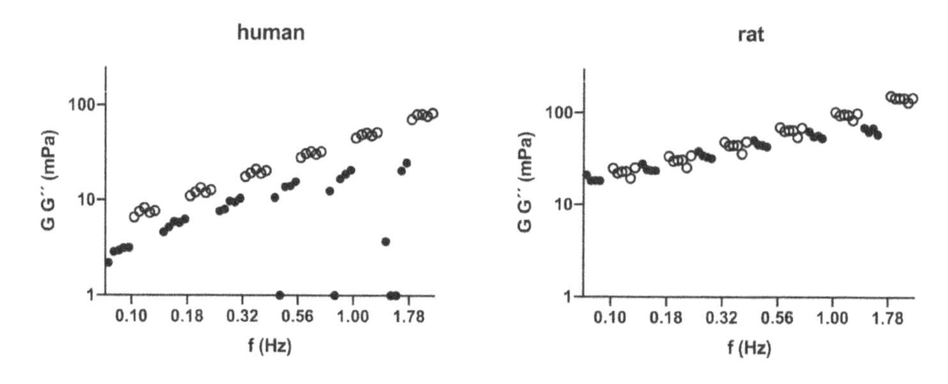

FIGURE 12.9
Frequency dependency of storage moduli (G', in mPa) and loss moduli (G", in mPa) obtained in linear mode of rat and human whole blood of 40% HCT, tested at 37°C. G' in rat blood is almost fourfold increased compared to human blood (see also Table 12.1). The stability of the blood suspension at quasi-static conditions in rat is indicated by the low loss factor at frequencies below 1Hz.

throughout the suspension. The exception is camel blood. In the blood of this extraordinary species G' was equal to 0 up to a HCT of 70%. Beyond this value, G' increased, but the loss factor was always higher than 1, which shows that a gel was not formed in camel blood, even at a high concentration. In the other species, G'' predominated over G' at HCT values below 70% indicating blood as a viscoelastic liquid even at pathologically high hematocrit, except in pig, where the blood appeared as a gel at hematocrit (> 60%) with $G'>G''$ [224]. Therefore pig and camel are opposing species concerning the elastic behavior of blood in quasi-static situations. Pig blood is used in the forensic discipline of bloodstain pattern analysis as a substitute for human blood during re-enactment studies due to its good availability from slaughtered animals. But the EDTA-blood of slaughtered animals is very different from the EDTA-blood that is withdrawn from conscious pigs. It has a higher viscosity and is more prone to form a gel, and this makes it obviously not suitable if the exact spatter must be simulated to human blood that was shed during a crime scene (Figure 12.9).

From comparative measurements, blood viscosity seems to result from the sum of scattered properties. For example, membrane strengthening in camels - in order to resist the osmotic variations in plasma during dehydration and rehydration [278, 279] - must be so important that the resulting blood suspension plays an underlying role (more information on camel blood is found below). Likewise, a reliable hemostasis seems to be mandatory for small animals with low total blood volume. The rate of clot formation is high in the rat [81] compared to human, which can be attributed to the high PLT count in rat blood. The blood suspension is stabilised and WBV is upshifted. In the following some mammalian species are described that have typical blood properties.

12.6 SPECIFIC ANIMAL SPECIES

12.6.1 Species with high RBC aggregability

12.6.1.1 Horse, Weddell seal

Horse shows high RBC aggregability within all animals tested so far ([9], refer to Table 12.1). Another species with very high red cell aggregability is the Weddell seal (Leptonychotes weddelli), but the difficulty of obtaining blood from free ranging animals limits the number of studies on Weddell seal blood. It is worth noting that this pinniped impresses with its high HCT (60% during dives), high MCV (150 fL) and high MCH (> 50pg), features needed for long and deep dives [43, 191, 203]. Images released to the public (PolarTREC webpage 2016) show the presence of frequent parachutes in a hemocytometer slit, indicating a flow behavior comparable to human cells. Information on RBC stiffness is lacking.

In the horse, the high aggregability occurs despite a poor cell elongational deformability [182, 220]. This is not intuitive, since rigidified human RBCs lose their ability to aggregate. But it must be questioned whether rigidity alone hinders aggregation, or whether the impaired surface properties due to the aldehyde treatment causes the loss of aggregability. It also demonstrates that the hemorheological parameters of animal species must be seen in a species-specific context. For the horse, the low RBC flexibility is a "normal" circumstance. If this physiological feature is associated with the deficiency of band 4.2 protein in RBC membranes [16], or with the lack of other proteins (band 6 and 8) [144] cannot be ruled out at present.

As a result of high RBC aggregation, sedimentation occurs in the viscometer gap during rheological experiments at very low shear rates, which reduces the stress response to the applied strain. A plateauing viscosity therefore does not indicate zero-viscosity. The movement of cells or cell clusters into the center between rotating parallel plates has also been described for blood [20]. If a steady state would be expected, the readings underestimate the true blood viscosity; rather it would display the viscosity value after completed phase separation. Such problems can occur when horse blood is measured in Couette, cone-plate, and double gap cylinder geometries. Once the cell fraction exceeds 70%, these time-dependent effects seem to be hindered, and typical flow curves for pseudoplastic fluids can be recorded.

But we also made another strange observation in association with the temperature at which the blood was measured. When a horse blood sample is exposed to rotating shear flow to test the stress response (at shear rates between 0.1 and 10 s^{-1}), the influence of temperature on viscosity depends on the temperature at which the first measurement was done. This phenomenon can be observed when series of experiments are performed at different temperatures. For instance, if the first experiment is performed at low temperature and subsequent experiments are performed at increasing temperatures in a stepwise manner, a sudden decrease of shear thinning behavior is observed between 22 and 27°C. In fact, the suspension suddenly has a Carreau-like behavior. This effect might indicate the onset of RBC aggregation at that temperature, associated with the beginning of phase separation, as explained above. If the same series of tests is started at the physiologic body temperature followed by stepwise decreases of temperature in equal intervals like before, this phase separating effect occurs during the first experiment and the subsequent curves are nicely upward shifted in parallel manner. The structure is frozen. When fibrinogen is added to human blood the same effect can be observed, but the behavior change occurs at lower temperatures. This observation indicates that temperature non-uniformly affects in-vitro measurements of blood viscosity when RBC aggregation occurs. It must be questioned whether the classical macroscopic viscometric or rheometric approaches can be used in such circumstance. At least, one must be aware of artifacts and carefully identify which settings can be considered valid and which protocols are only of limited use.

Horse RBCs easily form echinocytes through a loss of their intracellular volume [261] and through intracellular ATP depletion [37]. Serum electrolyte reduction and systemic diseases like colitis [79, 259] or more local signs of Clostridium perfringens infection [262] are the predominant clinical causes for echinocytosis. Interestingly, competitive races have triggered echinocytosis [66] as well, and circulating echinocytes were even persistent after the exercise.

The hemorheological test outcomes associated with equine echinocytosis are striking. First, a sample containing 30-50% echinocytes was filtered through 3μm pores. Filterability was not impaired [259], which might be the result of the lower cell volume of echinocytes. Second, treadmill-exercising horses received furosemide to reduce the pulmonary capillary pressure, which is a classical treatment in horses. The treatment aggravated echinocytosis, and the blood sample showed reduced RBC sedimentation rate, indicating partial loss of cell-cell interaction. However, the whole blood viscosity value was unchanged, which can be a matter of the method used [260].

Exercise-induced pulmonary hemorrhage is a frequent observation in racehorses that can shorten their competitive lifespan [184]. During high cardiac output and blood flow velocity, the pulmonary blood-air barrier can be ruptured and erythrocytes will be then able to invade the airways. Long-term pulmonary fibrosis can occur. Although the influence of blood viscosity on pulmonary vascular resistance is well recognized by the racehorse community (the HCT can dramatically increase during physical and psychical stress in the horse), the flow behavior of single equine echinocytes or cell clusters is not a focus of research. But we could visualize the flow and estimate the relevance of RBCs on vascular hindrance and subsequently on pulmonary blood pressure in the horse if we knew how equine RBCs behave in small vessels. In human, echinocytosis is accompanied by an increase of blood viscosity [195], an effect that could be expected in horse, as well. But in horse, RBC aggregability is so high that even spiculated cells can be incorporated into rouleaux [17] and the effect could be masked. The cell-free gliding layer might mask the effect as well. Further, a cluster containing a significant percentage of echinocytes might be even stabile. If this is the case, such a circulating cluster may not have a symmetrical shape and its behavior in flow will therefore be difficult to predict. If rigidified cell clusters play a role in exercise-induced pulmonary hemorrhage, the use of furosemide should be revisited.

12.6.2 Species with low RBC aggregability

12.6.2.1 Camel

Dromedary camel blood offers the unique possibility to study a nearly Newtonian blood suspension without the need to harden the red cells. Freshly withdrawn camelid blood appears brighter red than other mammalian blood [159,275]. Slow agitation of camel blood in an EDTA tube even makes the impression of a somehow "granular" fluid (personal observation). This observation with the naked eye is the first indicator of "something different" in camel blood.

Camelid RBCs are non-nucleated, ellipsoidal cells [1, 59, 155] with the long axis in the range of $7 - 8$ μm, and average small axes of $3.8 - 4.4$ μm [282], depending on the camelid species (Figure 12.10). Since the typical central dimple of mammalian RBCs is absent, camelid RBCs have a constant thickness of $1.0 - 1.1$ μm [173, 244, 282]. Their surface-to-volume ratio is only one third of human red cells [110]. Their small volume and the constant thickness of the cell allow high packing during centrifugation [243]. Therefore HCT values measured by automatic analyzers and PCV values measured by centrifugation are not the same: centrifugation displays a lower hematocrit [11].

One might think about the molecular structures that ensure an ellipsoidal cell shape and if this shape can be maintained when the cell passes through narrow networks. Generally, the symmetry and/or the function of the network must be different along the equator of the cell compared to its lateral faces on both sides. Differences also include the association of the cytoskeleton with the bilayer, especially on the cell poles. Until now we do not know the protein distribution and their function in local parts of the camel RBC membrane, but

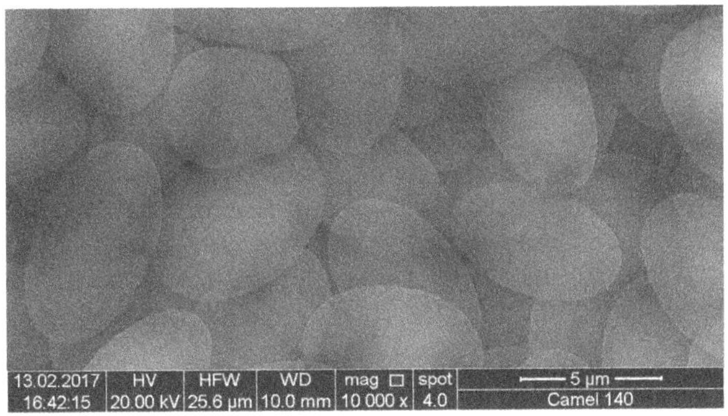

FIGURE 12.10
Intact ghosts from dromedary camel blood. A blood drop was vacuum dried for SEM investigation (10000x magnification). Cooperative work with Prof. H. Lichtenegger (University of Natural Resources and Life Sciences, Austria)

we know that the whole membrane is resistant to solubilisation by Triton X-100 and does not become spiculated at various treatments or conditions like ATP depletion [173]. But some treatments are indeed able to affect the cell, and crenation only occurred circularly around the equator, but not on the faces [219]. Therefore the distribution of structures connected to the cell membrane cannot be random. The structures also allow that the dromedary camel RBCs can become spherical [278] with an increase of intracellular volume. The possibility to stabilise the shape similarly to non-mammalian RBCs became a matter of discussion as soon as the marginal band of camel RBCs was first reported in 1966 [13]. In non-mammalian vertebrates the band stabilises the rim of the cell, and the number of microtubules is associated with the cell size. But since camelid RBCs are small, it was stated that a marginal band is unnecessary in this species to maintain the ellipsoid [82]. However, a band could stabilise the form nicely, thus explaining the extraordinary resistance of the cell against lysis. Cohen and coworkers [52] looked at camel RBC ghosts by TEM and detected a sub-membranous tubule system in a high number of cells. Under phase-contrast, the band appeared as a dense intracellular region that varied in size and thickness. But such structures were not found in every cell. Also Abdo and coworkers [1] saw the tubulin structure in many but not all dromedary camel RBCs. Other researchers did not find a marginal band and proposed that a band is only needed as a template for the development of the ellipsoidal shape and that mature camelid RBCs do not contain a marginal band [82]. But a high number of immature cells in previous observations would indicate significant blood loss or red cell destruction in many animals that gave their blood. Such events are not reported. It is more likely that there can be two populations of enucleated hemoglobin-carrying cells (both called "red cells") in blood samples from clinically healthy camels, and the relative amount of these cells is highly variable between the studied groups, but also among individual animals. This high variability is perhaps the most important result of all findings. One underlying reason for the difference in the studied groups might be the selection of animals. A phenotype can change over time in response to a confined environment (like camels only living in the Sahara or Arabian desert compared to those in a zoo or wildlife park in a moderate climate), or to a breeding program (e.g. for race performance or milk production). Alternatively, RBCs with a marginal band are indeed immature cells, released upon demand to quickly provide circulating oxygen carriers, and camels demonstrate clinical signs of diseases only when they are severely ill. A conjecture can be proposed: camel is an intrigue species. It uses several mechanisms when it has to deal with excessive heat exposure and low water intake [78]. Most important, several properties (e.g. metabolic rate, heterothermy, selective brain cooling, RBC lifetime) only come to play when the animal is dehydrated [36,60,211,277]. They vanish as soon as the animal has unrestricted access to water and are restored again when the camel is stressed. It is not astounding that a large mammal able to survive in harsh regions while no shelter can protect it from the sun uses every option to be prepared for "hard times". A marginal band could support animals freely ranging in the desert, whereas it may be unnecessary in animals with unlimited access to fresh water. The variability might therefore indicate the individual history of the animal, and the number of cells with a band within the same camel might vary with time. If a dromedary camel frequently experiences dehydration it could have a higher population of red cells with the band because microtubules favour intracellular transport processes. This would support intracellular water storage during rehydration. The good diffusional permeability to water [24] and the high hydrophilicity of the hemoglobin molecule also favour the process. In parallel, the tubules would stabilise the membrane against osmotic stress. Dromedary camels can restore their body weight after fasting within very short time [210]. We might assume that this is also the time needed to "refill" the cells with water. The instantaneous reduction in plasma osmolality generates a very high stress to the RBC membrane, and a tubule system would effectively protect RBCs. On the other hand, ad libitum watered camels may not need the band. A fortification of their RBC membrane might even hinder blood flow by the impediment of red cell deformation. This could be important if animals regularly perform exercise. It would be very interesting to see if animals, whose red cells do not possess a marginal band would gradually develop the second population of red cells (those with the band) if they were periodically depleted of water and vice versa.

Whatever the reason for the ellipsoidal shape, cell elongation cannot be measured by rotational ektacytometry. Indeed, instead of being elongated with increasing shear stresses, the elongational indices decrease above a shear stress threshold. Micropipette aspiration studies have shown that camelid RBC membranes are stiff [253]. This could be the result of the high band 3 protein concentration in the membrane (threefold compared to human RBCs in relation to the cell surface) [59, 110, 251], the limited rotation of the cytoplasmic domain of band 3 [111, 192] and its tight connection to ankyrin [147]. By probing camel RBCs with a cantilever no membrane tethers could be extracted (personal observation), which indicates that the enclosed bilayer area between the membrane proteins is low and cohesion between membrane elements is good. Camels' red cells have a significant number of sialic acid groups on their surface [110], which explains the significant adhesion that we observed, but appears contradictory in concern of its lack of aggregation [11].

In comparison to the similarly sized cells of sheep, the dromedary camel RBC (MCV: $25-30$ fL) carries more hemoglobin (MCH: $13-15$ pg), which results in a substantial MCHC of $45-55$ gdL^{-1} ([201], as well as reference values of CVR laboratory, Dubai, UAE). However, we have to consider that hematological values can vary in respect to the above-mentioned influences (e.g. environment and breeding program). Plasma viscosity at 37°C is very low; in fact it is only little higher than water (1.04(1.00/1.07) mPa.s), a result of the low plasma total protein concentration ($55-70$ gL^{-1}). Since the viscosity of a hemoglobin solution rises progressively with the hemoglobin concentration [264], the viscosity ratio across the RBC membrane must be significant. The cell free layer width is related to RBC aggregability [31, 112, 115] and deformability [285], but RBC aggregation does not occur and camelid RBCs are stiff [253]. How do the ellipsoidal cells with high stiffness and high viscosity contrast behave in flow?

Since dromedary camel blood is almost Newtonian, no viscosity decrease at high shear rates can occur. Intravascular shear rates typically increase at high cardiac output. One clinical problem in racing camels is the occurrence of anemia associated with a decrease of the plasma iron concentration [235]. The question arises whether the observed reduction in

RBC count in racing dromedaries is a physiologic effect to facilitate peripheral perfusion or to protect the vascular wall from energy input due to poor alignment of RBCs in the flow. Viscosity ratio and stiffness exclude typical cell motions [2–4, 54, 67, 71, 122, 138]. It is very unlikely that camels' RBCs fold into parachute shapes when they pass through capillaries. If camel RBCs do not align or orient in the flow, any increase of hematocrit affects the cardiac workload more severely than in other mammals. Microfluidic experiments with camelid blood that would help to understand the behavior of these cells in the microvasculature are highly warranted. Possibly, the observed anemia in racing camels reflects a mechanism to relieve the microvasculature from effects generated by a high RBC count.

On the other hand, no viscosity increase at blood stagnation zones will occur if blood is Newtonian. The lack of 2D clusters will keep the viscosity low in the veins. Extrapolating this finding to patho-physiological circumstances, it predicts that blood is always purely viscous, even when the flow approaches zero. This is a clear benefit for the animal and could explain the working ability of dromedary camels even when they are dehydrated.

12.6.2.2 Domestic ruminants

RBC volume of cow is comparable to horse (40 – 60 fL), but domestic sheep and goat have the smallest cells among domestic mammals (goat: 15 – 25 fL; sheep: 25 – 40 fL). Caprine cells are even smaller than camelid red cells. Small cells are more spherical, which restrains their aggregation (sphericity index obtained from the cell volume and the surface area in sheep: 0.70 and in human: 0.58) [238]. The contacting surface area of two small RBCs may be too small to support rouleaux, e.g., sheep RBCs show only minute aggregation indices [270]. Cow cells show slightly higher aggregation indices, which could be the result of the larger cells, but also a matter of the flexible assembly of the two domains of band 3 protein [104]. Oxygen affinity of hemoglobin is lower in ruminants than in horse, rat, and human [38].

Goat RBCs spiculate when the animals are anemic. This observation was explained by an imbalance in hemoglobin synthesis, resulting in HbC-associated poikilocytosis [137]. The presence of acanthocytes is surprising because goat RBC membranes are rich in glycoproteins [216] with the gp155 being tightly bound to band 3 protein and ankyrin [102], which should stabilise the connection between the cytoskeleton and the membrane. Goat platelets are as large as red cells, and their number per unit blood volume is one fourth the RBC number. The presence of small, stiff cells with different shapes affects the behavior of the blood suspension. There is very little knowledge from viscometry, but it was shown in an early study that HCT standardised samples (45%) showed low shear thinning at an overall low viscosity in goat compared to human, elephant, and dog. But when the cells became packed (at 85% HCT), the viscosity was much higher in the goat compared to the other species and shear thinning effects increased ([237]). This shows that the resulting material is less congested when packed RBCs are flexible.

Sheep RBCs carry less hemoglobin and their plasma viscosity is higher compared to dromedary camel [68]. Viscosity ratio is therefore not that high, which is underlined by the non-Newtonian behavior of sheep blood. However, shear thinning is low. Elongation and aggregation indices of sheep belong to the lowest that have been obtained from mammals [77, 182, 207, 270]. This could result at least in part from the good linkage between cytoskeleton and bilayer by the high amount of ankyrin [144]. A small number of sheep blood samples was adjusted to different HCT values and tested for their viscoelasticity in sinusoidal shear flow [274]. At 40% HCT no linear viscoelastic range could be detected (Figure 12.6). However, when the HCT was increased to 60%, the suspension stabilised and yielded; the yield stress was much higher than in horse and human at equal HCT. Like in the goat, viscosity is low at native HCT, but a HCT increase in these animals elevates the viscosity disproportionately [229, 237] and is therefore not warranted from a clinical point of view.

12.6.2.3 Rodent species

Rat Rats are often used for biomedical research and they are the rodent species of choice in toxicological tests within the OECD normative, since it is easier to withdraw blood quantities from rat than from mouse. Phenotypic differences exist between wildtype rat strains when considering RBC elongation [273]. Mutant strains can express a higher impairment of blood rheology (e.g. higher RBC aggregation indices, higher whole blood and plasma viscosity) than classical wildtype strains, even when they are clinically healthy. It is logical that genetically modified animals would show changes in blood rheology when the membrane protein profile is affected. In specific cases it may be necessary to analyse the hemorheological profile prior to the use of new or unknown rat strains in research.

Cats and dogs were used decades ago, but rodents are the experimental models of choice in these days to study the microcirculation. Especially the rat has been used to investigate the influence of blood viscosity and RBC properties on the cardiovascular function [18, 19, 29–31, 39, 103, 112, 114, 163, 174, 222]. Occasionally, guinea pigs and hamsters are used as well [27, 280, 281]. Although size and handling of an animal, as well as legal and ethical issues, are important for the selection of the animal - and many valuable species must be excluded for that reason - it cannot be omitted to note that the rheology of native rat blood is considerably different from human blood. Rat blood shows the highest apparent viscosity across a number of domestic animal species including human, but lacks a substantial RBC aggregation. In order to mimic human circumstances, aggregation must be initiated by the addition of pro-aggregating dextrans or poloxamer (e.g. Pluronic F98). The low molecular weight dextran 70 is not effective in the rat, but protocols to use other standard media have been described [15]. By having the minute aggregability in mind, the strong pseudoplastic behavior of the native bulk suspension is astounding. Even at long times the suspension is stabilised (Figure 12.9). G'-values at the tan δ-minimum of the suspension (obtained in linear mode) are fourfold higher than human values (see Table 12.1). Also at higher frequencies the suspension is stable, which may be of some clinical importance regarding the high physiological heart rate of rats. Plasma total protein concentration, fibrinogen concentration, and plasma cholesterol are comparable to human reference values [273], and may not be considered causative for such a suspension stability through an increase of plasma viscosity. However, platelet count is high in rats, which enables them to play a role as bulk elements. Due to their small volume (rat PLTs are about six times smaller than rat RBCs), platelets enhance the viscosity of the suspending phase, and this can stabilise the suspension.

Mouse Since the close concordance between the mouse and human proteome has been identified [178, 179], transgenic and knockout mouse models became a convenient model to study hematological diseases [153]. Several mouse models are described, which lack erythroid membrane proteins. Among them are the following (the list does not claim to include all current models): β-adducin [158, 187], α-adducin [179], band 3 [181], protein 4.1R [205], aquaporin1 [283], ankyrin-1 [101, 193], α-spectrin [248, 249], β-spectrin [33], and tropomodulin-1 [156]. β-adducin knockout mice RBC showed spherocytosis in association with decreased deformability, increased osmotic fragility, decreased MCV, increased MCHC, and decreased cation content [80]. Experiments comparing normal mouse RBCs to α-adducin-deficient RBCs suggest that roughly 33% of the band 3 molecules were immobilised by adducins, which favour a larger complex. The remaining band 3 dimers (30%) should be diffusing freely in the lipid bilayer [179]. Erythrocytes from 4.1R$^{-/-}$ mice [205] showed a large diminution of actin accompanied by a loss of cytoskeletal lattice structure, with formation of bare areas of membrane. Band 3$^{-/-}$ deficient mice were found to be completely devoid of glycophorin A [91]. The authors also concluded a putative "chaperone-like" function of band 3. The role of band 3 for a stabile biogenesis of the membrane cytoskeleton was confirmed by Southgate [223]. The AQP1/UT-B double knockout mouse was used as a model to study RBC water permeability [283]. Peters et al. [180] described the relevance of the band 3-spectrin-ankyrin-protein 4.2 interactions in the pathogenesis of spherocytosis in the mouse. But also the normal mouse has been investigated as a potential model for band 3 diffusion [117]. The site of binding of the cytoplasmatic domain of band 3 to ankyrin was identified in mice homozygous for the β-hairpin loop on band 3 [227]. The development of the membrane function during erythropoiesis was tested in RBCs of wildtype mice [256]. The importance of cholesterol in the arrest of erythroid development was tested in the $SR - BI$-/-/apoE -/- mouse [99]. Sepsis decreased RBC deformability in glucose-6-phosphate dehydrogenase-deficient mutant mice more effectively than in normal mice [225]. Hem1$^{-/-}$ mice showed defects in the spectrin-actin assembly. Loss of Hem-1 (hematopoietic protein 1) resulted in abnormal F-actin condensation, altered representation and phosphorylation of junctional complex proteins, and generated modified erythrocyte morphology [44]. Mice were described to have a preferred incidence for thrombosis when deficiencies in α-spectrin are present. This effect was related to the presence of adhesion molecules on RBCs [249, 250]. Mouse RBC ageing was investigated [157], and the ultrastructure of the mouse red cell cytoskeleton was investigated by cryo-electron tomography. The thickness was between 54 and 110 nm and the spectrin meshwork after removal of the bilayer showed an edge length of approximately 46 nm [164]. In contrast, the plasmalemma of dromedary camel RBC membrane was tested as 6.25 nm [1]. A microfluidic cytometer was developed for mouse blood [106], and wildtype as well as transgenic mice were even sent to space, in order to study the change of RBC membrane phospholipids during a space flight [196].

Despite the presence of these models, there is very limited knowledge on the flow behavior of genetically modified RBCs. This is presumably due to the little blood volume that can be harvested from mice (about 0.5 mL, associated with death of the animal), and the difficulty to get severely modified (e.g., homozygous) mice into adulthood. A blood sample from newborns to test the flow behavior is even more difficult to obtain. However, since many models can be accessed commercially or are preserved at academic institutions, pooled blood samples from the litter might offer enough volume for microfluidic tests.

12.7 ADAPTATION TO ENVIRONMENTAL STRESSORS AND LIFESTYLE

Living in cold climates modifies the bilayer composition, obviously to provide membrane fluidity at the low temperature. A change of the fatty acid profile can be generated even in a short time scale. RBC membranes obtained from a rat group that swam in cold water each day showed an increase of unsaturated fatty acids within several weeks. The corresponding elongation indices were higher compared to a group of rats swimming in warm water [233]. The same effect was found in fish swimming in cold or warm water [83, 130]. Trout RBCs showed a reduced temperature coefficient of the anion transport system compared to mammalian RBC ([197]). This helps to maintain the cellular function at changing water temperatures. There was also reduced temperature dependence in considering whole blood viscosity in Antarctic birds and bowhead whale. The authors explain that a reasonable blood viscosity might be maintained in the cold-exposed parts of a body if blood viscosity increases only sparsely with temperature lowering [62, 87].

For heterothermic camels adapted to hot climates, smaller fluctuations in blood viscosity in association with diurnal changes of body temperature reduce the need for the endothelium to periodically modulate shear stresses [268]. This could reflect another mechanism for conserving energy, which is vital for free-ranging dromedary camels to avoid heat generating processes [211]. Animals living in the harsh deserts must possess mechanisms to withstand fluctuations in plasma osmolarity. Membrane proteins play an important role in protecting the cell against osmotic stress, obviously on the basis of the band 3 - ankyrin complex: ankyrin is quantitatively increased in the Oryx antelope, and camelids have a strong binding of ankyrin to band 3 [110, 111, 173]. On the other hand every membrane reinforcement will affect red cell elongation. This was observed for the scimitar oryx antelope [268].

Adaptation to high external pressure and limited oxygen supply in diving animals may include modifications of the coagulation system as well, since the diving response [61] reduces the perfusion of all organs except brain and heart, and it is unclear what keeps the blood fluid in the other tissues during a 20 minute dive. The relevance of RBC aggregation for diving animals is not clarified, as well. Seals show very diverse RBC aggregability. Whereas ringed seal RBCs do not aggregate, the cells of Weddell seal show strong aggregability [43, 149]. Diving depths and aerobic dive limits do not correlate with RBC aggregability [43, 94, 149]. Hydrostatic pressure itself raises RBC aggregation [47] On the other hand, RBCs of human volunteers, who frequently dived to a depth of 300 feet showed higher aggregation indices than a control group [232]. Comparison of RBC aggregability between free-ranging seals and a group of captive animals of the same species, showed RBC aggregability was higher in the free-ranging animals [267]. Surprisingly, a change of lifestyle can be enough to change an intrinsic RBC property in regard to aggregability.

Throughout this chapter we wondered whether blood viscosity was a controlled parameter or the result of scattered properties. Blood from diving mammals and camels are examples that support the latter hypothesis. For these animals, the priority was to fulfil a specific a cellular-based requirement, and the viscosity thus results. For example, the priority of RBCs of diving animals is to be large, to contain as much hemoglobin as possible in order to store oxygen. Their MCHC is high but still in a "reasonable" range so that the cytoplasmic viscosity is not high enough to change cell deformability (see Table 12.2). Hematocrit at rest was higher than its theoretical "optimal" value [269]. But securing oxygen supply thanks to a high red cell count is of highest priority than issues in a high blood viscosity that results from high HCT. In the dromedary camels the ability of RBCs to withstand huge fluctuations of osmolality appears to be of higher priority than the ability of the suspension to have a shear thinning behavior.

12.8 A BOTTOM-UP APPROACH TO EXPLORE ANIMAL BLOOD SUSPENSIONS

Generally, systematic approaches must be carefully designed since it is impossible to test every animal species. Although the selection of species includes concerns regarding availability of animals and feasibility of blood sampling, there should be a clear rationale for choosing a given species. When size and shape are the criteria, it might be useful to include horse and cow, because these species show contrasting RBC aggregability for the same RBC size. Among the elliptic cells, camel could serve as reference for all non-mammalian species, since its RBCs are enucleated and possess a mammalian-type of cytoskeleton. When membrane proteins are the criterion, the band 3 - ankyrin complex or the protein 4.1 - actin complex might be a matter of research. In this context, the reader is also invited to read a review of the spectrin-ankyrin-4.1-adducin membrane skeleton of vertebrates [12]. The influence of membrane proteins on the quality of the suspension will be best analysed with the available genetically modified mouse models. But it is worth noting that RBCs from wildtype mouse strains do not form rouleaux. Therefore it is unclear if these models can shed light on the membrane proteins involved in RBC aggregability. Species selection might also arise from more "applied" questions, e.g. to compare exercising animals, or to identify the best working animal with respect to blood fluidity, to mention only a few possibilities.

We have several tools in hand to analyse blood suspensions. One is to measure RBC elastic moduli. The full toolbox of single cell spectroscopy techniques can be found in a recent textbook [4]. Although the interest in animal membrane elasticity started decades ago ([123, 253, 254], see Section 12.2) this research is still in its infancy. Recently, RBC stiffness was tested in three non-mammalian species by contact mode AFM [231]. The highest stiffness was found in an amphibian, and not in the fish. Interestingly, the largest RBCs and capillaries are also found in amphibians [10, 202]. The application of beads for tweezer experiments may be technically difficult for small RBCs, e.g. for the ovine and caprine families. It may be necessary to test two cell axes to investigate ellipsoidal cells. Most likely, test protocols to probe single red cells must be adapted for each species individually.

The in-vitro behavior of individual red blood cells and whole blood suspensions of animal species in microflows using microfluidic tools is another very interesting but not yet very developed area of research. Surprising behaviors may be observed in large nucleated flowing cells. The ellipsoidal shape and the presence of a nucleus certainly affect the alignment of the cells and their transport in the vascular network. The use of microfluidics requires knowledge of the capillary dimensions for each studied species. There is a significant diversity of vascular geometry within one individual, but also between species [40,42,58,103,118,127, 128,185,188,230]. Frog and some, but not all, fishes show much larger capillary diameters than species of the avian and mammalian class [25, 145, 146, 202, 230]. It is interesting to note that the average capillary diameter of species with nucleated RBCs is comparable to the mammalian one. For example, the average capillary diameter in finches is below 4 μm whereas the ellipsoidal RBCs of these species are in the range of 12×6 μm [146, 206]. Blood capillaries of sharks range between 8 and 12 μm [25], whereas the cell volume of their RBCs is often far beyond 600 fL [45]. Interesting features can be found in nature. The size of the RBC of Amphiuma tridactilum is 70 μm long and 40 μm wide and represents the largest red blood cell identified so far [10]. Specific Antarctic icefishes have few or almost no circulating red cells at all [263]. In addition, not all fishes have nucleated RBCs. Maurolicus mulleri has small ellipsoidal 6×3 μm sized red cells without a nucleus [276] and this species is thus an exception among fishes. Salamander species (e.g. Batrachoseps attenuates) have also non-nucleated red cells (called erythroplastids) [246], and some RBCs of frogs can also be non-nucleated [10]. Rheometry can provide some information on shear thinning, on the development of sudden normal forces in the gap (especially at lower gap width), and the presence of shear elasticity in a sample. The low viscosity (e.g. below 3 mPa.s at 10 s^{-1} in chicken at 30% HCT), the very weak pseudoplastic behavior of camel blood, and the lack of a G'-value of the bloods at 1 mPa stress indicate low cohesion of the suspension. This indicates low ability of two adjacent RBCs to interfere with each other.

Knockout animal models can help to identify proteins associated with membrane stabilization/destabilisation. In addition to mouse models, zebrafish has been cloned to investigate defects of RBC protein 4.1 and β-spectrin [129, 215]. Protein data banks can give information about the amino acid sequences of membrane proteins, e.g. α-spectrin [204]. There is knowledge on the qualitative appearance of erythroid membrane proteins from SDS-PAGE (see Figure 12.5). In the non-mammalian vertebrates a second protein-based filamentous system, the marginal band that circumscribes the cell membrane in the horizontal plane [50,51], and intermediate filaments [85], stabilise the membrane. Information on erythroid tubulin among the species has raised some interest in the past [23]. Its structure appears to be rather preserved [136] and red cell tubulin has been described to be homogenous unlike the tubulins in other somatic cells [199]. Polymerisation and self-assembly of microtubules in a circulating red cell can be one mechanism to stabilise the shape when a nucleus must be co-transported. The nucleus of chicken RBCs can be found in the center of the cell only at "rest", e.g. when the cell adheres to the glass slip in a blood smear, but it is able to move when the cell is indented (personal observation), or when the cell enters a glass capillary [76]. Almost nothing is known about the dynamics of nucleated RBCs in shear flow, but it appears logical that the alignment and orientation of such a cell in a vessel will be alleviated if the nucleus is stabilised to achieve a minimum of flow disturbance of the continuous plasma phase.

12.9 FUTURE CHALLENGES

This chapter presents some extraordinary blood suspensions with behaviors that would be very pathologic for human blood. It gives also an overview of the most important properties that determine a particular behavior, and to assist the reader in selecting the appropriate research model. Native animal whole blood of healthy individuals offers properties for fairly every situation by its rheological "fingerprint": e.g. high RBC aggregability and thixotropy (horse), no aggregability but high cohesion (rat), low RBC aggregability and low cohesion (sheep), almost Newtonian behavior (camel), and blood with RBCs with specific changes in the protein composition from genetic modified strains. Comparative studies can be useful because they allow testing different circumstances in fairly coherent study groups. Sometimes it is difficult to find such homogenous groups in cohorts of human patients due to the presence of co-morbidities.

Maybe it is imprudent or even unnecessary to try to find out the plan of nature behind the sometimes-huge differences in blood suspensions of animal species within one single working life span. Maybe it is easier to say that the picture is very complex and the diversity beautiful. But it is always worth spending more time to identify the physical and biochemical processes in RBC membranes that generate aggregability in the view of potential applications [90] . The physiological degree of RBC aggregability has to be known since it impacts flow disturbances, flow profile, the wall shear, and microvascular hematocrit. The dynamic structuration of blood flow is able to temporarily clog vessels and increase vascular resistance. Most likely, the whole vascular system of a species (including human), whether presenting aggregability or not, is adjusted to its species-specific aggregability.

Finally, the reader should not forget that cells are living systems and RBC is an active object[26, 28]. Energy-consuming processes that affect deformability [143] could also be involved in physicochemical modifications on the RBC surface, thus affecting the process of aggregation. Active processes may also be involved in the regulation of the amazing camel RBC properties. Understanding of such mechanisms can be helpful if desertification of land proceeds worldwide.

TABLE 12.1

Viscometric data and RBC aggregation in husbandry and companion mammalian species. Apparent viscosity of blood plasma and whole blood, as well as shear moduli of whole blood (G′, G″) obtained by rheometry, and aggregation indices (M0, M1) obtained by light transmission. "Own results" (OR) show HCT adjusted (40%) samples; the HCT may vary in other references. Viscosity is displayed at low shear rate (WBV LS) and at high shear rate (WBV HS). G′ and G″ were obtained in linear mode. Human data are included at the bottom of the list as a reference. Since the used shear rates are not uniform, they are indicated below each viscosity value. All viscosities were measured at 37°*C*, except some plasma samples that were tested at ambient temperature (indexed by *). Aggregation indices were measured at ambient temperature throughout. The column "n" indicates the sample size.

Species	n	PV	WBV LS (mPa.s)	WBV HS (mPa.s)	G′ at 1Hz (mPa)	G″ at 1Hz (mPa)	M0	M1	Ref
Horses	10	1.16 (1.11/1.32)	11.1 (10.9/12.2) at 1 s^{-1}	2.98 (2.83/3.05) (at 1000 s^{-1})	18.90 (15.7/20.35)	59.10 (57.10/59.60)	-	-	OR
	20	1.65 ± 0.08*	47.08 ± 17.42 (at 0.7 s^{-1})	6.58 ± 0.78 (at 94 s^{-1})	-	-	16.5 ± 2.8	55.9 ± 13.4	[228]
	32	-	8.33 ± 2.6 (at 5.7 s^{-1})	4.40 ± 0.38 (at 115 s^{-1})	-	-	-	-	[9]
	42	1.39 ± 0.47	9.28 ± 1.73 at 5.7 s^{-1}	3.94 ± 0.57 (at 115 s^{-1})	-	-	-	-	[221]
	19	1.10 ± 0.05	-	4.70 ± 0.49 (at 180 s^{-1})	-	-	-	-	[56]
	6	1.49 ± 0.06	-	-	-	-	-	-	[22]
	4	1.53 ± 0.24	88 ± 6 (at 0.28 s^{-1})	5.29 ± 0.42 (at 128 s^{-1})	-	-	-	-	[105]
	40	1.66 (1.57/1.84)*	38.17 (31.66/42.54) (at 0.7 s^{-1})	5.17 (4.79/5.51) (at 94 s^{-1})	-	-	13.2 (11.7/14.5)	57.6 (44.0/67.7)	[270]
	8	-	-	-	-	-	∼36	∼60	[17]
	6	-	-	-	-	-	∼37	∼60	[16]
	-	-	-	-	-	-	41.33 ± 0.95	-	[15]
	-	55.3 at 0.28 s^{-1}	-	-	-	-	-	-	[186]
Sheep	10	1.09 (1.04/1.12)	5.51 (5.11/6.37) at 1 s^{-1}	3.12 (3.05/3.32) (at 1000 s^{-1})	0.83 (0.44/1.28)	12.15 (11.8/12.3)	0 (0/0)	8.5 (5.6/7.8)	OR
	6	1.22	-	-	-	-	-	-	[48]
	5	-	5.8 ± 0.3 (at 11.5 s^{-1})	3.8 ± 0.2 (at 70 s^{-1})	-	-	-	-	[57]
	5	1.35 ± 0.05	8.2 ± 2.9 (at 0.7 s^{-1})	4.2 ± 0.3 (at 128 s^{-1})	-	-	-	-	[105]
	40	1.49 (1.44/1.57)*	6.59 (5.85/7.19) (at 0.7 s^{-1})	4.37 (4.09/4.68) (at 94 s^{-1})	-	-	0 (0/0)	6.6 (3.5/8.8)	[270]
	-	-	9.1 (at 0.28 s^{-1})	-	-	-	-	-	[186]
Goat	6	1.28	-	-	-	-	-	-	[48]
	5	1.61 ± 0.6	7.9 ± 1.0 (at 0.28 s^{-1})	5.52 ± 0.5 (at 128 s^{-1})	-	-	-	-	[105]
Cattle, Cow	10	1.36 (1.34/1.36)	6.12 (5.81/6.74) at 10 s^{-1}	3.73 (3.71/3.88) (at 1000 s^{-1})	0.35 (0.003/2.05)	34.06 (32.41/36.40)	-	-	OR
	6	1.95 ± 0.05	-	-	-	-	-	-	[22]
	7	-	8.51 ± 0.5 (at 0.28 s^{-1})	4.95 ± 0.13 (at 128 s^{-1})	-	-	-	-	[105]
	40	1.72 (1.63/1.94)*	6.55 (6.01/7.35) (at 0.7 s^{-1})	4.79 (4.48/5.14) (at 94 s^{-1})	-	-	1.7 (1.1/2.2)	11.6 (8.6/14.4)	[270]
	-	-	11.5 (at 0.28 s^{-1})	-	-	-	-	-	[186]
Camel	10	1.04 (1.00/1.06)	3.74 (3.59/3.95) (at 10 s^{-1})	3.44 (3.39/3.61) (at 1000 s^{-1})	0 (0/0)	27.17 (26.69/28.55)	-	-	OR

Continued on next page

TABLE 12.1

– continued from previous page

Species	n	PV	WBV LS (mPa.s)	WBV HS (mPa.s)	G′ at 1Hz (mPa)	G″ at 1Hz (mPa)	M0	M1	Ref
	11	-	6.3 (4.6/9.6) (at 11.6 s^{-1})	2.7 (2.2/3.5) (at 1000 s^{-1})	-	-	0 (0/0)	0.7 (0.1/0.8)	[11]
Pig	10	1.16 (1.10/1.21)	21.00 (18.62/22.65) (at 1 s^{-1})	2.95 (2.86/3.11) (at 1000 s^{-1})	41.9 (37.20/46.55)	62.85 (57.62/65.72)	-	-	[224]
	23	-	-	-	-	-	2.66 ± 1.15	16.77 ± 4.0	[116]
	5	-	8.9 ± 1.1 (at 11.5 s^{-1})	4.2 ± 0.8 (at 70 s^{-1})	-	-	-	-	[57]
	40	1.58 (1.49/1.69)*	24.69 (17.2/31.03) (at 0.7 s^{-1})	4.94 (4.33/5.50) (at 94 s^{-1})	-	-	3.7 (2.2/4.6)	26.2 (20.9/31.4)	[270]
Dog	13	-	15.89 ± 4.16 (at 1 s^{-1})	3.72 ± 0.40 (at 1280 s^{-1})	-	-	-	-	[98]
	40	1.61 (1.57/1.69)*	22.88 (18.82/27.75) (at 0.7 s^{-1})	14.58 (12.36/17.47) (at 94 s^{-1})	-	-	3 (2.7/4.0)	23.5 (21.1/27.9)	[270]
	121	1.30 (1.21/1.41)	33 (21.2/44) (at 0.22 s^{-1})	4.2 (3.6/4.35) (at 128 s^{-1})	-	-	-	-	[34]
	6	1.13	-	-	-	-	-	-	[48]
	-	-	42.6 (at 0.28 s^{-1})	-	-	-	-	-	[186]
Beagle dog	15	-	-	-	-	-	3.47 ± 1.85	15.95 ± 5.93	[186]
Cat	40	1.71 (1.62/1.78)*	30.19 (26.89/33.97) (at 0.7 s^{-1})	4.44 (4.18/4.67) (at 94 s^{-1})	-	-	5.2 (4.7/6.3)	29.6 (22.8/40.1)	[270]
Rabbit	3	1.11 ± 0.07	22 ± 4 (at 0.28 s^{-1})	4.5 ± 0.16 (at 128 s^{-1})	-	-	-	-	[105]
	40	1.30 (1.24/1.41)*	8.31 (6.16/10.24) (at 0.7 s^{-1})	4.04 (3.75/4.37) (at 94 s^{-1})	-	-	0 (0/0)	13.3 (8.2/17.2)	[270]
HL/LE rat	7	1.08 (1.07/1.10)	37.4 (36.70/39/70) (at 1 s^{-1})	3.49 (3.47/3.50) (at 1000 s^{-1})	54.35 (44.42/56.47)	94.65 (86.45/95.87)	-	-	OR
Lewis rat	20	1.22 (1.19/1.24)*	30.81 (23.45/35.93) (at 0.7 s^{-1})	5.84 (5.18/6.17) (at 94 s^{-1})	-	-	0.9 (0.5/2.2)	5.6 (3.5/8.2)	[273]
Sprague-Dawley rat	15	-	-	-	-	-	1.41 ±0.71	5.51 ± 2.25	[116]
	10	1.03 ± 0.04	-	2.78 ± 0.24	-	-	-	-	[236]
ZDF rat	10	1.48 (1.39/1.59)*	42.30 (37.05/48.68) (at 0.7 s^{-1})	7.56 (7.06/8.13) (at 94 s^{-1})	-	-	3.0 (1.5/5.0)	13 (9.5/15.0)	[273]
Rat	40	1.59 (1.55/1.65)*	35.40 (26.33/40.90) (at 0.7 s^{-1})	6.29 (5.88/6.92) (at 94 s^{-1})	-	-	0.7 (0.3/1.2)	2.4 (1.5/4.2)	[270]
	6	-	-	-	-	-	~1	~5	[16]
Mouse	16	-	-	-	-	-	3.19 ±1.72	6.55 ±3.44	[116]
	40	1.31 (1.29/1.34)* (at 0.7 s^{-1})	13.37 (10.69/16.57) (at 94 s^{-1})	4.88 (4.51/5.34)	-	-	0.2 (0.1/0.3)	0.6 (0.5/0.9)	[270]
Guinea pig	-	-	-	-	-	-	-	1.56 *pm* 0.78	[15]
Human	5	1.27 (1.24/1.28)	11.80 (11.45/12.60) (at 1 s^{-1})	2.95 (2.90/3.11) (at 1000 s^{-1})	14.65 (13.57/15.72)	49.40 (47.45/50.35)	5.8 ±0.5	28.5 ±6.6	OR
	5	-	6.5 ± 0.7 (at 11.5 s^{-1})	4.1 ± 0.3 (at 70 s^{-1})	-	-	-	-	[57]
	25	1.34 ± 0.08	44.6 ± 5.5 (at 0.28 s^{-1})	4.7 ± 0.3 (at 128 s^{-1})	-	-	-	-	[105]

TABLE 12.2

Mechanical characterization of animal RBCs . Average elongation (EI) of mammalian RBCs out of a laser diffraction pattern ($EI = (L - W)(L + W)$) of a RBC suspension sheared in a Couette or plate-plate geometry. Maximum elongation index of RBCs (EI_{max}) and shear stress to obtain half of the maximum elongation ($SS_{1/2}$, in Pa) was calculated out of the EI/SS-relationship [21]. The EI/SS curves are shown in Figure 12.5. Shear elastic modulus (in dyn cm^{-1}) was obtained by micropipette aspiration [253, 254]. Nine to 28 RBCs were tested from each individual sample."OR" indicates "Own Results"

Species	n	Shear modulus (dyn/cm^{-1})	DI_{max} ekta--cytometer	EI_{max} (Rheodyn SSD)	$SS_{1/2}$ (Rheodyn SSD)	EI_{max} (LORCA)	$SS_{1/2}$ (LORCA)	Ref
Horse	34	-	-	21.48*	12.68*	0.679 (0.59/0.88)	12.12 (10.2/15.25)	[182] OR
Sheep	36	-	-	5.65 (5.40/6.90)*	0.86 (0.69/1.36)*	0.47 (0.43/0.52)	3.61 (2.87/4.17)	[182] OR
Cattle	10	-	-	28.9 /28.5*	6.02 /5.79*	0.587 (0.581/0.595)	1.7 (1.44/1.86)	OR
Llama	1	0.16 ± 0.05	-	-	-	-	-	[253]
Pig	3	-	-	-	-	0.63 (0.61/0.7)	3 (2.88/4.54)	OR
Dog	13	-	-	-	-	0.6 ±0.08	3.0 ±0.8	[98]
	39	-	0.755 ±0.04	-	-	0.76 (0.71/0.84)	6.57 (5.21/8.2)	[253] [109] OR
Rabbit	1/7	0.0074 ±0.0008	0.66 ±0.032	-	-	0.61 (0.59/0.61)	1.87 (1.82/1.94)	[253] [109] OR
Opossum	-	0.015 ±0.004 0.008 ±0.002	-	-	-	-	-	[254] [253] OR
Lewis rat	1/20	0.007 ±0.001	0.75 ±0.04	-	-	0.67 (0.66/0.69)	2.84 (2.54/3.39)	[253] [109] [273]
ZDF rat	10	-	-	-	-	0.64 (0.63/0.64)	2.85 (2.61/3.35)	[273]
Dahl/SS rat	10	-	-	-	-	0.63 (0.63/0.64)	2.07 (1.86/2.20)	[273]
Balb/c mouse	2/10	-	0.82 ±0.04	-	-	0.56/0.64	2.29/1.49	[109] OR
Yellow necked mouse	4	-	-	-	-	0.67 (0.66/0.68)	3.03 (2.6/3.36)	OR
Chinese hamster	8	-	-	-	-	0.61 (0.60/0.613)	1.84 (1.55/2.06)	OR
Syrian hamster	15	-	0.80 ±0.04	-	-	0.64 (0.63/0.66)	2.7 (2.12/3.68)	[109] OR
Guinea pig	6	-	0.70 ±0.04	-	-	-	-	[109]
African elephant	4	-	-	-	-	0.95 (0.86/1.14)	21.03 (17.71/27.27)	[272]
Oryx antelope	4	-	-	22.64 (19.16/25.33)	3.67 (2.52/4.95)	-	-	[268]
Cape fur seal	4	-	-	27.88 (24.79/31.44)	1.45 (1.28/1.65)	-	-	[269]
Bottle--nose dolphin	2	-	-	24/31.5	1.4/2.1	-	-	[269]
Vervet monkey	8	-	-	37.43 (34.63/40.61)	1.7 (1.5/2.1)	-	-	OR
Turkey	2	0.081 ±0.008 0.041 ±0.004	-	-	-	-	-	[254] [253]
Pond turtle	1	0.240 ±0.082	-	-	-	-	-	[254]
Western painted turtle	2	0.194 ±0.018 0.120 ±0.041	-	-	-	-	-	[254] [253]
King snake	1	0.208 ±0.037	-	-	-	-	-	[254]
Conga snake	2	0.136 ±0.028 0.104 ±0.019	-	-	-	-	-	[254] [253]

Continued on next page

TABLE 12.2
–continued from previous page

Species	n	Shear modulus (dyn/cm^{-1})	DI_{max} ekta- -cytometer	EI_{max} (Rheodyn SSD)	$SS_{1/2}$ (Rheodyn SSD)	EI_{max} (LORCA)	$SS_{1/2}$ (LORCA)	Ref
Bull- -frog	2	0.0732 ±0.013 0.037 ±0.007	-	-	-	-	-	[254] [253]
Amphi- -uma	1	0.068 ±0.014	-	-	-	-	-	[254] [253]
Toad- -fish	2	0.067 ±0.023 0.034 ±0.012	-	-	-	-	-	[254] [253]
Trout	5	0.045/0.030 (2-18°C)	-	-	-	-	-	[123]
Human	3/5	0.0161 ±0.0044 0.0091 ±0.0021	0.68 ±0.034	-	-	0.618 (0.612/0.623)	1.44 (1.33/1.49)	[254] [253] [109] OR

*Smaller data set for Rheodyn SSD tests: 1 horse, 9 sheeps, 2 cattles

Bibliography

[1] M.S. Abdo, A.M. Ali, and P.F. Prentis. Fine structure of camel erythrocytes in relation to its functions. *Zschr. Mikr. Anat. Forsch.*, 104:440–448, 1990.

[2] M. Abkarian, M. Faivre, and A. Viallat. Swinging of red blood cells under shear flow. *Phys. Rev. Lett.*, 98:188302, 2007.

[3] M. Abkarian and A. Viallat. Vesicles and red blood cells in shear flow. *Soft Matter*, 4:653–657, 2008.

[4] M. Abkarian and A. Viallat. *RSC Soft Matter No. 4*, Chapter 10: On the importance of red blood cells deformability in blood flow. The Royal Society of Chemistry, London, UK, 2016.

[5] H.A. Abramson and L.S. Moyer. The electrical charge of mammalian red blood cells. *J. General. Physiol.*, pages 19601–607, 1936.

[6] N. Adili, M. Melizi, and H. Belabbas. Species determination using the red blood cells morphometry in domestic animals. *Veterinary World*, 9:960–963, 2016.

[7] M.R. Amaiden, V.S. Santander, N.E. Monesterolo, A.N. Campetelli, J.F. Rivelli, G. Previtali, C.A. Arce, and C.H. Casale. Tubulin pools in human erythrocytes: altered distribution in hypertensive patients affects na+, k+-atp activity. *Cell Mol. Life Sci.*, 68:1755–1768, 2011.

[8] J.L. Ambrus, J.M. Anain, S.M. Anain, P.M. Anain, J.M. Anain Jr., S. Stadler, P. Mitchell, J.A. Brobst, B.L. Cobert, and J.P. Savitzky. Dose-response effects of pentoxyphylline on erythrocyte filterability: clinical and animal studies. *Clin. Pharmacol. Ther.*, 48(1):50–56, 1990.

[9] F.A. Andrews, N.L. Korenek, W.L. Sanders, and R.L. Hamlin. Viscosity and rheologic properties of blood from clinically normal horses. *Am. J. Vet. Res.*, 53:966–971, 1991.

[10] H. Arikan and K. Cicek. Haematology of amphibians and reptiles: a review. *North-western Journal of Zoology*, 10:190–209, 2014.

[11] R. Auer, A. Gleiss, and U. Windberger. Towards a basic understanding of the properties of camel blood in response to exercise. *Emir J. Food Agric.*, 27:302–311, 2015.

[12] A.J. Baines. The spectrin-ankyrin-4.1-adducin membrane skeleton: adapting eukaryotic cells to the demands of animal life. *Protoplasma*, 244(1-4):99–131, 2010.

[13] M.E. Barclay. Marginal bands in duck and camel erythrocytes. *Anat. Rec.*, 154:313, 1966.

[14] O.K. Baskurt. Deformability of red blood cells from different species studied by resistive pulse shape analysis technique. *Biorheology*, 33:169–179, 1996.

[15] O.K. Baskurt, M. Bor-Kucukatay, O. Yalcin, and H.J. Meiselman. Aggregation behavior and electrophoretic mobility of red blood cells in various mammalian species. *Biorheology*, 37:417–428, 2000.

[16] O.K. Baskurt, R.A. Farley, and H.J. Meiselman. Erythrocyte aggregation tendency and cellular properties of horse, human, and rat: a comparative study. *Am. J. Physiol.*, 273:H2604–H2612, 1997.

[17] O.K. Baskurt and H.J. Meiselman. Susceptibility of equine erythrocytes to oxidant-induced rheologic alterations. *Am. J. Vet. Res.*, 60:1301–1306, 1999.

[18] O.K. Baskurt and H.J. Meiselman. Analyzing shear stress - elongation index curves: Comparison of two approaches to simplify data presentation. *Clin. Hemorheol. Microcirc.*, 31:23–30, 2004.

[19] O.K. Baskurt and H.J. Meiselman. RBC aggregation: more important than RBC adhesion to endothelial cells as a determinant of in vivo blood flow in health and disease. *Microcirculation*, 15:585–590, 2008.

[20] O.K. Baskurt, B. Neu, and H.J. Meiselman, editors. *Red Blood Cell Aggregation*. CRC Press, Boca Raton, FL, 2012.

[21] O.K. Baskurt, O. Yalcin, S. Ozdem, J.K. Armstrong, and H.J. Meiselman. Modulation of endothelial nitric oxide synthase expression by red blood cell aggregation. *Am. J. Physiol.*, 286:H222–H229, 2004.

[22] H. Bäumler, B. Neu, R. Mitlohner, R. Georgieva, H.J. Meiselman, and H. Kiesewetter. Electrophoretic and aggregation behavior of bovine, horse and human red blood cells in plasma and in polymer solutions. *Biorheology*, 38:39–51, 2001.

[23] O. Behnke. A comparative study of microtubules of disk-shaped blood cells. *J. Ultrastructure Research*, 31:61–75, 1970.

[24] Gh. Benga, S.M. Grieve, B.E. Chapman, C.H. Gallagher, and P.W. Kuchel. Comparative NMR studies of diffusion water permeability of red blood cells from different species. x. camel (Camelus dromedarius) and alpaca (Lama pacos). *Comp. Haematol. Int.*, 9:43–48, 1999.

[25] D. Bernal, C. Sepulveda, O. Mathieu-Costello, and J.B. Graham. Comparative studies of high performance swimming in sharks. I. Red muscle morphometrics, vascularization and ultrastructure. *Journal of Experimental Biology*, 206:2831–2843, 2003.

[26] A. Bernheim-Groswasser, N.S. Gov, S.A. Safran, and S. Tzlil. Living matter: mesoscopic active materials. *Advanced Materials*, 30:1707028, 2018.

[27] S. Bertuglia. Increased viscosity is protective for arteriolar endothelium and microvascular perfusion during severe hemodilution in hamster cheek pouch. *Microvasc. Res.*, 61:56–63, 2001.

[28] T. Betz, M. Lenz, J.F. Joanny, and C. Sykes. ATP-dependent mechanics of red blood cells. *PNAS*, 106:15320–15325, 2009.

[29] J.J. Bishop, P.R. Nance, A.S. Popel, M. Intaglietta, and P.C. Johnson. Relationship between erythrocyte aggregate size and flow rate in skeletal muscle venules. *Am. J. Physiol.*, 286:H113–H120, 2004.

[30] J.J. Bishop, A.S. Popel, M. Intaglietta, and P.C. Johnson. Effects of erythrocyte aggregation and venous network geometry on red blood cell axial migration. *Am. J. Physiol.*, 281:H939–H950, 2001.

[31] J.J. Bishop, A.S. Popel, M. Intaglietta, and P.C. Johnson. Effect of aggregation and shear rate on the dispersion of red blood cells flowing in venules. *Am. J. Physiol.*, 283:H1985–H1996, 2002.

[32] A. Biswas. Body fluid and haematological changes in a poikilothermic animal on cold exposure: some physical changes. *Biorheology*, 3:385–387, 1976.

[33] M.L. Bloom, T.M. Kaysser, C.S. Birkenmeier, and J.E. Barker. The murine mutation jaundiced is caused by replacement of an arginine with a stop codon in the mrna encoding the ninth repeat of b-spectrin. *PNAS*, 91:10099–10103, 1994.

[34] A.R. Bodey and M.W. Rampling. A comparative study of the haemorreology of various breeds of dog. *Clin. Hemorheol. Microcirc.*, 18(4):291–298, 1998.

[35] I.N. Bojesen and E. Bojesen. Sheep erythrocyte membrane binding and transfer of long-chain fatty acids. *J. Membr. Biol.*, 171:141–149, 1999.

[36] H. Bouaouda, M.R. Achaaban, M. Ouassat, M. Oukassou, M. Piro, E. Challet, K. El Allali, and P. Pevet. Daily regulation of body temperature rhythm in the camel (Camelus dromedarius) exposed to experimental desert conditions. *Physiol. Reports*, 2:e12151, 2014.

[37] J.H. Boucher. The equine spleen: source of dangerous red blood cells. *J. Equine Vet. Sci.*, 7:140–142, 1987.

[38] H.F. Bunn. Regulation of hemoglobin function in mammals. *Amer. Zool.*, 20:199–221, 1980.

[39] M. Cabel, H.J. Meiselman, A.S. Popel, and P.C. Johnson. Contribution of red blood cell aggregation to venous vascular resistance in skeletal muscle. *Am. J. Physiol.*, 272:H1020–H1032, 1997.

[40] P.B. Canham, R.F. Potter, and D. Woo. Geometric accommodation between the dimension of erythrocytes and the caliber of heart and muscle capillaries in the rat. *J. Physiol.*, 347:697–712, 1984.

[41] P.B. Canham, R.F. Potter, and D. Woo. Theoretical model of blood flow autoregulation: roles of myogenic, shera-dependent, and metabolic responses. *Am. J. Physiol.*, 295:H1572–H1579, 2008.

[42] G. Casotti and E.J. Braun. Structure of the glomerular capillaries of the domestic chicken and desert quail. *J. Morphol.*, 224:57–63, 2005.

[43] M. Castellini, R. Elsner, O.K. Baskurt, R.B. Wenby, and H.J. Meiselman. Blood rheology of Weddell seals and bowhead whales. *Biorheology*, 43:57–69, 2006.

[44] M.M. Chan, J.M. Wooden, M. Tsang, D.M. Gilligan, D.K. Hirenallur-S, G.L. Finney, E. Rynes, M. Maccoss, J.A. Ramirez, H. Park, and B.M. Iritani. Hematopoietic protein-1 regulates the actin membrane skeleton and membrane stability in murine erythrocytes. *PLoS One*, 8:e54902, 2013.

[45] C.A. Chapman and G.M.C. Renshaw. Hematological responses of the grey carpet shark (Chiloscyllium punctatum) and the epaulette shark (Hemiscyllium ocellatum) to anoxia and re-oxygenation. *J. Exp. Zool.*, 311A:422–438, 2009.

[46] D. Chen and D.K. Kaul. Rheologic and hemodynamic characteristics of red cells of mouse, rat, and human. *Biorheology*, 31:103–113, 1994.

[47] S. Chen, B. Gavish, G. Barshtein, Y. Mahler, and S. Yedgar. Red blood cell aggregability is enhanced by physiological levels of hydrostatic pressure. *Biochim. Biophys. Acta*, 1192:247–252, 1994.

[48] S. Chien, S. Usami, R.J. Dellenback, and C.A. Bryant. Comparative hemorheology - hematological implications of species differences in blood viscosity. *Biorheology*, 8:35–57, 1971.

[49] M.R. Clark, N. Mohandas, and S.B. Sohet. Osmotic gradient ektacytometry: comprehensive characterization of red cell volume and surface maintainance. *Blood*, 61:899–910, 1983.

[50] W.D. Cohen, M.F. Cohen, C.H. Tyndale-Biscoe, J.L. VandeBerg, and G.B. Ralston. The cytoskeletal system of mammalian primitive erythrocytes: studies in developing marsupials. *Cell Motil. Cytoskeleton*, 16:133–145, 1990.

[51] W.D. Cohen, Y. Sorokina, and I. Sanchez. Elliptical versus circular erythrocyte marginal bands: isolation, shape conversion, and mechanical properties. *Cell Motil. Cytoskeleton*, 40:238–248, 1998.

[52] W.D. Cohen and N.B. Terwilliger. Marginal bands in camel erythrocytes. *J. Cell Sci.*, 36:97–107, 1979.

[53] G.R. Cokelet and H.J. Meiselman. Rheological comparison of hemoglobin solutions and erythrocyte suspensions. *Science*, 162:275–277, 1968.

[54] D. Cordasco and P. Bagchi. Dynamics of red blood cells in oscillating shear flow. *J. Fluid Mech.*, 800:484–516, 2016.

[55] E.A. Davidson, C.A. Pickens, and J.I. Fenton. Increasing dietary epa and dha influence estimated fatty aid desaturase activity in systemic organs, which is reflected in the red blood cell in mice. *Int. J. Food Sci. Nutr.*, 12:1–9, 2017.

[56] L. Dintenfass and L. Fu-Lung. Plasma and blood viscosities and aggregation of red cells in racehorces. *Clin. Phys. Physiol. Measures*, 3:293–301, 1982.

[57] J.J. Durussel, J. Dufaux, A. Laurent, L. Penhouet, A.L. Bailly, M. Bonneau, and J.J. Merland. Comparative viscosity and erythrocytes disaggregation shear stress of blood-radiographic contrast media mixtures in three mammalian species. *Clin. Hemorheol. Microcirc.*, 17:261–269, 1997.

[58] S. Egginton and L.A. Johnston. An estimate of capillary anisotropy and determination of surface and volume densities of capillaries in skeletalmuscles of conger eel (conger conger l). *Q. J. Exp. Physiol.*, 68:603–617, 1983.

[59] A. Eitan, B. Aloni, and A. Livne. Unique properties of the camel erythrocyte membrane. II. Organization of membrane proteins. *Biochim. Biophys. Acta*, 426:647–658, 1976.

[60] A.O. Elkhawad. Selective brain cooling in desert animals: the camel (Camelus dromedarius). *Comp. Biochem. Physiol.*, 101A:339–344, 1992.

[61] R. Elsner. *The Biology of Marine Mammals*, chapter: Cardiovascular Adjustments to Diving, pages 117–145. New York, Academic Press, 1969.

[62] R. Elsner, H. Meiselman, and O.K. Baskurt. Temperature-viscosity relations of bowhead whale blood: a possible mechanism for maintaining cold blood flow. *Marine Mammal Science*, 20:339–344, 2004.

[63] R.L. Engen and C.L. Clark. High performance liquid chromatography determination of erythrocyte membrane phospholipid composition in several animal species. *Am. J. Vet. Res.*, 51:577–580, 1990.

[64] K.G. Engstrom and L. Ohlsson. Morphology and filterability of red blood cells in neonatal and adult rats. *Pediatr. Res.*, 27:220–226, 1990.

[65] E.H. Eylar, M.A. Madoff, O.V. Brody, and J.L. Oncley. The contribution of sialic acid to the surface charge of the erythrocyte. *J. Biol. Chem.*, 237:1992–2000, 1962.

[66] M.R. Fedde and S.C. Wood. Rheological characteristics of horse blood: significance during exercise. *Respir. Physiol.*, 94:323–335, 1993.

[67] D.A. Fedosov, B. Caswell, and G.E. Karniadakis. A multiscale red blood cell model with accurate mechanics, rheology, and dynamics. *Biophys. J.*, 98:2215–2225, 2010.

[68] B.F. Feldman, J.G. Zinkl, and N.C. Jain, editors. *Schalm's Veterinary Hematology. 5th Edition.* Lippincott Williams and Wilkins, 2000.

[69] H.P. Fernandes, C.L. Cesar, and M.L. Barjas-Castro. Electrical properties of the red blood cell membrane and immunohematological investigation. *Rev. Bras. Hematol. Hemoter.*, 33:297–301, 2011.

[70] K.D. Fink. *Microfluidic analysis of vertebrate red blood cell characteristics.* PhD thesis, University of California, Berkeley and University of California, San Francisco, 2017.

[71] T.M. Fischer and R. Korzeniewski. Effects of shear rate and suspending medium viscosity on elongation of red cells tank-treading in shear flow. *Cytometry Part A*, 79A:946–951, 2011.

[72] W.G. Frasher, H. Wayland, and H.J. Meiselman. Viscometry of circulating blood in dogs. 1: heparin injection; 2: platelet removal. *J. Appl. Physiol.*, 25:751–760, 1968.

[73] E. Friederichs, R.A. Farley, and H.J. Meiselman. Influence of calcium permeabilization and membrane-attached hemoglobin on erythrocyte deformability. *Am. J. Hematol.*, 41:170–177, 1992.

[74] P. Gaehtgens, H.J. Meiselman, and H. Wayland. Erythrocyte flow velocities in mesenteric microvessels of the cat. *Microvasc. Res.*, 2:151–162, 1970.

[75] P. Gaehtgens, F. Schmidt, and G. Will. Comparative rheology of nucleated and non-nucleated red blood cells. 1. Microrheology of avian erythrocytes during capillary flow. *Pflugers Arch.*, 390:278–282, 1981.

[76] P. Gaehtgens, G. Will, and F. Schmidt. Comparative rheology of nucleated and non-nucleated red blood cells. 2. Rheological properties of avian red cells suspensions in narrow capillaries. *Pflugers Arch.*, 390:283–287, 1981.

[77] J. Galehr. *System comparison, laser assisted optical rotational cell analyzer (LORCA) and laser shear stress diffractometer.* PhD thesis, Fachhochschule Technikum Vienna, 2008.

[78] J.B. Gaughan. Which physiological adaptation allows camels to tolerate high heat load - and what more can we learn? *J. Camelid Sci.*, 4:85–88, 2011.

[79] R.J. Geor, E.M. Lund, and D.J. Weiss. Echinocytosis in horses: 54 cases. *Am. J. Vet. Med. Assoc.*, 202:976–980, 1993.

[80] D.M. Gilligan, L. Lozovatsky, B. Gwynn, C. Brugnara, N. Mohandas, and L.L. Peters. Targeted disruption of the b-adducin gege (Add2) causes red blood cell spherocytosis in mice. *PNAS*, 96:10717–10722, 1999.

[81] V. Glanz. *Darstellung der Thrombusbildung mithilfe viskoelastischer Messmethoden.* PhD thesis, Fachhochschule Technikum Vienna, 2017.

[82] L. Goniakowska-Witalinska and W. Witalinski. Evidence for a correlation between the number of marginal band microtubules and the size of vertebrate erythrocytes. *J. Cell Sci.*, 22:397–401, 1976.

[83] A.S. Goryunov, A.G. Borisova, and S.P. Rozhkov. Structural changes in the erythrocyte membrane of the trout salmo irideus at seasonal acclimatization. *J. Evol. Biochem. Physiol.*, 42:559–565, 2006.

[84] B.L. Granger and E. Lazarides. Membrane skeletal protein 4.1 of avian erythrocytes is composed of multiple variants that exhibit tissue-specific expression. *Cell*, 37:595–607, 1984.

[85] Bruce L Granger, Elizabeth A Repasky, and Elias Lazarides. Synemin and vimentin are components of intermediate filaments in avian erythrocytes. *The Journal of Cell Biology*, 92(2):299–312, 1982.

[86] M.I. Gregersen, C.A. Bryant, S. Chien, R.J. Dellenback, V. Magazinovic, and S. Usami. Species differences in the flexibility and deformation of erythrocytes (RBC). *Bibl. Anat.*, 10:104–108, 1968.

[87] C.L. Guard and D.E. Murrish. Effects of temperature on the viscous behavior of blood from antarctic birds and mammals. *Comp. Biochem. Physiol.*, 52A:287–290, 1975.

[88] G. Gulliver. On the size and shape of red corpuscles of the blood of vertebrates, with drawings of them to a uniform scale, and extended and revised tables of measurements. *Proceedings of the Zoological Society of London*, pages 474–495, 1875.

[89] T.S. Hakim and A.S. Macek. Effect of hypoxia on erythrocyte deformability in different species. *Biorheology*, 25:857–868, 1988.

[90] X. Han, C. Wang, and Z. Liu. Red blood cells as smart delivery systems. *Bioconjugate Chem.*, 29:852–860, 2018.

[91] H. Hassoun, T. Hanada, M. Lutchman, K.E. Sahr, J. Palek, M. Hanspal, and A.H. Chishti. Complete deficiency of glycophorin A in red blood cells from mice with targeted inactivation of the band 3 (AE1) gene. *Blood*, 91:2146–2151, 1998.

[92] C.M. Hawkey. The value of comparative haematological studies. *Comp. Haematol. Int.*, 1:1–9, 1991.

[93] I. Hayam, U. Cogan, and S. Mokadi. Dietary oxidized oil enhances the activity of Na+K+ -ATPase and acetycholinesterase and lowers the fluidity of rat erythrocyte membrane. *Res. Commun.*, 4:563–568, 1993.

[94] M.S. Hedrick, D.A. Duffield, and L.H. Cornell. Blood viscosity and optimal haematocrit in a deep-diving mammal, thc northcrn elephant seal (Mirounga angustirostris). *Can. J. Zool.*, 64:2081–2085, 1986.

[95] R.S. Hetem, W.M. Strauss, L.G. Fick, S.K. Maloney, L.C.R. Meyer, M. Shobrak, A. Fuller, and D. Mitchell. Variation in the daily rhythm of body temperature of free-living arabian oryx (Oryx leucoryx): does water limitation drive heterothermy? *J. Comp. Physiol. B*, 180:1111–1119, 2010.

[96] S.S. Hillman, P.C. Withers, M.S. Hedrick, and P.B. Kimmel. The effects of erythrocythemia on blood viscosity, maximal systemic oxygen transport capacity and maximal rates of oxygen consumption in an amphibian. *J. Comp. Physiol. B*, 155:577–581, 1985.

[97] R. Hochmuth. *Handbook of Bioengineering*, Chapter 12: Properties of Red Blood Cells. McGraw-Hill Book Comp., New York, 1987.

[98] N. Hofbauer, U. Windberger, I. Schwendenwein, A. Tichy, and E. Eberspacher. Evaluation of canine red blood cell quality after processing with an automated cell salvage device. *Vet. Emerg. Crit. Care*, 26:373–383, 2016.

[99] T.M. Holm, A. Braun, B.L. Trigatti, C. Brugnara, M. Sakarmoto, M. Krieger, and N.C. Andrews. Failure of red blood cell maturation in mice with defects in the high-density lipoprotein receptor sr-bi. *Blood*, 99:1817–1824, 2002.

[100] R.J. Howard, P.M. Smith, and G.F. Mitchell. Identification of differences between the surface proteins and glycoproteins of normal mouse (balb/c) and human erythrocytes. *J. Memb. Biol.*, 49:171–198, 1979.

[101] H. Huang, P.X. Zhao, K. Arimatsu, K. Tabeta, K. Yamazaki, L. Krieg, E. Fu, T. Zhang, and X. Xin Du. A deep intronic mutation in the ankyrin-1 gene causes diminished protein expression resulting in hemolytic anemia in mice. *G3: Genes, Genomes, Genetics*, 3:1687–1695, 2013.

[102] M. Inaba and Y. Maede. A new major transmembrane glycoprotein, gp155, in goat erythrocytes. isolation and characterization of its association to cytoskeleton through binding with band 3-ankyrin complex. *J. Biol. Chem.*, 263:17763–17771, 1988.

[103] J.H. Jeong, Y. Sugii, M. Minamiyama, and K. Okamoto. Measurement of rbc deformation and velocity in capillaries in vivo. *Microvasc. Res.*, 71:212–217, 2006.

[104] J. Jiang, N. Magilnick, K. Tsirulnikov, N. Abuladze, I. Atanasov, P. Ge, N. Mohandas, A. Pushkin, Z.H. Zhou, and I. Kurtz. Single particle electron microscopy analysis of the bovine anion exchanger 1 reveals a flexible linker connecting the cytoplasmic and membrane domains. *PLOS one*, 8:e55408, 2006.

[105] H. Johnn, C. Phipps, S. Gascoyne, C. Hawkey, and M.W. Rampling. A comparison of the viscometric properties of the blood from a wide range of mammals. *Clin. Hemorheol. Microcirc.*, 12:639–647, 1992.

[106] Y. Ju, J. Song, Z. Geng, H. Zhang, W. Wang, L. Xie, W. Yao, and Z. Li. A microfluidics cytometer for mice anemia detection. *Lab Chip*, 12:4355–4362, 2012.

[107] M. Kaibara. Rheological behaviors of bovine blood forming artificial rouleaux. *Biorheology*, 20:583–592, 1983.

[108] L.B. Kasarov. Degradation of the erythrocyte phospholipids and hemolysis of the erythrocytes of different animal species by leptospirae. *J. Med. Microbiol.*, 3:29–37, 1970.

[109] L.N. Katyukhin, A.M. Kazenov, N. Maslove, and Y. Matskevich. Rheological properties of mammalian erythrocytes: relationship to transport atpases. *Comp. Biochem. Physiol. B*, 120:493–498, 1998.

[110] J.K. Khodadad and RS. Weinstein. The band 3-rich membrane of ilama erythrocytes: studies on cell shape and the organisation of membrane proteins. *J. Membrane Biol.*, 72:161–171, 1983.

[111] J.K. Khodadad and RS. Weinstein. Band 3 protein of the red cell membrane of the llama: crosslinking and cleavage of the cytoplasmic domain. *Biochem. Biophys. Res. Commun.*, 130:493–499, 1985.

[112] S. Kim, R.L. Kong, A.S. Popel, M. Intaglietta, and P.C. Johnson. Temporal and spatial variations of cell-free layer width in arterioles. *Am. J. Physiol.*, 293:H1526–H1535, 2007.

[113] S. Kim, M. Magendantz, W. Katz, and F. Solomon. Development of a differentiated microtubule structure: formation of the chicken erythrocyte marginal band in vivo. *J. Cell Biol.*, 104:51–59, 1987.

[114] S. Kim, A.S. Popel, M. Intaglietta, and P.C. Johnson. Aggregate formation of erythrocytes in postcapillary venules. *Am. J. Physiol.*, 288:H584–H590, 2005.

[115] S. Kim, O. Yalcin, M. Intaglietta, and P.C. Johnson. The cell-free layer in microvascular blood flow. *Biorheology*, 46:181–189, 2009.

[116] F. Kiss, E. Toth, K. Miszti-Blasius, and N. Nemeth. The effect of centrifugation at various g force levels on rheological properties of rat, dog, pig and human red blood cells. *Clin. Hemorheol. Microcirc.*, 62:215–227, 2016.

[117] G.C. Kodippili, J. Spector, J. Hale, K. Giger, M.R. Hughes, K.M. McNagny, C. Birkenmeier, L. Peters, K. Ritchie, and P.S. Low. Analysis of the mobilities of band 3 populations associated with ankyrin protein and junctional complexes in intact murine erythrocytes. *J. Biol. Chem.*, 287:4129–4138, 2012.

[118] A.G. Koutsaris. Volume flow estimation in the precapillary mesenteric microvasculature invivo and the principle of constant pressure gradient. *Biorheology*, 42:479–491, 2005.

[119] F.A. Kuypers, E.W. Easton, R. van den Hoven, T. Wensing, B. Roelofsen, J.A.F. Op den Kamp, and L.L.M. van Deenen. Survival of rabbit and horse erythrocytes in vivo after changing the fatty acyl composition of their phosphatidylcholine. *Biochim. Biophys. Acta*, 819:170–178, 1985.

[120] D. Kuzman, S. Svetina, R.E. Waugh, and Z. Zeks. Elastic properties of the red blood cell membrane that determine echinocte deformability. *Eur. Biophys. J.*, 33:1–15, 2004.

[121] B.L. Langille and B. Crisp. Temperature dependence on the viscosity of frogs and turtles: effect on heat exchange with environment. *Am. J. Physiol.*, 239:R248–R253, 1980.

[122] L. Lanotte, J. Mauer, S. Mendez, D. Fedosov, J.M. Fromental, V. Claveria, F. Nicoud, G. Gompper, and M. Abkarian. Red cells dynamic morphologies govern blood shear thinning under microcirculatory flow conditions. *PNAS*, 113:13289–13294, 2016.

[123] T. Lecklin, G.B. Nash, and S. Egginton. Do fish acclimated to low temperature improve microcirculatory perfusion by adapting red cell rheology? *J. Exp. Biol.*, 198:1801–1808, 1995.

[124] B.K. Lee, T. Alexi, R.B. Wenby, and H.J. Meiselman. Red blood cell aggregation quantitated via myrenne aggregometer and yield shear stress. *Biorheology*, 44:29–35, 2007.

[125] K. Lee, E. Shirshin, N. Rovnyagina, F. Yaya, Z. Boujja, A. Priezzhev, and C. Wagner. Dextran adsorption onto red blood cells revisited: single cell quantification by laser tweezers combined with microfluidics. *Biomedical Optics Express*, 9:2755, 2018.

[126] J. Lenard. Protein components of erythrocyte membranes from different animal species. *Biochemistry*, 9:5037–5040, 1970.

[127] A.M. Lewis, O. Mathieu-Costello, P.J. McMillan, and R.D. Gilbert. Effects of long-term, high-altitude hypoxia on the capillarity of the ovine fetal heart. *Am. J. Physiol.*, 277:H756–H762, 1999.

[128] K. Ley, A.R. Pries, and P. Gaehtgens. Topological structure of rat mesenteric microvessel networks. *Microvasc. Res.*, 32:315–332, 1986.

[129] E.C. Liao, B.H. Paw, L.L. Peters, A. Zapata, S.J. Pratt, C.P. Do, G. Lieschke, and L.I. Zon. Hereditary spherocytosis in zebrafish riesling illustrates evolution of erythroid beta-spectrin structure, and function in red cell morphogenesis and membrane stability. *Development*, 127:5123–5132, 2000.

[130] O. Lie. *Fish Quality-Role of Biological Membranes*, chapter: Fatty acid composition of membranes of fish influence diet and temperature. Tema Nord, Copenhagen, Denmark, 1995.

[131] I. Linhartova, B. Novotna, V. Sulimenko, E. Draberova, and P. Draber. Gamma-tubulin in chicken erythrocytes: changes in localization during cell differentiation and characterization of cytoplasmic complexes. *Developmental Dynamics*, 223:229–240, 2002.

[132] H. Lipowsky and S.B.W. Kovalcheck. The distribution of blood rheological parameters in the microvasculature of cat mesentery. *Circularion Res.*, 43:738–749, 1978.

[133] H. Lipowsky, S. Usami, and S. Chien. Methods for the simultaneous measurement of pressure differentials and flow in single unbranched vessels of the microcirculation for rheological studies. *Microvasc. Res.*, 14:345–361, 1977.

[134] H. Lipowsky, S. Usami, and S. Chien. In vivo measurements of "apparent viscosity" and microvessel hematocrit in the mesentery of the cat. *Microvasc. Res.*, 19:297–319, 1980.

[135] H.H. Lipowsky and B.W. Zweifach. Network analysis of microcirculation of cat mesentery. *Microvasc. Res.*, 7:73–83, 1974.

[136] M. Little and T. Seehaus. Comparative analysis of tubulin sequences. *Comp. Biochem. Physiol. B*, 90:655–670, 1988.

[137] P.A. Lorkin. Fetal and embryonic haemoglobins. *J Med Genetics*, 10:50–64, 1973.

[138] C. Loubens, J. Deschamp, F. Edwards-Levy, and M. Leonetti. Tank-treading of microcapsules in shear flow. *J. Fluid Mech.*, 789:750–767, 2016.

[139] P.S. Low. Structure and function of the cytoplasmic domain of band 3: center of erythrocyte membrane - peripheral protein interaction. *Biochim. Biophys. Acta*, 864:145–167, 1986.

[140] A.Y. Malkin. Non-newtonian viscosity in steady-state shear flows. *J. Non-Newton. Fluid Mech.*, 192:48–65, 2013.

[141] C.P. Mangum, editor. *Advances in Comparative and Environmental Physiology.* Springer-Verlag, Berlin, Heidelberg, New York, 1992.

[142] C. P. Mangum. Invertebrate blood oxygen carriers, *Comprehensive Physiology*, pp. 1097–1135, 2010.

[143] S. Manno, Y. Takakuwa, and N. Mohandas. Modulation of erythrocyte membrane mechanical function by protein 4.1 phosphorylation. *J. Biol. Chem.*, 280:7581–7587, 2005.

[144] H. Matei, L. Frentescu, and G. Benga. Comparative studies of the protein composition of red blood cells membranes from eight mammalian species. *J. Cell Mol. Med.*, 4:270–276, 2000.

[145] O. Mathieu-Costello. Morphometry of the size of the capillary-to-fiber interface in muscles. *Adv. Exp. Med. Biol.*, 345:661–668, 1994.

[146] O. Mathieu-Costello, P.J. Agey, E.S. Quintana, K. Rousey, L. Wu, and M.H. Bernstein. Fiber capillarization and ultrastructure of pigeon pectoralis muscle after cold acclimation. *J. Exp. Biol.*, 201:3211–3220, 1998.

[147] R.A. McPherson, W.H. Sawyer, and L. Tilley. Band 3 mobility in camelid elliptocytes: implications for erythrocyte shape. *Biochemistry*, 32:6696–6702, 1993.

[148] H.J. Meiselman, O.K. Baskurt, S.O. Sowemimo-Coker, and R.B. Wenby. Cell electrophoresis studies relevant to red blood cell aggregation. *Biorheology*, 36:427–432, 1999.

[149] H.J. Meiselman, M.A. Castellini, and R. Elsner. Hemorheological behaviour of seal blood. *Clin. Hemorheol.*, 12:657–675, 2001.

[150] I. Miko, N. Nemeth, V. Sogor, F. Kiss, E. Toth, K. Peto, A. Furka, E. Vanyolos, L. Toth, J. Varga, K. Szigeti, I. Benk, A.V. Olah, and I. Furka. Comparative erythrocyte deformability investigations by filtrometry, slit-flow and rotational ektacytometry in a long-term follow-up animal study on splenectomy and different spleen preserving operative techniques. *Clin. Hemorheol. Microcirc.*, 66:83–96, 2017.

[151] N. Mohandas. Red cell membrane disorders. *Int. J. Lab. Hematol.*, 39:47–52, 2017.

[152] N. Mohandas, M.R. Clark, M.S. Jacobs, and S.B. Shohet. Analysis of factors regulating erythrocyte deformability. *J. Clin. Invest.*, 66:563–573, 1980.

[153] N. Mohandas and P. Gascard. What do mouse gene knockouts tell us about the structure and function of the red cell membrane? *Baillieres Best Pract. Res. Clin. Haematol.*, 12:605–620, 1999.

[154] R. Monahan-Earley, A.M. Dvorak, and W.C. Aird. Evolutionary origins of the blood vascular system and endothelium. *J. Thromb. Haemost.*, 11(1):46–66, 2013.

[155] Moore D. M. (2000). Haematology of camelid species: ilamas and camels, in Schalm's Veterinary Haematology, 5th Edn, eds Feldman B. F., Zinkl J. G., Jain N. C., editors. (Philadelphia: Lippincott Williams & Wilkins;), 1184–1190.

[156] J.D. Moyer, R.B. Nowak, N.E. Kim, S.K. Larkin, L.L. Peters, J. Hartwig, F.A. Kuypers, and V.M. Fowler. Tropomodulin 1-null mice have a mild spherocytic elliptocytosis with appearance of tropomodulin 3 in red blood cells and disruption of the membrane skeleton. *Blood*, 79:492–499, 2010.

[157] T.J. Muller, C.W. Jackson, M.E. Dockter, and M. Morrison. Membrane skeletal alterations during in vivo mouse red blood cell aging. *J. Clin. Invest.*, 79:492–499, 1987.

[158] A.F. Muro, M.L. Marro, S. Gajovic, F. Porro, L. Luzzatto, and F.E. Baralle. Mild spherocytic hereditary elliptocytosis and altered levels of a- and y-adducins in b-adducin-deficient mice. *Blood*, 95:3978–3985, 2000.

[159] Grint N. and A. Dugdale. Brightness of venous blood in south american camelids: implications for jugular catheterization. *Vet. Anaesth. Analg.*, 36:63–66, 2009.

[160] M. Nakao. New insight into regulation of erythrocyte shape. *Curr. Opin. Hematol.*, 9:127–132, 2002.

[161] M. Nakao, T. Nakao, and S. Yamazoe. Atp and maintenance of shape of the human red cells. *Nature*, 187:945–946, 1960.

[162] M. Nakao, Y. Jinbu, S. Sato, Y. Ishigami, T. Nakao, E. Ito-Ueno, and K. Wake. "Structure and function of red cell cytoskeleton." *Biomedica biochimica acta*, 46(2-3): S5–9, 1987.

[163] B. Namgung, Y.C. Ng, H.L. Leo, J.M. Rifkind, and S. Kim. Near-wall migration dynamics of erythrocytes in vivo: effects of cell deformability and arteriolar bifurcation. *Frontiers Physiol.*, 8:963, 2017.

[164] A. Nans, N. Mohandas, and D.L. Stokes. Ultrastructure of the red cell cytoskeleton by cryo-electron tomography. *Biophys. J.*, 101:2341–2350, 2011.

[165] G.B. Nash and S. Egginton. Comparative rheology of human and trout red blood cells. *J. Exp. Biol.*, 174:109–122, 1993.

[166] G.J. Nelson. Lipid composition of erythrocytes in various mammals. *Biochim. Biophys. Acta*, 144:221–232, 1967.

[167] N. Nemeth, V. Sogor, F. Kiss, and P. Ulker. Interspecies diversity of erythrocyte mechanical stability ar various combinations in magnitude and duration of shear stress, and osmolality. *Clin. Hemorheol. Microcirc.*, 63:381–398, 2016.

[168] D.P. Nguyen, K. Yamaguchi, P. Scheid, and J. Piiper. Kinetics of oxygen uptake and release by red blood cells of chicken and duck. *J. Exp. Biol.*, 125:15–27, 1986.

[169] N. Mikko, Vertebrate Red Blood Cells, series Zoophysiology, Springer-Verlag Berlin Heidelberg, 1990.

[170] D. Obrist, B. Weber, A. Buck, and P. Jenny. Red blood cell distribution in simplified capillary networks. *Phil. Trans. R. Soc. A*, 368:826–834, 2010.

[171] K. Ohta, F. Gotoh, M. Tomita, N. Tanahashi, M. Kobari, T. Shinohara, Y. Tereyama, B. Mihara, and H. Takeda. Animal species differences in erythrocyte aggregability. *Am. J. Physiol.*, 262:H1009–H1012, 1992.

[172] O.O. Olumuyiwa-Akeredolu, P. Soma, A.V. Buys, L.K. Debusho, and E. Pretorius. Characterizing pathology in erythrocytes using morphological and biophysical membrane properties: relation to impaired hemorheology and cardiovascular function in rheumatoid arthritis. *Biochim. Biophys. Acta Biomembr.*, 1859:2381–2391, 2017.

[173] S.A. Omorphos, C.M. Hawkey, and C. Rice-Evans. The elliptocyte: a study of the relationship between cell shape and membrane structure using the camelid erythrocyte as a model. *Comp. Biochem. Physiol.*, 94B:789–795, 1989.

[174] P.K. Ong, S. Jain, and S. Kim. Spatio-temporal variations in cell-free layer formation near bifurcations of small arterioles. *Microvasc. Res.*, 83:118–125, 2012.

[175] J. Oulevey, E. Bodden, and O.W. Thiele. Quantitative determination of glycosphingolipids illustrated by using erythrocyte membranes of various mammalian species. *Eur. J. Biochem.*, 79:265–267, 1977.

[176] N.M. Palenske and D.K. Saunders. Blood viscosity and hematology of American bullfrogs (rana catesbeiana) at low temperature. *J. Thermal Biology*, 28:271–277, 2003.

[177] L. Pan, R. Yan, W. Li, and K. Xu. Super-resolution microscopy reveals the native ultrastructure of the erythrocyte cytoskeleton. *Cell Reports*, 22:1151–1158, 2018.

[178] E.M. Pasini, M. Kirkegaard, D. Salerno, P. Mortensen, M. Mann, and A.W. Thomas. Deep coverage mouse red blood cell proteome. *Mol. Cell Proteomics*, 7:1317–1330, 2008.

[179] E.M. Pasini, H.U. Lutz, M. Mann, and A.W. Thomas. Red blood cell (rbc) membrane proteomics - Part ii: comparative proteomics and rbc patho-physiology. *J. Proteomics*, 73:421–435, 2010.

[180] L. Peters and J. Barker. Spontaneous and targeted mutations in erythrocyte membrane skeleton genes: mouse models of hereditary spherocytosis, *Hematopoiesis: a developmental approach*, pp. 582–608, 2001.

[181] L.L. Peters, R.A. Swearingen, S.G. Andersen, B. Gwynn, A.J. Lambert, R. Li, S.E. Lux, and G.A. Churchill. Identification of quantitative trait loci that modify the severity of hereditary spherocytosis in wan, a new mouse model of band-3 deficiency. *Blood*, 103:3233–3240, 2004.

[182] R. Plasenzotti, B. Stoiber, M. Posch, and U. Windberger. Red blood cell deformability and aggregation behaviour in different animal species. *Clin. Hemorheol. Microcirc.*, 31:105–111, 2004.

[183] R. Plasenzotti, U. Windberger, F. Ulberth, W. Osterode, and U. Losert. Influence of fatty acid composition in mammalian erythrocytes on cellular aggregation. *Clin. Hemorheol. Microcirc.*, 37:237–243, 2007.

[184] D.C. Poole, T.S. Epp, and H.H. Erickson. Exercise-induced pulmonary haemorrhage (eiph): mechanistic bases and therapeutic interventions. *Equine Vet. J.*, 39:292–293, 2007.

[185] D.C. Poole and O. Mathieu-Costello. Relationship between fiber capillarization and mitochondria volume density in control and trained rat soleus and plantaris muscles. *Microcirculation*, 39:175–186, 1996.

[186] A.S. Popel, P.C. Johnson, M.V. Kameneva, and M.A. Wild. Capacity for red blood cell aggregation is higher in athletic mammalian species than in sedentary species. *J. Appl. Physiol.*, 77:1790–1794, 1994.

[187] F. Porro, L. Costessi, M.L. Marro, F.E. Baralle, and A.F. Muro. The erythrocyte skeletons of b-adducin deficient mice have altered levels of tropomyosin, tropomodulin and ecapz. *FEBS Letters*, 576:36–40, 2004.

[188] R.F. Potter and A.C. Groom. Capillary diameter and geometry in cardiac and skeletal muscle studied by means of corrosion casts. *Microvasc. Res.*, 25:68–84, 1982.

[189] A.R. Pries, D. Neuhaus, and P. Gaehtgens. Blood viscosity in tube flow: dependence on diameter and hematocrit. *Am. J. Physiol.*, 263:H1770–H1778, 1992.

[190] E. Quist and P. Powell. Polyphosphoinositides and the shape of mammalian erythrocytes. *Lipids*, 20:433–438, 1985.

[191] J. Qvist, R.D. Hill, R.C. Schneider, K.J. Falke, G.C. Liggins, M. Guppy, R.L. Elliot, P.W. Hochachka, and W.M. Zapol. Hemoglobin concentrations and blood gas tensions of free-diving weddell seals. *J. Appl. Physiol.*, 61:1560–1569, 1986.

[192] G.B. Ralston. Protein components of the camel erythrocyte membrane. *Biochim. Biophys. Acta*, 401:83–94, 1975.

[193] G. Rank, R. Sutton, V. Marshall, R.J. Lundie, J. Caddy, T. Romeo, K. Fernandez, M.P. McCormack, B.M. Cooke, S.J. Foote, B.S. Crabb, D.J. Curtis, D.J. Hilton, B.T. Kile, and S.M. Jane. Novel roles for erythroid ankyrin-1 revealed through an enu-induced null mouse mutant. *Blood*, 113:3352–3362, 2009.

[194] M. Rasia and A. Bollini. Red blood cell shape as a function of medium's ionic strength and ph. *Biochim. Biophys. Acta*, 1372:198–204, 1998.

[195] S.A. Reinhart, T. Schulzki, P.O. Bonetti, and W.H. Reinhart. Studies on metabolically depleted erythrocytes. *Clin. Hemorheol. Microcirc.*, 56:161–173, 2014.

[196] A.M. Rizzo, P.A. Corsetto, G. Montorfano, S. Milani, S. Zava, S. Tavella, R. Cancedda, and B. Berra. Effects of long-term space flight on erythrocytes and oxidative stress of rodents. *PLOS one*, 7:e32361, 2012.

[197] L. Romano and H. Passow. Characterization of anion transport system in trout red blood cell. *Am. J. Physiol.*, 246:C330–C338, 1984.

[198] P.D. Ross and A.P. Minton. Hard quasispherical model for the viscosity of hemoglobin solutions. *Biochem. Biophys. Res. Comm.*, 76:971–976, 1977.

[199] M. Rudiger and K. Weber. Characterization of the post-translational modifications in tubulin from the marginal band of avian erythrocytes. *Eur. J. Biochem.*, 218:107–116, 1993.

[200] Henon S., G. Lenormand, A. Richert, and F. Gallet. A new determination of the shear modulus of the human erythrocyte membrane using optical tweezers. *Biophys. J.*, 76:1145–1151, 1999.

[201] A. Saaed and M.M. Hussein. Change in normal haematological values of camels (Camelus dromedaries): influence of age and sex. *Comp. Clin. Pathol.*, 17:263–266, 2008.

[202] R.G.A. Safranyos, C.G. Ellis, K. Tyml, and A.C. Groom. Heterogeneity of capillary diameters in skeletal muscle of the frog. *Microvasc. Res.*, 26:151–156, 1983.

[203] K.O. Sakamoto, K. Sato, Y. Naito, Y. Habara, M. Ishizuka, and S. Fujita. Morphological features and blood parameters of weddell seal (Leptonychotes weddellii) mothers and pups during the breeding season. *J. Vet. Med. Sci.*, 71:341–344, 2009.

[204] M. Salomao, X. An, X. Guo, W.B. Gratzer, N. Mohandas, and A.J. Baines. Mammalian αI-spectrin is a neofunctionalized polypeptide adapted to small highly deformable erythrocytes. *PNAS*, 103:643–648, 2006.

[205] M. Salomao, X. Zhang, Y. Yang, S. Lee, J.H. Hartwig, J.A. Chasis, N. Mohandas, and X. An. Protein 4.1R-dependent multiprotein complex: new insight into the structural organization of the red blood cell membrane. *PNAS*, 105:8026–8031, 2008.

[206] B.B. Saxena. Unterschiede physiologischer konstanten bei Finkenvogeln aus verschiedenen Klimazonen. *Zschr. Vgl. Physiol.*, 40:376–396, 1957.

[207] K. Schmeidl. *Hamorheologische Eigenschaften des afrikanischen Nguni-Rindes und der Vergleich zum Osterreichischen Fleckvieh.* PhD thesis, Veterinary University Vienna, Austria, 2010.

[208] G.W. Schmid-Schoenbein, Y.C. Fung, and B.W. Zweifach. Vascular endothelium-leucocyte interaction. *Circ. Res.*, 36:173–184, 1975.

[209] G.W. Schmid-Schoenbein and B.W. Zweifach. RBC velocity profiles in arterioles and venules of the rabbit omentum. *Microvasc. Res.*, 10:153–164, 1975.

[210] K. Schmidt-Nielsen. The physiology of the camel. *Scientific American*, 201(6):40–51, 1959.

[211] K. Schmidt-Nielsen, E.C. Crawford, A.E. Newsome, K.S. Rawson, and H.T. Hammel. Metabolic rate of camels: Effect of body temperature and dehydration. *Am. J. Physiol.*, 212:341–6, 1967.

[212] K. Schmidt-Nielsen and C.R. Taylor. Red blood cells: why or why not? *Science*, 162:274–275, 1968.

[213] G.V.F. Seaman and G. Uhlenbruck. The surface structure of erythrocytes from some animal species. *Arch. Biochem. Biophys.*, 100:493–502, 1963.

[214] J. Seki and H.H. Lipowsky. In-vivo and in-vitro measurements epifluorescence of red cell velocity under microscopy. *Microvasc. Res.*, 38:110–124, 1989.

[215] E. Shafizadeh, B.H. Paw, H. Foott, E.C. Liao, B.A. Barut, J.J. Cope, L.I. Zon, and S. Lin. Characterization of zebrafish merlot/chablis as non-mammalian vertebrate models for severe congenital anemia due to protein 4.1 deficiency. *Development*, 129:4359–4370, 2002.

[216] S. Sharma and S.M. Gokhale. Sialoglycoproteins of mammalian erythrocyte membranes: a comparative study. *Asian-Aust. J. Anim. Sci.*, 24:1666–1673, 2014.

[217] S. Sharma, V. Punjabi, S.M. Zingde, and S.M. Gokhale. A comparative protein profile of mammalian erythrocyte membranes identified by mass spectrometry. *J. Membrane Biol.*, 247:1181–1189, 2014.

[218] N.M. Shpakova, N.V. Orlova, E.E. Nipot, and D.I. Aleksandrova. Comparative study of mechanical stress effect on human and animal erythrocytes. *Fiziol. Zh.*, 61:75–80, 2015.

[219] J.E. Smith, N. Mohandas, A.C. Greenquist, and S.B. Shohet. Deformability and spectrin properties in three types of elongated red cells. *Am. J. Hematol.*, 8:1–13, 1980.

[220] J.E. Smith, N. Mohandas, and S.B. Shohet. Variability in erythrocyte deformability among various mammals. *Am. J. Physiol.*, 5:H725–H730, 1979.

[221] C.S. Sommardahl, F.M. Andrews, A.M. Saxton, D.R. Geiser, and P.L. Maykuth. Alterations in blood viscosity in horses competing in cross country jumping. *Am. J. Vet. Res.*, 55:389–394, 1994.

[222] M. Soutani, Y. Suzuki, N. Tateishi, and N. Maeda. Quantitative evaluation of flow dynamics of erythrocytes in microvessels: influence of erythrocyte aggregation. *Am. J. Hematol.*, 268:H1959–H1965, 1995.

[223] C.D. Southgate, A.H. Chishti, B. Mitchell, S.J. Yi, and J. Palek. Targeted disruption of the murine erythroid band 3 gene results in spherocytosis and severe haemolytic anemia despite a normal membrane skeleton. *Nature Genetics*, 14:227–230, 1996.

[224] A. Sparer. *Porcine blood in forensic bloodstain pattern analysis: hemorheological issues.* PhD thesis, Fachhochschule Technikum Vienna, 2016.

[225] Z. Spolarics, M.R. Condon, M. Siddiqi, G.W. Machiedo, and E.A. Deitch. Red blood cell dysfunction in septic glucose-6-phosphate dehydrogenase-deficient mice. *Am. J. Physiol. Heart Circ. Physiol.*, 286:H2118–H2126, 2004.

[226] M. Stefanivoc, E. Puchulu-Campanella, G. Kodippili, and P.S. Low. Oxygen regulates the band 3-ankyrin bridge in the human erythrocyte membrane. *Biochem. J.*, 449:143–150, 2013.

[227] M. Stefanovic, N.O. Markham, E.M. Parry, L.J. Garrett-Beal, M.P. Cline, P.G. Gallagher, P.S. Low, and D.M. Bodine. An 11-amino acid b-hairpin loop in the cytoplasmic domain of band 3 is responsible for ankyrin binding in mouse erythrocytes. *PNAS*, 104:13972–13977, 2007.

[228] B. Stoiber, C. Zach, B. Izlay, and U. Windberger. Whole blood, plasma viscosity, and erythrocyte aggregation as a determining factor of competitiveness in standard bred trotters. *Clin. Hemorheol. Microcirc.*, 32:31–41, 2005.

[229] H.O. Stone, H.K. Thompson Jr, and K. Schmidt-Nielsen. Influence of erythrocytes on blood viscosity. *Clin. Hemorheol. Microcirc.*, 214:913–918, 1968.

[230] B. Stottinger, M. Klein, B. Minnich, and A. Lammetschwandtner. Design of cerebellar and nontegmental rhombencephalic microvascular bed in the sterlet, acipenser ruthenus: a scanning electron microscope and 3d morphometry study of vascular corrosion casts. *Microvasc. Microanat.*, 12:1–14, 2006.

[231] F. Tang, X. Lei, Y. Xiong, R. Wang, J. Mao, and X. Wang. Alteration young's moduli by protein 4.1 phosphorylation play a potential role in the deformability development of vertebrate erythrocytes. *Journal of Biomechanics*, 12:1–14, 2014.

[232] W.F. Taylor, S. Chen, G. Barshtein, D.E. Hyde, and S. Yedgar. Enhanced aggregability of human red blood cells by diving. *Undersea Hyperb. Med.*, 25:167–170, 1998.

[233] A. Teleglow, Z. Dabrowski, A. Marchewka, Z. Tabarowski, J. Bilski, J. Jaskiewicz, J. Gdula-Argasinska, J. Glodzik, D. Lizak, and M. Kepinska. Effects of cold water swimming on blood rheological properties and composition of fatty acids in erythrocyte membranes of untrained older rats. *Folia Biologica*, 59:203–209, 2011.

[234] Y.M. Tian, M.J. Cai, W.D. Zhao, S.W. Wang, Q.W. Qin, and H.D. Wang. The asymmetric membrane structure of erythrocytes from crucian carp studied by atomic force microscopy. *Chin. Sci. Bull*, 59:2582–2587, 2014.

[235] Y.M. Tian, M.J. Cai, W.D. Zhao, S.W. Wang, Q.W. Qin, and H.D. Wang. *The Camel: the Animal of the 21th Century.* Camel Publishing House, Bikaner, India, page 89ff, 2017.

[236] S. Toplan, N. Dariyerli, S. Ozdemir, M.C. Akyolcu, H. Hatemi, and G. Yigit. The effects of experimental hypothyroidism on hemorheology and plasma fibrinogen concentration. *Endocrine*, 28:153–156, 2005.

[237] S. Usami, S. Chien, and M.I. Gregersen. Viscometric characteristics of blood of the elephant, man, dog, sheep and goat. *Am. J. Physiol.*, 217:884–890, 1969.

[238] S. Usami, V. Magazinovic, S. Chien, and M.I. Gregersen. Viscosity of turkey blood: rheology of nucleated erythrocytes. *Microvasc. Res.*, 2:489–499, 1970.

[239] Hartenstein V. Blood cells and blood cell development in the animal kingdom. *Annu. Rev. Cell Dev. Biol.*, 22:677–712, 2006.

[240] L.L.M. Van Deenen and J. deGier. Chemical composition and metabolism of lipids in red cells of various species, The Red Blood Cell, pages 243–307. New York, Academic Press, 1964.

[241] E. Van den Akker, T.J. Satchwell, R.C. Williamson, and A. Toye. Band 3 multiprotein complexes in the red cell membrane; of mice and men. *Blood Cells, Molecules, and Diseases*, 45:1–8, 2010.

[242] M.A.P. Van den Boom, M.G. Wassink, B. Roelofsen, N.J. de Fouw, and J.A.F. Op den Kamp. The influence of a fish oil-enriched diet on the phospholipid fatty acid turnover in the rabbit red cell membrane in vivo. *Lipids*, 31:285–293, 1996.

[243] D. Van Houten, M.G. Weiser, L. Johnson, and F. Garry. Reference hematologic values and morphologic features of blood cells in healthy adult ilamas. *Am. J. Vet. Res.*, 53:1773–1775, 1992.

[244] L. Vap and A.A. Bohn. Hematology of camelids. *Vet. Clin. Exot. Anim.*, 18:41–49, 2015.

[245] J.F. Vernberg. Hematological studies on salamanders in relation to their ecology. *Herpetologica*, 11:129–133, 1955.

[246] M. Villolobos, P. Leon, S.K. Sessions, and J. Kezer. Enucleated erythrocytes in plethodontid salamanders. *Herpetologica*, 44:234–250, 1988.

[247] G. Viscor, J.R. Torrella, V. Fouces, and T. Pages. Hemorheology and oxygen transport in vertebrates. a role in thermoregulation? *J. Physiol. Biochem.*, 59:277–286, 2003.

[248] N.J. Wandersee, C.S. Birkenmeier, D.M. Bodine, N. Mohandas, and J.E. Barker. Mutations in the murine erythroid a-spectrin gene alter spectrin mrna and protein levels and spectrin incorporation into the red blood cell membrane skeleton. *Blood*, 101:325–330, 2003.

[249] N.J. Wandersee, S.C. Olson, S.L. Holzhauer, R.G. Hoffmann, J.E. Barker, and C.A. Hillery. Increased erythrocyte adhesion in mice and humans with hereditary spherocytosis and hereditary elliptocytosis. *Blood*, 103:710–716, 2004.

[250] N.J. Wandersee, A.N. Roesch, N.R. Hamblen, and et al. Defective spectrin integrity and neonatal thrombosis in the first mouse model for severe hereditary elliptocytosis. *Blood*, 97:543–550, 2001.

[251] M. Warda, A. Prince, H.K. Kim, N. Khafaga, T. Scholkamy, R.J. Linhardt, and H. Jin. Proteomics of old world camelid (camelus dromedarius): better understanding of the interplay between homeostasis and desert environment. *J. Adv. Res.*, 5:219–242, 2014.

[252] M. Warda and R. Zeisig. Phospholipid- and fatty acid-composition in the erythrocyte membrane of the one-humped camel and its influence on vesicle properties prepared from these lipids. *Dtsch. Tieraerztl. Wschr.*, 107:349–388, 2001.

[253] R.E. Waugh. Red cell deformability in different vertebrate animals. *Clin. Hemorheol.*, 12:649–656, 1992.

[254] R.E. Waugh and E.A. Evans. Viscoelastic properties of erythrocyte membranes of different vertebrate animals. *Microvasc. Res.*, 12:291–304, 1976.

[255] R.E. Waugh and E.A. Evans. Thermoelasticity of red blood cell membrane. *Biophys. J.*, 26:115–132, 1979.

[256] R.E. Waugh, Y.S. Huang, B.J. Arif, R. Bauserman, and J. Palis. Development of membrane mechanical function during terminal stages of primitive erythropoiesis in mice. *Exp. Hematol.*, 41:398–408, 2013.

[257] R.E. Weber, W. Voelter, A. Fago, H. Echner, E. Campanella, and P.S. Low. Modulation of red cell glycolysis: interaction between vertebrate hemoglobins and cytoplasmic domains of band 3 red cell membrane proteins. *Am. J. Physiol.*, 287:R454–R464, 2004.

[258] R.I. Weed, P.L. LaCelle, and E.W. Merrill. Metabolic dependence of red cell deformability. *J. Clin. Invest.*, 48:795–809, 1969.

[259] D.J. Weiss and R.J. Geor. Clinical and rheological implications of echinocytosis in the horse: a review. *Comp. Haematol. Int.*, 3:185–189, 1993.

[260] D.J. Weiss, R.J. Geor, and C.M. Smith. Effects of echinocytosis on equine hemorheology and exercise performance. *Am. J. Vet. Res.*, 55:204–210, 1994.

[261] D.J. Weiss, R.J. Geor, C.M. Smith, and C.B. McClay. Furosemide-induced electrolyte depletion associated with echinocytosis in horses. *Am. J. Vet. Res.*, 53:1769–1772, 1992.

[262] D.J. Weiss and A. Moritz. Equine immune-mediated hemolytic anemia associated with clostridium perfringens infection. *Veterinary Clinical Pathology*, 32:22–26, 2003.

[263] R. Wells, J.A. MacDonald, and G. Diprisco. Thin blooded antarctic fishes: a rheological comparison of the haemoglobin-free icefishes Chionodraco kathleenae and Cryodraco antarcticus with a red-blooded nototheniid, Pagothenia bernacchii. *J. Fish Biol.*, 36:595–609, 1990.

[264] R. Wells and H. Schmid-Schonbein. Red cell deformation and fluidity of concentrated cell suspensions. *J. Appl. Physiol.*, 27:213–217, 1969.

[265] X. Weng, G. Cloutier, P. Pibarot, and L.G. Durand. Comparison and simulation of different levels of erythrocyte aggregation with pig, horse, sheep, calf and normal human blood. *Biorheology*, 33:365–377, 1996.

[266] C. Whitfield, L.M. Mylin, and S.R. Goodman. Species-dependent variations in erythrocyte membrane skeleton proteins. *Blood*, 61:500–506, 1983.

[267] L.L. Wickham, D.P. Costa, and R. Elsner. Blood rheology of captive and free-ranging northern elephant seals and sea otters. *Can. J. Zool.*, 68:375–380, 1990.

[268] U. Windberger, R. Auer, Eloff S., R. Plasenzotti, and J.A. Skidmore. Temperature dependency of whole blood viscosity and red cell properties in desert ungulates: studies on scimitar-horned oryx and dromedary camel. *Clin. Hemorheol. Microcirc.*, 69:533–543, 2018.

[269] U. Windberger, R. Auer, and M.J. Smale. Blood rheology in cape fur seals and bottlenose dolphins: implications for muscle perfusion. *J. Biodivers. Endanger Species*, 3:1000149, 2015.

[270] U. Windberger, A. Bartholovitsch, R. Plasenzotti, K.J. Korak, and G. Heinze. Whole blood viscosity, plasma viscosity and erythrocyte aggregation in nine mammalian species: reference values and comparison of data. *Exp. Physiol.*, 88:431–440, 2003.

[271] U. Windberger and O.K. Baskurt. *Comparative Hemorheology*, chapter: Comparative Hemorheology. IOS press, Amsterdam, 2007.

[272] U. Windberger, R. Plasenzotti, and T. Voracek. The fluidity of blood in African elephants (loxodonta africana). *Clin. Hemorheol. Microcirc.*, 33:321–326, 2005.

[273] U. Windberger, K. Spurny, A. Graf, and H.J. Thomae. Hemorheology in experimental research: is it necessary to consider blood fluidity differences in the laboratory rat? *Lab Anim*, 49:142–152, 2015.

[274] U. Windberger, B. Stoiber, and R. Poschl, C.and van den Hoven. A comparative approach to measure elasticity of whole blood by small amplitude oscillation. *Rheology: Open Access*, 1:1, 2017.

[275] Ursula Windberger, Roland Auer, Monika Seltenhammer, Georg Mach, and Julian A Skidmore. Near-newtonian blood behavior-is it good to be a camel? *Frontiers in Physiology*, 10:906, 2019.

[276] K.G. Wingstrand. Non-nucleated erythrocytes in a teleostean fish Maurolicus Mulleri (gmelin). *Zeitschrift f§r Zellforschung*, 45:195–200, 1956.

[277] R. Yagil, S. Sod-Moriah, and N. Meyerstein. Dehydration and camel blood. I. Red blood cell survival in the one-humped camel Camelus dromedarius. *Am. J. Physiol.*, 226:298–300, 1974.

[278] R. Yagil, S. Sod-Moriah, and N. Meyerstein. Dehydration and camel blood. II. Shape, size, and concentration of red blood cells. *Am. J. Physiol.*, 226:301–304, 1974.

[279] R. Yagil, S. Sod-Moriah, and N. Meyerstein. Dehydration and camel blood. III. Osmotic fragility, specific gravity, and osmolality. *Am. J. Physiol.*, 226:305–308, 1974.

[280] O. Yalcin, F. Aydin, P. Ulker, M. Uyuklu, F. Gungor, J.K. Armstrong, H.J. Meiselman, and O.K. Baskurt. Effects of red blood cell aggregation on myocardial hematocrit gradient using two approaches to increase aggregation. *Am. J. Physiol.*, 290:H7665–H771, 2006.

[281] O. Yalcin, M. Uyuklu, J.K. Armstrong, H.J. Meiselman, and O.K. Baskurt. Graded alterations of RBC aggregation influence in vivo blood flow resistance. *Am. J. Physiol.*, 287:H2644–H2650, 2004.

[282] K. Yamaguchi, K.D. Jurgens, H. Bartels, and J. Piiper. Oxygen transfer properties and dimensions of red blood cells in high-altitude camelids, dromedary camel and goat. *J. Comp. Physiol.*, 157:1–9, 1987.

[283] B. Yang and A.S. Verkman. Analysis of double knockout mice lacking aquaporin-1 and urea transporter ut-b. Evidence for ut-b-facilitated water transport in erythrocytes. *J. Biol. Chem.*, 277:36782–36786, 2002.

[284] Y.Z. Yoon, J. Kotar, G. Yoon, and P. Cicuta. Non-linear mechanical response of the red blood cell. *Phys. Biol.*, 5:036007, 2008.

[285] J. Zhang, P.C. Johnson, and A.S. Popel. Effects of erythrocyte deformability and aggregation on the cell free layer and apparent viscosity of microscopic blood flows. *Microvasc. Res.*, 77:265–272, 2009.

Index

Note: Page numbers followed by "*fn*" indicate footnotes

N

R

S

T

Y

Z

CPSIA information can be obtained
at www.ICGtesting.com
Printed in the USA
JSHW020025040220
3992JS00002B/83